Albert Einstein's Special Theory of Relativity

Emergence (1905) and
Early Interpretation (1905–1911)

Albert Einstein as a patent clerk in Bern. (Copyright © by Lotte Jacobi)

Albert Einstein's Special Theory of Relativity

Emergence (1905) and
Early Interpretation (1905–1911)

Arthur I. Miller

Department of Physics
Harvard University

Department of Physics and Applied Physics
University of Lowell, Lowell, Massachusetts

1981

Addison–Wesley Publishing Company, Inc.
Advanced Book Program
Reading, Massachusetts

London · Amsterdam · Don Mills, Ontario · Sydney · Tokyo

Library of Congress Cataloging in Publication Data

Miller, Arthur I
  Albert Einstein's special theory of relativity.
  Bibliography: p.
  Includes index.
  1. Relativity (Physics)—History. 2. Physics—
History. 3. Einstein, Albert, 1879–1955.
4. Physicists—Biography. I. Title.
QC173.52.M54.   530.1′1   79-27495
ISBN 0–201–04680–6

Printed in the United States of America

ABCDEFGHIJK–HA–8987654321

To
Marlyn
Lori
Scott

CONTENTS

PREFACE

Midway into the most productive period of his life, 1902–1909, Albert Einstein wrote his first paper on electrodynamics entitled, "On the Electrodynamics of Moving Bodies." In 1905 the young scientist's intent, however, was not to develop electrodynamics as it was customarily interpreted, but to criticize it and then to propose a new basis for it. This he later referred to as the relativity theory, and the paper on electrodynamics became known as the relativity paper. Page for page Einstein's relativity paper is unparalleled in the history of science in its depth, breadth and sheer intellectual virtuosity. Whereas the consequences of the special relativity theory changed mankind's very view of its relation to the cosmos, this occurred neither immediately upon its publication, nor for several years afterward. This book is a biography and analysis of the relativity paper set into its historical context.

First as a student of physics, and then as a physicist, what struck me forcefully about the relativity paper was how Einstein had developed one of the most far-reaching theories in physics in a literary and scientific style that was parsimonious, yet not lacking in essentials; in a pace that, whenever necessary, possessed a properly slow cadence, yet was not devoid of crescendos and *tours de force*. It seemed as if Einstein's seminal paper on the relativity theory contained virtually everything the advanced undergraduate student needed for learning the basics of relativity theory – e.g., the relativity of simultaneity. On another level, for the mature physicist it could draw out in strong relief, and tie together, the fundamentals of the relativistic approach to mechanics and electrodynamics that more advanced texts clothe in elegant mathematics. My transition, some years ago, from physicist to historian of science owed itself in large part to my desire to know more about the relativity paper and its author. Through pursuance of the themes that underlay my research in the history of science – the nature of scientific discovery and the origin of scientific concepts – I discovered that there was more between the lines of Einstein's relativity paper than I had ever realized. It also became clear to me that there were no historical analyses available that attempted to connect the relativity paper in its entirety with the physics of its time. This lacuna was another impetus for my writing this book. Indeed, it is now strange to note, but prior to 1960 there were very few serious historical studies of Einstein's work on special relativity theory. The historical studies of Gerald Holton and Martin J. Klein were the first to begin filling in the pre-1905 background. As T. Hirosige and others have noted, it was Holton's 1960 essay, "On the Origins of the Special Theory of Relativity," that started the new interest in historical research on the special relativity theory, and that raised many of the modern questions, which Holton has continued to pursue in his subsequent essays. Klein, starting in

1962, has written extensively, and chiefly, on the effect of thermodynamics and statistical mechanics on Einstein's view of physical theory.

I have developed the history, philosophy and physics with an eye toward presenting to the widest possible audience the state of fundamental physical theory during 1890–1911. Against the background of Einstein's paper, set in its proper historical context, the book contains a detailed discussion of the internal structure of the paper using mathematics and physics at the sophomore or junior undergraduate level; an investigation of the genesis of Einstein's thoughts on relativity theory; and discussions of the philosophical and scientific (experimental and theoretical) currents of the times. Indeed, we shall see that Einstein presented the special relativity theory using no advanced mathematics at all. He performed only one integral in order to solve the rather straightforward problem of calculating the kinetic energy of an electron moving in an electrostatic field; later in 1905 he realized the deep meaning of this result – the equivalence of mass and energy. On the other hand, certain more advanced notions of mathematics and electrodynamics enter into my discussions of the state of electrodynamics in 1890–1905 and the early interpretation of relativity theory. With a view toward keeping the text self-contained, whenever necessary I have developed in footnotes the requisite mathematics and electrodynamics.

The analysis of Einstein's relativity paper is intended as a step toward improving scientific, philosophical and historical analyses of what Einstein did or did not accomplish or assert in 1905, and the multidisciplinary approach of modern historical scholarship may help to show the place this fascinating and complex episode has in the history of ideas.

Acknowledgements

My greatest intellectual debt I owe to my teacher Professor Gerald Holton. Without his generous assistance, encouragement and example as a scholar my transformation from physicist to historian of science would have been incomplete. I owe to Professor Holton the concept of this book. In 1972 he suggested that we analyze the internal structure of Einstein's 1905 relativity paper. There resulted a jointly written draft manuscript of 1972 which was circulated privately. During the ensuing years, the course of my own research expanded the plan for the full-fledged product, and this book is the result. Professor Holton graciously read most of the chapters in one form or another, and I am grateful for his comments.

I have benefitted much from Professor Martin J. Klein's writings and from our conversations. Dr. Rudolf Morf's critical comments on a version of my translation of Einstein's special relativity paper were most helpful. Among colleagues who contributed to the manuscript's final stages I mention Professors Edward M. Purcell, Silvan S. Schweber, and Kenneth Brecher. I thank Andrei Ruckenstein, and particularly Peter Galison, for checking details in portions of the manuscript.

Most of the research and writing for this book took place in the Department of Physics, Harvard University, to which I am most grateful for providing me with a "secular cloister."

It is a pleasure to acknowledge permission from the Algemeen Rijksarchief, The Hague, and the Estate of Henri Poincaré to reproduce correspondence between H. A. Lorentz and Henri Poincaré and to quote from other letters in their possession; from the American Philosophical Society (Archive for History of Quantum Physics) to quote from a letter from Walter Kaufmann to Arnold Sommerfeld; and from The Estate of Albert Einstein to quote from Einstein correspondence. I am grateful to the Poincaré family for opening to me for the first time their collection of Henri Poincaré's letters and manuscripts. The richness of this collection is made manifest in the two letters that I reproduced here, which shed new light on physics during the first decade of the 20th century. I thank Dr. Spencer R. Weart, Director, Center for History of Physics, for providing funds from the Friends of the Center for History of Physics of the American Institute of Physics that enabled me to arrange to microfilm the Poincaré archival materials.

For the travel and research funds essential for historical research I acknowledge grants from the Section for History and Philosophy of Science of the National Science Foundation, the Centre National de la Recherche Scientifique, the Lorentz Fund, Leiden, and the American Philosophical Society. A Fellowship from the John Simon Guggenheim Memorial Foundation affords me the luxury of completing this book while moving on to other studies.

For editorial assistance I have had the good fortune to have availed myself of the services of Dr. Marcel Chotkowski La Follette and Mrs. S. S. Stevens.

It is especially important in a book as complex as this to emphasize that old caveat that I alone am responsible for any textual errors and all conjectures.

ARTHUR I. MILLER

AUTHOR'S NOTES TO THE READER

In order to avoid a nesting of footnotes that contain no information other than a page number, I use abbreviations of the sort [Einstein (1907c)] which means the paper listed in the Bibliography (pp. 417–440) under Einstein and dated 1907c. Papers written by an author in a single year are listed in the Bibliography in the chronological order of their publication. Wherever necessary in the narrative I give the date a paper was received or its date of presentation as a lecture.

References to quotations from Einstein's 1905 special relativity paper, "On the Electrodynamics of Moving Bodies," are keyed to the line numbers of my translation in the Appendix (pp. 391–415). For example, [§1, 1–5] refers to lines 1–5 of §1. For convenient reference, certain of Einstein's equations in this paper are noted by a code; for example, [§8.15] indicates the fifteenth equation in §8 of the special relativity paper.

Attempting to retain the essence of the mathematical notation used in 1905, while excluding both Gothic letters then widely used and certain symbols that might confuse the modern reader (e.g., V for the velocity of light in vacuum c), has prevented me from standardizing notation. The remaining redundancies should not be confusing. For example, in this book when the letter E is used for the electric field, it is not likely to be confused with the quantity energy which is written as E, E^e, E_T, E_0, E_0^e or combinations thereof. I have taken care to define symbols such as E whenever they appear.

CAST OF CHARACTERS

The *dramatis personae* of fundamental physical theory during the late 19th and early 20th centuries are depicted as they were at that time.

By 1911 many physicists agreed that Isaac Newton's mechanics, as well as his views of space and time, from the 1687 *Principia* had to be revised. Of Newton Einstein wrote, "You found the only way which, in your age, was just about possible for a man of highest thought and creative power." (Courtesy of AIP Niels Bohr Library)

Augustin Fresnel's deduction in 1818 of the dragging coefficient was one of the most important results of 19th-century optics. (From A. Fresnel, *Oeuvres*)

Michael Faraday's 1831 law of electromagnetic induction served as a cornerstone for theories of electromagnetism. In 1851 Faraday investigated a particular sort of electromagnetic induction—unipolar induction, whose explanation became a subject of intense research, particularly by engineers and scientists in German-speaking countries. By the end of the 19th century Faraday's law of induction had ushered the Western world into the age of technology. (Courtesy of AIP Niels Bohr Library/Argonne National Library)

Newton's mechanics of 1687 had unified terrestrial and extraterrestrial phenomena. The next great synthesis occurred not quite 200 years later, during the 1860s, when James Clerk Maxwell unified electromagnetism and optics. (Courtesy of AIP Niels Bohr Library)

J. Clerk Maxwell

While a student at the ETH, Einstein learned a great deal about Maxwell's electromagnetic theory from independent study of August Föppl's text, *Einführung in die Maxwell'sche Theorie der Elektricität*. (From *Beiträge zur technischen Mechanik und technischen Physik, August Föppl* [Berlin: Springer-Verlag, 1924, courtesy of Springer-Verlag])

A. Föppl

Although in 1895 Einstein failed the entrance examination to the ETH, Einstein's high scores on mathematics and physics led one of the school's illustrious professors, Heinrich Friedrich Weber, to encourage him to try again. By 1900 their relationship had deteriorated, owing chiefly to a clash of personalities. (From *Schweizerische Bauzeitung*)

Werner von Siemens was the great German industrialist and scientist who founded the vast Siemens electrical industries. In 1867 von Siemens discovered the principle of the electrical dynamo and, in that year, with the assistance of his friend Kirchhoff, designed and built the first unipolar dynamo. In 1886 von Siemens donated funds toward establishing an institute for science and engineering at the ETH, which was to be directed by another friend, Heinrich Friedrich Weber. Einstein studied at this institute during the period 1896–1900. (From W. von Siemens, *Personal Recollections*)

Hermann von Helmholtz, Ludwig Boltzmann, and Gustave Kirchhoff were three of the masters of physics that Einstein studied at home during his ETH period.

(From H. von Helmholtz, *Wissenschaftliche Abhandlungen* [Leipzig: Barth, 1895])

(From L. Boltzmann, *Gesammelte Schriften* [Leipzig: Barth, 1909], courtesy of J. A. Barth, Publishers)

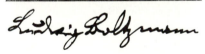

(From G. Kirchhoff, *Gesammelte Abhandlungen* [Leipzig: Barth, 1882])

In 1888 Heinrich Hertz empirically verified an important prediction of Maxwell's electromagnetic theory—electromagnetic waves. Hertz's fundamental theoretical research on Maxwell's theory influenced both Lorentz and Einstein. (From H. Hertz, *Miscellaneous Papers*)

The British physicist and electrical engineer, Oliver Heaviside, calculated the electromagnetic fields originating from a charged body in uniform linear motion. These results were fundamental toward theoretical research on the velocity-dependence of the electron's mass. (From the Heaviside Centenary Volume, courtesy of Institution of Electrical Engineers)

Einstein admired greatly the "incorruptible skepticism" of the philosopher-scientist Ernst Mach. (Courtesy of AIP Niels Bohr Library/Burndy Library)

Henri Poincaré was France's greatest living mathematician and among the first ranks of scientists and philosophers. His philosophical position led him to resist Einstein's views of space and time. (Courtesy of AIP Niels Bohr Library)

This photograph depicts H. A. Lorentz circa 1890. Of Lorentz, Einstein wrote, "for me personally, he meant more than all the others I have met on my life's journey." (From H. A. Lorentz, *Collected Papers*, vol. 1, courtesy of Martinus Nijhoff, Publishers)

Conrad Habicht, Maurice Solovine, and Einstein were the charter members of the "Olympia Academy," which met informally in Bern during 1902–1904.

Marcel Grossmann's meticulous notes at the ETH permitted Einstein the luxury of cutting classes in order to study at home such subjects as Maxwell's electromagnetic theory. In 1902 Grossmann's father was instrumental in obtaining for Einstein a job at the Patent Office in Bern, and in 1912 Grossmann taught Einstein the mathematics necessary for Einstein's final push toward the 1915 generalized theory of relativity. (Courtesy of ETH Bibliothek, Zürich)

Michele Besso was Einstein's life-long friend. In the special relativity paper Einstein thanked Besso for "several valuable suggestions." (Courtesy of the AIP Niels Bohr Library/Besso Family)

Max Planck made two great discoveries in his lifetime—the quantum of energy and Albert Einstein. (Courtest of AIP Niels Bohr Library/Burndy Library)

The photograph shows J. J. Thomson sitting in his study in a chair that had been used by Maxwell. Thomson's experiments at Cambridge University in the period 1897–1899 demonstrated conclusively that cathode rays were composed of negatively-charged "corpuscles," i.e., electrons. Lord Rayleigh wrote that Thomson's "attitude towards relativity was that of a looker-on." (From Lord Rayleigh, *The Life of Sir J. J. Thomson*, courtesy of Cambridge University Press)

During 1901–1903 at Göttingen, Walter Kaufmann's experiments on high-velocity electrons, i.e., β-rays, established the velocity-dependence of the electron's mass. Kaufmann's data supported the first field-theoretical description of an elementary particle, which was proposed in 1902 by his colleague, Max Abraham. Kaufmann maintained that his 1905 experiments, performed at Bonn, supported only the electron theories of Abraham and Alfred Bucherer, and that the results were "not compatible with the Lorentz-Einstein fundamental assumption." (Courtesy of Niedersächsische Staats- und Universitätsbibliothek, Göttingen)

Max Abraham was one of the most important physicists of the first decade of the 20th century. Although Abraham never accepted the special or general theories of relativity, Einstein greatly admired his ability as a physicist and paid close attention to his often sharply put criticisms. In their 1923 obituary notice for Abraham, Max Born and Max von Laue wrote that he "loved his absolute ether, his field equations, his rigid electron as a young man loved his first flame whose memory later experiences could not extinguish...in the history of physics he will be remembered as the completor of classical electrodynamics." (Courtesy of Niedersächsische Staats- und Universitätsbibliothek, Göttingen)

For the most part, during 1907–1911, Einstein left the defense of the relativistic viewpoint to Paul Ehrenfest—that is, after having set Ehrenfest straight in 1907. (From M.J. Klein, *Paul Ehrenfest*, courtesy of North-Holland Publishing Company)

Hermann Minkowski was among Einstein's teachers at the ETH. Minkowski's fame outside of mathematics is due largely to his 1907–1908 geometrical interpretation of the work of Einstein and Lorentz within the context of the electromagnetic world-picture. (Courtesy of the Houghton Library, Harvard University)

H. Minkowski

Max Born's 1909 Habilitationsschrift at Göttingen attempted to complete Hermann Minkowski's version of the electromagnetic world-picture. (From C. Reid, *Hilbert*, courtesy of Springer-Verlag)

In 1906 the Göttingen experimentalist, Adolf Bestelmeyer, measured the charge-to-mass ratio of the electron from data on the behavior of slow electrons, i.e., cathode rays, in an experimental arrangement that was an improvement over Kaufmann's. Bestelmeyer's results were the first to cast doubt on Kaufmann's 1905 data. (Courtesy of Niedersächsische Staats- und Universitätsbibliothek, Göttingen)

In 1906 the Bonn experimentalist, Alfred Bucherer, applied Bestelmeyer's experimental arrangement to measuring the charge-to-mass ratio of β-rays. His results supported the Lorentz-Einstein theory of the electron. (Courtesy of Stadtarchiv und Wissenschaftliche Stadtbibliothek, Bonn)

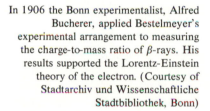

During the period 1907–1911 Max von Laue made several important contributions to the special theory of relativity, to which he had been introduced in late 1905 by Max Planck. In 1914 von Laue was awarded the Nobel Prize for his discovery of the diffraction of x-rays from crystals, which Einstein called one of the most beautiful in physics. In 1919 von Laue joined Einstein at the University of Berlin. Einstein had great respect for von Laue's courageous resistance to the Nazi regime. In reply to the physicist, P. P. Ewald, who had inquired during the mid-1930s if he could deliver any messages for Einstein in Germany, Einstein replied, "Greet Laue for me." Ewald named others, inquiring whether Einstein might wish to include them too in his greetings. Einstein's answer was simply to repeat, "Greet Laue for me." [quoted from Beyerchen (1977)]. (Courtesy of the AIP/Meggers Gallery of Nobel Laureates)

By the fall of 1911 the special relativity theory had been separated from Lorentz's theory of the electron. During the period 1905–1911 Einstein had progressed from an unknown physicist, who was a Patent Clerk, third class, to a Professor at the German University in Prague and a participant in the first summit meeting of physics—The Solvay Conference of 1911 whose participants are shown here. Seated, left to right: Nernst, Brillouin, Solvay, Lorentz, Warburg, Perrin, Wien, Mme. Curie, Poincaré. Standing: Goldschmidt, Planck, Rubens, Sommerfeld, Lindemann, de Broglie, Knudsen, Hasenöhrl, Hostelet, Herzen, Jeans, Rutherford, Kamerlingh Onnes, Einstein, Langevin. (Courtesy of AIP Niels Bohr Library/Institut International de Physique, Solvay)

INTRODUCTION

> He [Einstein] gave me [Shankland] his ideas on historical writing in science: "Nearly all historians of science are philologists and do not comprehend what physicists were aiming at, how they thought and wrestled with these problems." A means of writing must be found which conveys the thought processes that led to discoveries. Physicists have been of little help in this because most of them have no "historical sense." Mach's Science of Mechanics, however, he considered one of the truly great books and a model for scientific historical writing. He said, "Mach did not *know* the real facts of how the early workers considered their problems," but Einstein felt that Mach had sufficient insight so that what he said is very likely correct anyway. The struggle with their problems, their trying everything to find a solution which came at last often by very indirect means, is the correct picture.
>
> Report of an interview with A. Einstein by R. S. Shankland (4 February 1950).

Albert Einstein had definite ideas on research in the history of science. He preferred historical analyses that probed "what physicists were aiming at, how they thought and wrestled with these problems." With these words Einstein, himself, sets the motif for my book which uses his 1905 electrodynamics paper, *"Zur Elektrodynamik bewegter Körper"* (On the Electrodynamics of Moving Bodies) — known to us today as the relativity paper — as a window through which the modern reader can glimpse the intense intellectual struggles of physicists in the first decade of the 20th century. The drama of science in the making centered on the interplay between empirical data and physical theory, the clinging to notions that were neither always articulable nor clearly testable, the great investment of effort in empirical data and in already existing physical theories, and the irresistible drive toward the long-sought-after unification of the sciences.

These were the chief concerns in the clash among the dramatis personae of this book. Men like the fiery young theorist from Göttingen, Max Abraham; Europe's premier experimentalist on the characteristics of high-velocity electrons (i.e., β-rays), Abraham's colleague, Walter Kaufmann; the revered Dutch physicist, H. A. Lorentz; the great French mathematician-philosopher-scientist, Henri Poincaré; the leader of physics in Berlin, Max Planck; such rising young physicists as Poincaré's student Paul Langevin and Planck's student Max von Laue; and, of course, Einstein the Patent Clerk in Bern who, by 1905, had concluded that they all were "out of [their] depth."

Einstein's relativity paper was daring in its style and content. In 1905 the format of a journal paper in theoretical physics was much the same as its modern counterpart: these papers are essentially in three parts — the statement of a puzzle concerning empirical data; the proposal to modify an existing theory to explain the data; the deduction of further predictions. Imagine, for the moment, that you are on the editorial board of a prestigious physics journal and that you receive a paper from a little-known author that is unorthodox in style and format; its title has little to do with most of its content; it has no citations to current literature; a significant portion of its first half seems to be philosophical banter on the nature of certain basic physical concepts taken for granted by everyone; the only experiment explicitly discussed, adequately explainable using current physical theory, is not even considered by most physicists to be of fundamental importance; and the author proposes to discard some of the most widely-used notions (the ether in Einstein's relativity paper). True, with a minimum of mathematics, the little-known author deduces from his assumptions exactly a result that has heretofore required several drastic approximations. This is how Einstein's manuscript for the relativity paper may have struck the editor of the *Annalen der Physik* in 1905. The *Annalen*'s editor Paul Drude — known for his book on optics and the ether — had recently come to Berlin from Giessen University: "Until just $2\frac{1}{2}$ months earlier, the *Annalen* requested that all manuscripts be sent to Drude at Giessen. One can only speculate what the fate of the manuscript might have been there" [Holton (1979)].

As far as we know the editorial policy of the *Annalen* was that an author's initial contributions were scrutinized by either the editor or a member of the Curatorium; subsequent papers may have been published with no refereeing.[1] Einstein's having appeared in print in the *Annalen* five times by 1905, his relativity paper was probaby accepted on receipt. Planck apparently was the person who received Einstein's manuscript from the *Annalen*'s editorial office, and immediately recognized the fecundity of Einstein's work. During the winter term of 1905/1906 Planck asked his student von Laue to present a colloquium on it, and also himself wrote articles and lectured on it.

The kind of title Einstein had given his paper customarily signaled a discussion of the properties of moving bulk matter, either magnetic or dielectric. Einstein analyzed neither of these topics. In fact, the paper's first quarter contains a philosophical analysis of the notions of time and length. The paper's second half dispatches quickly certain problems of such fundamental importance that they generally rate separate papers — for example, the characteristics of radiation reflected from a moving mirror, and he concludes with certain results from the dynamics of electrons that generally appear at the beginning of papers where electrons are discussed. The only experiment developed in detail is at the paper's beginning and concerns the generation of current in a closed circuit as a result of the circuit's motion relative to a magnet; everyone knew this because electromagnetic induction had been discovered in 1831 by Faraday. The phenomenon of electromagnetic induction had grown since 1831, for, being fundamental to electrical dynamos, it had ushered the Western world into the age of technology. Everyone knew dynamos worked, but there remained fundamental problems concerning their operation — that is, problems considered to have been fundamental mostly by electrical engineers.

The physicists of 1905 who considered themselves to be on the cutting edge of basic scientific research, indeed who had defined the frontiers of physics, were busily formulating a field-theoretical description of all matter in motion. In their opinion this quest could benefit little from problems concerning how the current arises in electrical dynamos. Einstein thought otherwise.

If one read Einstein's paper very carefully, one saw that the young author had not only discovered a means to resolve the problems concerning electromagnetic induction, but even had discovered the connection between huge electrical dynamos, radiation, moving electrical bulk media, and the nature of space and time. To show this connection is one part of the book's plot.[2] Another part is how the physics community interpreted Einstein's paper on electrodynamics during 1905–1911 — for example, why until 1911 Einstein's relativity paper was interpreted only as a contribution to current research on electrodynamics, with many physicists confusing Einstein's work with Lorentz's, and referring to a "Lorentz-Einstein theory."

THE METHOD OF ANALYSIS

> I promise you four papers ... the first ... deals with radiation and the energy characteristics of light and is very revolutionary.... The second work is a determination of the true size of the atom from the diffusion and viscosity of dilute solutions of neutral substances. The third proves that assuming the molecular theory of heat, bodies whose dimensions are of the order of 1/1000 mm, and are suspended in fluids, should experience measurable disordered motion, which is produced by thermal motion. It is the motion of small inert particles that has been observed by physiologists, and called by them "Brown's Molecular motion [*Brownsche Molekularbewegung*]." The fourth paper exists in first draft and is an electrodynamics of moving bodies employing a modification of the doctrine of space and time; the purely kinematical part of this work will certainly interest you.
>
> A. Einstein to C. Habicht, written early Spring, 1905, in Seelig (1954).

Einstein completed the papers on the "energy characteristics of light," on "Brown's Molecular motion," and on an "electrodynamics of moving bodies" in intervals of approximately eight weeks. They constitute the famous triad of papers that appeared in Vol. 17 of the *Annalen der Physik*. [Einstein's (1905b) is dated 17 March 1905, and was received in Berlin on 18 March 1905; (1905c) is dated May 1905, and was received in Berlin on 11 May 1905; (1905d) is dated June 1905, and was received in Berlin on 30 June 1905.] But in 1905, and for a long time afterwards, their unity and their intended meaning was missed. Each paper seemed to be different from the next, and of very different value. Einstein's proposal that in certain instances it is useful as a heuristic device to assert that light behaves like a particle, i.e., a light quantum, was indeed "very revolutionary" in the sense that it was contrary to Maxwell's electromagnetic theory of light and without a fundamental base. The light quantum was in fact heatedly resisted until, in 1927, it found its place within a new atomic theory;

by this time Einstein had long since abandoned the light quantum as having not even a sufficiently heuristic significance. Soon after Einstein's Brownian motion theory, another one was published in 1906 by the well-known Polish physicist Marian von Smoluchowski, whose methods, while far less deep, most scientists found more comprehensible than Einstein's [Brush (1976), Teske (1975)]. Einstein's paper on electrodynamics was at the time widely considered to be a generalization of someone else's theory, i.e., Lorentz's (whose most recent version Einstein had not even read). Einstein did not refer to the electrodynamics paper as "revolutionary" and he usually avoided the term [Klein (1975)]. Yet the electrodynamics paper of course turned out to be the most far-reaching in physics.

Nor was the merit of Einstein's papers immediately recognized everywhere. The second paper that Einstein mentioned in the letter to Habicht was submitted to the University of Zurich as a possible Ph.D. thesis (1905a), and was rejected. (Einstein added another sentence and it was accepted. A previous thesis had been rejected in 1901 [Hoffmann (1972)].) In the fall of 1907 the relativity paper was rejected by the University of Bern as his *Habilitationsschrift*. One experimentalist wrote, "I cannot at all understand what you have written" [Seelig (1954)].[3]

The 1905 relativity theory was presented in thirty-odd pages of print, and Einstein developed it almost as an essay. Written at white heat [in "five or seven weeks" (Seelig (1952))], it is pristine in form, and yet in its way as complete as was Newton's theory presented in the book-length *Principia*. Testimony enough for the fecundity of Einstein's paper is that its content is a cornerstone of every subsequent fundamental physical theory. It spawned scientific treatises. Some of them served to embellish the content of Einstein's original paper, and others contain differing interpretations of the relativity theory. In addition the theory still affects the philosophical thought of the 20th century.

How did this all come about? What was the situation in physics in 1905? What feats of imagination did it take to make this advance? Questions of this sort invite a line-by-line analysis of Einstein's relativity paper just as they would in literature. Like the works of great literary masters, Einstein's relativity paper contains a high density of original ideas, often expressed in succinct loaded sentences whose full thrust and drama cannot be discerned outside of their historical context.

While the works of literary authors are commonly considered to be an integral part of our cultural milieu, this distinction is generally not granted to scientific authors. It is difficult, for example, to imagine a teacher of English who has never read one of Shakespeare's plays. But few people today, including physics researchers, teachers of physical science or philosophers of science have carefully read Einstein's relativity paper of 1905, although it is brief, requires little mathematics, but had immense effects on intellectual and societal pursuits in the 20th century. While many different in-depth analyses of the works of high literature are available to humanistic scholars, physicists have virtually no access to analyses that guide the reader through the real and apparent complexities of a major scientific work, placing the work in its proper historic context, and developing its technical content so as to make the terminology and symbols understandable to the modern scientist.

The historical analysis to be given focuses on the period 1890–1911. Chapter 1 is a detailed discussion of research in the electrodynamics and optics of moving bodies during 1890–1905, and portions of Chapter 3 discuss contemporaneous fundamental problems in electromagnetic induction. These analyses set the stage for Einstein's thinking toward the relativity theory. It will be noted that a main thrust of basic research in 1905 is similar to that of today — toward the unification of physics within a field-theoretical framework. In the first decade of the 20th century many physicists believed that reaching this goal was imminent, and their researches possessed an elegance and significance that later successes have tended to obscure. I conclude Chapter 1 with a Note on Sources to describe my method for the reconstruction of this development and of the early interpretations of the special relativity theory.

Chapter 2 analyzes Einstein's philosophical presuppositions before 1905, principally those concerning the origin of scientific concepts, and the structure and assessment of scientific theories.

Einstein divided his first relativity paper into an untitled introductory section and two main parts each comprised of five subsections. [An English-language version of the 1905 relativity paper is in the Appendix (pp. 391–415).] Chapters 3–12 provide an analysis of these sections, and trace their interpretation and development of their contents after publication until 1911.

A result of a section-by-section analysis of Einstein's relativity paper is that the key episodes in the discovery and early interpretation of relativity theory assume naturally a life of their own. For example, Chapters 1, 7 and 12 trace the developments surrounding Walter Kaufmann's experiments on the mass of β-rays. Chapters 4 and 7 discuss Einstein's discovery of the relativity of simultaneity and its early interpretation. Chapters 1, 3, 9, 10, 11 and 12 demonstrate the stunning power of Einstein's axiomatic method for electrodynamics and optics, vis-à-vis the constructive methods of the other physicists of 1905.

The cutoff date for commentary is set by the fact that after 1911 Einstein's viewpoint was seen more generally to be clearly demarcated from Lorentz's, and Einstein's own post-1911 research focused on generalizing the special theory of relativity.

The book concludes with an epilogue that picks up the strands of footnote discussions on Einstein's post-1905 researches, particularly on a generalization of the special theory of relativity.

To recapitulate, this book addresses the following topics:

(1) the state of the electrodynamics and optics of moving bodies from 1890–1905;

(2) the fundamental problems of electromagnetic induction in 1905;

(3) Einstein's thinking toward the special theory of relativity;

(4) an analysis of the internal structure of the special relativity paper;

(5) the developments in experimental and theoretical physics that determined the interpretation of Einstein's 1905 paper on electrodynamics during 1905–1911.[4]

A BIOGRAPHICAL SKETCH: 1879–1905

Your mere presence in the class destroys the respect of the students.

An instructor at the Luitpold Gymnasium [Frank (1947)].

If I have the good fortune to pass my examinations successfully I shall go to the Zurich Polytechnic. I shall remain there four years in order to study mathematics and physics. I dream of becoming an instructor in those branches of the natural sciences, specializing in the theoretical portion . . . . There is also a certain independence in the scientific profession that greatly pleases me.

A. Einstein [circa 1896, Aarau, in Hoffmann and Dukas (1979)].

You are an intelligent young man Einstein, a very intelligent young man. But you have one major fault – one cannot tell you anything.

H. F. Weber [circa 1900, Seelig (1952)]

Prior to Einstein's Bern period, 1902–1909, his life was in many ways unlike that of any other groundbreaking scientist. For one thing [as we know from Einstein (1946) and Frank (1947)] Einstein's school career as a youth in Germany was stormy.[5] He seems to have made no attempt at hiding his innate repulsion from the Prussian militarism of the staff at the Luitpold Gymnasium, Munich, and so the school authorities were only too happy to permit him to leave in 1894, without a diploma. After almost a year spent traveling in Italy (in 1894 Einstein's family had moved from Munich to Milan), Einstein took the entrance exam for the Eidgenössische Technische Hochschule (ETH), Zurich, and failed. Owing to his high grades in mathematics and physics he was advised to complete one year at the Kantonschule at Aarau, whereupon he could be matriculated at the ETH. Only the fastidious notetaking of his friend Marcel Grossmann saved the day at the ETH (1896–1900). Finding the curriculum too narrow, e.g., Maxwell's electrodynamics was omitted, Einstein had cut many classes in order to study the works of major 19th-century physicists independently at home. On the final exams he scored 4.91 out of a possible six points – well, but not superlatively. Einstein later recalled that at Zurich he had been unorderly, distant, unfriendly and dreamy, yet he possessed a small circle of good friends with whom he often met for iced coffee at the Café Metropole. Normally, if a student desired, the ETH diploma led to his becoming an assistant at the school, and so it was for Einstein's good friends Grossmann, Jakob Ehrat, and Louis Kollros. However, the animosity of the reigning power in Einstein's field of study, Heinrich Friedrich Weber, toward Einstein's independent attitude grew into an "instinctive antipathy," owing in part no doubt to Einstein's habit of addressing Weber as "Herr Weber" instead of "Herr Professor."[6] Weber had the last word because he selected others as his assistants. This was a "bitter pill" for Einstein, and feeling "suddenly abandoned" and helpless, he wandered through a succession of odd teaching positions and unemployment, until Grossmann's father helped to secure for him a position at the Swiss Federal Patent Office, Bern, as a technical expert, third class. Working dutifully at his job Einstein reserved his iconoclasm for evenings and weekends which he devoted to physics. In his Annus Mirabilis of 1905 Einstein published three papers in Vol. 17 of the *Annalen der Physik* each of which showed — in retrospect – a genial mind at work. In addition, in that year he wrote and published a doctoral thesis, and published a result that had escaped his notice in the relativity paper — the mass-energy equivalence.

FOOTNOTES

1. I thank Mr. Allan Needell for this information at which he arrived as a result of studying the Planck-Wien correspondence. In 1905 the *Annalen*'s editor was Paul Drude and the Curatorium was composed of F. Kohlrausch, M. Planck, G. Quincke, W. C. Röntgen, E. Warburg, "with the participation of the German Physical Society and especially M. Planck" (from the title page of Vol. 17).

2. In the book's subtitle I have used the term emergence owing to its neutrality. I was tempted to use discovery or invention; however, in a letter of 6 January 1948 to his life-long friend Michele Besso, Einstein distinguished explicitly between discovery and invention, preferring invention: "[Mach's] weakness, as I see it, lies in the fact that he believed more or less strongly, that science consists only of putting experimental results in order; that is, he does not recognize the free constructive element in the creation of a theory. He thought that theories are somehow the result of a *discovery* and not of an *invention*" (italics in original). But Hoffmann (1972) recollects Einstein's reply to someone who suggested to Einstein that Beethoven was a greater composer than Mozart as: "He [Einstein] said that Beethoven created his music, but Mozart's music was so pure that it seemed to have been ever-present in the universe, waiting to be discovered by the master." I considered using the term creation, but Einstein also employed it in two different ways. In "Physics and Reality" (1936) he described "thinking" as the "creation and use of definite functional relations between [concepts]…" — i.e., creation is synonymous with invention; and in "Reply to Criticisms" (1949) he described "concepts and theories as free inventions of the human spirit (not logically derivable from what is empirically given)." On the other hand, in the Hoffmann recollection Einstein described Beethoven as only creating music while Mozart, whom Einstein held in higher esteem, plucked out of the air music that was "ever-present in the universe." We shall see in Chapters 2 and 3 where we discuss Ernst Mach's philosophy of science and what Einstein meant by "concepts," that it is reasonable to conjecture that in 1905 Einstein had begun to liberate himself from the constraints of sense data, and moved toward a Leibnizian notion of a harmonious world that existed independently of the scientist as observer. Thus, as Einstein

emphasized in his later years, just as Mozart discovered music, the creative scientist plucked theories from the objective reality beyond the appearences, i.e., beyond Mach's sense data, through dazzling and unquantifiable feats of invention. In summary, although in his later years Einstein expressed his philosophical position with great clarity, his occasional interchangeable use of the terms creation, discovery and invention led me to choose for the book's subtitle the term emergence. In the text I shall use the more conventional terminology of creation or discovery.

3. Prior to being considered for a professorship at a Germanic university, one had to habilitate to the position of *Privatdozent*, to which there was no analogous rank elsewhere in academia. Habilitation was achieved through the faculty's acceptance of the *Habilitationsschrift* which was an original piece of work beyond the Doctoral Thesis. The rank of *Privatdozent* carried no remuneration from the university, rather a small fee was paid by the *Dozent's* students. Generally only those with outside means of support could afford to be a *Privatdozent*. Einstein had his job at the Patent Office.

In the fall of 1907, Einstein's colleague at the Patent Office, Joseph Sauter, brought Einstein to the attention of Paul Gruner who was the Professor of Theoretical Physics at the University of Bern. With Gruner's agreement Einstein presented the 1905 special relativity paper as his *Habilitationsschrift*. Gruner, however, recalled to Seelig (1952) that at the time he had found the special relativity theory to have been "very problematical." The curt assessment of Einstein's *Habilitationsschrift* that I quoted in the text was by a professor of experimental physics, Aimé Forster. In a letter to Gruner dated 11 February 1908 [Seelig (1952), (1954)], Einstein wrote that as a result of a recent conversation with Gruner at the town library, and on the advice of friends, that he would try again to habilitate at Bern. This time his *Habilitationsschrift* was accepted. Unfortunately we do not know its topic because, Seelig (1952) writes, this "can no longer be ascertained from the protocols" of the University. During the Wintersemester 1908/1909 at Bern, Einstein taught a course entitled "Theory of Radiation" which had a regular attendance of four: Einstein's good friend Besso, Lucien Chavan another acquaintance who, in 1905, had taken private lessons with Einstein in electricity, and two others named Schenker and Max Stern.

By late 1907 Alfred Kleiner, Director of the Physical Institute at the University of Zurich, had become interested in having Einstein on the faculty. In 1901 and 1905 Kleiner had been involved in rejecting two of Einstein's Ph.D. theses, but in the interim had changed his opinion of Einstein. Kleiner paid a surprise visit to Einstein's class and criticized his teaching. Einstein agreed, and then, as he related to Besso [Seelig (1952)], he "countered craftily, that you are not required to employ me as a teacher." In a more recent biography, Hoffmann (1972) writes that Kleiner had been in contact with Einstein since January of 1908, and had urged Einstein to try again to habilitate at Bern in order to ease the way for a position as professor at Zurich.

Early in 1909 a position opened at Zurich which was offered to a former classmate of Einstein's at the ETH, Friedrich Adler. In an act of great generosity, rooted in his esteem for Einstein, Adler refused the offer. On 7 May 1909 Einstein was elected to the rank of Professor Extraordinarius of Theoretical Physics (equivalent to Associate Professor).

4. Having discussed the book's intent, I should like to mention two topics that I chose to exclude: (1) Science in its social context — With great interest I await results from scholars engaged in studying the cultural and political influences on developments

in electrodynamics and special relativity during the period 1890–1911. (2) Gestalt switches, methods offered by philosophers of science for understanding the dynamics of scientific discovery and scientific progress, paradigms and scientific revolutions – Thus far I have found these notions to be inapplicable to the subject matter of this book. Elsewhere I have discussed paradigms, gestalt switches and the methodology of scientific research programs [see Miller (1974), (1975)]. As for scientific revolutions we shall see that it took six years for Einstein's view to be disentangled from Lorentz's, and even then there were prominent holdouts.

5. Especially useful biographies of Einstein are: Bernstein (1973), Frank (1947), (1953), Hoffmann (1972), Reiser (1930) which was Einstein's son-in-law Rudolf Kayser's pseudonym, and Seelig [(1952), (1954)]. Holton's (1973) book contains some new biographical material.

6. Weber's lectures were considered generally to have been lively and clear but old-fashioned, i.e., he never discussed Maxwell's theory [Kollros (1956), Reiser (1930), Einstein (1946)]. For further biographical information on Weber see Weiss (1912) and "Professor Dr. H. F. Weber," *Revue Polytechnic Schweizerische Bauzeitung*, *59*, 299 (1912). It was reported in 1897 that Weber experimented on himself in order to determine the highest shock that a human could comfortably absorb from alternating current. He found that a person standing on wet ground with dry shoes could absorb comfortably shocks of 1,000 V of 50cps alternating current [*Elektro. Z.*, *18*, 615–616 (1897) and *Electrical Review*, *41*, 526 (1897)]. One wonders how Einstein and his friends interpreted such experiments.

CHAPTER 1

ELECTRODYNAMICS: 1890—1905

> [The special relativity theory is the] coalescence of H. A. Lorentz's theory and the principle of relativity.
>
> A. Einstein (1907e)

> There is no doubt, that the special theory of relativity, if we regard its development in retrospect, was ripe for discovery in 1905. Lorentz had already observed that for the analysis of Maxwell's equations the transformations which later were known by his name are essential, and Poincaré had even penetrated deeper into these connections. Concerning myself, I knew only Lorentz' important work of 1895 – "La théorie électromagnétique de Maxwell" and "Versuch einer Theorie der elektrischen und optischen Erscheinungen in bewegten Körpern" – but not Lorentz' later work, nor the consecutive investigations by Poincaré. In this sense my work of 1905 was independent.
>
> A. Einstein to C. Seelig, 19 February 1955, quoted from Born (1969)[1]

For the purpose of this book I focus on developments in the electrodynamics and optics of moving bodies that influenced Einstein's thinking toward the special relativity theory.[2]

1.1. HERTZ'S ELECTROMAGNETIC THEORY OF 1890: THE MAXWELL-HERTZ THEORY

> And now, to be more precise, what is it that we call the Faraday-Maxwell theory? Maxwell has left us as the result of his mature thought a great treatise on Electricity and Magnetism; it might therefore be said that Maxwell's theory is the one which is propounded in that work. But such an answer will scarcely be regarded as satisfactory by all scientific men who have considered the question closely.
>
> H. Hertz (1892)

A foundational analysis of electromagnetic theory led experimentalist-theorist par-excellence, Heinrich Hertz, to propose (1890b) that the electrodynamics of moving bodies could be described by means of four equations which he considered axiomatic, i.e., he made no attempt to derive them:[3]

$$\mathbf{V} \times \boldsymbol{E} = -\frac{1}{c}\frac{\partial \boldsymbol{B}}{\partial t},$$

$$\tag{1.1}$$

$$\nabla \times H = \frac{1}{c}\frac{\partial D}{\partial t} + \sigma E, \tag{1.2}$$

$$\nabla \cdot D = 4\pi\rho_{\text{true}}, \tag{1.3}$$

$$\nabla \cdot B = 4\pi m_{\text{true}}, \tag{1.4}$$

where (using Hertz's nomenclature), $E(H)$ was the electric (magnetic) force, $D(B)$ was the electric (magnetic) polarization, σ was the specific conductivity, ρ_{true} was the density of true electricity (i.e., the amount of electricity that is entrained completely by a moving body, except for conduction currents), and m_{true} was the density of true magnetism (i.e., in order to treat electric and magnetic phenomena analogously, Hertz introduced a true magnetic volume charge density which was entrained completely by moving matter); $(1/c)(\partial D/\partial t)$ was the displacement current; σE was the conduction current; c was the velocity of light; and for consistency of notation I use Gaussian (cgs) units. Hertz related the quantities D, E, H, and B by the equations

$$D = \varepsilon E, \tag{1.5}$$

$$B = \mu H, \tag{1.6}$$

where ε was the dielectric constant and μ was the magnetic permeability (in general these quantities could be direction dependent); Hertz considered E and B to be fundamental quantities. In order to remove ambiguities concerning the notion of force, Hertz proposed an in-principle operational definition for the electric force that could be extended easily to the magnetic force:

> The phenomenon by which we define the electric force is the mechanical force which a certain electrified body experiences in empty space under electrical stress.

In Hertz's opinion speculations on the constitution of matter impeded progress, i.e., they were "illegitimate" questions, so he discussed the nature of neither electricity nor the ether.[4]

Equations (1.1)–(1.4) Hertz wrote relative to the reference system S that was fixed in the ether, and so they satisfied the basic assumption of any wave theory of light: relative to the free ether (space devoid of any ponderable matter, and so $\mu = \varepsilon = 1$ and $\sigma = 0$) electromagnetic disturbances were propagated with a velocity that was constant and independent of the source's motion. That is, in the free ether Eqs. (1.1)–(1.4) yield the wave equations for an electric (or a magnetic) disturbance

$$\left(\nabla^2 - \frac{1}{c^2}\frac{\partial^2}{\partial t^2}\right)E = 0 \tag{1.7}$$

and this equation was independent of the source's motion because the source's velocity and acceleration were nowhere present in Eq. (1.7).

Extending his theory to moving bodies required Hertz to decide whether the ether within a ponderable body participated totally, partially, or not at all in the body's motion. He chose the first alternative, believing that a partial ether drag necessitated two sets of electric and magnetic forces (E and H) for the dragged and motionless parts of the ether; besides, a total ether drag permitted him to maintain "the principle of action and reaction." Hertz was honestly open about his doubts on the "arbitrary" assumption of a totally dragged ether because "the few existing indications as to the nature of the motion of the ether lead us to suppose" that this was not the case. One of Hertz's "few existing indications" must have been Hippolyte Fizeau's (1851) experiment, which supported Augustin Fresnel's hypothesis of a partial ether drag. Hertz continued in 1890 by emphasizing that his theory of the electrodynamics of moving bodies represented "electromagnetic phenomena in a narrower sense" because he focused on those electromagnetic phenomena that he considered independent of any assumptions concerning the ether's motion. Hertz extended Eqs. (1.1)–(1.4) to moving bodies by assuming axiomatically their validity in another reference system R in arbitrary motion relative to S:

$$\nabla \times E = -\frac{1}{c}\frac{dB}{dt}, \tag{1.8}$$

$$\nabla \times H = \frac{1}{c}\frac{dD}{dt} + \sigma E; \tag{1.9}$$

Hertz assumed that Eqs. (1.3) and (1.4) and the term σE remained unchanged because only their instantaneous fields were considered. By the derivative d/dt Hertz meant the rate of change with respect to time of a vector field taken relative to the system R; this time derivative could be related to the change of the vector quantity calculated in S by the well-known relationship established by von Helmholtz (1874)[5]:

$$\frac{dU}{dt} = \frac{\partial U}{\partial t} + v(\nabla \cdot U) - \nabla \times (v \times U) \tag{1.10}$$

where U was any vector field and the right-hand side was evaluated in S. Equation (1.10) is another way of writing the convective derivative $dU/dt = \partial U/\partial t + (v \cdot \nabla)U$. So Eqs. (1.8) and (1.9) become in S

$$\nabla \times E = -\frac{1}{c}\frac{\partial B}{\partial t} - \frac{v}{c}(\nabla \cdot B) + \nabla \times \left(\frac{v}{c} \times B\right), \tag{1.11}$$

$$\nabla \times H = \frac{1}{c}\frac{\partial D}{\partial t} + \frac{v}{c}(\nabla \cdot D) - \nabla \times \left(\frac{v}{c} \times D\right) + \sigma E. \tag{1.12}$$

But, continued Hertz, the system S need not have been absolutely fixed in space, because v could be taken as the velocity relative to R. Consequently, "absolute motion" had no effect on electromagnetic processes. Hertz did not delve further into the subject of appropriate transformation equations for the physical quantities between S and R;[6] rather, his axiomatic assertion of the form invariance of the electromagnetic field equations [or "covariance" as Hermann Minkowski (1908a) described this mathematical property] led Hertz to predict new effects whose empirical confirmation could in turn serve to confirm his axiom of covariance. For example, from Eq. (1.12) Hertz predicted that a moving dielectric was the source of a magnetic field. Two years before Wilhelm Konrad Röntgen (1888) had observed such an effect. A. Eichenwald's (1903) more precise measurements disconfirmed Hertz's prediction, and supported instead Lorentz's elaboration of the Maxwell-Hertz theory. [For brevity, I use Abraham's (1902–1903) nomenclature, "Maxwell-Hertz theory."]

1.2. THE OPTICS OF MOVING BODIES

> The aberration of light, which according to the emission theory results directly from the composition of two rectilinear motions, is explained more easily by the wave theory.
>
> H. A. Lorentz (1886)

Nineteenth-century research on optical phenomena concerned chiefly the following overlapping problems [Hirosige (1976)]: (1) stellar aberration and its meaning for the wave theory of light, beginning with the experimental work of François Arago (1810); (2) studies, mostly by astronomers, of the manner in which light waves propagated, and their goal was to determine the proper motion of the solar system (corrections due to stellar aberration were important); (3) the relation between ponderable matter and the ether, particularly the detection of motion relative to the ether. Lorentz's 1886 review paper "On the Influence of the Earth's Motion on Luminiferous Phenomena," was the impetus for physicists to investigate the third problem, and it served also to emphasize the importance of an adequate explanation of stellar aberration to the optics of moving bodies.

In 1728 James Bradley reported that instead of observing the parallax of the "fixed" star γ Draconis, he had discovered an effect that depended both on changes in the earth's motion about the sun, and that the velocity of light was finite. Bradley explained stellar aberration by applying the Newtonian addition law of velocities to light corpuscles, as was the prevalent notion of light. If, in the geocentric system, c was the light velocity from a star, $-v$ was the star's velocity relative to the earth (i.e., $v = 30$ km/sec which is the earth's velocity relative to the sun), and u_r was the velocity of starlight relative to a terrestrial observer, then Newton's mechanics provided the relationship between these quantities as

$$u_r = c - v; \tag{1.13}$$

and Bradley calculated the aberration angle δ as

$$\sin \delta = \frac{v}{c} \sin \theta, \tag{1.14}$$

where θ (the angle between u_r and $-v$) defined the line of sight for a telescope's eyepiece to receive light corpuscles (see Fig. 1.1).

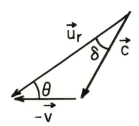

FIG. 1.1. The stellar aberration of light from a fixed star that is observed in the geocentric system. The telescope must be pointed along the direction of $-u_r$, which is the apparent position of the star. The star's true position is along $-c$.

From observations of the stellar aberration of several stars, Bradley concluded that the velocity of light was independent of a star's distance from the earth and was the same for every star. His measurement of the velocity of light ($c = 3.04 \times 10^{10}$ cm/sec) was in fair agreement with the one obtained in 1675 by the Danish astronomer Ole Römer ($c = 2.2 \times 10^{10}$ cm/sec).

François Arago wrote in (1810) that the determination of the velocity of light was one of the "most beautiful results of modern astronomy." Yet according to the Newtonian addition law for velocities, the velocity of light should depend on the motions of its emitter and of the measuring instrument. Arago noted that Bradley's apparatus did not possess the accuracy sufficient for displaying variations in the velocity of light. For this purpose Arago covered half of his telescope with an achromatic prism. He found that the aberration angle was independent of whether light passed through the prism, and was the one from Eq. (1.14). In order to save the corpuscular theory Arago proposed that the stars imparted to the light corpuscles an infinite number of velocities, but the eye could resolve only one among these velocities, or at best a small interval of velocities.

Thomas Young's optical researches on interference at the beginning of the 19th century brought about renewed interest in a wave theory of light. Young (1804) went on to suggest that stellar aberration could be explained by a wave theory provided that the "luminiferous ether pervades the substance of all material bodies with little or no resistance, as freely perhaps as the wind passes through a grove of trees." By 1815 Arago felt disposed toward a wave theory of light, particularly due to Young's researches, the French engineer Augustin Fresnel's 1815 investigation of diffraction, and the results of his own optical experiments, e.g., his 1811 discovery of chromatic polarization [see Whittaker (1973), vol. I and Hahn (1970)]. Arago's explanation for the 1810 observations of stellar aberration, based on the corpuscular theory of light, had not sat well with a scientist of his stature. In 1818 Arago turned to Fresnel to see whether,

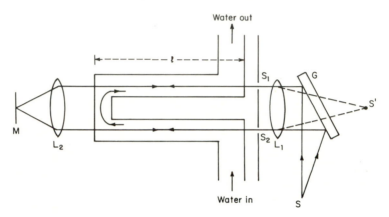

FIG. 1.2. A schematic to illustrate Fizeau's 1851 measurement of the velocity of light. Monochromatic light from a source S at rest on the earth was reflected from a glass plate G and then rendered parallel by the lens L_1. The slits S_1 and S_2 selected two rays that traversed the U-shaped glass tube. Water could flow in the tube in the direction indicated in the figure. The mirror M, at the focus of the lens L_2, interchanged the two rays so that one of them always traversed the glass tube in the direction of the running water, while the other ray moved opposite to the water's direction of flow. After having traversed the tube's length twice, the two light rays were combined to produce interference fringes at S', which was at the focus of the lens L_1. With the water at rest, Fizeau measured no shift of fringes from the case in which the tube was empty. To order v/c, where v was the earth's velocity relative to the ether, he expected the result because it had been obtained already by Arago (1810). But what if, continued Fizeau, the water moved with a velocity v relative to the laboratory? He considered three possibilities: (1) the water totally dragged the ether, i.e., $\kappa = 1$; (2) there was no ether dragging at all, i.e., $\kappa = 0$; (3) Fresnel's hypothesis according to which part of the ether was dragged, i.e., $\kappa = 1 - 1/N^2$. From Eq. (1.15), Fizeau calculated the optical path difference between the two rays that traversed the U-shaped tube with running water to be

$$\Delta = 2lc \left[\frac{1}{c/N - \kappa v} - \frac{1}{c/N + \kappa v} \right],$$

where l was the length of each arm of the U-shaped glass tube. To accuracy v/c, the expected fringe shift $\delta = \Delta/\lambda$ was

$$\delta = \frac{4N^2 lv}{c\lambda} \kappa.$$

Fizeau used yellow light of wavelength $\lambda = 5.26 \times 10^{-7}$ m, $l = 1.487$ m, $N = 1.33$ and $v = 7.059$ m/sec. He measured an average fringe shift $\delta_{observed} = 0.23$. Using $\kappa = 1 - 1/N^2$, the predicted fringe shift was $\delta_{calculated} = 0.2022$. Fizeau concluded: "These two values are nearly identical."

as Fresnel (1818) wrote, the "observations could be reconciled more easily with the theory in which light is considered as being vibrations in a universal fluid."

Fresnel (1818) reminded Arago that little was known about the physiology of vision, and that a constraint in the wave theory of light was that the wave velocity could depend only on the medium through which it traveled and not on the source's motion. With the words, "However extraordinary this hypothesis may appear at first sight," Fresnel proposed to explain Arago's result by assuming that relative to an observer in the ether the velocity of the light propagating through a transparent body that was moving relative to the ether with the velocity v, was altered by a fraction of the body's velocity — namely, $(1 - 1/N^2)v$, where N was the body's refractive index.

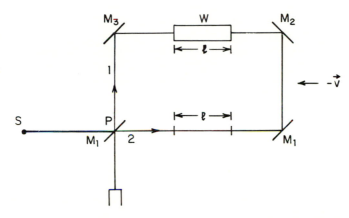

FIG. 1.3. A schematic to illustrate Hoek's 1868 experiment. Monochromatic light from the source S was split by the half-silvered mirror P into two rays that traversed the closed path $PM_1M_2M_3P$ in opposite directions, where M_1, M_2 and M_3 were mirrors. The quantity $-v$ was the ether's velocity relative to the laboratory. Part of the path M_2M_3 contained a glass tube W filled with water. After having traversed their closed paths, ray 1 was reflected by the plate P and combined with the part of ray 2 that was refracted by P, to produce interference fringes in the telescope T. On rotating the apparatus 180° so that ray 2 traveled opposite to earth's motion through the ether, Hoek found no shift in the interference fringes. According to Hoek's optical arrangement, the path difference between rays 1 and 2 could only have arisen from the portions of the closed circuit of length l, as indicated in the diagram. From Eq. (1.15), or Eq. (1.17) below, the time for ray 1 to traverse the lengths l of water and then of air was

$$t_1 = \frac{l}{c + v} + \frac{l}{(c/N)v + \kappa v - v},$$

and the time t_2 for ray 2 to circulate in the opposite direction was

$$t_2 = \frac{l}{c - v} + \frac{l}{c/N - \kappa v + v},$$

where $\kappa = 1$ if there were a complete dragging of the ether by the water, $\kappa = 0$ if no ether were dragged by the water, and $\kappa = 1 - 1/N^2$ if there were a partial dragging. Since to accuracy v/c, $t_1 = t_2$ then $\kappa = 1 - 1/N^2$.

Fresnel interpreted this result to mean that the moving body dragged the excess of ether within it, and the excess was $(1 - 1/N^2)$. Consequently, Fresnel's mathematical basis for Young's 1804 suggestion to explain stellar aberration required a partial dragging of the ether.

According to Fresnel's hypothesis the velocity of light did not obey strictly the Newtonian rule for the addition of velocities. Relative to a reference system fixed in the ether, the velocity of light c' was

$$c' = c/N + \kappa v, \tag{1.15}$$

where the quantity κ was Fresnel's dragging coefficient

$$\kappa = 1 - 1/N^2, \tag{1.16}$$

and if $N = 1$, then $c' = c$ as expected for the free ether. Fresnel's "extraordinary" hypothesis turned out to be one of the most important statements of 19th-century physics. Hippolyte Fizeau (1851) achieved the first experimental verification of Fresnel's dragging coefficient, in which c' was the velocity of light relative to the laboratory and v was the measured velocity in the laboratory of the moving water through the light passed (see Fig. 1.2). In a null interferometer experiment, Hoek (1868) took v as the earth's velocity relative to the ether. To order v/c, he explained the absence of any fringe shift when his apparatus was rotated by 180° by invoking the Fresnel dragging coefficient (see Fig. 1.3). Albert A. Michelson and Edward W. Morley (1886) performed a more exact repetition of Fizeau's experiment and confirmed his result.

1.3. LORENTZ'S, "ON THE INFLUENCE OF THE EARTH'S MOTION ON LUMINIFEROUS PHENOMENA" (1886)

> To what degree the ether participates in the motion of bodies that traverse it ... is of interest not only for the theory of light. It has acquired a more general importance since the ether probably plays a role in electric and magnetic phenomena.
>
> H. A. Lorentz (1886)

Owing primarily to the researches of Arago, Fresnel and Young, by the 1830's the wave theory of light was predominant. By 1886 there were two widely-used explanations of the results of optical experiments accurate to order v/c, where v was the velocity of ponderable matter relative to a fixed reference system: the theories of Fresnel (1818) and of George Gabriel Stokes (1845), (1846). Stokes, among others, strongly objected to the notion of a massive body, such as the earth, moving through the ether without disturbing it. In 1845 Stokes introduced a wave theory which assumed complete dragging of the ether at the earth's surface, that decreased to zero with increasing distance from the earth; and Stokes could deduce Fresnel's dragging coefficient. But Lorentz (1886) demonstrated the inconsistency of Stokes'

assumption of the existence of a velocity potential with no relative velocity
between the earth and the ether over the earth's surface. Lorentz devoted the
major part of his (1886) to the comparison of a theory composed of elements of
Stokes' and Fresnel's theories, i.e., containing a velocity potential and
Fresnel's dragging coefficient, with a pure Fresnel theory. It was Lorentz's
experiences with atomistic theories of electricity that, by 1886, may well have
influenced him to interpret a Fresnel theory as one in which the alteration of
the absolute velocity of light by a moving transparent body was due to the
influence of the body's constituent molecules on the ether in their immediate
vicinity. Consequently, by 1886 Lorentz was tending toward the opinion that
in a pure Fresnel theory there was no dragging at all.

   Lorentz (1886) used Huygens' principle and Fresnel's hypothesis to deduce
the velocity u_r of light that traversed a medium of refractive index N that was at
rest on the earth as (see Fig. 1.4)

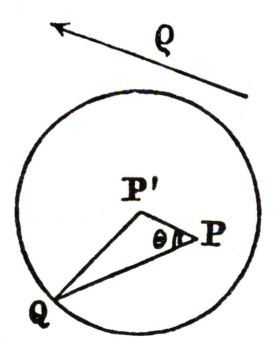

FIG. 1.4.   Lorentz's (1886) derivation of Eq. (1.17). At the time t in the geocentric system there
is a point P on a spherical wave front, and the wave is traversing a medium of refractive index N
that is at rest on the earth. According to Huygens' principle at the time $t + dt$ later pQ would have
been the distance to the next wave front if the ether were at rest. Since the ether moves relative to P
with the velocity ρ, along PP', then after the time dt the center of the next wave front is at P' instead
of P. Equation (1.17) follows from taking account of the directed distances $\boldsymbol{PP'} = \boldsymbol{\rho}\, dt = -\boldsymbol{v}/N^2 dt$
(Fresnel's hypothesis), $\boldsymbol{PQ} = \boldsymbol{u_r}\, dt$, and $\boldsymbol{P'Q} = c/N\, dt$. (Reprinted from Lorentz's (1886) in Vol. 4
of H. A. Lorentz, *Collected Papers*, with the permission of Martinus Nijhoff Publishers.)

$$u_r = \frac{c}{N} - \frac{v}{N^2}, \tag{1.17}$$

where the source could have been either on the earth or in the ether. For $N = 1$, Eq. (1.17) reduced to Eq. (1.13), and for $N \neq 1$, Eq. (1.17) explained Arago's experiment and an equivalent one by George Bidell Airy (1871) — see Fig. 1.5.

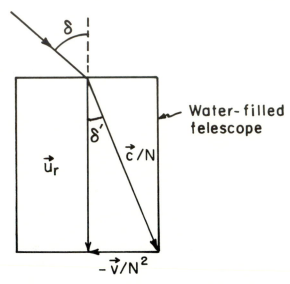

FIG. 1.5. Consider, in the geocentric system, a water-filled telescope whose line of sight to a star is normal to the direction of the star's velocity relative to the earth which is $-v/N^2$ (according to Fresnel's hypothesis). The law of sines yields $\sin \delta' = v/(cN)$. Since the starlight is refracted on entering the water then δ' is not the aberration angle. Using Snel's law to relate v and δ', i.e., $\sin \delta = N \sin \delta'$, we obtain $\sin \delta = v/c$. This derivation is based on the ones of Veltmann (1873), Lorentz (1886) and Drude (1900). The notion of seeking deviations from stellar aberration in air by using a water-filled telescope had been suggested by Boscovich in 1766, and was mentioned by Fresnel (1818) who predicted no change because this experiment was equivalent to Arago's. Airy (1871) carried out the experiment and found no change in the aberration angle.

Lorentz continued (1886), by noting that from the viewpoint of the geocentric system we could say that "the waves are entrained by the ether" according to the amount $-v/N^2$. For consistency with the nomenclature of the time Lorentz defined u_r as the velocity of the "relative ray" and c/N as the velocity of the "absolute ray." For example, in order to view the light from a fixed star, a telescope, or a system of aligned slits, at rest on the earth had to be oriented in the direction of the relative ray because the relative ray was the direction in which energy was transported (see Fig. 1.5). On the other hand, an observer at rest in the ether measured the velocity of the light that was propagating through the medium at rest on the moving earth to be

$$c' = u_r + v. \tag{1.18}$$

Equation (1.18) is identical to Eq. (1.15). Lorentz (1886) noted that the ether-fixed observer could interpret Eq. (1.18) as the "entrainment of the light waves by the ponderable matter." Consequently, although the phenomenon of stellar aberration depended on only the relative velocity between the earth and the star, ether-based theories of optics described it in two different ways depending on whether the source or observer were in motion. As we shall see in Chapters 3 and 10, Einstein considered redundancies of this sort as "asymmetries which do not appear to be inherent in the phenomena." In summary Fresnel's hypothesis of a dragging coefficient explained: (1) the dependence of the velocity of light on the velocity of the medium through which it propagated; (2) to first order in v/c, where v was the earth's velocity relative to the ether, optical phenomena were unaffected by the earth's motion. On the second point, Lorentz (1886) cited Wilhelm Veltmann's essentially kinematical proof that used Fermat's principle and Fresnel's hypothesis to show that, to order v/c, optical phenomena were unaffected by the earth's motion. In particular Veltmann proved that, to order v/c, interferometer experiments could not detect the earth's motion relative to the ether, i.e., Hoek's result was a foregone conclusion.[7]

Lorentz did not elaborate on Veltmann's important work because his goal was a more general one; namely, Lorentz (1886) set out to formulate a consistent theory for first-order optical phenomena that contained elements of Stokes' and Fresnel's theories. Then, using Fermat's principle, Lorentz demonstrated the consistency of the hybrid theory with empirical data accurate to order v/c, and concluded: "Briefly, everything occurs as if the earth were at rest, and the relative rays were the absolute rays." He went on to emphasize that only stellar aberration and the accompanying optical Doppler effect could reveal the earth's irregular motion relative to the sun. Lorentz deduced these phenomena as follows. The fixed stars were far enough away so that he could assume that their light reached the earth as plane waves. The phase of a plane wave was $\Phi = 2\pi v[t - d/c]$ where v was the wave's frequency and d was the distance from the source to the observer. Lorentz wrote the quantity d as $d = \hat{n} \cdot \mathbf{r}$, where $\hat{n}$ was the unit vector in the direction of the wave normal and $\mathbf{r}$ was the directed distance from source to observer. A plane wave was characterized by its amplitude, frequency, direction (i.e., orientation of $\hat{n}$) and, in ether-based theories of light, its velocity. Lorentz's method of relating the frequencies and velocities between ether-fixed and inertial reference systems was based on assuming that the phase Φ had the same numerical value in every reference system.[8] If the phase of plane waves emitted from a fixed star was

$$\Phi = 2\pi v[t - \hat{n} \cdot \mathbf{r}/c], \tag{1.19}$$

then on the moving earth the phase was also Φ, i.e.,

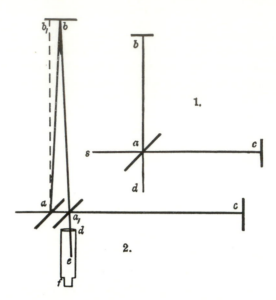

Fig. 1.6. I take the liberty of using the schematics from the "classic" 1887 Michelson-Morley experiment because the apparatus' of 1881 and 1887 were essentially the same. Furthermore I employ Michelson's description that was based on an ether-fixed reference system; consequently, the earth moves through the ether with a velocity v to the right. A monochromatic source of yellow light ($\lambda = 5.89 \times 10^{-7}$ m) was split by a half-silvered mirror into two beams. These beams traversed equal distances to mirrors b and c where they were reflected back to a and then combined to produce interference fringes. The optical paths $ac = ab = l = 11$ m (in 1881 the optical paths had been 1.2 m, and therefore the 1887 data were more accurate). On rotating the apparatus by 90° Michelson found no displacement of the interference fringes that could be expected owing to the ray along ab traveling parallel to the earth's motion through the ether, and the ray along ac traveling normal to this direction. Consequently, the times for the rays to have traversed to-and-fro the same optical path lengths were equal, regardless of whether they moved along or normal to the direction of the earth's motion through the ether. The time $t_{\parallel}$ for a ray to twice traverse the distance $ac = l$ was

$$t_{\parallel} = \frac{l}{c+v} + \frac{l}{c-v} = \frac{2l}{c(1-v^2/c^2)}.$$

In 6(b) the interferometer moved a distance aa_1, and so the ray ab was reflected by the moving mirror b into the path aba_1. If the beam of light ab left at a at time $t = 0$ and arrived at a_1 at time $t_{\perp}$, then since $aa_1 = l$

$$l^2 = \left(\frac{c}{2}t_{\perp}\right)^2 - \left(\frac{vt_{\perp}}{2}\right)^2$$

or

$$t_{\perp} = 2l/c\sqrt{1 - v^2/c^2}.$$

(In 1881 Michelson had erroneously calculated that $t_{\perp} = 2l/c$.) The expected fringe shift upon rotating the instrument by 90° was

$$2\delta = (2c/\lambda)(t_{\parallel} - t_{\perp}),$$

$$\Phi = 2\pi v_r \left[t - \hat{n} \cdot r_r / u_r\right], \tag{1.20}$$

where quantities subscripted with r were measured by observers on the moving earth which was assumed to be an inertial reference system. Newtonian mechanics related the distances in the two reference systems (one fixed to the star and the other to the earth) through the equations

$$r = r_r + vt, \tag{1.21}$$

$$t = t_r. \tag{1.22}$$

Substitution of Eq. (1.21) into Eq. (1.19) yielded

$$v_r = v(1 - \hat{n} \cdot v / u_r) \tag{1.23}$$

which, wrote Lorentz, "is precisely the modification of the number of vibrations which can be deduced from Doppler's principle," and to first order in v/c the quantity u_r in Eq. (1.23) was

$$u_r = c - \hat{n} \cdot v, \tag{1.24}$$

which, to this order of accuracy, was Eq. (1.13).

Lorentz went on to point out that the dependence of the refractive index on the frequency of light could lead to deviations from the laws of optics in reference systems at rest. Arago's experiment did not reveal any discrepancies because he had used achromatic prisms. It was conceivable, continued Lorentz, that spectroscopic investigations of starlight through dispersive media might reveal the earth's motion relative to the ether. Lorentz avoided this topic because "dispersion in a moving prism cannot be studied theoretically until we have formed some idea of the mechanism of light propagating through ponderable matter that is in motion relative to the ether."

Lorentz concluded his analysis of first-order phenomena by focusing on Fresnel's theory. He presented a well-known argument for demonstrating the impossibility of confining the ether between solid walls: the displacement of ether by mercury rising to the top of an inclined barometer suggests that the ether in the Torricellian vaccum must have flowed freely through the tube's

---

Fig. 1.6 (continued)

or to order $(v/c)^2$,

$$2\delta = 2lv^2/\lambda c^2.$$

In 1887, for a comparison reading, Michelson and Morley took v to be earth's orbital velocity about the sun, i.e., $v = 30$ km/sec, and so the expected fringe shift was $2\delta = 0.4$ fringe. They observed a fringe shift of "certainly less than the twentieth of this, and probably less than the fortieth part." See Silberstein (1914) and Richardson (1916) for discussions of the subtleties in the calculation of $t_\perp$ that concerned the reflection of light from a moving mirror – for example, only to accuracy v/c is the triangle aba_1 an isosceles triangle. Certain subtleties in Michelson's 1887 calculation of $t_\perp$ had been mentioned by Hicks (1902) who, however, drew incorrect conclusions [see Swenson (1972)]. (From *Am. J. Sci.*)

walls. Lorentz continued by stating a preference for Fresnel's theory because it seemed to be the simplest manner of explaining experiments accurate to order v/c, particularly stellar aberration, as well as the demonstration with the mercury-filled barometer. For assuming ponderable matter to have been constituted of extended atoms which affected the ether only in their immediate vicinity, then the ether freely penetrated bodies. Lorentz emphasized, however, that "in my opinion we cannot permit ourselves to be guided in such an important problem by considerations concerning the degree of probability or simplicity of one hypothesis or the other, but to address ourselves to experiment in order to ascertain the state of rest or motion of the ether at the earth's surface." He knew only of two experiments accurate to second order in v/c that addressed the question of whether the ether at the earth's surface was at rest or in motion: (1) Fizeau's (1860) claim to have observed an effect of the earth's motion through the ether on the change of the azimuth of the plane of polarization of the polarized light that was incident obliquely on a pile of glass plates; although this experiment supported Fresnel's theory, Lorentz, along with others, doubted its accuracy.[9] (2) Michelson's interferometer experiment (1881) which Michelson had claimed supported only Stokes' theory, but Lorentz (1886) had shown that Stokes' theory was internally inconsistent. Using the Newtonian rule for the addition of velocities, Michelson had predicted that, after rotation of his interferometer by 90°, there should occur a fringe shift proportional to $(2l/c) \cdot (v^2/c^2)$, where l was the length of the interferometer's arm, and v was the earth's velocity relative to the ether (this result was accurate to second order in v/c); upon finding no fringe shift, Michelson set $v = 0$ which he offered as empirical proof for Stokes' theory (see Fig. 1.6). Besides the mathematical difficulties with Stokes' theory, Lorentz pointed out a calculational error committed by Michelson in his data analysis: Michelson had calculated the time required for the light ray to traverse the interferometer arm normal to the direction of the earth's motion to be $2l/c$, instead of $2l/c + lv^2/c^3$ [the exact result was $(2l/c)(1/\sqrt{1 - v^2/c^2})$]. The extra term, Lorentz continued, reduced the calculated fringe shift by a factor of two, thereby placing any effect beyond Michelson's experimental accuracy[10]; so Michelson's data ruled out neither Fresnel's theory nor the hybrid theory composed of elements of Fresnel's and Stokes' theories.

In 1887 Michelson, in collaboration with E. A. Morley at the Case Western Reserve University, repeated the interferometer experiment of 1881 taking into account Lorentz's critical comments; this was the "classic" Michelson-Morley experiment.[11] Although their result was of great concern to Lorentz, he did not mention it in the seminal paper (1892a) on his electromagnetic theory. The only data discussed by Lorentz in support of the new theory were from first-order experiments: stellar aberration, Fizeau's 1851 measurement of the velocity of light in moving water, and its 1886 repetition by Michelson and Morley. Lorentz chose not to follow the empiricist course of selecting theories that he had set down in 1886 because, as he wrote (1892b), theories intermediate between those of Stokes and Fresnel "being more complicated, are less worthy of consideration."[12]

1.4. LORENTZ'S ELECTROMAGNETIC THEORY OF 1892: THE MAXWELL-LORENTZ THEORY

> It is right to mention that it is only after having read [Hertz's (1890b)] mémoire that I have undertaken the study of bodies in motion. Thus I have the advantage of knowing in advance the necessary results.
>
> H. A. Lorentz (1892a)

1.4.1. Perspective[13]

Straightaway Lorentz wrote that the goal of his (1892a) was "a theoretical deduction of the 'coefficient of entrainment' that Fresnel introduced in the theory of aberration." To achieve this goal would require some dazzling assumptions, for which Lorentz had been prepared by his earlier researches on electromagnetic theory and on the molecular theory of gases. Lorentz's principal assumption was that the sources of electromagnetic disturbances were microscopic charged particles that moved about in an all-pervasive absolutely resting ether, and so his notion of an ether was a slight variant of Fresnel's, which permitted the relative motion of its parts. Lorentz asserted axiomatically that the state of the ether at each point of space was described by the equations:

$$\nabla \times \boldsymbol{E} = -\frac{1}{c}\frac{\partial \boldsymbol{B}}{\partial t}, \tag{1.25}$$

$$\nabla \times \boldsymbol{B} = \frac{1}{c}\frac{\partial \boldsymbol{E}}{\partial t} + \frac{4\pi}{c}\rho\boldsymbol{w}, \tag{1.26}$$

$$\nabla \cdot \boldsymbol{E} = 4\pi\rho, \tag{1.27}$$

$$\nabla \cdot \boldsymbol{B} = 0, \tag{1.28}$$

where ρ was the particle's charge density; $\boldsymbol{w}$ was the particle's velocity relative to the ether; $(4\pi/c)\rho\boldsymbol{w}$ was the convection current of Rowland; Lorentz ensured conservation of "total current," i.e., conservation of charge, by proposing that the particles moved like "rigid bodies."[14] (I use vector notation and Gaussian (cgs) units instead of Lorentz's 1892 units and component notation; Abraham (1902–1903) was the first to emphasize the usefulness of this system of units, as well as coining the term "Maxwell-Lorentz field equations.") Hereafter I shall refer to $\boldsymbol{E}$ and $\boldsymbol{B}$ as the electric and magnetic fields, instead of using Lorentz's nomenclature of "dielectric displacement" and "magnetic force." Lorentz took the force density exerted by the ether on a charged particle as

$$f = \rho E + \rho \frac{w}{c} \times B; \tag{1.29}$$

he referred to Eqs. (1.25)–(1.29) as the "fundamental equations." These equations were written relative to a reference system S fixed in the ether.

As Emil Wiechert wrote (1900): "H. A. Lorentz was the first to successfully utilize the distinction between ether and matter in Maxwell's theory." That distinction enabled Lorentz, for example, to describe electromagnetic phenomena using two field quantities instead of Hertz's four, and this great simplification appealed also to Einstein's view of what constituted good theoretical physics [Einstein (1957)].

Lorentz initially developed his electromagnetic theory to discuss the electromagnetic properties of bulk matter. He assumed that bulk matter was composed of molecules that contained a particle of charge density ρ and mass m. In order to explain optical phenomena such as dispersion and the velocity of light in moving matter, Lorentz assumed that the charged particles could perform oscillations in response to incident light. He obtained the electromagnetic and optical properties of bulk matter by suitably averaging over the electromagnetic fields arising from the charged vibrating microsopic particles.[15] In addition to providing a theoretical basis for the dielectric constant, Lorentz's atomistically-based theory avoiding the notions of true electricity and of true magnetism, i.e., magnetic poles. Consequently, as Lorentz wrote (1892a), he could "simplify" Hertz's Eq. (1.4) to Eq. (1.28). By 1894, the notion of true magnetism was excluded from many influential expositions of Hertz's theory; for example, August Föppl's text (1894) cited as a reason for its exclusion the absence in nature of magnets with one pole. A theory of the magnetic permeability eluded Lorentz. Consistent with the theme of this book I take the liberty of discussing only the fields arising from the charged microscopic particles, unless indicated otherwise.

Emphasizing the continuity in the work of Maxwell, von Helmholtz and Hertz, Lorentz pointed out that, in the absence of charges, the theories of Hertz and Lorentz offered the same description for the state of the free ether because, when $\rho = 0$, then Eqs (1.25)–(1.28) with macrosopic fields, were formally identical with Eqs. (1.1)–(1.4); thus, Lorentz's equations describing radiation in the absence of charges were

$$\nabla \times E = -\frac{1}{c}\frac{\partial B}{\partial t}, \tag{1.30}$$

$$\nabla \times B = \frac{1}{c}\frac{\partial E}{\partial t}, \tag{1.31}$$

$$\nabla \cdot E = 0, \tag{1.32}$$

$$\nabla \cdot B = 0, \tag{1.33}$$

and the wave equation Eq. (1.7) followed; consequently in Lorentz's theory, relative to the ether, the velocity of light was a constant that was independent of the source's motion. However, Fresnel's dragging coefficient predicted direction-dependent effects which had not been observed.

1.4.2. Lorentz's Optics of Moving Bodies: 1892

The first five-sixths of Lorentz's 179 page treatise set the foundation for the results in the concluding chapter, "Propagation of Light in a Moving Ponderable Dielectric," where he discussed optical processes as they occurred in inertial reference systems. Lorentz's first step for treating problems in the optics of moving bodies was transforming the wave equations for radiation with its sources (i.e., the inhomogeneous wave equations deduced from Eqs. (1.25)–(1.28)) from the ether-fixed system S to an inertial reference system S_r moving relative to S with velocity $v = v\hat{i}$. In S these inhomogeneous wave equations were

$$\left(V^2 - \frac{1}{c^2}\frac{\partial^2}{\partial t^2}\right)E = 4\pi V\rho + \frac{4\pi}{c^2}\frac{\partial}{\partial t}(\rho w), \tag{1.34}$$

$$\left(V^2 - \frac{1}{c^2}\frac{\partial^2}{\partial t^2}\right)B = \frac{4\pi}{c}V \times (\rho w), \tag{1.35}$$

where $w = v + u$, with u the source's velocity relative to S_r; for brevity Eqs. (1.34) and (1.35) can be written as

$$\left(V^2 - \frac{1}{c^2}\frac{\partial^2}{\partial t^2}\right)a_1 = a_2. \tag{1.36}$$

The wave Eqs. (1.34) and (1.35) described an electromagnetic disturbance propagating through the ether with a velocity c that was independent of the source's motion.

Assuming that every molecule in the dielectric moved with the velocity of an inertial reference system S_r (i.e., $w_x = v$, $w_y = w_z = 0$), Lorentz transformed Eq. (1.36) from S to S_r using a set of transformation equations that were mathematically equivalent to the "Galilean transformations"[16] from Newtonian mechanics, i.e.,

$$x_r = x - vt, \tag{1.37}$$

$$y_r = y, \tag{1.38}$$

$$z_r = z, \tag{1.39}$$

$$t_r = t, \tag{1.40}$$

and the convective derivative for uniform linear translation obtainable from Eq. (1.10) as

$$\left(\frac{\partial}{\partial t}\right)_{S_r} = \left(\frac{\partial}{\partial t}\right)_{S} + \boldsymbol{v} \cdot \boldsymbol{\nabla}_r. \tag{1.41}$$

Consequently, in S_r the wave equation Eq. (1.36) became

$$\left[V_r^2 - \frac{1}{c^2}\left(\frac{\partial}{\partial t_r} - v\frac{\partial}{\partial x_r}\right)^2\right]K = J, \tag{1.42}$$

where K and J were functions of $(x_r, y_r, z_r, t_r = t)$, and $\boldsymbol{V} = \boldsymbol{V}_r$. But since Eq. (1.42) did not have the proper form for describing wave motion, Lorentz proposed an additional coordinate transformation on the inertial coordinates (x_r, y_r, z_r, t_r) in order that Eq. (1.42) possessed the proper form of a wave equation, i.e., that of Eq. (1.36):

$$x' = \gamma x_r, \tag{1.43}$$

$$y' = y_r, \tag{1.44}$$

$$z' = z_r, \tag{1.45}$$

$$t' = t - (v/c^2)\gamma^2 x_r, \tag{1.46}$$

where $\gamma = (1/\sqrt{1 - v^2/c^2})$. (I call the primed reference system Q'.) Lorentz considered the transformation from S_r to Q' as a purely mathematical coordinate transformation – for example, he introduced x' as "a new independent variable," and similarly for t'. In Q', Eq. (1.42) became

$$\left[V'^2 - \frac{1}{(c^2 - v^2)}\frac{\partial^2}{\partial t'^2}\right]K' = J', \tag{1.47}$$

where K' and J' were functions of (x', y', z', t'), and Eq. (1.47) described a disturbance propagating with a velocity $c\sqrt{1 - v^2/c^2}$.[17] Consequently, whereas a Galilean transformation from S to S_r failed to yield a proper wave equation, a further transformation from S_r to Q' resulted in a wave equation for a disturbance that depended on the emitter's motion, thereby violating an ether-based wave theory of light. Although Lorentz did not comment explicitly on this result for Q', we can assume that he noticed it because he wrote that calculations in the remainder of (1892a) were only to first-order accuracy in v/c, because this approximation facilitated further calculations, and it led to a "*théorème générale.*" To first order in v/c the equations for the electromagnetic field quantities of the molecules constituting matter had the same form in S as in a reference system connected with S_r through the equations:

$$x' = x_r, \tag{1.48}$$

$$y' = y_r, \tag{1.49}$$

$$z' = z_r, \tag{1.50}$$

$$t' = t - (v/c^2)x_r. \tag{1.51}$$

I refer to Q' in the v/c limit as R', in which the velocity of light is c. Hence, to order v/c the mathematical coordinate system Q' becomes in its spatial coordinates identical with the spatial Galilean coordinates, and the time coordinate mixes the Galilean absolute time t_r ($=t$) with the Galilean spatial coordinate x_r.

Lorentz concluded his (1892a) by deriving the Fresnel dragging coefficient from the inhomogeneous wave equations in S_r. Lorentz's derivation depended on the restoring forces of the vibrating constituent charged particles, and the number of charged particles per unit volume of the material through which the light was propagating.[18] Since his ether was absolutely at rest he attributed a dynamical interpretation to his coefficient: the cause of the Fresnel dragging coefficient was the interaction between light and the harmonically bound charged particles which Lorentz assumed constituted the moving medium.[19]

In summary, Lorentz's electromagnetic theory predicted a Fresnel dragging coefficient of $\kappa = 1 - 1/N^2$; Hertz predicted $\kappa = 1$; J. J. Thomson's (1880) version of Maxwell's theory the value $\kappa = \frac{1}{2}$. Lorentz's and Thomson's predictions were accurate to order v/c, but only Lorentz's agreed with the data of Michelson and Morley (1886).

1.5. THE HYPOTHESIS OF CONTRACTION

> Fresnel's hypothesis taken conjointly with his coefficient $1 - 1/N^2$, would serve admirably to account for all the observed phenomena were it not for the interferential experiment of Mr. Michelson, which has, as you know, been repeated after I published my remarks on its original form, and which seems decidedly to contradict, Fresnel's views. I am totally at a loss to clear away this contradiction, and yet I believe if we were to abandon Fresnel's theory, we should have no adequate theory at all, the conditions which Mr. Stokes has imposed on the movement of aether being irreconcilable to each other.
>
> Can there be some point in the theory of Mr. Michelson's experiment which has as yet been overlooked?
>
> H. A. Lorentz to Lord Rayleigh, 18 August 1892, quoted from Schaffner (1972)

Lorentz's concern over being unable to explain the 1887 Michelson-Morley experiment was clear from a letter that he wrote to Lord Rayleigh on 18 August 1892, shortly after having completed "La théorie..." [quoted in Schaffner (1972)]. Later in 1892 Lorentz found a way to explain "Mr. Michelson's experiment," which he described in "The Relative Motion of the Earth and the Ether" (1892b). From the velocity addition rule of Newton's mechanics, Lorentz deduced that, to second order in v/c, where v was the earth's velocity relative to the ether, the difference in times δt for the rays to traverse the interferometer's arms parallel ($t_{\parallel} = 2l/[c(1 - v^2/c^2)]$) and normal ($t_{\perp} = 2l/c\sqrt{1 - v^2/c^2}$) to its motion was

$$\delta t = t_{\parallel} - t_{\perp} = \frac{lv^2}{c^3} \; ; \tag{1.52}$$

δt was a measure of the fringe shift that Michelson and Morley had expected as a result of rotating their interferometer by $90°$; but within experimental error none was found, and so somehow $t_{\parallel} = t_{\perp}$. Whereas Michelson had assumed that $v = 0$, Lorentz thought otherwise (1892b):

> This experiment [Michelson and Morley] has been puzzling me for a long time, and in the end I have been able to think of only one means of reconciling its result with Fresnel's theory.

So he proposed what was clearly a physics of desperation: Suppose that the length of the interferometer arm in the direction of the earth's motion was $l(1 - \alpha)$, then to order $(v/c)^2$, $\delta t = 0$, if $\alpha = v^2/2c^2$ and the Michelson-Morley experiment was explained. That is, the quantity l in the equation for $t_{\perp}$, deduced from the Newtonian addition law for velocities, had to be replaced by $l\sqrt{1 - v^2/c^2}$. The shrinking of dimensions only in the direction of motion became known as the Lorentz contraction. In (1892b) Lorentz continued by giving what the Göttingen theorist Max Abraham later referred to as (1904b) as a "plausibility" argument for it. In a tone similar to that of his model Fresnel, Lorentz emphasized that such a phenomenon was "not inconceivable."[20] In this highly speculative vein he next posited yet another hypothesis: "It is not far-fetched to suppose" that the molecular forces, which after all determined the shape of a body, "act by intervention of the ether." This hypothesis required Lorentz to prove that molecular forces were affected by motion through the ether in such a way that they influenced the shape of a moving object by at least a term of second order in v/c. But, he continued, "since the nature of the molecular forces is entirely unknown to us, it is impossible to test this hypothesis." Consequently, one more bold hypothesis was necessary — that the "influences of the motion of ponderable matter on electric and magnetic forces" was the same for molecular forces. Lorentz's next step was to calculate the Lorentz force (i.e., Eq. (1.29) integrated over the charged-body's volume) in S_r, which he had done soon after publication of (1892a). Using the fictitious reference system Q' to transform a problem of electrodynamics into one of electrostatics, Lorentz obtained to second order in v/c:

$$F'_{x'} = F_{xr}, \tag{1.53}$$

$$F'_{y'} = F_{yr}(1 + v^2/2v^2), \tag{1.54}$$

$$F'_{z'} = F_{zr}(1 + v^2/2c^2), \tag{1.55}$$

where primed quantities referred to Q'. Lorentz hypothesized that molecular forces also transformed in this manner. According to Eqs. (1.53)–(1.55) a system in equilibrium in S_r was also in equilibrium in Q', which was taken as at rest in the ether. Lorentz next argued that this could be the case only if dimensions in S_r in the direction of motion were shortened by a factor $(1 - v^2/2c^2)$, and he obtained this result by rewriting Eq. (1.43) as $x_r = x'/\gamma$, then retaining only terms of second order in v/c. Thus hypothesizing that molecular forces transformed like the electromagnetic force permitted Lorentz to give a plausibility argument for the Lorentz contraction. As if to emphasize this point, Lorentz went on to write:

> One may not of course attach much importance to this result; the application to molecular forces of what was found to hold for electric forces is too venturesome for that.

He continued by asserting that the Lorentz contraction might not be the correct explanation of the Michelson-Morley experiment, because other deformations "would answer the purpose equally well"; this more general hypothesis I refer to as the hypothesis of contraction.[21]

Lorentz listed the two methods available for detecting the contraction of moving bodies: (1) "By juxtaposition of two rods," i.e., by congruence, but with both rods undergoing the same change of dimensions; (2) "comparing the lengths of two rods at right angles to each other" by means of interferometry, "but in this way we should come back once more to the Michelson-Morley experiment." He concluded forthrightly that the hypothesis of contraction led to no observable consequences.

In summary, Lorentz explained the Michelson-Morley experiment of 1887, accurate to second order in v/c, by proposing the hypothesis of contraction, not from reasoning based on the new electromagnetic theory, but from the Newtonian addition law for velocities. However, since this law predicted a dependence of the velocity of light on the source's motion, it had to be modified with Fresnel's hypothesis for use in electromagnetic theory. Furthermore, applications of the Newtonian addition law for velocities to stellar aberration were to first order accuracy in v/c.[22] Then, to provide a plausibility argument, that would also give the aura that the Lorentz contraction was connected with his electromagnetic theory, Lorentz proposed honestly stated interconnected assumptions – for example, if molecular forces transformed like the electromagnetic force, the transformation equations from S_r to Q' could be invoked for obtaining the required change of dimensions. Furthermore, neither the Lorentz contraction, nor the consequences of the more general hypothesis of contraction could be observed. Thus, the hypothesis of contraction was proposed to explain a single experiment; it was obtained from reasoning unconnected with the electromagnetic theory; its connection with the electromagnetic theory was at best tenuous – in short, the hypothesis of contraction was ad hoc. In his classic treatise, *Versuch einer Theorie der elektrischen und optischen Erscheinungen in bewegten Körpern* (1895) Lorentz discussed the hypothesis of contraction as he had in (1892b), but at the end, and in a chapter entitled "Experiments which Cannot be Explained without Further Ado." I refer to this treatise as the *Versuch*.[23]

1.6. LORENTZ'S VERSUCH (1895)

1.6.1. A Method for Solving Electrodynamical Problems

In the *Versuch* Lorentz deduced all results from (1892a) systematically and succinctly, and then went on to generalize them, in particular, the *théorème générale*.

In the chapter "Application to Electrostatics," Lorentz presented the following procedure for transforming electrodynamical problems into electrostatics problems. In a system S_r in which the ions constituting bulk matter were at rest (in 1895 Lorentz referred to the sources of the electromagnetic field as ions), their electromagnetic properties would be constant, and so the convective derivative of the ion's electromagnetic quantities would vanish, i.e., from Eq. (1.41)

$$\left(\frac{\partial}{\partial t}\right)_s = -\boldsymbol{v} \cdot \boldsymbol{\nabla}_r. \tag{1.56}$$

Lorentz transformed Eqs. (1.34) and (1.35) to S_r by use of Eq. (1.56) and the Galilean transformations Eqs. (1.37)–(1.40):

$$\left[\boldsymbol{\nabla}_r^2 - \left(\frac{\boldsymbol{v} \cdot \boldsymbol{\nabla}_r}{c}\right)^2\right]\boldsymbol{E} = 4\pi\boldsymbol{\nabla}_r\rho - \frac{4\pi}{c^2}\boldsymbol{v} \cdot \boldsymbol{\nabla}_r(\rho\boldsymbol{v}), \tag{1.57}$$

$$\left[\boldsymbol{\nabla}_r^2 - \left(\frac{\boldsymbol{v}}{c} \cdot \boldsymbol{\nabla}_r\right)^2\right]\boldsymbol{B} = \frac{4\pi}{c}\boldsymbol{\nabla}_r \times (\rho\boldsymbol{v}), \tag{1.58}$$

where Lorentz evaluated the fields $\boldsymbol{E}$ and $\boldsymbol{B}$ in S, although all other quantities were expressed in the inertial reference system S_r (Lorentz assumed that $\rho = \rho_r$). Next, in order to facilitate solving Eqs. (1.57) and (1.58), Lorentz introduced the scalar potential ϕ through the equation

$$\left[\boldsymbol{\nabla}_r^2 - \left(\frac{\boldsymbol{v}}{c} \cdot \boldsymbol{\nabla}_r\right)^2\right]\phi = -4\pi\rho \tag{1.59}$$

where

$$\boldsymbol{E} = -\boldsymbol{\nabla}_r\phi + \frac{\boldsymbol{v}}{c}\left(\frac{\boldsymbol{v}}{c} \cdot \boldsymbol{\nabla}_r\phi\right), \tag{1.60}$$

$$\boldsymbol{B} = \frac{\boldsymbol{v}}{c} \times \boldsymbol{E}.^{24} \tag{1.61}$$

Taking $\boldsymbol{v} = v\hat{\imath}$, Lorentz could rewrite Eq. (1.59) as

$$\left[\left(1 - \frac{v^2}{c^2}\right)\frac{\partial^2}{\partial x_r^2} + \frac{\partial^2}{\partial y_r^2} + \frac{\partial^2}{\partial z_r^2}\right]\phi = -4\pi\rho, \tag{1.62}$$

and it did not escape his attention that Eq. (1.62) was almost a Poisson equation, because he next introduced a coordinate transformation from S_r to a reference system S'' which was obtained from S_r "through increase of all dimensions which lie along the x_r-axis (thus also the corresponding dimensions of the ion) in the ratio $\sqrt{1 - v^2/c^2}$ to c." These transformation equations were a variant of the ones from S_r to Q':

$$x_r = x''\sqrt{1 - v^2/c^2}, \tag{1.63}$$

$$y_r = y'', \tag{1.64}$$

$$z_r = z'', \tag{1.65}$$

$$t_r = t''. \tag{1.66}$$

With Eqs. (1.63)–(1.66), Eq. (1.62) became Poisson's equation:

$$V''^2\phi'' = -4\pi\rho''$$

where ϕ'' and ρ'' were functions of x'', y'', z''; from demanding charge conservation in S'',[25] Lorentz deduced that

$$\rho'' = \rho\sqrt{1 - v^2/c^2} \tag{1.67}$$

and therefore

$$\phi'' = \phi\sqrt{1 - v^2/c^2}. \tag{1.68}$$

Lorentz's demonstration that in S'' the ion's field characteristics could be obtained from a Poisson equation, completed the reduction of an electrodynamical problem to an electrostatic one. For Lorentz, S'' was a purely mathematical coordinate system that was obtained from S_r by stretching dimensions along the common x-axes of S_r and S'', with all other dimensions remaining unchanged. This was a purely mathematical process, i.e., a mapping, because by definition Lorentz's ions were rigid bodies.

Since in S''

$$E'' = -V''\phi'' \tag{1.69}$$

then Lorentz deduced from Eqs. (1.60), (1.61), (1.63), (1.66) and (1.68) the forces in S_r and S'' as (the forces in S and S_r are equal since these reference systems are related by the Galilean transformation)

$$F_{x_r} = F_{x''}, \tag{1.70}$$

$$F_{y_r} = F_{y''}''\sqrt{1 - v^2/c^2}, \tag{1.71}$$

$$F_{z_r} = F_{z''}''\sqrt{1 - v^2/c^2} \tag{1.72}$$

(or equivalently from Eq. (H) of footnote 24).[26] Thus if the system of ions was in equilibrium in S'', it was also in equilibrium in S_r. These results transcended merely the solution of electrodynamical problems, for they meant also that electrostatic phenomena, i.e., experiments performed using apparatus at rest in the laboratory (S_r) could be affected by the earth's motion only to second order in v/c. Thus, for example, with these results Lorentz explained the failure of Des Coudres (within his experimental accuracy) to detect effects of the earth's motion on the mutual induction of two current-carrying loops at rest in the laboratory.

Lorentz's deduction of the uniformly moving charge's electromagnetic field from the value of these quantities in the charge's rest system S'' should not be interpreted in the sense of the relativity theory where we *begin* the calculation from the viewpoint of a system such as S''. Lorentz's approach to this sort of calculation was diametrically opposite to the relativity theory because his interpretation of covariance was a constructive one that was based on transforming a problem in the electrodynamics or optics of moving bodies into a statics problem — that is, into a reference system having all the properties of being at rest in the ether. This was manifestly clear in Lorentz's Chapter V, "Application to Optical Phenomena," where he proposed a method for explaining systematically every experiment in the optics of moving bodies accurate to first order in v/c.

1.6.2. A Method for Solving Problems in the Optics of Moving Bodies

Analysis of this class of problems precluded Lorentz's taking S_r as the ion's rest system; consequently he returned to equations of the form of Eq. (1.42) which, neglecting certain terms of second order in v/c, he rewrote as

$$\left[\left(\frac{\partial}{\partial x_r} + \frac{v_x}{c^2}\frac{\partial}{\partial t_r}\right)^2 + \left(\frac{\partial}{\partial y_r} + \frac{v_y}{c^2}\frac{\partial}{\partial t_r}\right)^2 + \left(\frac{\partial}{\partial z_r} + \frac{v_z}{c^2}\frac{\partial}{\partial t_r}\right)^2 - \frac{1}{c^2}\frac{\partial^2}{\partial t_r^2}\right]K = J,$$

$$(1.73)$$

where, for the sake of generality, the relativity velocity $\boldsymbol{v}$ between S and S_r is in an arbitrary direction. He transformed Eq. (1.73) into a form suitable for describing waves traveling with velocity c by introducing "a new independent variable"

$$t_L = t - \frac{\boldsymbol{v}}{c^2} \cdot \boldsymbol{r}_r \qquad (1.74)$$

which he referred to as the "local time" coordinate, in order to distinguish it from the "universal time" t, and $r_r = \sqrt{x_r^2 + y_r^2 + z_r^2}$. This interpretation of t_L was consistent with Lorentz's life-long belief that the true or universal space and time coordinates were either those relative to the ether (x, y, z, t) or those relative to S_r ($x_r, y_r, z_r, t_r = t$). Eq. (1.73) became

$$\left(\nabla_r^2 - \frac{1}{c^2}\frac{\partial^2}{\partial t_L^2}\right)K'' = J'' \tag{1.75}$$

and so, to order v/c, electromagnetic disturbances propagated with a velocity c that was independent of the source's motion when the electromagnetic quantities were expressed as functions of the set of coordinates:

$$x_r = x - vt, \tag{1.76}$$

$$y_r = y, \tag{1.77}$$

$$z_r = z, \tag{1.78}$$

$$t_L = t - \frac{v}{c^2}x_r, \tag{1.79}$$

where K'' and J'' were functions of (x_r, y_r, z_r, t_L). (Since the local time t_L is a function of x_r, then Eq. (1.75) follows from Eq. (1.73) for the case of $\boldsymbol{v} = v\hat{\imath}$ when $\partial/\partial x_r$ is replaced with $\partial/\partial x_r - v/c^2\,\partial/\partial t_L$, $\partial/\partial y_r$, and $\partial/\partial z_r$ are unchanged, and $\partial/\partial t_r$ is replaced with $\partial/\partial t_L$.)

This set of modified Galilean transformations was the transformation to the reference system R' from 1892. Lorentz found that only in charge-free space, i.e., the free ether, did the electromagnetic field equations have the same form in S_r as in S.[27] Thus, Lorentz specialized to the case of bulk matter that was nonmagnetic, nondielectric, nonconducting, and neutral, so that under the modified Galilean transformation Eqs. (1.76)–(1.79), the electromagnetic field Eqs. (1.30)–(1.33) — written with macroscopic fields, assumed the same form in S_r as they did in S. These equations were covariant under the modified Galilean transformations:

$$\boldsymbol{\nabla}_r \times \boldsymbol{E}_r = -\frac{1}{c}\frac{\partial\boldsymbol{B}_r}{\partial t_L}, \tag{1.80}$$

$$\boldsymbol{\nabla}_r \times \boldsymbol{B}_r = \frac{1}{c}\frac{\partial\boldsymbol{E}_r}{\partial t_L}, \tag{1.81}$$

$$\boldsymbol{\nabla}_r \cdot \boldsymbol{E}_r = 0, \tag{1.82}$$

$$\boldsymbol{\nabla}_r \cdot \boldsymbol{B}_r = 0, \tag{1.83}$$

where

$$\boldsymbol{E}_r = \boldsymbol{E} + \frac{v}{c} \times \boldsymbol{B}, \tag{1.84}$$

$$\boldsymbol{B}_r = \boldsymbol{B} - \frac{v}{c} \times \boldsymbol{E}, \tag{1.85}$$

and Lorentz introduced $\boldsymbol{E}_r$ and $\boldsymbol{B}_r$ merely as "new" vectors without any further explanation; so, to order v/c, using t_L instead of t_r the laws of optics were the same in S_r as in S. Lorentz referred to this result as the "theorem of corresponding states": if the state of a system characterized by E and B as functions of (x, y, z, t) existed in S, then there existed in S_r a corresponding state characterized by E_r and B_r as functions of (x_r, y_r, z_r, t_L). From Eqs. (1.84)–(1.85) areas of brightness and darkness were the same in S and S_r, so that the lateral dimensions of a light bundle (composed of light rays) remained unchanged.

Lorentz noted that there was no need to discuss Veltmann's (1873) result because the theorem of corresponding states replied more generally to the absence of any effects of the earth's motion on optical phenomena in experiments accurate to order v/c. Very likely two of Lorentz's chief reasons for this statement were: (1) while Veltmann had assumed the Fresnel dragging coefficient, Lorentz could derive it; (2) while Veltmann's proof was essentially kinematical, the theorem of corresponding states offered a dynamical explanation for empirical phenomena.

1.6.3. Lorentz's 1895 Theory of the Optics of Moving Bodies

The section of Chapter V that contained the theorem of corresponding states, Lorentz aptly entitled, "Reduction to a Resting System," because according to Eqs. (1.80)–(1.83) radiation in S_r was analyzed as if S_r were at rest relative to the source. For example, suppose that in S there was a source of radiation at a point Q, P was an observation point at rest relative to Q, and the radiation that reached P could be characterized as plane waves. Then the plane wave that reached P at the time t had been emitted at the earlier time $t - \hat{n} \cdot r/c$. Suppose now that P was at the origin of S_r. Then according to the theorem of corresponding states, an observer at P calculated the characteristics of the radiation, i.e., frequency, amplitude and direction, as if S_r were at rest in the ether, and the radiation had a phase

$$\Phi_r = (2\pi/T_r)(t_L - \hat{n}_r \cdot r_r/c), \qquad (1.86)$$

where $\hat{n}_r$ was the wave normal in the direction of the relative ray. Lorentz referred to the period T_r as the "relative period." Consequently, the radiation that reached P at the local time t_L had been emitted at an earlier time $t_L - \hat{n}_r \cdot r_r/c$, which was characteristic of the solution to the inhomogeneous wave Eq. (1.75).

Since the physical time was t, i.e., the time from the Galilean transformations, Lorentz used Eq. (1.79) to rewrite Eq. (1.86) as

$$\Phi_r = (2\pi/T_r)[t - (\hat{n}_r + v/c) \cdot r_r/c]. \qquad (1.87)$$

Eq. (1.87) described plane waves with the wave normal

$$\hat{n}' = \hat{n}_r + v/c \qquad (1.88)$$

where for $v = 0$, $\hat{n}'$ and $\hat{n}_r$ were identical. For the case of stellar aberration in the geocentric system, Lorentz rewrote Eq. (1.88) as

$$\hat{n}_r = \hat{n}' - v/c \tag{1.89}$$

which, to order v/c, is identical with Eq. (1.13). Whereas the lateral dimensions of a light bundle remained unchanged, the wave normal deviated from the direction of the ray. Consequently, on the moving earth a telescope must be oriented along $-\hat{n}'$ which was the apparent direction of the star, whose true direction was $-\hat{n}$.

In order to calculate the optical Doppler effect, Lorentz made the following replacement

$$\hat{n}_r + \frac{v}{c} = \hat{n}'\left(1 + \hat{n}_r \cdot \frac{v}{c}\right), \tag{1.90}$$

then the phase Φ_r in Eq. (1.87) became

$$\Phi_r = \frac{2\pi}{T_r}\left[t - \left(1 + \hat{n}_r \cdot \frac{v}{c}\right)\frac{n' \cdot r_r}{c}\right] \tag{1.91}$$

where ratios of velocities to that of c^2 were ultimately neglected. Lorentz defined the quantity

$$T = T_r\bigg/\left(1 + \hat{n}_r \cdot \frac{v}{c}\right) \tag{1.92}$$

as the radiation's "absolute period," i.e., its period measured in the source's rest system. To order v/c, Eq. (1.92) became

$$T = T_r\left(1 - \hat{n}_r \cdot \frac{v}{c}\right) \tag{1.93}$$

which was Lorentz's 1895 version of Eq. (1.23). Eq. (1.23) could be obtained from Eq. (1.93) through the replacements $T = 1/v$, $T_r = 1/v_r$ and $u_r = c$.

We recall that Lorentz's (1886) derivation of the optical Doppler effect and stellar aberration was strictly Galilean, and so it used different values for the velocity of light in S and S_r. The 1895 theorem of corresponding states permitted Lorentz to use the same light velocities in S and S_r; thereby he derived stellar aberration as a result of the deviation of the wave normal in S_r. But the 1895 deduction of the optical Doppler effect and stellar aberration required two steps. Why? Had Lorentz used the invariance of the phase of a plane wave in 1895, he would have proceeded thus:

$$\frac{1}{T}\left(t - \hat{n} \cdot \frac{r}{c}\right) = \frac{1}{T_r}\left(t_L - \hat{n}_r \cdot \frac{r_r}{c}\right)$$

$$= \frac{1}{T_r}\left[t\left(1 + n_r \cdot \frac{v}{c}\right) - \left(\frac{v}{c} + \hat{n}_r\right) \cdot \frac{r}{c}\right] \tag{1.94}$$

and obtained Eq. (1.88) and the result that $\hat{n}$ was the quantity $\hat{n}'$ in Eq. (1.90). We can conjecture that in 1895 Lorentz did not use the invariance of phase because he interpreted the theorem of corresponding states not as a principle of relativity that related phenomena in two equivalent reference systems, but as a means to perform calculations and to explain experiments by reducing phenomena in a moving reference system to phenomena occurring in a reference system at absolute rest. Support for this conjecture is Lorentz's title to Chapter V, "Reduction to a Resting System." More support accrues as we move through the development of electrodynamics circa 1905.

Suppose now that in S_r there was a medium of refractive index N through which light was propagating. Then Lorentz replaced Eq. (1.86) with

$$\Phi_r = \frac{2\pi}{T_r}\left(t_L - \frac{\hat{n}_r \cdot r_r}{W_r}\right) \tag{1.95}$$

where $W_r = c/N$. Then Eq. (1.88) became

$$\frac{\hat{n}'}{W'} = \frac{\hat{n}_r}{W_r} + \frac{v}{c^2} \tag{1.96}$$

where W' was the velocity of light relative to S_r. Lorentz neglected terms of order $(v/c)^2$, and obtained from Eq. (1.96)

$$W' = W_r - \frac{v}{N^2}\cdot\hat{n}_r \tag{1.97}$$

and the light velocity relative to the ether was $W'' = W' + \hat{n}_r \cdot v$, or

$$W'' = W + (1 - 1/N^2)\,v\cdot\hat{n}_r. \tag{1.98}$$

For light traveling through a dispersive medium Lorentz demonstrated that a term $-(v\cdot\hat{n}_r/N)\,T_r(dN/dT_r)$ should be added to Eq. (1.98). Zeeman (1914–1915) experimentally verified this extra term. Thus, Lorentz's 1895 derivation of the Fresnel dragging coefficient was more general than the one from his (1892a) because it avoided assumptions on the constitution of matter. The local time coordinate permitted Lorentz to avoid assumptions of this sort.

As we shall see in Chapters 3 and 10, Einstein's realization that Lorentz's local time should be the physical time, and his bold extension of a fundamental postulate of ether-optics to every inertial system — that in special relativity the velocity of light is the same in every inertial reference system and is independent of the source's motion — removed the necessity for the notions of relative and absolute rays. Consequently, Einstein reduced stellar aberration and the optical Doppler effect to purely kinematical phenomena that depended on only measureable quantities.

For the case in which both source and observer move with velocities v_1 and v_2, respectively, relative to the ether, Lorentz's prediction for the optical Doppler effect suffered from a defect indigenous to every ether-based theory of light: it contained separately the unknown velocities v_1 and v_2, that is,

$$T_r = T\left[\frac{1 - \hat{n}_r \cdot \boldsymbol{v}_2/c}{1 - \hat{n}_r \cdot \boldsymbol{v}_1/c}\right] \tag{1.99}$$

or, to order v/c,

$$T_r = T\left(1 - \hat{n}_r \cdot \frac{(\boldsymbol{v}_1 - \boldsymbol{v}_2)}{c} - 2\left(\hat{n}_r \cdot \frac{\boldsymbol{v}_1}{c}\right)\left(\hat{n}_r \cdot \frac{\boldsymbol{v}_2}{c}\right)\right), \tag{1.100}$$

and only the relative velocity $\boldsymbol{v}_1 - \boldsymbol{v}_2$ is known. This shortcoming was emphasized in Paul Drude's widely-read *Theory of Optics* (1900):

> In the above it is assumed that the source A is at rest and the point of observation B in motion. The considerations also hold in case both A and B move. It is then the relative velocity of B with respect to A measured in the direction of the propagation of the light. In this case the rigorous calculation shows that the actual period T and the relative period T' observed at B stand to each other in the ratio $T : T' = \omega - v' : \omega - v$, in which v is the absolute velocity of B, v that of A in the direction of the ray, and ω that of the light in the medium between A and B. It is only when v' and v are both small in comparison with ω that this rigorous equation reduces to that given in the text, i.e. to the customary form of Doppler's principle. Now we know nothing whatever about the absolute velocities of the heavenly bodies: hence in the ultimate analysis the application of the usual equation representing Doppler's principle to the determination of the relative motion in the line of sight of the heavenly bodies with respect to the earth might lead to errors. Attention was first called to this point by Moessard (C. R. 114, p. 1471, 1892).

1.6.4. Summary State of Lorentz's Theory of the Electrodynamics and Optics of Moving Bodies in 1895

The *Versuch* displayed Lorentz's theory as a dynamical theory which explained the positive and null experiments in optics and electrodynamics as the result of the interaction between the ether and the oscillator ions of which Lorentz assumed matter to be composed.

Consequently, in the *Versuch* Lorentz postulated two different sets of space and time transformations for use in two different sets of problems: the transformation Eqs. (1.63)–(1.66) were used to reduce electrodynamical problems to electrostatic problems, and the transformation Eqs. (1.76)–(1.79) to reduce problems in the optics of moving bodies to this class of problems for bodies at rest. This signaled that Lorentz's theory was still in a developmental stage for, after all, it purported to be a unified description of electromagnetism and optics. The hypothesis of the local time coordinate was the basis for the theorem of corresponding states that permitted systematic explanation of all optical phenomena to an accuracy of first order in v/c. Although both Lorentz and Hertz took the electromagnetic field equations as axioms, their approaches to covariance differed: Lorentz's was constructive because he attempted to derive covariance using space and time transformation equations, as well as suitable transformation equations for the electromagnetic field quantities to which he gave no physical interpretation; Hertz took covariance as a basic property of his theory and then deduced its consequences, e.g., the magnetic field of a moving dielectric. As for the status of Lorentz's hypothesis of contraction in 1895, it had been suggested by a law that did not always apply to optics, Newton's addition law for velocities; it had been cooked up to rescue the electromagnetic theory from the results of a single experiment; it was explained as arising from the interaction between the ether and the oscillator ions, which permitted molecular forces to transform so as to cause contraction, provided molecular forces, as yet unknown, transformed like electromagnetic forces; it predicted an unobservable effect; its plausibility argument turned on a set of transformation equations to a nonphysical coordinate system, and moreover these transformations had been posited for use in electrodynamical problems, and not in optical problems. In short, this hypothesis marred the magnificent edifice that was Lorentz's electromagnetic theory, and this point was emphasized in the epistemological-physical criticism of that titan of international science, Henri Poincaré.[28]

1.7. HENRI POINCARÉ

> Il faut donc renoncer à developper une theórie parfaitement satisfaisante et s'en tenir provisoirement à la moins défectueuse de toutes qui paraît être celle de Lorentz.
>
> H. Poincaré (1895)

1.7.1. Poincaré's Criticisms of Lorentz: 1895–1902

Poincaré's first published critique of Lorentz's theory was "*À propos de la théorie de M. Larmor*" (1895), in which he proposed three conditions that an electrodynamical theory of moving bodies should satisfy: (1) Fizeau's 1851 experiment; (2) the principle of conservation of electricity and magnetism; (3) the principle of action and reaction.[29] Of the theories that Poincaré discussed (those of von Helmholtz, Hertz, Larmor, Lorentz and Thomson), only Lorentz's satisfied the first two criteria, and so Poincaré considered it to be the "least defective" of the theories. From the failure thus far to detect any effects of the earth's motion on optical phenomena, Poincaré went on to assert that the only motion observable was of ponderable matter relative to ponderable matter, and not to the ether.

During his 1899 lectures at the Sorbonne, Poincaré (1901) commented again on Lorentz's theory, focusing on the *Versuch*. He declared the hypothesis of contraction to be a "*coup de pouce*," for preventing the detection, to second order in v/c, of absolute motion; should we expect "a new *coup de pouce*, a new hypothesis" to explain the expected null results to each higher order in $(v/c)^2$? No, he replied, because the true theory of the electrodynamics of moving bodies had to be rigorously consistent with the principle that only the relative motion of ponderable bodies was observable.

Perhaps, continued Poincaré, "without very profound modifications" Lorentz's theory may ultimately be satisfactory.

Poincaré, however, did not consider Lorentz's hypothesis of a local-time coordinate as an ad hoc hypothesis, almost certainly because of the central role it played in the theorem of corresponding states which explained every known experiment accurate to first order in v/c.

By the end of the 19th century new developments in the various branches of physics, especially in electromagnetism, necessitated the extension of the conventions of classical mechanics into what Poincaré (1902) referred to as the "physical sciences," i.e., electromagnetic theory. Here, due to lack of sufficient empirical data, Newton's principles, the principle of conservation of energy and the principle of relative motion could no longer be considered as conventions. Nevertheless, their usefulness in the domain of classical mechanics, for establishing relationships among phenomena, suggested strongly the possibility of their validity in the physical sciences.

By 1902, Poincaré could write in his best-selling reprint volume *Science and Hypothesis* that Lorentz's theory was the "most satisfactory" in the microscopic domain because it "best explains the known facts"; in particular, it explained the null results of first-order optical experiments and this was important to Poincaré, whose philosophy and physics excluded absolute motion; but he still remained critical of the ad hoc hypothesis of contraction, and he wrote in *Science and Hypothesis* (1902):

> Experiments have been made that should have disclosed the terms of the first order; the results were nugatory. Could that have been by chance? No one has admitted this; a general explanation was sought, and Lorentz found it. He showed that the terms of the first order should cancel each other, but not the terms of the second order. Then more exact experiments were made, which were also negative; neither could this be the result of chance. An explanation was necessary, and was forthcoming; they always are; hypotheses are what we lack the least.

However, at this time, it was the Lorentz theory's violation of the principle of action and reaction that most concerned Poincaré because this principle was a convention.

1.7.2. The Principle of Action and Reaction

In the *Festschrift* celebrating the 25th anniversary of Lorentz's doctorate, Poincaré (1900b) proposed a means for reconciling Lorentz's electromagnetic theory with the principle of action and reaction. Let us first review the theoretical situation in 1900. Lorentz's assumption of an absolutely resting ether meant that the ether acted on matter, but not the reverse; consequently, the violation of Newton's third law was built into Lorentz's electromagnetic theory, and Lorentz discussed that issue in the *Versuch*. Using the electromagnetic field equations to replace ρ and ρw in the force density Eq. (1.29) Lorentz obtained

$$f = \nabla \cdot \ddot{\mathbf{T}} - \frac{d\mathbf{g}}{dt} \qquad (1.101)$$

where the tensor-dyadic $\ddot{\mathbf{T}}$ is Maxwell's stress tensor,

$$\ddot{\mathbf{T}} = \frac{1}{4\pi}(\mathbf{EE} + \mathbf{BB}) - \ddot{\mathbf{I}}w, \qquad (1.102)$$

where $\ddot{\mathbf{I}}$ is the identity dyadic, w is the electromagnetic energy density

$$w = (E^2 + B^2)/8\pi \qquad (1.103)$$

and

$$\mathbf{g} = \frac{1}{4\pi c}(\mathbf{E} \times \mathbf{B}) \qquad (1.104)$$

which Lorentz identified as related to "Poynting's energy current."[30]

Lorentz calculated that the net force F_N on every charged particle contained in a volume V is

$$F_{\text{Net}} = \int f \, dV = \oint \ddot{\mathbf{T}} \cdot \hat{n} \, d\sigma - \frac{d}{dt} \int \mathbf{g} \, dV \qquad (1.105)$$

where he used Gauss' law for writing $\int \nabla \cdot \ddot{\mathbf{T}} \, dV = \oint \ddot{\mathbf{T}} \cdot \hat{n} \, d\sigma$, thereby expressing the stress tensor in the form intended by Maxwell to signify that the medium external to the enclosed volume V exerts forces across its surface, and $\hat{n}$ is the outward normal to the element of surface area $d\sigma$.

For stationary systems $\mathbf{g}$ was time independent and so Eq. (1.105) became

$$F_{\text{Net}} = \int f \, dV = \oint \ddot{\mathbf{T}} \cdot \hat{n} \, d\sigma, \qquad (1.106)$$

which was familiar from Maxwell's electromagnetic theory.

Lorentz focused upon the case of no charges within the volume V; then $F_{\text{Net}} = \mathbf{0}$,[31] and Eq. (1.105) becomes

$$\oint \ddot{\mathbf{T}} \cdot \hat{n} \, d\sigma = \frac{d}{dt} \int \mathbf{g} \, dV \qquad (1.107)$$

from which he concluded that whenever Poynting's energy current varied with time, the ether was not in equilibrium. Von Helmholtz (1894b) had also noticed this point and had sought equations characterizing the ether's flow. Such speculations were antithetical to Lorentz's own electromagnetic theory in which the Maxwell stresses were "fictional stresses" and were relegated to only calculational purposes. Thus Lorentz (1895) did not feel compelled "to raise [Newton's third law] to a fundamental law of unrestricted validity." Saying no more about this matter, he went on to develop the consequences of the new electromagnetic theory.

Poincaré (1900b), however, could not permit such a blemish on Lorentz's theory; his reasons were twofold: (1) According to classical mechanics Newton's third law led to the important law of conservation of momentum, and consequently violation of Newton's third law vitiated this conservation law as well. (2) Poincaré demonstrated that Newton's third law was linked with two other conventions – the principles of relative motion and energy – and the principle of relative motion imposed itself "*impérieusement à l'esprit*": "the mutual action between two bodies depends only on the relative position and relative velocity, but not on their absolute position and absolute velocity."

Poincaré analyzed the case in which the volume V contained a charged body. Extending the integral in Eq. (1.105) out to infinity, he obtained for the net force F_{Net}

$$F_{\text{Net}} = - \frac{d}{dt} \int g \, dV, \tag{1.108}$$

since the external electromagnetic fields vary as $1/r^2$; thus the contribution from the Maxwell stresses vanished. Then setting $F_{\text{Net}} = d/dt(mv)$, where m was the inertial mass of the charged particle, Poincaré obtained

$$\frac{d}{dt}\left[mv + \frac{1}{4\pi c} \int E \times B \, dV \right] = 0. \tag{1.109}$$

In order to rescue the principle of reaction, Poincaré drew from Eq. (1.109) the conclusion that the principles of reaction, and of conservation of momentum, were no longer applicable only to matter with mass; rather, electromagnetic energy, which he likened to a "*fluid ficitif*" with mass, had a "momentum" G given by

$$G = \frac{1}{4\pi c} \int E \times B \, dV; \tag{1.110}$$

the equation for the conservation of momentum for a system consisting of an emitter with mass and unidirectional radiation was the constancy of the quantity $mv + G$; thus, the momentum G compensated for the recoil of an emitter of unidirectional radiation. In other words, even after the surface enclosing the charged particle was allowed to become infinite in extent, thereby enclosing also the sources of the external field, the net force on the entire system still did not vanish; in order to simulate an isolated closed system, Poincaré interpreted the quantity dG/dt as a force F' so that $F_{Net} + F' = 0$, thereby satisfying Newton's third law.

Poincaré next considered the case of an emitter and absorber of radiation in relative inertial motions. Here the principle of relative motion was imperfectly satisfied, because its application necessitated a "compensating mechanism" – the local time.

Applying the principle of relative motion (with the local time) and the principle of conservation of energy to the case under investigation, Poincaré encountered the need for yet another hypothesis so that Newton's third law could be satisfied. In order to free the theory of certain offending terms, Poincaré postulated an "apparent complementary force" of the ether on the emitter and absorber *separately*; therefore Newton's third law could now be satisfied separately by the emitter and absorber.[32]

In a summary presentation in *Science and Hypothesis*, Poincaré (1902) succinctly described how Lorentz's theory could be altered to preserve Newton's third law:[33]

> According to Lorentz, we do not know what the movements of the ether are; and because we do not know this, we may suppose them to be movements compensating those of matter, and re-affirming that action and re-action are equal and opposite.

Consequently, in Poincaré's view, it was permissable to multipy hypotheses in order to save a theory that explained adequately a wide range of phenomena, yet whose structure violated a part of classical mechanics. However, this procedure was invalid if it was a matter of explaining a single piece of experimental data; on this point Poincaré was adamant, for, in the same essay of 1900, he harshly criticized Lorentz for the hypothesis of contraction – "hypotheses are what we lack least."

In a lengthy letter of 20 January 1901, Lorentz assessed Poincaré's valiant attempt at saving Newton's third law: "But must we, in truth, worry ourselves about it?" Lorentz explained that the result of Michelson and Morley most concerned him, and that it was "impossible" to alter the electromagnetic theory in order to include Poincaré's apparent complementary force because the theory was based upon the premise of a motionless ether: "I would not have been led to that theory if the phenomena of aberration had not forced me to it." Furthermore, Lorentz continued, just as the Maxwell stresses had no physical meaning, the quantity G could not be interpreted physically as a momentum for this could be construed as meaning that the ether was moving. For Lorentz the key issue concerning Newton's third law was that in an ether-based theory the action of an emitter of unidirectional radiation could not be simultaneously compensated for by the reaction of the absorber, and Poincaré had dodged this point [see Miller (1979a)].

But developments subsequent to 1900 in a new discipline, named by Wiechert in 1896 "*Elektronentheorie*," emphasized that the principle of action and reaction was transcended by the law of conservation of momentum.

1.8. THE ELECTROMAGNETIC WORLD-PICTURE

By 1900 attempts to deduce the laws of electromagnetism from increasingly complex mechanical models of the ether (i.e., a mechanical world-picture) paled before the successes of Lorentz's electromagnetic theory which took the equations of electromagnetism as axiomatic — for example, Lorentz's theorem of corresponding states that, to order v/c, could explain optical phenomena in moving bodies. But as Lorentz's theory stood in 1895 little was known of the charge and mass of its fundamental particle, the ion. These quantities were absorbed in Lorentz's procedure for averaging over the ion's fields, and in dispersion theory they occurred in the form e^2/m.

This situation was changed in 1896 when Pieter Zeeman, a *Privatdozent* at Leiden and a former student of Lorentz, discovered an effect that had eluded Faraday himself — the splitting of the spectral lines from a sodium flame that was situated in a magnetic field. Zeeman (1897) recalled that Lorentz had immediately explained the splitting. According to Lorentz's electromagnetic theory the molecules that constituted matter were composed of ions that were capable of being displaced, and a fixed part. Lorentz's explanation of the Zeeman effect was based on his assuming, as he had done in (1892a), that the mobile ions underwent oscillations. Lorentz's fit to the frequencies of the observed spectral lines, required that the charge-to-mass ratio of the mobile ions was of the order of 10^7 emu/gm, and that these ions were charged negatively. (In the Gaussian (cgs) system of units the unit for the elementary electric charge e is an electrostatic unit (esu) or statcoulomb. Circa 1905, experimentalists used the absolute electromagnetic system of units in which the unit for the elementary charge ε is an electromagnetic unit (emu) or abcoulomb. The relation between e and ε is $\varepsilon = e/c$. Max Abraham, and other theorists who employed the Gaussian (cgs) units, preferred to compare their calculations directly with the experimental data instead of multiplying these data with a factor of c in order to convert from (emu/gm) to (esu/gm). Consequently, many theorists designated the charge to mass ratio as (e/cm) or as (ε/m). Hereafter for historical accuracy I use the units of the charge-to-mass ratio as (emu/gm).) This charge-to-mass ratio was 10^3 less than that of the positively charged hydrogen ion that was known from electrolysis. Combining ε/m from the Zeeman effect with empirically determined quantities from dispersion, Lorentz showed that $m/m_H = 1/350$, where m_H was the mass of a hydrogen atom, and so $\varepsilon/m_H \simeq 10^4$. From electrolysis it was known that $\varepsilon'/m_H \simeq 10^4$, where ε' was the charge on the hydrogen ion active in electrolysis. Thus, with some confidence, Lorentz could write that the charge on the hydrogen ion was of the "same order of magnitude" as the charge on the hydrogen atom's lighter constituent. By the time that Lorentz published his results (1898), Thomson (1897) had found empirically that the charge-to-mass ratio for the cathode ray, i.e., the free electron, was also of the order of 10^7. Thomson (1899) went on to determine that the charge on cathode rays was the same as that on the hydrogen ion from electrolysis, i.e., of the order of 10^{-10} esu. The particle elementary to Lorentz's theory had at last been found.

With these accomplishments in view, in addition to Lorentz's impressive 1900 speculations on reducing gravitational interactions to those of elec-tromagnetism,[34] Wilhelm Wien proposed in the Lorentz *Festschrift* "the possibility of an electromagnetic foundation for mechanics." Wien had in mind a research effort whose goal was to deduce mechanics, and then all of physical theory, from the equations of Lorentz's electromagnetic theory. This electromagnetic world-picture was a bold suggestion because, as Wien wrote, it was "diametrically opposite" to Hertz's (1894) well-developed proposal for a thus far unsuccessful mechanical world-picture. Wien went on to set the stage for this research program.[35]

A far-reaching implication of an electromagnetic foundation for mechanics was that, if matter could ultimately be reduced to electrical charges and an ether, then mass had its origin in the electromagnetic field, as Thomson (1881) had suggested, although without any intent or desire to reduce mechanics to electromagnetism. By means of a hydrodynamic analogy, Thomson had demonstrated that the inertia of a charged body increased when the body was set in motion in the ether, because moving charged matter constituted a current that gave rise to a magnetic field acting back upon the charge. This self-induction effect would add to the moving body's inertia or mass, and Thomson calculated it to be a constant quantity.

Heaviside (1889) improved upon Thomson's calculation of the elec-tromagnetic field due to a uniformly moving charge and obtained the nonconstancy of a moving charged body's mass.[36] Wien emphasized Searle's (1897) results for the energy of charged moving spheres and Heaviside ellipsoids that had succeeded in clarifying the work of Thomson (1889) and Heaviside (1889). For example, Searle's expression for the total energy of a spherical body of radius R with a uniform surface distribution of charge in uniform straight-line motion was

$$E^e = \frac{e^2}{2R}\left[\frac{1}{\beta}\ln\left(\frac{1+\beta}{1-\beta}\right) - 1\right], \qquad (1.111)$$

(where $\beta = v/c$) which demonstrated that the body's energy and therefore its mass increased as a result of its motion. (*N. B. Hereafter, all quantities superscripted with e are derived from the charged body's self-electromagnetic fields.*)[37]

Wien went on to point out that Philipp Lenard's recent measurements of ε/m for cathode rays that had been accelerated to $c/3$ displayed an increase of mass with velocity [Lenard (1898), (1900)]. However, continued Wien, "these quantitative measurements are not regarded as conclusive." Wiens's reason was that Lenard's communications contained three data points for ε/m as a function of increasing velocity. The middle value for ε/m was less than the other two. Wien concluded (1900) by emphasizing that "the possibility for or against [an electromagnetic foundation for mechanics] is to be decided by experiment." Although Lorentz did not immediately throw his support behind Wien's program, in (1901b) he underscored the importance of accurate measurements for deciding between "real and apparent masses," where real meant mechanical and apparent referred to the electromagnetic mass.

1.9. WALTER KAUFMANN'S EXPERIMENTS OF 1901–1902

> And here we come to a question which deeply affects the structure of matter in general. If an electric atom only in virtue of its electrodynamic properties behaves like an inert particle, is it then possible to regard all masses as only apparent? May we not, instead of the sterile efforts to reduce electrical to mechanical phenomena, attempt the reverse process of reducing mechanics to electrical principles? Here we return to views already cultivated by Zöllner 30 years ago and lately improved upon by H. A. Lorentz, J. J. Thomson and W. Wien: *if all material atoms consist of conglomerates of electrons, then their inertia results as a matter of course* (italics in original).
>
> W. Kaufmann (1901a)

Walter Kaufmann, a Göttingen experimentalist, decided to seek the empirical evidence on the electron's mass that Wien had called for in 1900. In a lecture delivered September 1901 to the 73rd *Naturforscherversammlung* at Hamburg,* Kaufmann reviewed the state of electromagnetic theory. He (1901a) caught the mood of the times when he said that efforts toward a mechanical world-picture were "sterile"; and that the coincidence of the charge-to-mass ratio for the particle responsible for the Zeeman effect and the cathode ray was a "stunning result," and constituted a "direct proof" of Lorentz's theory. Kaufmann emphasized excitedly the central role played by the electron in such wide-ranging fields of research as luminous vapors (e.g., Zeeman effect) dispersion, stellar aberration, electrolysis, the electron theory of metals, photoelectric effect, decay of "uranium compounds," and, further-more, the electron's charge obtained from Planck's theory of cavity radiation was nearly equal to the one from electrical experiments. Kaufmann proclaimed that unification was in the air: "Everywhere, therefore, in all states of aggregation the electron plays an important part in electrical and optical phenomena." Electrons, continued Kaufmann, were the "long-sought-for 'primordial atoms'," which Thomson (1897) had asserted were necessary to replace the hydrogen atom in order to render tenable Prout's hypothesis.

---

* The *Naturforscherversammlung*, whose name in full is *Versammlung deutscher Naturforscher und Ärzte* (Meeting of German Scientists and Physicians), was held annually in September. The proceedings of the science section, with discussion sessions, were published in the *Physikalische Zeitschrift*.

Along with Thomson, Wien and Lenard, Kaufmann was considered as one of the foremost experimenters on the characteristics of the electron. Kaufmann earned his doctorate at Munich in 1894 and was an assistant at the Physics Institute at the University of Berlin from 1896–1899 [(Campbell (1973) and Kossell (1947)]. In 1897, while at the University of Berlin, Kaufmann had performed several highly regarded experiments that served to further elucidate the nature of cathode rays, particularly their charge-to-mass ratio.[38]

By 1901, Kaufmann, among other physicists, had ascertained empirically that the rays emitted from radium salts behaved in electric and magnetic fields as if they were composed of negatively charged particles. From the results of these deflection experiments, performed separately for electric and magnetic fields, researchers concluded that the charge-to-mass ratio of the rays was of the same order of magnitude as that of cathode rays, i.e., 10^7. The rays from radium salts were at first referred to as Becquerel rays after Henri Becquerel who had discovered them in 1898, and then by the name given to them arbitrarily by Ernest Rutherford (1899), β-rays. In 1901 the terms β-ray and electron were used interchangeably because empirical data indicated the indentity of β-rays with the electrons in Zeeman effect and cathode rays. Since for the same magnetic field strength the β-rays were deflected less than cathode rays, then β-rays moved faster than cathode rays. Cathode rays had been accelerated to 0.3c, and β-rays were measured to move at velocities exceeding 0.9c. In fact, the high velocities of cathode rays had led the noted Australian experimentalist William Sutherland to proclaim (1899): "The electrons stream through the aether with nearly the velocity of light and yet provoke no noticeable resistance. What wonder, then, that any aethereal resistance to planetary motion has remained beyond our ken!"

The constancy of the charge-to-mass ratio for cathode rays, however, was not expected to persist for velocities approaching that of light, i.e., for β-rays. Thomson, himself, had mentioned this point in the description of his classic experiments of 1897 and again in (1899) [see also Sutherland (1899)]. We recall that Thomson (1881) had been the first to have emphasized in print the possibility of a velocity dependence of the mass of moving charged bodies. Kaufmann (1901a, b) wrote that the question of whether the electron's mass was "real" or "apparent" was much discussed in scientific circles. To Kaufmann in 1901 (now at Göttingen), β-rays were the ideal candidates for investigating the "true nature" of the electron, even though it was a "complete riddle" how radium salts could spontaneously emit particles with velocities that were almost impossible to obtain using electric forces.

Whereas the velocity spectrum of the cathode rays was discrete, the velocity spectrum of the β-rays was, Kaufmann wrote, "inhomogeneous," i.e., continuous. He considered this characteristic of the β-ray spectrum as an "inconvenience." In order to disentangle the various velocities in the β-ray spectrum, Kaufmann took recourse to a technique from optics—"Kundt's method of crossed spectra" [Kaufmann (1901b, c)]. Most likely because

August Adolph Kundt's experiments were so well known in 1901, Kaufmann did not elaborate on the direct connection between the design of his experimental apparatus on β-radiation and Kundt's optical experiments on light radiation. A review of Kundt's work reveals what could well have been

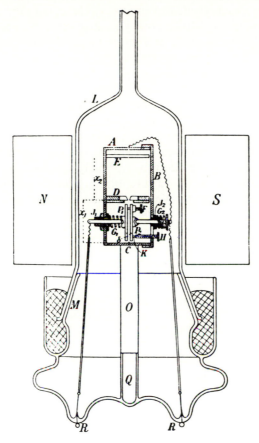

FIG. 1.7. Kaufmann's schematic of his experimental apparatus for demonstrating the velocity dependence of the electron's mass. A granule of radium bromide (1 mm long by 0.3 mm thick) was placed at C with its longer dimension normal to the plane of the diagram, and closely-spaced capacitor plates ($\delta = 0.1525$ cm and height 1.775 cm) served to form a collimated beam of β-rays

with velocities of the order of 0.9 c. The sense of the coordinate axes is $y \longrightarrow$. An average plate

voltage of 7,000 V was the source of an electric field $\boldsymbol{E} = -E\hat{j}$. The measurement chest A was of dimensions $2 \times 3 \times 4.5$ cm and was situated in an evacuated enclosure, which was between the poles of an electromagnet (labeled N and S) of average field strength 300 Gauss, and $\boldsymbol{B} = -B\hat{j}$. The β-ray beam passed through the combined field region of length $x_1 = 2.07$ cm. Then the beam emerged through the diaphragm D (0.5 mm diameter) that was carefully aligned with C. Along the distance $x_2 = 2$ cm the beam was acted on only by the magnetic field before striking the photographic plate E. (From *Göttinger Nachr.*)

the connection. In a series of experiments performed during 1871–1872, Kundt had adapted a method, due originally to Newton, to the investigation of anomalous dispersion. Kundt disentangled the wavelengths in the region of an absorption band by passing a collimated beam of sun light that was incident first on a prism which yielded normal dispersion, through a second prism which yielded anomalous dispersion. The second prism's refractive face was oriented normally to that of the first prism, hence the name method of crossed spectra. The resulting curve exhibited the discontinuous behavior of the refractive index in the neighborhood of an absorption band, i.e., anomalous dispersion. Before Kundt the phenomenon of anomalous dispersion was doubted owing to the difficulty in demonstrating it.

Kaufmann's electromagnetic analogue of Kundt's method of crossed spectra is in Fig. 1.7. The closely-spaced capacitor plates defined a collimated beam of β-rays. The different velocities of the ray's constituent electrons were disentangled by passing the electrons normally through parallel electric and magnetic fields, which spread out the inhomogeneous electron beam into a curve on a photographic plate. When Thomson had used electric and magnetic fields in combination, they were oriented normally to each other in order to form a spot on a fluourescent screen from which the cathode ray's velocity v and ε/m could be determined. Under the influence of Kundt's optical experiments, Kaufmann used parallel electric and magnetic fields in order to

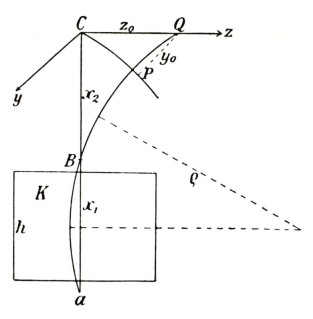

FIG. 1.8. Kaufmann's schematic for illustrating the electron's trajectory in the apparatus from Fig. 1.7. (From *Göttinger Nachr.*)

obtain a curve. We shall see in Chapter 12 that Kaufmann's opting for an analogue of Kundt's method over Thomson's crossed-field arrangement was a tactical error because analysis of the resulting complicated curve was fraught with too many possibilities for error. In 1907, Adolf Bestelmeyer applied Thomson's crossed-field arrangement to an inhomogeneous beam of cathode rays in order to select out a particular velocity, i.e., he invented the velocity filter.

Kaufmann used the schematic in Fig. 1.8 for describing his method of data analysis. The line aBC defined the x-axis as the direction of β-rays that were undeviated by the electric and magnetic fields that were oriented parallel to the negative y-axis. Owing to the distribution of velocities in the beam emerging from the diaphragm B, a curve was expected on a photographic plate in the y-z-plane. Electric and magnetic fields were necessary because Kaufmann sought to determine two unknowns from a point on the curve – the electron's velocity v and its charge-to-mass ratio. The electric field E and the magnetic field B caused an electron to be deflected into a trajectory whose projection on the x-z-plane was a curve of curvature ρ where

$$\frac{1}{\rho} = \frac{e}{\mu vc} B; \tag{1.112}$$

since the maximal deflection $y_0 = 0.198$ cm was much less than the measured distance that the electron traveled from the aperture labeled B in Fig. 1.8, then to 0.3% accuracy Kaufmann could neglect the y-component of the deflected electron's velocity and set $v_x = v = $ constant to obtain Eq. (1.112); Kaufmann's symbol for the electron's mass is μ. (Hereafter I shall designate the electron's mass with the symbol μ because this was the convention in German-language scientific papers during the first decade of the 20th century. The subscript o designates the electron's mass at low velocities compared to that of light. Thus Simon's measurement of the electron's charge-to-mass ratio is written as ε/μ_0.) From the projection of the β-ray's trajectory on the x-y-plane Kaufmann deduced that a β-ray emerged from the diaphragm labeled B in Fig. 1.8, at an angle θ, formed by the intersection of the tangent to the curve's projection in the x-y-plane, and the x-axis. It followed that[39]

$$\tan \theta = \frac{dy}{dx} = \frac{es_1}{\mu v^2} E \tag{1.113}$$

where s_1 was the point of maximum curvature, and Kaufmann set $v_x = v = $ constant. Consequently, a β-ray of velocity v suffered a deflection in the y-direction to the coordinate

$$y_0 = \frac{e}{\mu v^2} Es_1 s_2 \tag{1.114}$$

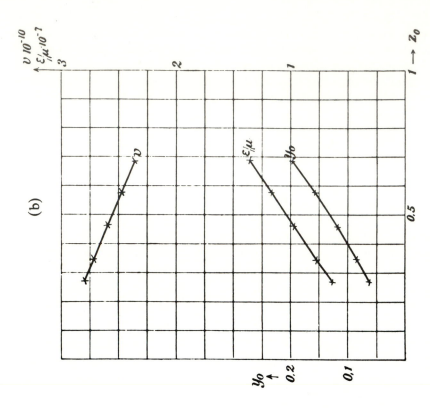

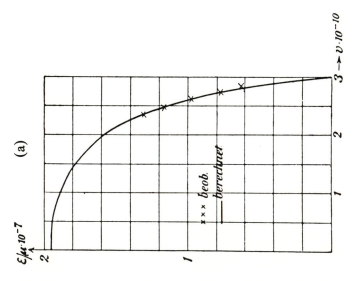

FIG. 1.9. Kaufmann's (1901c) experimental results for $\varepsilon/\mu_0^\varepsilon$ and v (*beob.* = observed and *berechnet* = calculated). In (b), y_0 and z_0 are the plate coordinates in cm. (From *Göttinger Nachr.*)

where s_2 was the projection on the x-y-plane of the β-ray's linear trajectory from B to y_0.

In summary, the magnetic field alone would have deflected an electron into the x-z-plane, and the electric field alone would have deflected an electron into the x-y-plane. With both fields in operation an electron was deflected into a curve whose projections on the x-z- and x-y-planes were Eqs. (1.112) and (1.114), respectively. The combined fields served to produce a curve on the photographic plate in the y-z-plane. Figure 1.8 contains a point P on the curve. In Section 1.11 we shall see that Kaufmann's (1902) mathematical analysis of the expected curve in terms of only y and z values, instead of the quantities s_1 and s_2, illustrated clearly that it was symmetric about the y-axis, i.e, $\varepsilon/\mu \sim z^2/y$, and thus was a parabola for a constant electron mass. Figure 1.8 shows that one-half of the curve was obtained for a particular orientation of the electric field. Kaufmann obtained the other half by reversing the direction of the electric field. (See Fig. 1.11.) Kaufmann used a micrometer to measure the quantities (y_0, z_0). The null point of the empirical curve was set by the "undeflected" Röntgen radiation from the radium bromide.

The cusp of the empirical curve at the intersection of the y- and z-axes designated the velocity $v = c$ because electrons arriving there would not have been deviated by the electric or magnetic fields, i.e., the trajectory curvature was $\rho = \infty$. The fastest β-rays populated the curve closest to the origin which was the region of interest to Kaufmann. From Eqs. (1.112) and (1.114) Kaufmann obtained

$$\beta = Es_1s_2/y_0\rho B. \tag{1.115}$$

Straightforward geometrical considerations permitted the calculation of ρ from the apparatus' dimensions and the z_0-coordinate on the plate.[40] Then the values for s_1 and s_2 were calculated. Kaufmann made two exposures over a period of 48 hours. From five data points (see Fig. 1.9) he found [see Eq. (1.112)] that ε/μ decreased as the β-ray's velocity approached that of light. Having established a velocity dependence of the electron's mass μ, Kaufmann turned next to the problem of what part of the mass was generated by the electron's self-electromagnetic fields. He took the electron's total mass μ as $\mu = m_0 + \mu^e$, where, like Lorentz, he called m_0 the "real mass" (i.e., the mechanical mass) and μ^e the "apparent mass" (i.e., the electromagnetic mass); he calculated the "apparent mass" from Searle's (1897) result for the energy of a moving sphere

$$\mu^e = \frac{1}{v}\frac{dE^e}{dv} = \frac{3}{4}\frac{\mu_0^e}{\beta^2}\left[-\frac{1}{\beta}\ln\left(\frac{1+\beta}{1-\beta}\right) + \frac{2}{1-\beta^2}\right] \tag{1.116}$$

where $\mu_0^e = \frac{4}{3}(e^2/2Rc^2)$.

Expressing the charge-to-mass ratio as

$$\varepsilon/\mu = \varepsilon/(m_0 + \mu_0^e\eta), \tag{1.117}$$

where m_0 is the mechanical mass, η is the velocity-dependent term in Eq. (1.116), Kaufmann used his five empirical values for velocities to obtain a least squares fit to the quantity ε/μ in Eq. (1.117). See Fig. 1.9 for Kaufmann's close agreement between the values for ε/μ calculated using Eq. (1.117) with those determined from the five data points.

From the least squares fit Kaufmann determined values for ε/μ that were accurate to within 0.8% of the measured ones, and "most probable values" of $\mu_0^e/\varepsilon = 0.122 \times 10^{-7}$ and $m_0/\varepsilon = 0.39$. Consequently, he calculated that for "very slow rays" the electrons that constituted β-rays had a charge-to-mass ratio of

$$\varepsilon/\mu_0 = 1.95 \times 10^7. \tag{1.118}$$

The value of ε/μ_0 in Eq. (1.118) was close to that obtained by Simon (1899) from the deflection of cathode rays moving at velocities of the order of 0.2c in a magnetic field. Simon's experiment had been performed at Berlin under the supervision of Kaufmann, and he had obtained

$$\varepsilon/\mu_0 = 1.865 \times 10^7 \tag{1.119}$$

where here μ_0 was the mass of an electron that constituted a cathode ray. Since $(\mu_0^e/\varepsilon)/(m_0/\varepsilon) = 0.313$ then, Kaufmann continued, approximately $\frac{1}{3}$ of a fast moving electron's mass was "apparent."

Kaufmann concluded his (1901b) report by stating how riddled with hypotheses was his data analysis — for example, that the electron was a sphere with a uniform surface charge distribution. But by early 1902 he thought otherwise of his data, for on 11 January 1902, Kaufmann's colleague at Göttingen, Max Abraham (1902a), proposed the first field-theoretical description of an elementary particle. Abraham assumed that the electron's mass originated entirely in its own electromagnetic field. He demonstrated the inappropriate use of "apparent mass" by Kaufmann, since it could be associated only with a force oriented along the electron's trajectory (which Abraham named the "longitudinal mass"), and not normally as in Kaufmann's experiments.[41] Abraham went on to calculate the electromagnetic mass that Kaufmann should have used, and he called it the "transverse mass." Kaufmann reanalyzed his 1901 data and found agreement with Abraham's prediction for the transverse mass [Kaufmann (1902a)]. At the 74th *Naturforscherversammlung* at Karlsbad, Kaufmann (1902b) proclaimed that the electron's entire mass was of electromagnetic origin. In a sequel paper, Abraham (1902b) reviewed his new theory and then announced the realization of an "electromagnetic mechanics." And so, he continued, "however continuously the ether manages to fill space, electricity has an atomistic structure." The coexistence of the continuous and the discrete was a hallmark of the new physics.

1.10. ABRAHAM'S THEORY OF THE ELECTRON

> The electron's inertia is produced exclusively by its electromagnetic field.
> Therefore: atomistic structure of electricity, but continuous space distribution of the
> ether! Let that be our solution.
>
> M. Abraham (1902b)

1.10.1. Abraham's Fundamental Assumptions

Abraham began his in-depth exposition of his new theory (1903) by emphasizing, as had Kaufmann, the fundamental role played by the electron in optical and electrical phenomena. This realization, continued Abraham, had led many physicists such as Wiechert [(1898), (1900)] to link Lorentz's "electromagnetic theory of light" to the cathode rays, i.e., free electrons. Abraham wrote: "For this electron theory of matter the problem of the dynamics of electrons is of fundamental meaning [in particular] the problem of whether the electron's inertia can be explained completely by the dynamical interaction of its electromagnetic field." Abraham next cited recent speculations that part of the electron's mass was electromagnetic in nature by such physicists as Sutherland (1899), Drude [e.g., (1900)], and Lorentz (1901b); Wien's (1900) bold suggestion of research toward an "electromagnetic foundation for all of physics"; and then there were Kaufmann's (1901c) data that confirmed the velocity dependence of the electron's mass. Abraham emphasized that Searle's result for the electromagnetic energy of a rigid sphere with a uniform surface charge distribution could produce only its longitudinal mass, because the particle's electromagnetic momentum was required for deducing its transverse mass. He then proceeded to develop his work on the "dynamics of the electron."

Accepting the electron's charge density as a fundamental quantity, Abraham set a goal in his papers of 1902–1903 of deriving the electron's purely electromagnetic mass from its self-electromagnetic fields, which he assumed could be described by the "Maxwell-Lorentz field equations" – Eqs. (1.25) to (1.28) – with E and B as the self-fields.[42] Abraham assumed the electron to be a rigid sphere, with either a uniform volume or a uniform surface charge distribution.

He chose a rigid electron because a deformable one could explode, owing to the enormous repulsive forces between its constituent elements of charge, unless nonelectromagnetic adhesive forces were postulated; but in that case, "the electromagnetic foundation of mechanics would be excluded from the outset." An indication of the effect of Hertz's foundational analysis toward clarifying the goals of physical theory was that Abraham, like Wien, emphasized the harmoniousness of the electromagnetic world-picture with Hertz's program of "tracing back every energy to the kinetic energy of moving masses, every force to kinematical connections." But Hertz's program defined a mechanical world-picture in which contiguous actions would have been explained by rigid connections among hidden masses, instead of through an ether.

For as Abraham emphasized particularly in his (1903), Hertz's goal of purifying mechanics could be achieved not through a mechanical world-picture but via an electrodynamical route in which the electron's mass was due to its self-induction; then the quantities force, torque, work, energy and momentum could be deduced from the electron's self-fields. Furthermore, in Abraham's view, Hertz's conditions for rigidity found their expression in the electromagnetic world-picture through postulation of the electron's invariable sphericity—that is, from the mechanical conditions that the total self and external forces (F^e and F_{ext}, respectively), and torques on the electron vanished:

$$\int (f^e + f_{ext})\, dV = 0, \qquad (1.120)$$

$$\int [r \times (f^e + f_{ext})]\, dV = 0, \qquad (1.121)$$

where he considered the force densities in Eqs. (1.120) and (1.121) as "auxiliary quantities (*Hülfsbegriffe*)" since they could be deduced from the electron's self-fields.

Thus, Abraham's mode of pursuing Hertz's goal of deducing kinematics from the motions of atoms permitted him to replace Hertz's complicated conditions of rigid connection with tractable rigidity conditions on the elementary particle itself. Abraham's fundamental equations were the Maxwell-Lorentz equations (which, he emphasized, were equivalent to the "Maxwell-Hertz equations" in the absence of charges), the rigidity conditions of Eqs. (1.120) and (1.121), and the mechanical condition that the electron moved like a rigid sphere, i.e.,

$$w = v + \omega \times r \qquad (1.122)$$

where w was the electron's velocity relative to the ether, and ω its angular velocity.

An important ingredient in Abraham's theory was the term g from Poincaré's Eq. (1.101) which Abraham referred to as the "electromagnetic momentum density." This was a bold move because, unlike Lorentz and Poincaré, Abraham took g to be a physical entity, thereby maintaining conservation of momentum at the expense of Newton's third law.

Toward deducing the electron's equations of motion, Abraham defined the quantities

$$G^e = \int g^e\, dV, \qquad (1.123)$$

$$\mathscr{L}^e = \int \boldsymbol{r} \times \boldsymbol{g}^e \, dV, \tag{1.124}$$

where $\boldsymbol{G}^e$ was the electron's momentum, and $\mathscr{L}^e$ was the angular momentum with $\boldsymbol{r}$ as radius vector from the ether-fixed system's origin to the electron's center of mass. Then, taking Newton's mechanics as a guide, he proposed equations of motion for the electron:

$$\frac{d\boldsymbol{G}^e}{dt} = \int \frac{\partial g^e}{\partial t} \, dV, \tag{1.125}$$

$$\frac{d\mathscr{L}^e}{dt} = -\int \boldsymbol{v} \times \boldsymbol{g}^e \, dV + \int \boldsymbol{r} \times \frac{\partial \boldsymbol{g}^e}{\partial t} \, dV. \tag{1.126}$$

Abraham explained that these were the equations of motion expected for a "rigid body in an ideal fluid." Since the fluid stressed the body, then the body's momentum need not be collinear with its velocity, thereby accounting for the unfamiliar first term on the right-hand side of Eq. (1.126). Although mechanics ensured that Eqs. (1.125) and (1.126) were solvable, in electromagnetism that was not the case; furthermore through the retarded potentials the electromagnetic-field quantities depended on the entire "prehistory of the electron's motion." Therefore, the situation was generally "hopeless" for solving Eqs. (1.125) and (1.126), except for a few special sorts of motions which Abraham proceeded to investigate.

Abraham's criterion for a motion to be a "distinguished motion" was that it could be described using the uniformly moving particle's rest system; thus, Abraham could use Lorentz's 1895 results for transforming the inhomo-geneous wave equation into a Poisson equation. (In the calculations to follow Abraham specialized to the distinguished motion of pure translation, i.e., $\boldsymbol{\omega} = \boldsymbol{0}$.) He interpreted Eqs. (1.63)–(1.66) as transforming the electron to a "resting system" in which dimensions along the x-axis were stretched in the ratio $1 : \sqrt{1 - v^2/c^2}$; and so the electron's "image" in S'' was an ellipsoid with axes $(\gamma R, R, R)$. Abraham emphasized that S'' was a "real system if $v/c < 1$"—that is, real in the mathematical sense of the term because Abraham's electron was a rigid sphere.

An important expression of "Newton's first axiom" in Abraham's theory was the inertial motion that could be maintained without any external forces balancing the electron's self-forces. This "force-free" motion satisfied the equations

$$d\boldsymbol{G}^e/dt = \boldsymbol{0}, \tag{1.127}$$

$$d\mathscr{L}^e/dt = \boldsymbol{0}. \tag{1.128}$$

But Eq. (1.128) did not guarantee force-free inertial motion unless the electron's velocity and momentum were collinear [see Eq. (1.126)]. Using a symmetry argument stunning in its modernity, Abraham investigated how the electron's electromagnetic fields transformed under inversions of the x-, y-, and z-axes, and inferred: the momentum and velocity of a moving charge are collinear if the charge's shape is symmetric about axes perpendicular to its direction of motion. Consequently, Abraham's real spherical electron could perform force-free motion and so also could the imaginary electron in S'' whose major axis was along the x''-axis. Abraham went on to mention one more condition that had to be satisfied in order for the rigid spherical electron to execute force-free motion: its velocity had to be smaller than that of light.

Abraham completed his analysis of uniform motion by setting his theory of the electron into a Lagrangian formulation. He defined the Lagrangian function in terms of the electron's self-electromagnetic fields as

$$L^e = \frac{1}{8\pi} \int (B^2 - E^2) \, dV \qquad (1.129)$$

which could be written in terms of the convection potential ψ as

$$L^e = -\tfrac{1}{2} \int \rho \Psi \, dV. \qquad (1.130)$$

Abraham verified the consistency of the Lagrangian formulation by showing that the electron's momentum calculated from its defining expression in Eq. (1.123) was the same as would be calculated from Eq. (1.129), as

$$\boldsymbol{G}^e = \partial L^e / \partial \boldsymbol{v}, \qquad (1.131)$$

and that the electron's total energy $E^e = (1/8\pi) \int (E^2 + B^2) \, dV$ followed from its customary expression in Lagrangian mechanics as

$$E^e = -L^e + \boldsymbol{v} \cdot \boldsymbol{G}^e. \qquad (1.132)$$

For Abraham the beauty of the Lagrangian formulation was that the Lagrangian L^e in Eq. (1.130) could be most simply calculated using the electron's self-electromagnetic potential, charge distribution, and volume element in S'', as

$$L^e = -\sqrt{1 - v^2/c^2} \; E_0^e \qquad (1.133)$$

where E_0^e was the electron's energy in its fictitious rest system S''. Abraham emphasized that Eq. (1.133) was a new result in electrodynamics. Searle (1897) had calculated E^e which was the total energy in the laboratory reference system of the rigid spherical electron [see Eq. (1.111)], but Abraham required E_0^e. In the German edition of Maxwell's *Treatise* Abraham found the electromagnetic energy of an ellipsoid with axes $(\gamma R, R, R)$, that possessed a uniform surface charge distribution, i.e., the E_0^e for the "image" in S'' of the rigid spherical electron; then the Lagrangian in Eq. (1.133) became

$$L^e = -\frac{e^2}{2R}\left(\frac{1-\beta^2}{2\beta}\right)\ln\left(\frac{1+\beta}{1-\beta}\right) \qquad (1.134)$$

(for a uniform volume distribution of charge the $\frac{1}{2}$ in $e^2/2R$ is replaced by $\frac{3}{5}$) and its momentum followed from Eq. (1.131)

$$G^e = \frac{e^2}{2Rc}\left[\frac{1+\beta^2}{2\beta^2}\ln\left(\frac{1+\beta}{1-\beta}\right) - \frac{1}{\beta}\right]. \qquad (1.135)$$

Abraham was struck by the validity in the microscopic regime of that most elegant and useful instrument of theoretical physics – Lagrangian mechanics: "[for velocities less than that of light] our electromagnetic foundation ... extends the sphere of influence of Lagrangian mechanics in a most remarkable manner." Abraham (1903) continued in a vein that indicated his attachment to the thinking of Ernst Mach: "This result is significant not only as a contribution to the theory of knowledge but also as a more economical expression because it traces back the dynamics of any sort of motion to the calculation of the Lagrangian function."

1.10.2. Abraham's Predictions for the Electron's Mass

Abraham's next step was calculating the electron's mass from the momentum of its self-fields [Eq. (1.131)]. The resultant force acting on the electron had two parts, one due to external fields (F_{ext}) and the other due to its own fields; thus

$$F^e + F_{ext} = m_0 a \qquad (1.136)$$

where m_0 was the mechanical (inertial) mass. Since in the electromagnetic world-picture $m_0 = 0$, then Eq. (1.136) becomes

$$F^e + F_{ext} = 0. \qquad (1.137)$$

Extending the integration region for the electron's self-fields to infinity, Abraham deduced from the "Lorentz-Poincaré transformation" [Eq. (1.101)] that

$$F^e = -dG^e/dt. \qquad (1.138)$$

Consequently, Eq. (1.137) became

$$F_{ext} = dG^e/dt, \qquad (1.139)$$

which was Abraham's basic equation for calculating the electron's mass. Abraham went on to show that unambiguous identification of the electron's mass as the coefficient of its acceleration required restricting the particle to "quasi-stationary motion": the electron's rate of change of momentum could be equated to its mass times acceleration if the electron's acceleration occurred during a time interval t that was much greater than the time taken for light to traverse the electron's radius R, i.e., $R/(ct) \ll 1$.[43] During this sort of acceleration the electron did not radiate. After establishing that the approximation of quasi-stationary motion could be applied to Kaufmann's experiment, Abraham replaced Eq. (1.139) with

$$F_{\text{ext}} = dG_L^e/dt + dG_T^e/dt \tag{1.140}$$

where

$$dG_L^e/dt = m_L a_L \tag{1.141}$$

and

$$dG_T^e/dt = m_T a_T, \tag{1.142}$$

m_L and m_T are the electron's "longitudinal" and "transverse" masses, respectively, which depend on the components of the external force along and normal to the electron's trajectory. From Eq. (1.135) Abraham calculated these masses as

$$m_L = \frac{e^2}{2R\beta^3 c^2}\left[\frac{2\beta}{1-\beta^2} - \ln\left(\frac{1+\beta}{1-\beta}\right)\right], \tag{1.143}$$

$$m_T = \frac{e^2}{4R\beta^3 c^2}\left[(1+\beta^2)\ln\left(\frac{1+\beta}{1-\beta}\right) - 2\beta\right]. \tag{1.144}$$

Thus, wrote Abraham, "the electromagnetic mass ... is a tensor of rotatory symmetry whose symmetry axis is defined by the direction of motion." Whereas Kaufmann's 1901 data were not in good agreement with Abraham's electron theory, Abraham emphasized the agreement of Kaufmann's (1902a) analysis with m_T from Eq. (1.144) to 1% in the region $0.6 < v/c < 0.9$.[44] Owing to the small deflections of the β-rays, the applied electromagnetic force could be taken as transverse to their trajectory, and hence to a good approximation Kaufmann measured only m_T.

For $\beta \ll 1$ the Eqs. (1.143) and (1.144) become

$$m_L = \frac{2}{3}\frac{e^2}{Rc^2}\left[1 + \frac{6}{5}\beta^2 + \frac{9}{7}\beta^4 + \cdots\right], \tag{1.145}$$

$$m_T = \frac{2}{3}\frac{e^2}{Rc^2}\left[1 + \frac{6}{3\cdot5}\beta^2 + \frac{9}{5\cdot7}\beta^4 + \cdots\right], \tag{1.146}$$

and in the limit of $\beta = 0$, the electromagnetic world-picture predicted that the electron had a mass

$$\mu_0^e = \frac{2}{3}\frac{e^2}{Rc^2}. \tag{1.147}$$

Abraham estimated the size of the electron's radius as follows: from Eq. (1.147) Abraham obtained

$$R = \frac{2}{3}\left(\frac{\varepsilon}{\mu_0^e}\right)\varepsilon.$$

He assumed that in the limit of $v = 0$ he could replace ε/μ_0^e with Simon's value in Eq. (1.119). From electrolytic experiments, and experiments such as Thomson's (1899), the electric charge e was known to be of the order of 10^{-10} esu. Consequently, Abraham estimated that the electron's radius was about 10^{-13} cm.

[Since the electrostatic energy E_0^e of a spherical electron with a uniform surface charge distribution is

$$E_0^e = e^2/2R, \tag{1.148}$$

then the electron's electrostatic, i.e., rest, mass is

$$m_0^e = \frac{e^2}{2Rc^2}. \tag{1.149}$$

The $\beta = 0$ limit of m_L and m_T is $\mu_0^e = (2/3)\, e^2/(Rc^2)$ or $\frac{4}{3}m_0^e$. The $\frac{4}{3}$-factor is discussed in Sec. 12.5.8.]

Abraham concluded his 1903 paper by proving that force-free motion is stable if the electron in S'' is ellipsoidal in shape, with its major axis along the direction of its motion.

In summary, by 1903 Abraham had apparently succeeded in explaining the electron's mechanical properties, albeit within the quasi-stationary approximation: mechanics had been absorbed into electromagnetism.

1.11. KAUFMANN'S 1902–1903 DATA

The importance of Kaufmann's data to the basic physics of his time requires that we survey his 1902–1903 experiments and data analysis.

Besides using the incorrect mass in 1901, Kaufmann's equation for determining the curvature ρ of the electron's trajectory in the magnetic field was also incorrect.[40] Kaufmann (1902a) took these corrections into account and then applied the empirical data from 1901 to Abraham's prediction of m_T as follows. From Eqs. (1.112) and (1.114), and identifying μ as m_T,

$$\frac{e}{m_T} = \frac{s_1 s_2}{y_0 \rho^2} \frac{Ec^2}{B^2}. \tag{1.150}$$

In order to isolate the velocity dependence of Abraham's m_T, Kaufmann rewrote it as

$$m_T = \tfrac{3}{4}\mu_0^e \psi(\beta) \tag{1.151}$$

where

$$\psi(\beta) = \frac{1}{\beta^2}\left[\frac{1+\beta^2}{2\beta}\ln\left(\frac{1+\beta}{1-\beta}\right) - 1\right] \tag{1.152}$$

where $\psi(0) = \frac{4}{3}$, and so Eq. (1.150) became

$$\frac{s_1 s_2}{y_0 \rho^2} \psi = \frac{4}{3} \frac{\varepsilon}{\mu_0^e} \frac{B^2}{Ec}. \tag{1.153}$$

Using empirical values for y_0 and z_0, Kaufmann could calculate s_1 and s_2, and thereby determine β [see Eq. (1.115)]. Then he calculated ψ. Kaufmann found that to 1.2% the left-hand side of Eq. (1.153) remained constant, and so he proclaimed: "*The electron's mass is purely electromagnetic in nature*" (italics in original).

Kaufmann encountered a problem in calculating ε/μ_0^e from Eq. (1.153), namely, that his empirical curve indicated that at $z_0 = 0$ the electron's velocity was 2.785×10^{10} cm/sec, instead of $c = 3 \times 10^{10}$ cm/sec. He assumed that this 7.2% error resided in observational errors, and perhaps in inhomogeneities in the electric field. In order to compensate for the 7.2% discrepancy, he multiplied values of v with the factor 3/2.785, which raised the value of ε/μ_0^e from Abraham's theory [Eq. (1.153)] to

$$\left(\frac{\varepsilon}{\mu_0^e}\right)_{\text{corrected}} = 1.84 \times 10^7, \tag{1.154}$$

which differed by 1.34% from Simon's. Kaufmann set his goal at improving his experimental procedures in order to remove the 7.2% error.

In September 1902 Kaufmann was hard at work with improved apparatus and a new method for data analysis. He described these new developments at the 74th *Naturforscherversammlung*, held in September at Karlsbad, but he was prepared to display only his earlier (1902b) result for $(\varepsilon/\mu_0^e)_{\text{corrected}}$. Perhaps it was Abraham's presence that imbued Kaufmann with the courage to assert that the 7.2% error would disappear in the new experiments, and that the "*mass of the electron is purely electromagnetic in nature*," and was described by Abraham's theory. In the ensuing discussion session, Kaufmann deferred a question pertaining to how ψ was calculated to Abraham. Abraham replied that he could offer only a theoretical derivation and let Kaufmann discuss the empirical determination of ψ. Abraham delivered the next paper (1902b).

On 7 March 1903, Kaufmann described his new series of experiments to the Göttingen Academy. The salient points were:

(1) Kaufmann borrowed from the Curies an extremely active radium chloride source that was only 0.2 mm thick, thereby allowing him to decrease the diaphragm to 0.2 mm (see Fig. 1.7). The new source permitted Kaufmann to decrease the exposure time to 20 hours for each branch of the empirical curve. An improved high-voltage battery increased Kaufmann's confidence in the electric field remaining constant over the exposure time of 40 hours. He assumed that any inhomogeneities in the electric field would emerge in the data analysis. Kaufmann claimed that these improvements permitted him to obtain the "finest possible curve." Admitting the possibility of inhomogeneities

in the magnetic field, Kaufmann used a value for B obtained by integrating over the apparatus a function that suitably described the mapped magnetic field variations as a function of x.

(2) By suitable limiting procedures, based on the apparatus' geometry, Kaufmann extrapolated measured values of the curve coordinates (y_0, z_0) closer to the curve's cusp. Kaufmann designated these new coordinates as y' and z' and referred to them as the "reduced electric and magnetic deflections." [Precisely they are $y' = y_0 h x_2 / 2 s_1 s_2$ and $z' = (x_2^2 + x_1 x_2)/2\rho$. The result for y' follows from Chapter 1, footnote 39. For infinitely small deflections $s_1 s_2 = (h/2) x_2$, in which case $y' = y_0$ with the additional assumption of a homogeneous electric field in order that the electron's orbit between the capacitor plates be symmetric about its point of maximum curvature. The expression for z' follows from the result in Chapter 1, footnote 40 for z_0 much smaller than ρ, x_1 and x_2. In his (1906) Kaufmann more systematically related y' and z' directly to the electric and magnetic fields (see Section 7.2).]

Kaufmann rewrote Eqs. (1.115) and (1.153) as

$$\beta = k_1 \frac{z'}{y'}, \tag{1.155}$$

$$k_2 = \psi \left(k_1 \frac{z'}{y'} \right) \frac{z'^2}{y'}, \tag{1.156}$$

$$\varepsilon/\mu_0^e = k_1 k_2 N_1, \tag{1.157}$$

where k_1, k_2 and N_1 were constants that depended on the apparatus' dimensions, the velocity of light and the electric and magnetic field strengths. Kaufmann referred to the quantities k_1 and k_2 as "apparatus constants." For a constant electron mass, the Eq. (1.156) for the "reduced curve" described a parabola symmetric about the y-axis.

Kaufmann planned to determine the apparatus constants from the empirical curve, and then to compare these results with the apparatus constants that he obtained from the apparatus' dimensions and the measured electric and magnetic fields; any discrepancies between these two sets of constants were a measure of the inhomogeneity of the electric field. [In his (1902b) Kaufmann had emphasized that the 7.2% discrepancy in v at $z_0 = 0$ applied also to the differences between the empirically determined and calculated values for k_1 using the 1902 data (see Eq. (1.155)).] Besides Eq. (1.156) being quadratic in z', the equation for the reduced curve had to be treated carefully because, in Kaufmann's opinion, it resisted the method of least squares. His reason was that Eq. (1.156) was a transcendental equation whose Taylor series expansion converged too slowly for β close to unity. Consequently Kaufmann took the following approach toward determining k_1 and k_2 from values of y' and z', to which he referred loosely as a least squares approximation. Using closely spaced values of y' and z' in Eq. (1.156), Kaufmann determined those series of closely spaced values of k_1 that resulted in a series of approximately equal values of k_2. The best k_2 of the series, continued Kaufmann, was the one which gave the minimal value for the expresssion $(k_2 - \bar{k}_2)^2$, where $\bar{k}_2$ was the arithmetic mean of the series of k_2. This process was repeated for each photographic plate. Then for each plate Kaufmann used the (k_1, k_2) pair determined for closely spaced values of (y', z') in order to check the constancy of the quantity $\psi z'^2/y'$ over the reduced curve. In Fig. 1.10 are Kaufmann's results from Plate 19 which, he wrote, was the "most clear error-free picture."

z_0	y_0	z'	y'	β	$\psi(\beta)$	$\psi(\beta)z'^2/y'$
0.15	0.04055	0.1495	0.04045	0.951	3.00	(1.66)	—
0.20	0.0531	0.199	0.0529	0.967	3.32	(2.49)	—
0.25	0.0682	0.247	0.0678	0.938	2.79	2.525	− 0.036
0.30	0.0842	0.296	0.0834	0.912	2.52	2.645	+ 0.084
0.35	0.1032	0.3435	0.1019	0.865	2.22	2.57	+ 0.009
0.40	9.1240	0.391	0.1219	0.825	2.04	2.56	− 0.001
0.45	0.1460	0.437	0.1429	0.787	1.915	2.565	+ 0.004
0.50	0.1710	0.4825	0.1660	0.747	1.82	2.55	− 0.011
0.55	0.1985	0.5265	0.1916	0.707	1.74	2.515	− 0.046

FIG. 1.10. Kaufmann's results from the Plate 19. The coordinates (y_0, z_0) are obtained by micrometer measurements from the plate, the calculated reduced coordinates are (y', z'), all coordinates are in cm, and $k_1 = 0.257$, $k_2 = 2.561$. The quantities in the last column are the differences $k_2 - N_1\psi z'^2/y'$, and their smallness led Kaufmann to proclaim that *"the shape of the curve is represented with satisfactory accuracy by Abraham's formula"* (italics in original).

The k_1's and k_2's determined from the empirical curve in Plate 19 were less than those calculated from the apparatus' dimensions by 0.1% and 3%, respectively. Thus Kaufmann was led to assert the constancy of the electric field for these runs. On the other hand, the previous problem of too *low* a velocity at $z_0 = 0$, appeared now in some of his other plates as too *high* a velocity near $z_0 = 0$, e.g., $v = 1.04c$. Kaufmann attributed this latest shortcoming to the difficulty in accurately determining y_0 and z_0 close to the cusp owing to the high intensity of electrons in that area.

Kaufmann used Eq. (1.157) to calculate ε/μ_0^e for each of the four runs. His properly weighted average for the four runs was

$$\varepsilon/\mu_0^e = 1.77 \times 10^7 \qquad (1.158)$$

which was 5% less than Simon's. Eq. (1.158) was 4% less than a value of ε/μ_0^e with a correction for taking into account that the slowest β-rays moved faster than Simon's cathode rays — that is the value of ε/μ_0^e that Kaufmann calculated from Abraham's theory to accuracy β^2 was

$$\varepsilon/\mu_0^e = 1.845 \times 10^7 \qquad (1.159)$$

(as we shall see in a moment this value turned out to have been calculated incorrectly). Kaufmann emphasized, however, that the value of ε/μ_0^e from his Plate 19 (1.80×10^7) differed by only 2.5% from Eq. (1.159). Needless to say, at this point Kaufmann preferred comparing Abraham's theory to Eq. (1.159) rather than to Simon's result. Ascribing these deviations to merely "observational errors," Kaufmann boldly asserted that "*not only Becquerel rays, but cathode rays as well, consist of electrons whose mass is purely electromagnetic*" (italics in original).

Kaufmann concluded the (1903) by mentioning the Plate 19 that contained "the by far the clearest and faultless picture... The accompanying figure is copied from Plate 19 without, however, approaching the clarity of the plate itself." Eager at last to see the empirical curves that had the physicists of 1901–1903 so in a tizzy, I hurriedly leafed through the installment of the *Göttingen Proceedings* containing Kaufmann's (1903). Plate 19 appeared with the errata to the (1903). Despite Kaufmann's cautionary remark concerning the quality of Plate 19's reproduction, a glance at Fig. 1.11 leads one to agree with the American physicist Gilbert N. Lewis, who was greatly interested in problems concerning the mass of a moving electron, (as we discuss in Chapter 7), who observed (1908) that "it seems incredible that [Kaufmann's data], which consisted of a somewhat hazy spot on a photographic plate, could have been determined with the precision claimed."

FIG. 1.11. The actual size of Kaufmann's Plate 19 from his best run of the 1902–1903 series of experiments. Using a micrometer, Kaufmann determined y_0 and z_0 from plates of this sort. The sense of the y- and z-axes is $\overset{z}{\underset{y}{\llcorner}}$. (From *Göttinger Nachr.*)

Kaufmann's errata stated also that the value for ε/μ_0^e in Eq. (1.159) contained numerical errors and should have been

$$\varepsilon/\mu_o^e = 1.885 \times 10^7, \tag{1.160}$$

which increased the deviation to 6% between Eq. (1.158) and Simon's value of the charge-to-mass ratio that was corrected for slow electrons—Eq. (1.159). [In a (1905) publication Kaufmann again used the value of ε/μ_0^e in Eq. (1.160) but in (1906) he changed it to 1.878×10^7. The earlier value, he noted (1906), resulted from his having forgotten to distinguish between the transverse and longitudinal masses in Abraham's theory.][45]

In an address to the Göttingen Academy on 31 October 1903, one of Kaufmann's former colleagues at Göttingen, Carl Runge, pointed out serious errors in Kaufmann's data analysis. [Kaufmann's (1903) was also his farewell address, and he went next to Bonn.] Runge noted that "in his excellent work on the electromagnetic mass of the electron, Mr. Kaufmann did not discuss his measurements according to the usual method of least squares." Kaufmann's nonstandard determination of k_1 and k_2, continued Runge, did not attribute proper weights to the curve coordinates. Runge emphasized that an "or-thodox" application of the method of least squares removed this difficulty. He used Eqs. (1.155) and (1.157) to set up a linear relationship between y' and z' of the form

$$ay' = k_1 f(az') \qquad (1.161)$$

where $a = 1/k_1 k_2$ and $az' = 1/\beta\psi$. Inserting into Eq. (1.161) closely spaced values of β with the calculated z' from a particular plate, Runge applied the method of least squares to find optimum values of a and k_1 that minimized the square of the difference between ay' and $k_1 f(az')$. Although Runge's and Kaufmann's k_1's agreed well (to 0.1%), Runge's k_2's differed from Kaufmann's on the order of 4% and the agreement between the calculated and observed values of y' was on the order of 0.3% (except for the smallest y' which was nearest the reduced curve's cusp). Runge's and Kaufmann's values for ε/μ_0^e were in best agreement for Plate 19 (Runge obtained 1.755×10^7), and their worst deviation was 8%. Runge's weighted average for ε/μ_0^e was $1.755 \pm 0.059 \times 10^7$, which was close to Kaufmann's. Runge concluded that the mean errors in Kaufmann's data could explain the deviation between Kaufmann's and Simon's values for the electron's charge-to-mass ratio, provided that Simon's value was free of error.

Abraham (1905) reproduced Runge's data analysis of Kaufmann's Plate 19 as evidence in favor of his electron theory. He went on to use the data from Kaufmann's often reproduced Plate 19 to prove that the quasi-stationary approximation was valid to one part in 10^{12} (see Chapter 1, footnote 43). For further support for his theory, Abraham cited H. Starke's (1903) experiment, at the Physics Institute of the University of Berlin, which "bridged over" the region between low- and high-velocity electrons. Using an evacuated tube similar to Thomson's in 1897, Starke accelerated electrons to 0.3c. Unlike Thomson, however, Starke used parallel electric and magnetic fields in order to perform a low-velocity version of Kaufmann's experiment. Starke concluded that, within experimental errors, the decrease of ε/m_T with velocity agreed with

Abraham's formula for m_T, and "*the assumption of a purely electromagnetic mass for electrons*" (italics in original).

Patterns emerge from Kaufmann's papers:

(1) The nth paper attempts to correct errors in theory and data analysis from the $(n-1)$st.

(2) Kaufmann's goal was to achieve agreement with a value of ε/μ_0^e from only one experiment, i.e., Simon's. Sometimes this is not a good psychological situation for an experimentalist. Indeed, we shall see that the lower values of ε/μ_0^e turned out to be nearer to the correct value of ε/μ_0^e. Then there was the human factor; namely, the colleagueship between Kaufmann and the resolute Abraham. We can well believe Lorentz who, in his lectures (1910–1912), wrote that Kaufmann "believed himself to be driven to the conclusion that the theory of the spherical electron fitted [his data] better."

1.12. LORENTZ'S THEORY OF THE ELECTRON

> I shall suppose that there is no other, no 'true' or 'material' mass.
>
> H. A. Lorentz (1904c)

1.12.1. Prelude: The Period 1899–1903

Lorentz (1899a) had speculated on the velocity-dependence of the oscillator ion's mass so that he could respond as completely as was then possible to a criticism of Liénard's (in 1899 he still referred to the basic constituent of his theory as an ion): if part of the path in Michelson's interferometer contained a medium of refractive index $N \neq 1$, then perhaps the hypothesis of contraction would be insufficient to explain a null effect. Lorentz's response required (1) still another set of space and time transformations, and then (2) the hypothesis that the ratio of the ion's mass in S_r to the mass in the new instantaneous rest system, along its direction of motion, was $\gamma^3 \varepsilon$, and that normal to this direction it was $\gamma \varepsilon$, where ε was an undetermined scale factor.[46] Like the hypothesis of contraction, Lorentz proposed this new hypothesis in a tentative vein ("such a hypothesis seems startling at first sight") referring to, but not citing, previous work in this direction by Thomson and Searle, among others.

During the period 1900–1903 Lorentz discussed the possibility that some of the free electron's mass could be of electromagnetic origin, although he was as yet somewhat cautious about supporting an electromagnetic world-picture [(1900), (1901a)]. Indicative of Lorentz's caution was that his (1901b) expression, in the $v \ll c$ limit, for the mass of a free "ion," i.e., $(1 + \frac{6}{5}\beta^2)$, was unrelated to the low-velocity limits of his (1899a) expressions for the mass of a bound ion. Abraham was somewhat mystified over Lorentz's (1901b) result for the free ion's mass, because Abraham (1903) wrote that Lorentz gave this result "without derivation" and it was inapplicable to β-ray phenomena which were in the high-velocity regime. Apparently Abraham missed the point that Lorentz did not have to derive his (1901b) expression for the free ion's mass because it was easily obtainable from Searle's (1897) expression for the electromagnetic energy of a rigid sphere with a uniform surface charge distribution [see Eq. (1.111), footnote 61 below and Eq. (1.145)]. Consequently, in 1899 Lorentz had not extended the hypothesis of contraction from bulk matter to the ion.

It could well have been Abraham's theory of the electron that caused Lorentz, in 1903, to extend his electromagnetic theory to include the motion of free electrons. His first full-scale efforts in this direction appeared in the review paper that he completed in December 1903 and published in 1904, *Weiterbildung der Maxwellschen Theorie. Elektronentheorie.* Besides reviewing Abraham's theory of the electron, Lorentz discussed further applications of his own electromagnetic theory to electrostatic and optical phenomena, and to problems concerning moving dielectric and magnetic media, as well as to the motion of free electrons; here Lorentz once again used the transformation techniques from the *Versuch.*[47] In the concluding portion he speculated on the mass of the electron. [In the (1904b) paper Lorentz discussed the heavier positively charged constituents of a molecule as if it were a specie of the negatively charged electron. This was consistent with the goal of an electromagnetic world-picture, toward which, as we shall see, Lorentz was already working.]

Lorentz, in addition to reviewing the experiment of Michelson and Morley, discussed two new second-order experiments: (1) The electromagnetic experiment of Trouton and Noble (1903) which attempted to measure the turning couple of a parallel-plate capacitor hung on a string at rest relative to the earth; charging the capacitor should lead to a turning moment when the capacitor adjusted its orientation to minimize the electromagnetic energy between its plates (Fig. 1.12). Trouton and Noble concluded: "There is no doubt that the result is a purely negative one." (2) Lord Rayleigh (1902) sought to detect whether an isotropic substance at rest on the moving earth exhibited double refraction owing to the strain originating in Lorentz's proposed hypothesis of contraction – that is, whether a moving isotropic body should respond differently to light propagating through it parallel and transverse to its direction of motion. Rayleigh found no double refraction and his data were accurate to one part in 10^{10}. In a repetition of Rayleigh's experiment, D. B. Brace (1904), an American physicist at the University of Nebraska, also found no isotropy[48] and his data were accurate to one part in 10^{13}. According to Lorentz, with the additional hypothesis of the elastic force in the suspension cord transforming like the electromagnetic force, then the Trouton-Noble experiment could be explained the same way as the Michelson-Morley experiment. But, continued Lorentz, the problem remained of exactly how these systems were deformed, and thus far the deformation problem had resisted solution because "the motion of molecules is ignored . . . consequently

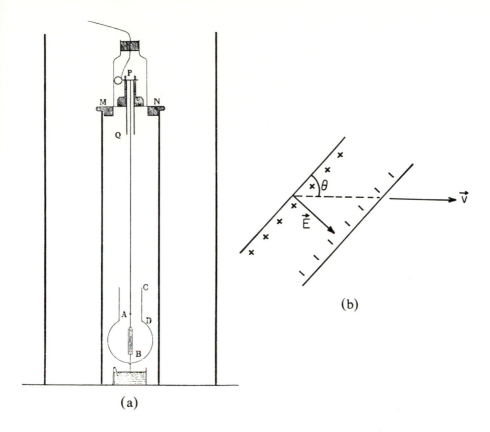

(b)

(a)

FIG. 1.12.   In (a) is Trouton's and Noble's schematic of their 1903 apparatus. From the 37 cm long phosphor bronze strip PA was hung a parallel-plate capacitor AB. The region between the capacitor plates contained mica as a dielectric, and the capacitor was charged to voltages of the order of 3000 V. The capacitor was hung in a closed glass tube which served to prevent effects due to air drafts, and the liquid contact at the tube's bottom served also to dampen any extraneous oscillations. A mirror attached to the capacitor was viewed through a telescope in order to detect any oscillations. Oscillations were expected for the following reason: the earth's motion through the ether caused the charges on the capacitor plates to constitute convection currents. Consequently, Trouton and Noble expected the capacitor's total energy density to increase beyond the electrostatic energy density $E^2/8\pi$ by the contribution from the magnetic field whose origin was the convection current. The additional energy density depended on the capacitor's orientation relative to the direction of the earth's motion through the ether. The capacitor tends to orient itself so as to minimize its energy, and consequently a torque acts on it. In order to allow the orientation of the capacitor relative to the earth's velocity through the ether to change, Trouton and Noble observed the orientation of the capacitor over a period of 9 days. The schematic (b) illustrates the principle of their experiment. E is the capacitor's electric field and v is the direction of the earth's motion through the ether. From the Biot-Savart law the contribution of the magnetic field to the capacitor's energy is proportional to $E^2(v^2/c^2)\cos^2\theta$. Consequently the capacitor's total energy is proportional to $E^2(1 + (v/c)^2 \cos^2\theta)$, and a torque of magnitude $E^2(v/c)^2 \sin 2\theta$ acts on the capacitor. For $\theta = \pi/2$, the capacitor's energy is a minimum, and so at this angle the torque vanishes. Thus the capacitor was expected to turn in order to maintain its plane perpendicular to the direction of the earth's motion through the ether. (From *Philos. Trans. R. Soc. London.*)

the removal of this defect seems to be a pressing task"; Lorentz immediately addressed himself to it by proposing another "conjecture (*Vermutung*)" – the "theory of the electromagnetic mass." Lorentz proposed to transfer the hypothesis of the change of dimensions from bulk matter to the constituent molecules. In a footnote to this passage, Lorentz mentioned the result of pushing this new conjecture to the conclusion that "perhaps the dimensions of the individual electrons become changed by translation." Continuing in the text, Lorentz wrote that taking this route could "probably" lead to a theory of the optics and electrodynamics of moving bodies, thereby explaining Rayleigh's experiment as well. These "further developments," Lorentz hoped, could signal "an increase in understanding and a unification of the description." On the other hand, Lorentz was sensitive to criticism of his accumulation of hypotheses, all for the express purpose of supporting the hypothesis of contraction, and in particular to Poincaré's pungent criticisms of 1900 that "hypotheses are what we lack least." Lorentz's response to Poincaré may well have been a gentle lesson to the great mathematician in how theoretical physicists often had to proceed when working on the frontiers of their subject:

> Poincaré with right, has criticized the theory (although we should be permitted the excuse, that we are entering into this domain cautiously groping forward) on the grounds that it has introduced an hypothesis contrived for the purpose of explaining just the experiment of Michelson, and that perhaps similar artifices [*Kunstgriffe*] would be necessary to account for new experimental results.

Lorentz continued by referring the reader to a forthcoming article in Dutch (published 23 April 1904) where he treated systematically second-order effects; the English version of this paper appeared on 27 May 1904, "Electromagnetic Phenomena in a System Moving with Any Velocity Less than that of Light," and it contained Lorentz's theory of the deformable electron.[49]

1.12.2. Lorentz's Fundamental Assumptions of his Own Theory of the Electron

In the paper on "Electromagnetic Phenomena..." Lorentz wrote that he was prompted to formulate a theory of the electron not only by new experiments, but also by Poincaré's foundational criticisms. Lorentz's proposal of a deformable electron was particularly courageous in the face of Abraham's assertion that such a theory ran counter to the goal of an electromagnetic world-picture. One of Lorentz's principal reasons for this move might have been the generality of his theory, which purported to explain second-order phenomena in the optics of moving bodies in addition to the electron's mass. Lorentz made more than 10 fundamental assumptions – for example, the hypothesis of contraction; all forces transform like electromagnetic forces; suitable transformation equations for the space and time coordinates; the velocity of light was the ultimate velocity. Nevertheless, these hypotheses satisfied Poincaré (1904) who found them "complementary to one another, permitting Lorentz to explain the results of several second-order experiments" – the experiments of Michelson and Morley, Rayleigh and Brace, Trouton and Noble, as well as those of Kaufmann which enabled him to fix the factor of ε from the space and time transformations of 1899.

As he had done since 1892, Lorentz first transformed the electromagnetic field equations form S to S_r, using both the convective derivative and the Galilean transformations. Then he introduced "a new set of independent variables" in order to transform the Galilean coordinates to a reference system Σ' with coordinates x', y', z', t':[50]

$$x' = \gamma l x_r, \tag{1.162}$$

$$y' = l y_r, \tag{1.163}$$

$$z' = l z_r, \tag{1.164}$$

$$t' = \frac{l}{\gamma} t - \gamma l \frac{v}{c^2} x_r. \tag{1.165}$$

Lorentz went on to demonstrate that if the Maxwell-Lorentz Eqs. (1.25)–(1.28) were to retain their form in Σ', then the relation between the fields, velocities, and charge densities were:[51]

$$E'_{x'} = \frac{1}{l^2} E_x, \tag{1.166}$$

$$E'_{y'} = \frac{\gamma}{l^2} \left(E_y - \frac{v}{c} B_x \right), \tag{1.167}$$

$$E'_{z'} = \frac{\gamma}{l^2} \left(E_z + \frac{v}{c} B_y \right), \tag{1.168}$$

$$B'_{x'} = \frac{1}{l^2} B_x, \tag{1.169}$$

$$B'_{y'} = \frac{\gamma}{l^2} \left(B_y + \frac{v}{c} E_x \right), \tag{1.170}$$

$$B'_{z'} = \frac{\gamma}{l^2} \left(B_z - \frac{v}{c} E_y \right), \tag{1.171}$$

$$u'_{x'} = \gamma^2 u_{x_r}, \tag{1.172}$$

$$u'_{y'} = \gamma u_{y_r}, \tag{1.173}$$

$$u'_{z'} = \gamma u_{z_r}, \tag{1.174}$$

$$\rho' = \rho/\gamma l^3, \tag{1.175}$$

where Eqs. (1.166)–(1.171) related field quantities between S and Σ', and Eqs. (1.172)–(1.175) related kinematical quantities between S_r and Σ', and in Eq. (1.175) Lorentz assumed that $\rho = \rho_r$.

The field transformation equations turned out to be correct, but those for the velocity and the charge density were not. The root of the problem, as Poincaré [(1905a), (1906)] showed, was Lorentz's two-step procedure which used S, S_r, and Σ', in addition to the convective derivative.

Indeed, Lorentz very likely deduced Eqs. (1.172)–(1.175) by assuming that $u_{x_r} \ll 1$.[52] Consequently Lorentz's paper of 1904 treated electromagnetic phenomena only in those where, relative to S_r, the electron moved with a velocity much less than that of light, and not with any velocity less than that of light.

Even with the transformation Eqs. (1.166)–(1.175), the Maxwell-Lorentz equations in Σ' were

$$\mathbf{V}' \times \mathbf{E}' = -\frac{1}{c}\frac{\partial \mathbf{B}'}{\partial t'}, \tag{1.176}$$

$$\mathbf{V}' \times \mathbf{B}' = \frac{1}{c}\frac{\partial \mathbf{E}'}{\partial t'} + \frac{4\pi}{c}\rho' \mathbf{u}', \tag{1.177}$$

$$\mathbf{V}' \cdot \mathbf{E}' = \left(1 - \frac{vu'_{x'}}{c^2}\right)\rho', \tag{1.178}$$

$$\mathbf{V}' \cdot \mathbf{B}' = 0. \tag{1.179}$$

According to Eq. (1.178) there exists a reference system where charge is not conserved.

Lorentz could ignore the incorrectness of Eq. (1.178) because he carried out all his calculations in an "electrostatic system, i.e., a system having no other motion but the translation with the velocity v." Consequently, since $\mathbf{u}' = \mathbf{u}_r = 0$, the system Σ' became the more exact version of the electron's rest system S'', from the *Versuch*, and thus satisfied Gauss' law, and agreed with the force equations in the *Versuch*. We refer to Σ' in the $\mathbf{u}' = \mathbf{0}$ case as S'. In summary, in the electron's rest system S', the offending terms in the fundamental equations vanished.[53]

As he had speculated in his (1904b), Lorentz applied the hypothesis of contraction to the electron, as follows. Consistent with the theorem of corresponding states, the electron should be spherical in S' because it was spherical in S. Since in S' the electron was stretched along the x-axis by the amount γl, and its dimensions normal to this direction were changed by l [in Lorentz's notation the electron's dimensions in S' were $(\gamma l, l, l)$], then according to Lorentz's (1904c) hypothesis of contraction the electron's "translation would produce [in S_r] the deformation $(1/\gamma l, 1/l, 1/l)$," in order that the electron retained its spherical shape in the "imaginary system" S'; that is, the distortion in the real system S_r canceled the distortion in the imaginary system S'. As he had done previously, Lorentz deduced the distortion in S_r by inversion of the transformation equations between S_r and S'. Lorentz's deformable electron can be likened to a balloon with a uniform surface charge distribution.

1.12.3. Lorentz's Predictions for the Electron's Mass

Lorentz's assumption that the electron was a sphere in S' enabled him to calculate with relative ease, compared to Abraham, the electron's electromagnetic momentum due to its self-fields. In the system S the electron's electromagnetic momentum is

$$G_x^e = \frac{1}{4\pi c}\int (E_y B_z - E_z B_y)\, dV. \tag{1.180}$$

Upon transforming the fields and volume element to S', Eq. (1.180) becomes

$$G_x^e = \frac{\gamma l v}{6\pi c^2}\left(\int E'^2\, dV'\right) \tag{1.181}$$

(where from spherical symmetry $E_{x'}'^2 = E_{y'}'^2 = E_{z'}'^2 = \frac{1}{3}E'^2$). Eq. (1.181) can be rewritten as

$$G_x^e = \frac{4}{3}\frac{\gamma l v}{c^2}\left[\frac{1}{8\pi}\int E'^2\, dV'\right]. \tag{1.182}$$

The integral in Eq. (1.182) is the electrostatic energy E_0^e of Lorentz's electron. Since in S' Lorentz's electron is a sphere with a uniform surface charge distribution then Lorentz could easily calculate E_0^e as

$$E_0^e = \frac{1}{8\pi}\int E'^2\, dV' = \frac{e^2}{2R}, \tag{1.183}$$

[whereas the "image" of Abraham's real electron was dilated in S'', and so Abraham's E_0^e was more difficult to calculate—compare Eqs. (1.133) and (1.134)]. Consequently,

$$\mathbf{G}^e = \left(\frac{2}{3}\frac{e^2}{Rc^2}\right)\gamma l\mathbf{v}. \tag{1.184}$$

Assuming quasi-stationary motion, Lorentz obtained the electron's longitudinal and transverse masses

$$m_L = \frac{2}{3}\frac{e^2}{Rc^2}\frac{d}{dv}(\gamma l v) \tag{1.185}$$

and

$$m_T = \frac{2}{3}\frac{e^2}{Rc^2}\gamma. \tag{1.186}$$

In order to evaluate the derivative in the longitudinal mass, Lorentz had to state explicitly the 1904 theorem of corresponding states: to a state in S characterized by E and B as functions of x, y, z, t, there corresponded a state in S' characterized by E' and B' as functions of x', y', z', t'; Lorentz could then calculate that $l = 1$,[54] and thus

$$m_L = \frac{2}{3}\frac{e^2}{Rc^2}\frac{1}{(1 - v^2/c^2)^{3/2}} \tag{1.187}$$

and

$$m_T = \frac{2}{3}\frac{e^2}{Rc^2}\frac{1}{\sqrt{1 - v^2/c^2}}. \tag{1.188}$$

We should bear in mind that the comparative simplicity of Lorentz's calculations and results resulted from his assuming the hypothesis of contraction.

Expanding Lorentz's prediction for m_T and m_L in terms of $\beta = v/c$ as

$$m_T = \frac{2}{3}\frac{e^2}{Rc^2}\left[1 + \frac{1}{2}\beta^2 + \frac{3}{8}\beta^4 + \cdots\right], \tag{1.189}$$

$$m_L = \frac{2}{3}\frac{e^2}{Rc^2}\left[1 + \frac{3}{2}\beta^2 + \frac{9}{8}\beta^4 + \cdots\right] \tag{1.190}$$

show how little they differed from Abraham's in the limit of low velocities. See Eqs. (1.146) and (1.145), respectively.

Lorentz went on to assert the necessity that all masses behaved like the electromagnetic mass of the electron in order that the theorem of corresponding states remain valid.

Aware that he was in competition with a theory the results of which "have been confirmed in a most remarkable way by Kaufmann's measurements of the deflexion of radium-rays in electric and magnetic fields," Lorentz went to compare his m_T with Kaufmann's data (1902b), (1903) and with Runge's (1903) analysis. Lorentz interpreted Runge's (1903) as showing that Kaufmann's calculation of k_1 and k_2 was a consistency check on Eqs. (1.155) and (1.156). Lorentz demonstrated that this was the case for his theory with Abraham's ψ replaced with $\psi = \frac{4}{3}(1 - \beta^2)^{-1/2}$. Unlike Abraham's theory, Lorentz's permitted the elimination of v between Eqs. (1.155) and (1.156), thereby relating y' and z' through the equation

$$y'^2 = (k_3)^2 z'^2 + (k_4)^2 z'^4, \tag{1.191}$$

where k_3 and k_4 are constants. Then using Kaufmann's measured values of z'

as input to Eq. (1.191), Lorentz deduced values of y' that agreed as closely with Kaufmann's data as did the ones from Runge's least square analysis.

However, Lorentz realized that he had constructed his theory at the price of piling up hypotheses:

> It need hardly be said that the present theory is put forward with all due reserve.
>
> $\cdots$
>
> Our assumption about the contraction of the electrons cannot in itself be pronounced to be either plausible or inadmissible. What we know about the nature of electrons is very little, and the only means of pushing our way farther will be to test each hypothesis as I have here made.

On 16 July 1904 Kaufmann wrote to Lorentz requesting a copy of his recently published "Electromagnetic Phenomena...."[55] [The library at Bonn, where Kaufmann was now a faculty member, did not stock the *Proceedings of the Royal Academy of Amsterdam.*]

Kaufmann reported that his "colleague Abraham" had derived Lorentz's prediction for the electron's transverse mass, and so he could compare immediately Lorentz's prediction with that of Abraham, but with the data at hand it was difficult to decide between the two theories; nevertheless, it seemed to Kaufmann that Lorentz's prediction agreed better with the data than Abraham's. "On account of the great importance of the whole problem," Kaufmann promised to repeat his measurements with increased accuracy. The results were not to please Lorentz. In short, Kaufmann claimed that his improved 1905 data disconfirmed what was by that time known as the Lorentz-Einstein prediction for m_T. I discuss this episode in Chapters 7 and 12.

1.13. ABRAHAM RESPONDS TO LORENTZ'S THEORY OF THE ELECTRON

> The proof is missing that an additional non-electromagnetic force could render stable the motion of a deformable ellipsoid.
>
> M. Abraham (1904a)

1.13.1. Abraham's 1904 Response

By the latter part of July 1904, Abraham had read a copy of Lorentz's (1904c) and he lost no time in sending off a critical note to the *Physikalische Zeitschrift*, "The Fundamental Hypotheses of the Electron Theory," where he compared the hypotheses of his theory of the rigid electron with Lorentz's.[56]

Delineating between Lorentz's electromagnetic theory and the theories of the electron of Abraham and Lorentz, Abraham (1904a) emphasized that special assumptions on the structure of the electron were required only for explaining data accurate to second order in v/c.

Abraham (1904a, b) and Lorentz (1904c), (1909) used their theories of the electron to interpret the results of Rayleigh and Brace as follows. Since the index of refraction depends on the mass of the isotropic substance's constituent electrons, then the index of refraction becomes double valued owing to the moving electrons developing transverse and longitudinal masses. Abraham's theory predicted a double refraction well within the limits of accuracy of Rayleigh and Brace, while Lorentz's 1904 theory of the electron predicted no double refraction because of hypotheses concerning how intermolecular forces transform and how moving matter contracts, among others.

With the intellectual honesty that was his trademark, Abraham admitted that his electron theory predicted a double refraction within the limits of accuracy of Rayleigh and Brace, and so his only recourse was to criticize Lorentz's hypothesis of contraction which was the key element in the incompatibility between the two theories of the electron. Abraham's principal points of contention were: (1) The deformable electron was not in the spirit of the electromagnetic world-picture, since it required nonelectromagnetic forces in order to maintain its equilibrium. (2) A rigid ellipsoid could execute force-free motion only if moving in the direction of its major axis: "proof is missing that an additional non-electromagnetic force can render stable the motion of a deformable ellipsoid." (3) Although the mathematics of an electron theory based on the real electron's assuming the shape of a "Heaviside ellipsoid" was simpler, "the physics is more complicated than the hypothesis of the rigid spherical electron." Besides, Abraham continued, Lorentz's theory of the electron precluded its moving with a hyperlight velocity since the factor $\sqrt{1 - v^2/c^2}$ which determined the real electron's shape became imaginary. Some physicists in 1905 were hesitant to accept the notion of an ultimate velocity, and Abraham pointed out that his student at Göttingen P. Hertz had written a doctoral dissertation demonstrating that a rigid spherical electron could be accelerated with a finite force to the velocity of light. Elsewhere in 1904 Sommerfeld (1904a, b, c and (1905)) had shown that a rigid spherical electron with a uniform volume charge distribution could execute force-free motion at, and beyond, the velocity of light, whereas a rigid spherical electron with a surface charge distribution could not. In another major paper of 1904, Abraham (1904b) had asserted that "consequently it is very probably compatible with the electron theory that the β-radiation of radioactive materials contain electrons moving with the velocity of light." Thus one of Abraham's reasons for considering his theory of the electron as provisional was that it excluded the possibility that the rigid electron could attain the velocity of light. However, he was inclined toward the opinion that a finite force could not accelerate bulk matter to the velocity of light.[57]

Abraham concluded his 1904 critique of Lorentz's theory of the electron by emphasizing that ultimately only empirical data could decide between the two theories, particularly further experiments by Kaufmann.[58]

In December 1904 Lorentz (1904d) discussed the state of his theory of the electron and expressed his great concern over the missing cohesive forces. If they could not be accounted for, he emphasized, then his electromagnetic theory could no longer explain the "complete independence of phenomena from the earth's translational velocity." Brace (1905a) in a comprehensive review of the state of ether-drift experiments wrote that without the additional cohesive forces "the hypothesis of Lorentz [i.e., of a deformable electron] is incomplete... Hence we must either abandon the contraction hypothesis or modify it."[59]

1.13.2. Abraham's 1905 Response: The Missing Energy

> [Lorentz's theory of the electron is incompatible with the] program of the electromagnetic world-picture.
>
> M. Abraham (1905)

In the first (1905) edition of his book, *Theorie der Elektrizität: II, Elektromagnetische Theorie der Strahlung*, Abraham discussed at length the fundamental hypotheses of the "dynamics of the electron and the electromagnetic world-picture." After acknowledging the wide explanatory powers of Lorentz's theory of the electron, Abraham again stressed its incompatibility with the "program of the electromagnetic world-picture," and now he had the calculations to support this criticism; in fact, he had already sent them to Lorentz in a letter of 26 January 1905.[60]

Abraham (1905) made the important observation that Lorentz's predictions for the electron's mass were derived from its momentum without considering the deformable electron's energy; taking the latter route, Abraham found that the m_L calculated from the deformable electron's energy E^e [$m_L = (1/v)(dE^e/dv)$][61] disagreed with the one calculated from its momentum ($m_L = dG^e/dv$), because the entire energy of Lorentz's electron could not be accounted for by electromagnetic forces alone. Abraham went right to the heart of the matter by showing that Lorentz's G^e, calculated from its defining equation in terms of electromagnetic fields [$G^e = (1/4\pi c) \int E \times B \, dV$], i.e.,

$$G^e = \frac{4}{3} \frac{E_0^e}{\sqrt{1 - v^2/c^2}} \frac{v}{c^2} \qquad (1.192)$$

was not the same as the one obtained from Lagrangian mechanics ($G^e = \partial L^e / \partial v$), i.e., from Eqs. (1.133) and (1.183)

$$G^e = \frac{E_0^e}{\sqrt{1 - v^2/c^2}} \frac{v}{c^2}. \qquad (1.193)$$

Abraham deduced the exact relationship between $(1/v)\,(dE^e/dv)$ and dG^e/dv in Lorentz's theory from Eq. (1.132) as (where $v = v\hat{\imath}$)

$$\frac{1}{v}\frac{dE^e}{dv} = \frac{dG^e}{dv} + \frac{1}{v}\left[G^e - \frac{dL^e}{dv}\right].$$ (1.194)

Then he substituted the G^e from Eq. (1.192), and the L^e from Eq. (1.133), with $E_0^e = e^2/2R$, to obtain

$$\frac{1}{v}\frac{dE^e}{dv} = m_L + \frac{m_T}{4},$$ (1.195)

instead of just m_L. Abraham's theory guaranteed that $G^e = dL^e/dv$. The root of the problem, continued Abraham, was that, according to Eqs. (1.133), and (1.192), the Lorentz electron had a total energy E_T of

$$E_T^e = E_0^e \sqrt{1 - v^2/c^2} + \frac{4}{3}\frac{E_0^e}{\sqrt{1 - v^2/c^2}}\frac{v^2}{c^2},$$ (1.196)

instead of $\frac{4}{3}E_0^e/\sqrt{1 - v^2/c^2}$.

Therefore, Abraham wrote, in order to maintain the energy principle, Lorentz should have included elastic forces which would have provided the proper energy counterterm for rendering $E_T^e = \frac{4}{3}E_0^e/\sqrt{1 - v^2/c^2}$. Abraham considered it "logically impossible" to amend Lorentz's theory of the electron in a "purely electromagnetic" manner in order to effect its necessary Lagrangian form; he continued, should further experimentation reveal Lorentz's m_T as correct, then the hope of formulating a world-picture based on the "electron as the smallest building block would thus be considered as an error."

Thus, circa 1905, Abraham and Lorentz claimed to be able to account for the dynamics of the electron on a purely electromagnetic basis. Fundamental to both theories was the asssumption that the electron's energy and momentum were identical with the (self-) electromagnetic energy and momentum. However, to "derive" Newton's second law, and thereby to calculate unambiguously the electron's mass, appeared to necessitate taking account of only quasi-stationary motion. In Abraham's theory G^e is calculated from the Lagrangian of the electromagnetic field on the assumption that the electron is a rigid sphere which undergoes an imaginary contraction in an imaginary coordinate system, while in Lorentz's theory G^e is calculated from its electromagnetic definition under the assumption that the deformable electron undergoes a real contraction in the direction of motion but maintains its spherical shape in an imaginary coordinate system. The portions of both theories amenable to experiment (m_T) were in agreement with the available data.

There were, however, two basic faults with Lorentz's theory, which was clearly the more ambitious of the two: (1) The Maxwell-Lorentz equations in general were not left unchanged under the transformation Eqs. (1.162)–(1.165). This was, as Poincaré discovered, a technical problem. (2) More serious was the instability of the deformable electron caused by the Coulomb repulsion of its constituent parts, a problem that manifested itself in the ambiguity in m_L. Henri Poincaré addressed himself to these problems.

1.14. POINCARÉ'S PENULTIMATE STATEMENT ON A LORENTZ ELECTRON-BASED ELECTROMAGNETIC WORLD-PICTURE

> This impossibility of experimentally demonstrating the absolute motion of the Earth appears to be a general law of Nature; it is reasonable to assume the existence of this law, which we shall call the *postulate of relativity* and to assume that it is universally valid. Whether this postulate, which so far is in agreement with experiment, be later confirmed or disproved by more accurate tests, it is, in any case, of interest to see what consequences follow from it (italics in original).
>
> H. Poincaré (1906)

The universal scope of Lorentz's theory and its theorem of corresponding states particularly interested Poincaré. At the Congress of Arts and Sciences, September 1904, St. Louis, Missouri, Poincaré elatedly discussed the main results of Lorentz's 1904 theory of the electron. Within the broader context of Lorentz's theory, Poincaré referred to the theorem of corresponding states as the "principle of relativity" (1904):

> The laws of physical phenomena must be the same for a stationary observer as for an observer carried along in a uniform motion of translation; so that we have not and can not have any means of discerning whether or not we are carried along in such a motion.

Three letters that I recently found from Poincaré to Lorentz, written during late 1904 to mid-1905 (as was customary among many French scientists, Poincaré's letters are undated), reveal that Poincaré had also discovered the problem in Lorentz's theory concerning the longitudinal mass, simultaneously and independently of Abraham. The results appeared in his classic paper, "*Sur la dynamique de l'électron*," a brief version of which was published in the *Comptes Rendus* (5 June 1905), and a longer version in the *Rendiconti del Circolo matematico* (1906).[62] Even if Einstein had read the *Comptes Rendus* version we can be sure that it could not have influenced his thinking toward his paper, "On the Electrodynamics of Moving Bodies," received for publication on 30 June 1905; however, for reasons to become clear in a moment, Poincaré's "*Sur la dynamique...*" was considered by some as evidence that Poincaré, more than anyone else in the late 19th and early 20th centuries anticipated Einstein's 1905 theory of relativity. (As far as I know Einstein had not read Poincaré's (1905a) prior to writing the special relativity paper — see the letter of Einstein to Seelig that I quoted from under the title to Chapter 1.) We shall review Poincaré's principal results for not only was his paper the penultimate effort of the electromagnetic world-picture, his mathematical methods were

important for further elaborations of the Lorentz-Einstein theory, in particular by Hermann Minkowski.

Poincaré, after stating his "postulate of relativity," and suggesting that
lengths should be compared using light rays, succinctly related the problem of
the calculation of m_L to the deformable electron's equilibrium as follows. The
deformable electron is in equilibrium if the Lorentz force density vanishes
there, i.e.,

$$f = \rho E = 0; \tag{1.197}$$

since $\rho \neq 0$, then $E = 0$. But from Gauss' law $\nabla \cdot E = 4\pi\rho$, and if $E = 0$, then
$\rho = 0$. The reason for this paradoxical situation was the electron's self-
Coulomb field that prevents the Lorentz force from vanishing in the electron's
rest system. That is, the electron could explode in S' owing to the repulsive
coulomb forces between its constituent parts. The problem Poincaré set for
himself was to find cohesive forces that would have the proper Lorentz
transformation properties; his mathematical tool was the principle of least
action.

But first Poincaré had to correct certain technical errors in Lorentz's
(1904c). He began by eliminating the reference system S_r, thereby relating the
imaginary system Σ' with the apparently unobservable ether-fixed system S
through what Poincaré referred to as the "Lorentz transformations"
[elimination of S_r enabled Poincaré to dispense with the convective derivative,
and instead to transform the time derivatives in the electromagnetic field
equations directly with Eq. (1.201) below]:

$$x' = \gamma l(x - vt), \tag{1.198}$$

$$y' = ly, \tag{1.199}$$

$$z' = lz, \tag{1.200}$$

$$t' = \gamma l\left(t - \frac{v}{c^2}x\right). \tag{1.201}$$

I can think of no better way to present Poincaré's proof that the Lorentz
transformations form a group than to refer the reader to the second of the three
letters that Poincaré wrote to Lorentz during late 1904 to mid-1905 (Fig.
1.13).[63] In the previous letter, Poincaré had written that Lorentz's (1904c)
proof that $l = 1$ was not conclusive. Most likely Poincaré's reason was that
Lorentz's (1904c) proof depended on transformation equations for forces and
accelerations that were valid only for the electron's instantaneous rest
system – Lorentz's proof lacked generality. Poincaré elegantly resolved this
problem by eliminating S_r, thereby writing the Lorentz transformation in a
form that revealed the symmetry between S and Σ' – that is, for Poincaré the
mathematical symmetry. Then, (see Fig. 1.13), Poincaré used a method that
has become part of the repertoire of every physics student to prove that two

Mon cher Collègue,

Merci de votre aimable lettre. Depuis que je vous ai écrit mes idées se sont modifiées sur quelques points. Je trouve comme vous $l = 1$ par une autre voie.

Soit $-\varepsilon$ la vitesse de Σ translation celle de la lumière étant prise pour unité.

$$k = (1-\varepsilon^2)^{-\frac{1}{2}}$$

On a la transformation

$$x' = kl(x+\varepsilon t), \quad t' = kl(t+\varepsilon x)$$
$$y' = ly, \quad z' = lz.$$

Ces transformations forment un groupe. ... deux transformations composantes correspondant à

$$k, \; l, \; \varepsilon$$

et

$$k', \; l', \; \varepsilon'$$

leur résultante correspondra à

$$k'', \; l'', \; \varepsilon''$$

ou :

$$k'' = (1-\varepsilon''^2)^{-\frac{1}{2}}, \quad l'' = ll', \quad \varepsilon'' = \frac{\varepsilon + \varepsilon'}{1+\varepsilon\varepsilon'}$$

Si nous voulons maintenant prendre

$$l = (1-\varepsilon^2)^m, \quad l' = (1-\varepsilon'^2)^m$$

nous n'aurons :

$$l'' = (1-\varepsilon''^2)^m$$

que pour $m = 0$

D'un autre côté je ne trouve d'accord entre le calcul des masses par le moyen des quantités de mouvement électromagnétique et par le moyen de la moindre action, et par le moyen de l'énergie que dans l'hypothèse de Langevin.

J'espère tirer bientôt au clair cette contradiction, je vous tiendrai au courant de mes efforts.

Votre très dévoué Collègue,

Poincaré

FIG. 1.13. A letter of Poincaré to Lorentz, written sometime during the period late 1904 to mid-1905, that contains Poincaré's proof that the Lorentz transformations form a group. Comparing Poincaré's equations for x', y', z', and t' with Eqs. (1.198)–(1.201) we see that Poincaré's relative velocity $\varepsilon = -v$, the quantity k is what I call γ, and he took $c = 1$. In Poincaré's equations for l, l' and l'', ε is the relative velocity between S and an inertial reference system Σ', and ε' is the relative velocity between Σ' and another inertial reference system Σ''. All relative velocities are along the common x-axes of S, Σ', Σ''. (Reproduced with permission from the Estate of Henri Poincaré.)

successive Lorentz transformations could be replaced by a single Lorentz transformation. Therefore, as Poincaré wrote in the letter to Lorentz, the Lorentz transformations "form a group." (Coincidentally, as we shall see in Chapter 8, Einstein used the same method to prove that the relativistic space and time transformations form a group.) The relative velocity of the single Lorentz transformation was the quantity ε'' (see Fig. 1.13), and the relationship of ε'' with ε and ε' constituted a new addition law for velocities that was independent of the value of l. Poincaré went on to prove that if l were an even function of ε of the form $(1 - \varepsilon^2)^m$ then $m = 0$, and $l = 1$. Poincaré's goal in his (1905a) and (1906) was to investigate every possible model of the electron, i.e., every connection between γ and l; consequently, he worked with the Lorentz transformations for $l \neq 1$, until he could give a physical reason for $l = 1$.[64]

Requiring the Lorentz covariance of the Maxwell-Lorentz equations and the Lorentz-force equation in Σ', Poincaré (1906) used Eqs. (1.198)–(1.201) to transform the space and time coordinates in these equations back to S. Thus he obtained the correct transformation equations for the electromagnetic field quantities and the force densities that were valid also in Σ'; he displayed them so as to emphasize that the quantities $(f, f \cdot v/c)$, $(\rho v/c, \rho)$ and (A, ϕ) transformed under the Lorentz transformations like the space and time coordinates (r, ct); from them he could form quantities that, if $l = 1$, were invariants of the Lorentz group, in analogy with the quadratic form $c^2 t^2 - x^2 - y^2 - z^2$. In the course of his (1906), Poincaré used this procedure systematically to search for other invariants of the Lorentz group. Then, in the concluding section, "Hypotheses concerning gravitation," Poincaré used an imaginary time coordinate ict in order to take advantage of the theory of invariants in a "4-dimensional space" with the positive definite metric $x^2 + y^2 + z^2 + \tau^2$ where $\tau = ict$.[65]

Toward a fundamental analysis of the electromagnetic world-picture, Poincaré deduced the Lorentz transformation property of the electron's Lagrangian as

$$L^e = l L_0^e \sqrt{1 - v^2/c^2} \qquad (1.202)$$

where $L_0^e = (1/8\pi) \int (B'^2 - E'^2)\, dV' - L_0^e$ is the Lagrangian in Σ'. From Eqs. (1.201) and (1.202) followed the Lorentz invariance of the principle of least action for any value of l, i.e.,

$$\int L^e\, dt = \int L_0^e\, dt'. \qquad (1.203)$$

Poincaré's reason for including factors of l was to propose a Lagrangian formulation general enough to include every possible model for an electron. He discussed the three models currently under investigation which he categorized as follows (at this point Poincaré specialized to the electron's rest system S'). (1) Abraham's theory in which the "real electron is spherical and [under the transformation Eqs. (1.63)–(1.66)] the imaginary electron will become an ellipsoid" whose major axis is along its direction of motion. (2) Theories of a deformable electron in which a real resting electron has a radius R. But since the "imaginary immobile electron will always be a sphere of radius R, the axes of the real [moving] electron will then be $(R/\gamma l, R/l, R/l)$." In Lorentz's theory, continued Poincaré, $l = 1$. In the theory that had been proposed independently by Poincaré's former student, Langevin (1905) and by Bucherer (1904), (1905), the real moving electron underwent a deformation that preserved its volume, and so $dV = dV'$, or $l = \gamma^{-1/3}$ (see Fig. 1.14). In the Langevin-Bucherer theory,

$$m_T = \frac{2}{3}\frac{e^2}{Rc^2}\frac{1}{[1 - (v^2/c^2)]^{1/3}} \quad \text{and} \quad m_L = \frac{2}{3}\frac{e^2}{Rc^2}\frac{1 - \frac{1}{3}(v^2/c^2)}{[1 - (v^2/c^2)]^{4/3}}.$$

Poincaré demonstrated, in general, that only the Langevin-Bucherer model could describe a deformable electron with a Lagrangian of the sort in Eq. (1.129) or equivalently Eq. (1.202), because no extra cohesive forces were necessary and so it avoided any discrepancy in the calculation of m_L.[66]

(a)

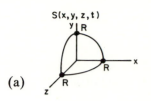

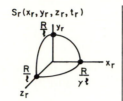

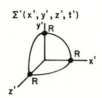

In the ether-fixed reference system S the real-sphere electron has axes: (R,R,R).	In the inertial reference system S_r the real-deformed electron has axes: $\left(\dfrac{R}{\gamma\ell}, \dfrac{R}{\ell}, \dfrac{R}{\ell}\right)$.	In the auxiliary reference system Σ' the imaginary-sphere electron has axes: (R,R, R).

(b)

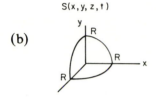

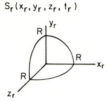

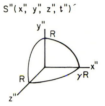

FIG. 1.14. In (a) are the three different reference systems employed in the electromagnetic world-picture for theories of deformable electrons—Bucherer-Langevin, and Lorentz. These theories transform according to the Lorentz transformation Eqs. (1.198)–(1.201). Figure (b) describes Abraham's theory in which the real electron is a rigid sphere in S and S_r. The imaginary electron in S'' is dilated in the direction of the relative motion of S'' and S, by an amount γ. Abraham's theory transforms according to Eqs. (1.63)–(1.66), i.e., it is not Lorentz covariant.

Poincaré posed the problem of what sort of term must be added to the Lagrangian in Eq. (1.202) which would permit a stable deformable electron with $l = 1$. Toward its solution he rewrote the principle of least action as

$$J = \int [L^e + L^c]\, dt \qquad (1.204)$$

where L^e was the Lagrangian containing the electron's self-electromagnetic fields [i.e., Eq. (1.202)] and L^c was the additional term permitting $l = 1$. Poincaré's demand that L^c be the "potential" arising from the "supplementary forces" serving to bind the electron caused L^c to be the product of a pressure and a volume, and he calculated the quantity L^c as

$$L^c = -E_0^c \sqrt{1 - v^2/c^2}. \qquad (1.205)$$

The correct Lagrangian $L_T = L^e + L^c$ for Lorentz's theory was

$$L_T = -(4/3) E_0^e \sqrt{1 - v^2/c^2}, \qquad (1.206)$$

which removed the discrepancy in the calculation of m_L. Toward clarifying the meaning of L^c, Poincaré rewrote the term E_0^c in Eq. (1.205) as

$$E_0^c = \frac{e^2}{8\pi R^4}(\tfrac{4}{3}\pi R^3) = \frac{E_0^e}{3} \tag{1.207}$$

where $\tfrac{4}{3}\pi R^3$ is the resting electron's volume, and E_0^c is therefore the energy that results when the electron is acted upon by an internal stress in S',

$$p_0^c = -\frac{e^2}{8\pi R^4} . \tag{1.208}$$

The quantity p_0^c, often referred to as the Poincaré stress, served to cancel the self-stress due to the electron's self-coulomb field in S', thereby preventing the deformable electron from exploding in its rest system,[67] and thus increasing the deformable electron's energy by an amount ΔE, where

$$\Delta E = -p^c V \tag{1.209}$$

$$= (E_0^e/3)\sqrt{1 - v^2/c^2}. \tag{1.210}$$

Thus was Abraham's criticism met.

Consequently instead of Eq. (1.196), the Lorentz electron's total energy E_T should have been written

$$E_T = E_T^e + \Delta E \tag{1.211}$$

or

$$E_T = \frac{4}{3}\frac{E_0^e}{\sqrt{1 - v^2/c^2}} \tag{1.212}$$

so that $(1/v)\,(dE^e/dv) = (dG^e/dv)$. Poincaré went on to prove that the pressure p_0^c was a Lorentz-invariant quantity, and that its ensuring the electron's stability held for any sort of motion, and not just for quasi-stationary motion.[68] Although Poincaré did not emphasize the point, the stress p_0^c served also to explain the contraction of a moving electron; since the Poincaré stress was Lorentz-invariant, it was necessary for it to be always normal to the electron's surface.

Poincaré continued in (1906) to prove that only $l = 1$ left unchanged the electron's equations for quasi-stationary motion. Consequently only $l = 1$ was consistent with the principle of relativity, and therefore also with the negative results of the ether-drift experiments that had been performed thus far.

Thus, concluded Poincaré:

> *If the electron's inertia is exclusively of electromagnetic origin, if it is subjected to no forces other than those of electromagnetic origin, or to the forces due to the supplementary potential* [L^c], *then no experiment can detect evidence of absolute motion* (italics in original).

Having proven that only $l = 1$ satisfied the principle of relativity, Poincaré could write the Eqs. (1.198)–(1.201) as

$$x' = \gamma(x - vt), \tag{1.213}$$

$$y' = y, \tag{1.214}$$

$$z' = t, \tag{1.215}$$

$$t' = \gamma\left(t - \frac{v}{c^2}x\right). \tag{1.216}$$

For Poincaré the property that permitted Eqs. (1.213)–(1.216) to be inverted by changing v to $-v$, and interchanging primes, was only a mathematical symmetry because the S system never moved.

The goal of most philosophically inclined scientists has always been the unification of the sciences, and so not unexpectedly Poincaré was "tempted to infer" a relation between the "causes giving rise to gravitation and those which give rise to the supplementary potential." Unlike Abraham, Poincaré insisted that Lorentz's theory of the electron was in the spirit of the electromagnetic world-picture, because Poincaré took Lorentz covariance to be his guide toward this goal; the Poincaré pressure ensured the Lorentz electron's consistency with the requirement of Lorentz covariance. In the final section of (1906), Poincaré formulated a Lorentz-covariant theory of gravity that reduced to Newton's theory when the velocity of light became infinite.

Poincaré (1906) had demonstrated that Lorentz convariance could not be satisfied in a theory that attempted to construct a particle with mass from only its self-electromagnetic fields. In 1911 Max von Laue used mathematical techniques that were in their nascent form in Poincaré's (1906), and developed subsequently by Minkowski (1907), to prove that the Poincaré stress ensured that the electron's energy and momentum transformed properly under the Lorentz transformation: This clarification of the role of kinematical quantities such as energy and momentum, was possible only within the special theory of relativity. I discuss von Laue's work in Chapter 12.

In summary, by mid-1905, there were three models for the electron, but only Lorentz's could adequately explain the hard-won empirical data accurate to second order in v/c. It explained the Fresnel dragging coefficient, the dependence of mass on velocity, and the isotropy of the velocity of light in every inertial system as effects all *caused* by the interaction between the electron's constituting bulk matter and the ether. Cast by Poincaré into the elegant Hamilton-Lagrange formalism, replete with such advanced mathematics and physics circa 1905 as group theory, four-dimensional vector spaces, and the quasi-stationary approximation, Lorentz's dynamical theory appeared to many physicists as the most likely candidate for the sledgehammer needed for the breakthrough to a unified field-theoretical description of matter – that is, to serve as the cornerstone for the electromagnetic world-picture.

It seemed as if all the actors were on stage for this momentous event. Or were they? For something unexpected occurred when, on 30 June 1905, Albert Einstein's "On the Electrodynamics of Moving Bodies" arrived in Berlin from Bern at the desk of the editor of the *Annalen der Physik*. By degrees Einstein's results would alter the course of the physics of the electron.

1.15. A NOTE ON SOURCES

> In discussing the approach of "nearly all historians" (perhaps somewhat too brusquely) Einstein accentuates the need to deal with the private phase of scientific effort — how a man thinks and wrestles with a problem. In discussing the physicists themselves (perhaps also too brusquely) Einstein accentuates the need for a particular kind of historical sense, one that largely intuits how scientists may have proceeded, even in the absence of "the real facts" about the creative phase. It is a challenging statement, a recommendation to adopt for research in the history of science a lesson Einstein had learned from his research in physics: *just as in doing physics itself*, Einstein here advises the historian of science to leap across the unavoidable gap between the necessarily too limited "facts" and the mental construct that must be formed to handle the facts. And in such a historical study, *as in physics itself*, the solution comes often "by very indirect means"; the best outcome that can be hoped for is not certainty but only a good probability of being "correct anyway" (italics in original).
>
> From G. Holton's (1973e) analysis of the Einstein-Shankland interview of 4 February 1950, from which I quoted in the epigraph to this book's Introduction.

There is little extant pre-1905 documentation describing Einstein's thinking as he made his way toward the relativity theory — for example, on electrodynamics and optics. Thus, much detective work must be done in order to discuss the origins of Einstein's ideas for the relativity paper and how they relate to the physics of 1905. One has to base oneself often on those conjectures that receive support from the widest number of sources. I can think of no better way to elaborate on this point than to quote from Holton's (1973e): "Einstein . . . advises the historian of science to leap across the unavoidable gap between the necessarily too limited 'facts' and the mental construct that must be formed to handle the facts. And in such a historical study, *as in doing physics itself*, the solution comes often 'by very indirect means'; the best outcome that can be hoped for is not certainty but only a good probability of being 'correct anyway'."

A source of immense value for any analysis of Einstein's work is his only substantial autobiography: the "Autobiographical Notes" of 1946 which Einstein sometimes called his "obituary." There he wrote that "the obituary can limit itself in the main to the communicating of thoughts which have played a considerable role in my endeavors." We shall take seriously Einstein's caveat that "today's person of 67 is by no means the same as one of 50, of 30 or of 20. Every reminiscence is colored by today's being what it is, and therefore by a deceptive point of view." Thus, quotations used from the "Autobiographical Notes" need support by documentation from the period to which it refers. This is in fact possible to do, so that the historian of science can only marvel at how successfully Einstein plumbed into the past of his own thinking.

1.15.1. On Einstein's Knowledge of Electrodynamics in 1905

Having examined the state of electrodynamics in 1905, we now shall look at Einstein in the setting of contemporaneous research. First we list some books, monographs and journals concerning electrodynamics to which Einstein (definitely, very probably, maybe) had access before sending off his relativity paper in late June 1905; the supporting evidence cited is enlarged on in succeeding chapters. During 1896–1900 Einstein could have availed himself of the library facilities at the ETH. The nature of Einstein's work at the Patent Office, e.g., assessing patents for dynamos, almost certainly led him to the library at the University of Bern at which time he could have perused the literature for his own research as well.

Books and Monographs
A. *Definitely read, and evidence*
1) Lorentz's (1892a), (1895): Letter written 19 February 1955 by Einstein to Seelig that is quoted in the epigraph to Chapter 1 [see Born (1969)]. Einstein mistakenly attributed both monographs to the year 1895.
2) Hertz (1892): See Holton (1973c) and Seelig (1954).
3) Föppl (1894): See Holton (1973c). As was the case with Hertz (1892), Einstein studied Föppl at home in the evenings during the ETH period.
4) Boltzmann, von Helmholtz and Kirchhoff: See Seelig (1954) and Frank [(1947), (1953); also Einstein's (1905c) cites Kirchhoff's (1876).

B. *Very probably*
1) Drude (1900): In any era there are certain books that are standard fare for the student of physics, and Drude's was one of them; however, Einstein does not mention it.
2) Abraham-Föppl (1904c): Like its predecessor, Föppl (1894), this book became an instant success. Having read Föppl, Einstein could have at least glanced at the version rewritten and brought up-to-date by the well-known physicist Abraham.

C. *Maybe*
1) Lorentz (1904a, b): Lorentz's widely-referenced monographs were the most complete expositions of the state of electromagnetic theory in 1904.
2) Lorentz *Festschrift* (1900): Einstein's (1906c) cites it.

Journals
A. *Definitely, and evidence*
1) *Annalen der Physik*: By 1905 Einstein had published five papers in this leading German physics journal. Since in his articles from 1901–1907 Einstein

cited only papers that had appeared in the *Annalen*, we can postulate that he read this journal. Indeed it is not likely that he would have published in a journal to which he had no access.

B. *Very probably*

1) *Physikalische Zeitschrift*: A well-distributed and widely-read journal that contained original communications, conference proceedings, German translations of important French and English papers, book reviews and general news about the German-speaking science community. Further reasons for Einstein's awareness of the *Physikalische Zeitschrift* are given with the next journal.

2) *Verhandlung der Deutschen Physikalischen Gesellschaft* (*Proceedings of the German Physical Society*): In order to support my conjecture that before 1905 Einstein very probably had access to the *Physikalische Zeitschrift* and the *Verhandlung*, I have found it necessary to turn to certain of Einstein's papers and correspondence from 1907. The reason is that in 1907 Einstein for the first time recalled his view of physics in 1905, and mentioned explicitly the works of others, although he still did not cite the sources. In what follows I shall present indirect evidence that in 1906–1907 Einstein could have had access to the *Physikalische Zeitschrift* and the *Verhandlung*, and then it is reasonable to conjecture that he also could have had access to these journals before 1906.

Einstein devoted a paragraph in his (1907d) to a plea for others to pursue basic problems in theoretical physics (that is, as Einstein defined them) from a "new viewpoint." He wrote that recent "circumstantial examination of the literature" had brought to his attention works by Planck and Kaufmann, both of whom Einstein thanked for mentioning his "first work on the principle of relativity." Planck's (1906a) lecture emphasized why Einstein's first paper on electrodynamics was an important generalization of Lorentz's theory of the electron. Kaufmann's paper could only have been his (1906). Einstein's ignoring Kaufmann's empirical results was already indicative of his opinion of the interplay between data and theory — Kaufmann had asserted that his empirical data definitely disconfirmed the "Lorentz-Einstein theory."

Consequently, it appears that in 1906 and 1907 Einstein had access to the *Proceedings of the German Physical Society*. Planck's (1906b) discussed Kaufmann's experiments and how the various electron theories fared under them; among these theories was the "Lorentz-Einstein" theory. Einstein's (1907d) could have been alluding only to Planck's (1906a), since there Planck focused exclusively on Einstein's work. But Einstein went on in (1907d), for the first time in print, to use the term "*Relativitätstheorie.*" It is a variant of a term "*Relativtheorie*" first used by Planck in (1906b); in the discussion session to Planck's lecture the experimentalist A. H. Bucherer used the term "*Relativitätstheorie.*" Therefore, Einstein could also have read Planck's (1906b) in the *Physikalische Zeitschrift*. Consequently, we can conjecture that in 1905 Einstein very probably had access to both the *Proceedings of the*

German Physical Society and to the *Physikalische Zeitschrift*. As further evidence to support my conjecture that before 1905 Einstein had access to the *Physikalische Zeitschrift*, I offer Einstein's recollection (1946) that before 1905 he had been aware of the instability of Lorentz's electron (see Section 2.3). Einstein very probably learned of this problem from Abraham's (1904a) in the *Physikalische Zeitschrift*.

It was not unusual for *Annalen* authors to skimp on references. Einstein pushed this style to the extreme in the relativity paper, while in the other *Annalen* papers he followed current practice.

We continue our discussion of Einstein's possible sources of scientific information with an exchange of letters in the fall of 1907 between Einstein and Johannes Stark, then the editor of the *Jahrbuch der Radioaktivität und Elektronik*. Stark had invited Einstein to write a review paper on the special relativity theory; on 25 September 1907 Einstein replied [see Hermann (1966)]. Einstein asked Stark to send him copies of important recent "theoretical works" because he was "unfortunately in no position" to obtain them since "in my free time the library is closed . . . other than my own works I know of only [the recent ones] of H. A. Lorentz [(1904c)], one of E. Cohn, one of K. von Mosengeil [(1907)], and two of Planck." From Einstein's knowledge of physics displayed in his relativity paper and in his papers of 1906–1907, we may assume that he had ample time to at least browse at the University Library. We can interpret Einstein's request to Stark to mean those works that Einstein may have missed during the "circumstantial examination of the literature" permitted by his job at the Patent Office. In my opinion, Einstein's letter cannot be interpreted to mean that he had neither the time nor the opportunity to visit libraries in Bern.

Stark responded by sending Einstein a copy of Planck's "latest major work" (1907a), a recent *Annalen* paper of von Laue, and a mention of Planck's (1907b).

On 1 November 1907 Einstein replied to Stark: "It is good that you called my attention to Planck's work on Kaufmann's experiment. I was unaware of any research of Mr. Planck on this subject." Apparently after Stark's letter, Einstein had read Planck's (1907b) in the *Proceedings*, and then read Planck's (1906b) which had appeared also in the *Physikalische Zeitschrift*. In the completed *Jahrbuch* essay (1907e), Einstein cited Planck's (1907b) and (1906b) papers from the *Proceedings*. In light of the fact that in his (1907d) Einstein used a variant of Planck's term "*Relativtheorie*," I was puzzled by the passage above from the 1 November 1907 letter to Stark. Anyone with access to the *Proceedings* is likely to have been able to peruse the *Physikalische Zeitschrift*. The *Proceedings* version of Planck's (1906b) omitted the discussion session, and so Einstein's (1907d) use of the term "*Relativitätstheorie*" could be taken as an indication of his having read the *Physikalische Zeitschrift* version of Planck's (1906b). The most likely answer is that a result of Einstein's "circumstantial examination" of Planck's (1906b) in the *Physikalische*

Zeitschrift was to pick out Planck's name for a theory emphasizing a principle of relativity. On the testimony of Einstein to Besso, and others, Einstein was not a voracious reader of journals; rather, he skimmed them and chose to read in more detail whatever struck his interest. In his (1907d) Einstein referred to neither Planck's papers (1906a, b) nor to Kaufmann's paper, even though it was in the *Annalen*. In 1907 Einstein continued his practice of citing only papers from the *Annalen*, and selectively at that.

In the course of this book, I shall have the occasion to note Einstein's awareness of the contents of certain scientific papers without having seen them. A scientist can find out about a particular paper from its description elsewhere in the literature. So, for example, it is unwarranted to conclude from Einstein's letter to Stark of 25 September 1907 that "the works of Poincaré were [still] unknown to Einstein" [Hermann (1966)]. Hermann meant Poincaré's classic papers of 1905–1906. Kaufmann's (1906), however, had described a principal result of Poincaré – the Poincaré pressure, and analysis of Einstein's (1907d) in Section 12.5.3 reveals that Einstein may well have included a portion of the energy resulting from the Poincaré pressure implicitly as a special case of the problem under consideration.

3) *Centralblatt für Elektronik* and *Elektrotechnische Zeitschrift*: The ETH curriculum's emphasis on applied electricity could well have served to introduce Einstein to these journals – for example, they contained many articles on the design and development of dynamos. Since Einstein's work at the Patent Office often concerned assessing patents for dynamos [Flückiger (1974)] he may often have perused these journals during 1902–1909.

From the list of materials to which Einstein could have had access before 1905, there emerges the picture of a scientist who possessed the potential of having been informed of at least the broad outline of developments in physics.[69]

Even supposing, for example, that before 1905 Einstein had read only the *Annalen*, he would have been aware of: Wien's (1900) as reprinted in the *Annalen* – proposal of the electromagnetic world-picture; Abraham (1903) – Abraham's most detailed exposition of his theory [Einstein's (1905d) uses Abraham's nomenclature of transverse and longitudinal masses]; Abraham (1904b) – the most up-to-date treatment of such problems in the optics of moving bodies as the reflection of light from a perfectly reflecting moving mirror [Einstein's (1905d) solved this problem and obtained a result "in agreement with experiment and with other theories"] – the only other theory was Abraham's and its results were mathematically identical to Einstein's; Abraham's (1904b) also addressed the problem of faster-than-light motion, particularly of Wien's assertion that electrons could not attain $v = c$; Wien's (1904b) reply to Abraham's (1904b) also sketched some important results of Lorentz's (1904c) theory of the electron; Wien (1904a) discussed the major theories of electrodynamics, e.g., those of Abraham, Cohn and Lorentz. On the basis of the material to which Einstein may have had access before 1905, he could well have been aware of a great deal of the extant empirical data – for example:

1) *Electromagnetic induction*: Einstein's (1905d) begins by analyzing the current generated in a conductor owing to the "relative motion of the conductor and magnet." In §6 of the (1905d) he applied the results of this case to "unipolar machines." Empirical data on problems concerning electromagnetic induction would have been available to Einstein from, for example, Föppl (1894), Arnold (1895), Lecher (1895), Weber (1895) and Grotrian (1901).

2) *Experiments designed to discover effects of the earth's motion on electrical and optical phenomena*: Einstein (1905d) refers to "unsuccessful attempts to discover any motion of the earth relatively to the 'light medium'" without listing any of them. Then he specialized to the class of experiments accurate to the "first order of small quantities" – that is, to measurements accurate to order v/c of the velocity of light in moving reference systems.

Experiments with this accuracy could be divided into two sorts:

(i) Positive experiments (these were not designed as ether-drift experiments): Observations of stellar aberration by Bradley (1728) and Arago (1810); the experiments of Fizeau (1851) and Michelson and Morley (1886).

(ii) Null experiments: For example, Hoek (1868), among others.

At the ETH Einstein can have become aware of the existence of these experiments through his readings in Lorentz's (1895).

The unnamed ether-drift experiments that Einstein grouped into "unsuccessful attempts" undoubtedly included those accurate to second order in v/c. By 1905 there were two such groups of experiments:

(i) Michelson (1881) and Michelson and Morley (1887): By 1905 Einstein would have known of the result of Michelson's experiment, since it was discussed in the last chapter of Lorentz's (1895). But its genetic role in fashioning the relativity theory is another matter. Holton (1973e) has argued cogently that we should take Einstein at his own word that the Michelson-Morley experiment had "at most an indirect effect" on Einstein's thinking toward the special relativity theory. Einstein recalled later that only after 1905 had he become interested in this experiment, that he may not even have known about it until "only after 1905" [Shankland (1963)], and he was convinced in any case that the result had to be so.

(ii) Rayleigh (1902) and Brace (1904): Abraham's (1904b) discussed these experiments.

3) *Experiments to determine the velocity dependence of the electron's mass*: Before 1905 Kaufmann had published papers on the mass of β-rays [see Kaufmann's 1901–1903 publications and Starke (1903) had published results on the velocity-dependence of the mass of fast cathode rays]. For reasons to be discussed in Chapter 12, Einstein chose judiciously not to compare his prediction for m_T with Kaufmann's data.

4) *Experiments to measure the pressure of radiation*: In §8 of the (1905d) Einstein checked his predictions for the pressure of radiation on a perfectly reflecting moving mirror by showing that for $v \ll c$ it was "in agreement with experiment and with other theories...". Abraham's (1904b) discussed the available experimental data of Lebedew (1901) and Nichols and Hull (1903), which had appeared in the *Annalen*.

FOOTNOTES

1. In the letter of Einstein to Seelig, Einstein mistakenly listed Lorentz's (1892a) as having been published in 1895.

2. Briefly, an overview of electromagnetic theory until Hertz is: If in 1870 Voltaire could have traveled once again between the continent and Britain, he would have changed his observation of 140 years earlier – he would have "left the world a vacuum and now he finds a plenum." By 1870, owing mainly to the influence of Laplace's elaboration of Newton's mechanics, continental physicists had become accustomed to applying Newtonian action-at-a-distance physics to electricity. In England a Cartesian viewpoint with an ether reigned supreme in the form given to it by James Clerk Maxwell's electromagnetic theory of light. A striking feature of Maxwell's theory was its unification of two hitherto distinct branches of physical theory – light and electromagnetism. Although by 1870 continental physicists considered that optical phenomena were mediated by an ether, this was not the case for Europe's most well-known electromagnetic theories – for example, those of Franz Ernst Neumann and Wilhelm Weber. Neumann's 1845 theory described electromagnetic induction between two closed circuits as due to the instantaneous effect of a vector potential. Weber's 1848 theory interpreted electrical current as the streaming of equal and opposite microscopic charges. He calculated that the force between two moving charges q_1 and q_2 is

$$F = \frac{q_1 q_2}{r^2}\left[1 - \frac{\dot{r}^2}{c^2} + 2\frac{r\ddot{r}}{c^2}\right]$$

where r is the distance between the charges, c is the velocity of light, and with this force law Weber could explain all known electrostatic and electrodynamical effects. As C. W. F. Everitt (1974) has written, Neumann's potential theory and Weber's force law provided the starting points for almost all work done in Europe on electromagnetic theory until the 1870's.

Weber had shown that electrostatic and electrodynamical units were related by a velocity which he and Kohlrausch had measured in 1856 to be 3.107×10^{10} cm/sec. Although in 1857 Kirchhoff noted the closeness of this quantity to the velocity of light, Maxwell in 1862 declared this occurrence to be not merely coincidental: just as optics could be based upon an ether, so could electromagnetic theory, and the two ethers were the same.

But Maxwell's theory added to what Hermann von Helmholtz [see Hertz (1894)] described as "the pathless wilderness" of electromagnetic theories: Maxwell's theory contained twenty equations in twenty unknowns (they included potentials and fields); it did not clearly define electricity; electrical disturbances were supposed to propagate as stresses and strains in an all-pervasive dielectric medium. Hertz (1892) wrote of "many a man . . . compelled to abandon the hope of forming for himself an altogether consistent conception of Maxwell's ideas." Henri Poincaré (1901) wrote of a well-known French scientist who said, "I understand everything in [Maxwell's Treatise (1873)] except what a charged sphere is."

In 1870 von Helmholtz set about putting some order into this situation by taking what Max Planck (1931) described as "a middle course." He formulated an electromagnetic theory with an adjustable parameter whose three possible values were the limiting cases for the theories of Neumann, Weber and Maxwell. Consequently, most continental physicists learned their Maxwellian electromagnetism within the context of an action-at-a-distance framework. Von Helmholtz's foundational analyses during the 1870's revealed in Weber's theory instabilities that did not occur in nature; therefore, the ultimate choice was between variants of a potential theory and Maxwell's theory. The decision had to be made empirically, and von Helmholtz suggested certain tests. The principal ones were: (1) In 1874 he proposed attempting to detect a magnetic field set up by a moving charge, i.e., a convection current; this phenomena was found by Rowland in 1876, with a magnitude consistent with Maxwell's theory. (2) Then in 1879 von Helmholtz proposed to seek an effect predicted only by Maxwell's theory – electromagnetic waves; this phenomenon was detected in 1888 by von Helmholtz's former student, Hertz. Consistent with an ether-based view of electromagnetism, Hertz's discovery of electromagnetic waves was heralded also as proof for the existence of an ether. Poincaré's Sorbonne lectures of 1888, 1890 and 1899 completed the process of convincing most everyone of the fecundity of ether-based electromagnetic theories.

For discussions of the history of electromagnetism until Hertz, see: Bork (1966a), Everitt (1974), Gillispie (1960), Heimann (1970), (1971), Hirosige (1966), (1969), Merz (1965, vol. 2), Rosenfeld (1952), Whittaker (1973, vol. 1), and Woodruff (1968).

3. Oliver Heaviside had arrived at a similar set of equations [see Whittaker (1973, vol. 1)]. The quantities true magnetism and true charge, as well as free charge (magnetism), whose sources are $E(B)$ did not serve to clarify the electrodynamics of bulk matter, and Lorentz excluded them from his theory of 1892. For biographical information on Heaviside see Süsskind (1972).

4. Other questions that Hertz (1894) deemed to have been illigitimate were: What is the nature of force? What is the nature of velocity? Hertz's philosophy of science as presented in his (1894) is discussed in Chapter 2.

5. There Helmholtz focused upon calculating the induced currents for a circuit moving through a magnetic field. Maxwell (1873) had extended only Faraday's law to moving bodies and proved that the magnitude and direction of the induced current was the same regardless of whether it was calculated in the moving circuit's rest system. For this purpose, Maxwell used the convective derivative. Evidently Hertz missed this point in Maxwell's writings – see the footnote Hertz added to (1890b) in *Electric Waves* (1892).

6. For example, Hertz may well have considered that for an observer at rest in R, Eq. (1.11) was

$$\nabla \times E' = -\frac{1}{c}\frac{\partial B}{\partial t} + \frac{u}{c}m_{\text{true}}$$

where

$$E' = E + \frac{u}{c} \times B$$

and $u = -v$.

7. Briefly [following Lorentz, (1886) and (1909)] two points A and B are on the line of length ds that intersects normally two wave fronts. Then the quantity $\int_A^B ds/u_r$, where u_r is the ray velocity, is less than that for any other line connecting A and B. The integral $\int_A^B ds/u_r$ is the time for a light ray to propagate from A to B, and the condition that the integral be a minimum for the ray velocity is a statement of Fermat's principle. From Eq. (1.17), neglecting terms of order $(v/c)^2$, we can write $1/u_r$ as

$$\frac{1}{u_r} = \frac{1}{c/N} + \frac{v}{c^2} \cos \theta \tag{A}$$

where θ is the angle between u_r and $-v/N^2$ (see Fig. 4). Then we have

$$\int_A^B \frac{ds}{u_r} = \int_A^B \frac{ds}{c/N} + \frac{v}{c^2} \int_A^B \cos \theta \, ds. \tag{B}$$

The integral in Eq. (B) containing θ is v/c^2 times the projection of the path AB on the direction of the relative velocity between the earth and the ether. This term is determined only by the position of the endpoints A and B, and so it is the same for every path connecting A and B. Consider next that at A there is a half-silvered mirror which splits a ray from a terrestrial light source into two rays that are recombined at B, i.e., an interference experiment. The difference in the optical paths of the two rays is independent of v because the term containing θ in Eq. (B) is the same for both paths. Therefore, to order v/c, interferometer experiments cannot detect the earth's motion relative to the ether. Since according to Huygens' principle all optical phenomena are cases of interference, then to order v/c all optical phenomena are independent of the earth's motion.

8. Consider that at the time $t = 0$ a wave crest passes by the origin 0 of an ether-fixed reference system S. At a time d/c later this wave crest reaches a point P at rest in the ether, where d is the distance between P and O. From that moment an observer at P begins counting the wave crests that pass by P until the time t, i.e., during the interval $t - d/c$. Thus in the time interval $t - d/c$, the observer at P counts $v(t - d/c)$ wave crests. Consider next an observer at the point P' which is the origin of an inertial reference system S_r. We assume that at $t = t_r = 0$ the coordinate origins of S and S_r coincided. The moving observer begins counting wave crests at the time $t = t_r = 0$, until the points P and P' coincide. Consequently, the observer at P' counts wave crests for an interval $t_r - d_r/u_r$, and counts the number of wave crests $v_r(t_r - d_r/u_r)$, where quantities subscripted with r refer to the reference system S_r. The spatial coordinates in S and S_r are related by the Galilean transformations of Newtonian mechanics. Although the plane wave's defining characteristics may differ when measured in S and S_r, the observers at P and P' must both have counted the same number of wave crests. The phase of an electromagnetic wave depends only on the relative separation of the points O and P or O and P', and not on the systems of reference.

9. Among the doubters were Maxwell (1878) and Drude (1900). Subsequent measurements detected no such effect – e.g., Brace (1905b).

10. Michelson and Morley (1887) wrote that this error had also been pointed out in 1881 by Potier.

11. It should be pointed out that Michelson and Morley (1887) neither obtained a null result nor considered their data as optimally accurate. In fact, they took only six hours of readings over a period of five days, and promised that "the experiment will therefore be repeated at intervals of three months, and thus uncertainty will be avoided." Although they never carried out this proposal Swenson (1972) shows that by the turn of the century, the Michelson-Morley experiment of 1887 became a null experiment. For further discussion of the Michelson-Morley experiment and in particular of its supposed role in physics *after* relativity as an *experimentum crucis*, for Einstein's thinking toward the special relativity theory, see Holton (1973b).

12. Lorentz often emphasized this quasi-aesthetic criterion, e.g., Lorentz (1897). See Miller (1974), (1979a) for discussion of this component of Lorentz's thinking.

13. H. A. Lorentz was one of the acknowledged scientific masters. The nine volumes of his *Collected Papers* reveal the range and depth of his research. Lorentz was at ease equally with theoretical problems of a type couched heavily in mathematics (as, for example, in his later work on Einstein's theory of gravitation) or of a very applied nature (what better example is there than his work in hydraulic engineering rendering possible the construction of an all-important enclosure dike in the Zuiderzee – Lorentz personally supervised the construction as well). Almost certainly Lorentz considered his electromagnetic theory to have been the crowning achievement of his life's work. In his doctoral thesis of 1875, and then in sequel papers during the period 1875–1880, Lorentz demonstrated a deep understanding of Maxwellian electrodynamics in the action-at-a-distance reformulation of von Helmholtz.
Hertz's discovery of electromagnetic waves in 1888 and Poincaré's lectures at the Sorbonne convinced Lorentz in 1891 of the advantages of a continuum-based theory of electromagnetism. Lorentz (1904b) stressed that his theory was a mixture of the theory of Maxwell-Hertz and the 1876 theory of Clausius, in which the force between two particles of electricity depended upon their velocities relative to an ether.
In 1902 Pieter Zeeman and Lorentz were awarded the Nobel Prize for the discovery and explanation of the splitting of the spectral lines of sodium by a magnetic field, i.e., the Zeeman effect (see Section 1.8). For discussion of Lorentz's electromagnetic theory, see Goldberg (1969), Hirosige (1966), (1969), Miller (1973), (1974), (1979a), and Schaffner (1969), (1972). For biographical information on Lorentz, see De Haas-Lorentz (1957) and McCormmach (1973).

14. Lorentz used Eqs. (1.26) and (1.27) to obtain $\partial\rho/\partial t + \mathbf{V}\cdot(\rho\mathbf{w}) = 0$, which is a form of the convective derivative of ρ; therefore, in Lorentz's theory ρ is independent of time in the microscopic particle's rest system.

15. In 1878 Lorentz had used the model of a charged harmonic oscillator within a molecule of matter for dispersion calculations in bodies at rest. For this purpose Lorentz had utilized von Helmholtz's action-at-a-distance electrodynamics [see Hirosige (1969)].

16. Since the ether-fixed system S never moves, then the process of inverting Eqs. (1.37)–(1.40) by interchanging the subscripts and v with $-v$ has no physical meaning. The nomenclature "Galilean transformation" appeared first in Frank and Rothe (1911).

17. This can be proven directly by using a device that Lorentz did not have at his disposal. In S,

$$x^2 + y^2 + z^2 = c^2 t^2. \tag{A}$$

Equation (A) can be transformed to Q' using the equations inverse to Eqs. (1.43)–(1.46) that relate S and Q':

$$x = \gamma x' + vt,$$

$$y = y',$$

$$z = z',$$

$$t = t' + \gamma \frac{v}{c^2} x',$$

and (A) becomes

$$x'^2 + y'^2 + z'^2 = c^2(1 - v^2/c^2)t'^2.$$

Hence in Q' the velocity of light is $c\sqrt{1 - v^2/c^2}$.

18. According to Lorentz's theory, in S_r the mean dipole moment P_y arising from harmonically bound charged particles that responded to a plane electromagnetic wave incident along the x-axis was

$$\left[V_r^2 - \frac{1}{c^2}\left(\frac{\partial}{\partial t_r} - v\frac{\partial}{\partial x_r} \right)^2 \right] P_y = -\frac{1}{Q}\frac{\partial^2 P_y}{\partial t_r^2}, \tag{A}$$

where the quantity Q depended on the restoring forces and the number of charged particles per unit volume of the material that was at rest in S_r, and through which the light was propagating [see Lorentz (1892a), (1909) or Drude (1900)]. Lorentz substituted into Eq. (A) the solution

$$P_y = A \cos \frac{2\pi}{T}(t_r - x_r/W_r) \tag{B}$$

where W_r was the velocity of the light relative to S_r. To order v/c, Lorentz deduced

$$W_r = \frac{c}{N} \pm \left(1 - \frac{1}{N^2} \right)v \tag{C}$$

where $N = \sqrt{1 - c^2/Q}$.

19. In this way, Lorentz's theory avoided a rather embarrassing shortcoming of Fresnel's original theory; namely, that the components of a light beam composed of different frequencies were dragged along at different velocities.

20. Lorentz's contraction hypothesis had been proposed simultaneously but independently by G. F. FitzGerald [see Bork (1966b), Brush (1967)].

21. Lorentz could well have used the following argument: from the exact expressions for $t_{\parallel}$ and $t_{\perp}$, i.e.,

$$t_{\parallel} = \frac{2l_{\parallel}/c}{(1 - v^2/c^2)},$$

$$t_{\perp} = \frac{2l_{\perp}/c}{\sqrt{1 - v^2/c^2}};$$

their ratio is

$$\frac{t_{\parallel}}{t_{\perp}} = \frac{l_{\parallel}}{l_{\perp}\sqrt{1 - v^2/c^2}}$$

and so $t_{\parallel}$ could equal $t_{\perp}$ for $l_{\parallel} = l[1 - v^2/c^2]^{n+1/2}$ and $l_{\perp} = l[1 - v^2/c^2]^n$ for any n; therefore the Lorentz contraction was one of many possibilities.

22. The reason is that increased accuracy in observations could reveal variations in the dimensions and orientation of the aberration ellipse owing to the earth's velocity relative to the ether. These effects were expected to be less than the square of the aberration constant, i.e., less than $(v/c)^2 = 10^{-8}$. Poincaré (1908a, b) discussed this fine point, and then went on to show that Lorentz's (1904c) theory of the electron explained why the earth's motion relative to the ether could never be observed "even though our instruments were ten thousand times as accurate." As we shall see later in Chapter 1, Poincaré mentioned such compensating effects as the contraction of measurement instruments on the moving earth.

23. The Michelson-Morley experiment of 1887 was one among three experiments that concerned Lorentz. The other two were the experiments of Fizeau (1860) – see Section 1.3 – and Mascart (1872). Lorentz used the theorem of corresponding states to prove that, to order v/c, Fizeau should have detected no change in the azimuth of the plane of polarization additional to the one expected from the laws of optics for systems at rest as follows. According to the theorem of corresponding states regions of brightness and darkness were unaltered by the earth's motion. Consequently, the orientation of a polarizer arranged for extinction was unchanged by the earth's motion. Several repetitions of Fizeau's experiment yielded a null result, and Brace's (1905b) experiment was considered as conclusive. Brace (1905c) emphasized the importance of a decision on the status of Fizeau's results owing to "the positive results of this experiment, which are the only ones ever obtained by experimental means on the problem of the 'aether drift'."

Mascart (1872) found that the rotary power of optically active substances, such as quartz, was the same as if they were at rest in the ether. Lorentz's 1895 theory of optics, however, predicted a first-order effect of one part in 10^4, unless a certain two terms conspired to cancel [see also Lorentz (1902)]. With a precision of better than one part in 10^6 Brace (1905b) demonstrated that the earth's motion through the ether had no effect on rotary polarization.

Separate publication in the Dover reprint volume of Lorentz's (1895) analysis of the Michelson-Morley experiment has led many physicists and historians and philosophers of science to emphasize unduly the role of this experiment in Einstein's thinking toward the special relativity theory. For further discussion of this point see footnote 7 of Chapter 1, and the Appendix which presents the history of the Dover reprint volume.

24. In Lorentz's theory the electric and magnetic fields are related to a scalar potential ϕ and a vector potential A through the equations:

$$E = -\nabla\phi - \frac{1}{c}\frac{\partial A}{\partial t},\tag{A}$$

$$B = \nabla \times A.\tag{B}$$

These potentials satisfy inhomogeneous wave equations that are deducible from Eqs. (A) and (B), and the Maxwell-Lorentz equations:

$$\left(\nabla^2 - \frac{1}{c^2}\frac{\partial^2}{\partial t^2}\right)A = -4\pi\rho\,\frac{v}{c},\tag{C}$$

$$\left(\nabla^2 - \frac{1}{c^2}\frac{\partial}{\partial t^2}\right)\phi = -4\pi\rho.\tag{D}$$

Since for a uniformly moving charge $(\partial/\partial t)_S = -v\cdot\nabla_r$ and $\nabla = \nabla_r$ then these wave equations become

$$\left[\nabla_r^2 - \left(\frac{v\cdot\nabla_r}{c}\right)^2\right]A = -4\pi\rho\,\frac{v}{c},\tag{E}$$

$$\left[\nabla_r^2 - \left(\frac{v\cdot\nabla}{c}\right)^2\right]\phi = -4\pi\rho,\tag{F}$$

and Eq. (E) could be reduced to (F) through the relation

$$A = (v/c)\phi.\tag{G}$$

If $v = v\hat{i}$, then, from Eqs. (1.60)–(1.61) Lorentz could write the force on another ion q' that moved in the electromagnetic fields E and B owing to a uniformly moving ion q as

$$F = -q'\nabla\Psi\tag{H}$$

where

$$\Psi = (1 - v^2/c^2)\phi.\tag{I}$$

Searle (1897) referred to Ψ as the "convection potential."

25. So far as I know, Thomson (1889) invented the mathematical technique involving the coordinate transformation Eqs. (1.63)–(1.66) for changing inhomogeneous wave equations for electromagnetic quantities into a Poisson equation. His goal was to provide an alternate derivation of Heaviside's (1889) formulae for the electromagnetic field of a uniformly moving charged body. Although Lorentz was aware of Thomson's 1881 paper (Lorentz's (1892a) referred to it) which was a precursor to the one of 1889, by 1892 and certainly by 1895 Thomson's mathematical trick may have been so well known that Lorentz felt it reasonable not to refer to Thomson's works; on the other hand, Lorentz may have discovered for himself the coordinate transformation to a reference system having the properties of the moving

charge's absolute rest system; furthermore, the conclusions which Lorentz drew from the Eqs. (1.63)–(1.66) went far beyond Thomson's.

26. Setting $v = v\hat{i}$, and using Eqs. (1.63)–(1.66), and (1.68), Lorentz's electric field Eq. (1.60) becomes

$$E_x = -\frac{\partial}{\partial x''}\,\phi'' = \frac{qx''}{r''^3}, \tag{A}$$

$$E_y = -\gamma\frac{\partial}{\partial y''}\,\phi'' = \gamma q\,\frac{y''}{r''^3}, \tag{B}$$

$$E_z = -\gamma\frac{\partial}{\partial z''}\,\phi'' = \gamma q\,\frac{z''}{r''^3}, \tag{C}$$

since $\phi'' = q/r''$ where $r'' = \sqrt{x''^2 + y''^2 + z''^2}$ – that is,

$$E = \frac{qr_r(1 - v^2/c^2)}{S^3} \tag{D}$$

where $S = \sqrt{x_r^2 + (1 - v^2/c^2)(y_r^2 + z_r^2)}$ which is Heaviside's (1889) result for the electric field of a uniformly moving point charge. The electric field is directed along a radius vector drawn from its resting system S_r, and is measured by an observer in S, i.e., in the ether. In applications S was cavalierly taken as the laboratory system – see, for example, Abraham (1905). The convection potential ψ can be written as

$$\psi = (1 - v^2/c^2)q/S \tag{E}$$

and so equipotential surfaces are

$$S^2 = x_r^2 + (1 - v^2/c^2)(y_r^2 + z_r^2) = \text{constant} \tag{F}$$

which are the family of ellipsoids

$$\frac{x_r^2}{(1 - v^2/c^2)} + y_r^2 + z_r^2 = \lambda \tag{G}$$

whose major axes are normal to their directions of motion; Searle (1897) referred to this family of ellipsoids as "Heaviside ellipsoids."

Eq. (D) is the electric field for either a point charge or far enough away from a charge distribution of arbitrary shape so that it can be taken to be a point charge; the electric field lines spread out radially from the source, and accumulate in the charge's direction of motion. Heaviside (1889) went on to assert that Eq. (D) was valid also for the case of a moving charged sphere with a uniform surface distribution of charge, and he assumed that the electric field lines were everywhere normal to the sphere's surface. In a reprint of the (1889) paper in Heaviside's (1892, vol. II), he noted Searle's disagreement with this extension because the electric field of a particle moving with a velocity close to that of light could not be deduced from the usual electrostatic potential. Heaviside agreed and showed that the conductor with a uniform surface distribution of charge q whose motion would yield at all points the radial electric field distribution of a point charge of magnitude q placed at its center was not a sphere but an ellipsoid, and E was not everywhere normal to its surface. Morton (1896) proved that the static charge distribution on a conductor of any shape was unaltered by its motion. He went on to calculate the potential for an ellipsoid with a surface charge distribution. Searle (1897) investigated the Heaviside ellipsoids in further detail and deduced the potentials and electromagnetic energies for surface charge distributions on such shapes as a Heaviside ellipsoid and a sphere.

The Heaviside ellipsoid is not the same shape as the electron in S'' whose major axis is along S_r's direction of motion. That no deformation of the electron is implied in describing its fields by means of the convection potential had to be emphasized as late as 1916 [see Richardson (1916)].

Poincaré (1906) deduced Eq. (D) from Lorentz's coordinate and time transformations of 1904 – the Lorentz transformations.

The electric field of a moving point charge was also deduced in 1898 by Liénard and then in 1900 by Wiechert by means of a more general method than direct solution of the inhomogeneous wave equation. The Liénard-Wiechert potentials were applicable also to the nonuniform motion of charged particles – that is, to problems concerning radiation.

27. The stumbling block was Gauss' law, which from Eqs. (1.76)–(1.79) was

$$\mathbf{V}_r \cdot \boldsymbol{E}_r = 4\pi \left[1 - \frac{v u_{x_r}}{c^2} \right] \rho$$

where $\boldsymbol{E}_r = \boldsymbol{E} + (\boldsymbol{v}/c) \times \boldsymbol{B}$ and $\boldsymbol{u}_r$ was the ion's velocity relative to S_r.

28. Henri Poincaré was France's greatest living mathematician, and among the first ranks of the physicists and philosophers of his day. On 15 December 1913 Gaston Darboux (1913) eulogized him as "à la fois mathématicien hors de pair, physicien pénétrant et profond philosophe." Poincaré was a member of l'Académie des Sciences (elected in 1887 and President in 1906) and of l'Académie française (elected in 1887 and Director in 1912). Darboux recalled that Poincaré had been the only member of l'Académie des Sciences whose researches qualified him to belong to every section – geometry, mechanics, astronomy, physics, geography and navigation. By the end of his life Poincaré had written in the neighborhood of 30 books and 500 papers. Poincaré's curriculum vitae is a 111 page book [see Lebon (1912)].

Studies of the development of Poincaré's philosophical viewpoint support the conjecture that it was precipitated by his research in non-Euclidean geometry; in particular, his discovery (1887) that any non-Euclidean geometry could be generated by infinitesimal rotations or translations in spaces of any number of dimensions [see Miller (1973), (1979b)]. According to the German philosopher Immanuel Kant not only was there no geometry other than Euclid's, but we even lack the intellectual capacity for visualizing non-Euclidean worlds. This Kantian viewpoint was dealt crushing blows when in 1827 Bolyai and Lobaschevsky independently formulated non-Euclidean geometries; then in 1854 von Helmholtz demonstrated how non-Euclidean worlds could be visualized; in 1868 the Italian geometer Beltrami proved the consistency of three dimensional non-Euclidean geometries in spaces with constant curvature (this proof rested upon the unproven assumption of the consistency of Euclidean geometry). Poincaré's keen appreciation for the role of organizing principles in Kant's viewpoint, led him to attempt to salvage this aspect of Kantian philosophy. Instead of the Kantian *a priori* intuitions of space and time, which asserted that we were forever trapped in a Euclidean prison, Poincaré proposed two new organizing principles: (1) the notion of groups of infinitesimal transformations; (2) the notion of mathematical induction. Consequently, according to Poincaré, as a result of studying displacements, i.e., the relationships between sensations, these organizing principles enabled all life forms to realize that of every possible geometry, only three dimensional Euclidean geometry applied to the world of sensations. Making the mental leap to ideal solids (e.g., triangles whose interior angles added up to exactly 180°), the laws of geometry became axioms that were of such generality that they were no longer experimentally verifiable – that is, they became "conventions." (Poincaré's philosophical view is often referred to as conventionalism.)

In Poincaré's viewpoint, the conventions of classical mechanics such as Newton's three principles, the conservation of energy and the principle of relative motion, according to which (1900c) "[t]he movement of any system whatever ought to obey the same laws, whether it is referred to fixed axes or to the movable axes which are implied in uniform motion in a straight line" were, like the geometrical axioms, obtained from data. For, as Poincaré wrote, experiment is (1900c) "the sole source of truth." The conventions of classical mechanics were obtained from empirical data by means of dazzling and unquantifiable mental leaps.

Poincaré had concluded as a result of analyzing the foundations of geometry and classical mechanics that only relative, not absolute, quantities should appear in these theories. He referred to this result in different terms: in geometry, in 1899, as the principle of relativity; in classical mechanics first as the law of relativity (1899) and then as the principle of relative motion (1900c). Thus, this principle was characteristic of his holistic view of science.

Whereas certain conventions could be discarded if no longer "fruitful," Poincaré assumed that those of Euclidean geometry and of relative motion could not because they were necessary to our very survival.

Consistent with his belief in the psychological origin of scientific concepts, Poincaré always agreed with Lorentz on the indispensability of the ether. His reason was that just as in the world of perceptions there was no action-at-a-distance, so should it be in science which offered a sophisticated description of this world – for example, the ether mediated forces and supported light in transit. Many of Poincaré's philosophical writings appear in his reprint volumes (1902), (1905b), (1908a), (1913). Other analyses of Poincaré's philosophical view are Goldberg (1970a) and Holton (1973b).

29. I have discussed Poincaré's foundational analysis of electromagnetic theory in Miller (1973), (1979a, b). For other views, see Goldberg (1967), (1969), (1970a), Holton (1973b), Kahan (1959). In Lorentz's theory conservation of electric charge follows from the vanishing of the convective derivative of the ion's charge density in its rest system (see Chapter 1, footnote 14). Conservation of magnetism follows from there being no real magnetic charges in Lorentz's theory. The law of conservation of charge turned out to be a Lorentz-invariant law, and this point was noted by Poincaré (1906) as a purely mathematical property of how the charge density and convection current transform according to Lorentz's transformations of 1904 (see Section 1.14).

30. The properties of the stress tensor useful for our subsequent analysis can be illustrated with the following example. The x-component of the force exerted on charges contained enclosed in a volume V is

$$\int f_x \, dV = \int \left(\frac{\partial T_{xx}}{\partial x} + \frac{\partial T_{xy}}{\partial y} + \frac{\partial T_{xz}}{\partial z} \right) dV - \frac{d\boldsymbol{g}}{dt} \tag{A}$$

$$= \oint (T_{xx} \cos{(\hat{n}, \hat{x})} + T_{xy} \cos{(\hat{n}, \hat{y})} + T_{xz} \cos{(\hat{n}, \hat{z})}) \, d\sigma - \frac{d\mathbf{g}}{dt} \qquad \text{(B)}$$

where $\hat{n}$ is the outward unit vector from the volume V.

The stress tensor's components are the total force per unit area exerted on a surface. Consider a perfectly absorbing surface in y-z-plane. If radiation in the form of plane waves propagating along the x-axis is incident on this surface from the left, then only T_{xx} contributes to Eq. (B). Since $\cos(\hat{n}, \hat{x}) = -1$ (where $\hat{n}$ is the normal outward from the surface), then $-T_{xx}$ is the resultant force per unit area on the plane surface. Since the plane wave propagating along the x-axis has electromagnetic quantities E_y and B_z, then from Eq. (1.102)

$$-T_{xx} = (E_y^2 + B_z^2)/8\pi. \qquad \text{(C)}$$

The average value of Eq. (C) reveals that the normal pressure per unit area of a beam of light on a perfectly absorbing surface is numerically equal to the electromagnetic energy density of the light beam [this example is from Lorentz (1909)].

Two other interesting properties of $\mathbf{\ddot{T}}$ are:
(1) $T_{xy} = (1/4\pi)(E_x E_y + B_x B_y) = T_{yx}$, i.e., the stress tensor is a symmetric tensor,
(2) $T_{xx} + T_{yy} + T_{zz} + \omega = 0$, i.e., the stress tensor is traceless.

31. The other case discussed by Lorentz was the stationary system in which $d\mathbf{g}/dt = \mathbf{0}$. An example of such a system is in footnote 30.

32. Poincaré restricted the demonstration to first order in v/c in order to avoid discussing "a certain complementary hypothesis" which could only have been the hypothesis of contraction.

33. Poincaré never again suggested this means of saving Newton's third law, and, in fact, he used against it Lorentz's observation of the nonsimultaneity of action and reaction in electromagnetic field theory [Poincaré (1904)].

34. The successes of his electromagnetic theory had spurred Lorentz on to take a step toward extending his theory to gravitation. He turned to an idea from the 1836 gravitational theory of Mossotti, that had been recently discussed by Wilhelm Weber and J. C. F. Zöllner, in which, wrote Lorentz (1900), "positive and negative charges differ from each other to a larger extent than may be expressed by the signs $+$ and $-$." Lorentz assumed that the positively charged constituent of matter was a specie of the negatively charged electron. Lorentz supposed that the positively and negatively charged electrons constituting ponderable matter acted on each other differently; the force between particles of unlike sign being greater than the force between charges of like sign. Including this modification into his electromagnetic theory, Lorentz deduced a gravitational force law that produced an advance for Mercury's perihelion in the right direction, but too small. Lorentz was not deterred, however, for he wrote that it "is sufficient to show that gravitation may be attributed to actions which are propagated with no greater velocity than that of light."

35. For other discussions of the electromagnetic world-picture, see McCormmach (1970a), Miller (1973), (1974), (1977c). For discussions of the mechanical world-picture, see Merz (1965), Schaffner (1972) and Whittaker (1973, vol. 1).

36. Heaviside (1889) obtained this result by interpreting the quantity $V^2 - 1/c^2 \cdot (\partial^2/\partial t^2)$, i.e., the D'Alembertian, in the inhomogeneous wave equation as an operator

on the vector potential, and then expanding its inverse in a series of derivatives which acted on the convection current. Perhaps horrified by this seemingly idiosyncratic procedure, Thomson almost immediately thereafter invented the mathematical method for directly solving the inhomogeneous wave equation for a uniformly moving charge, i.e., the transformation Eqs. (1.63)–(1.66), used in conjunction with the convective derivative.

37. Searle's work further impressed upon Wien the possibilities for an electromagnetic basis for gravitation that Lorentz (1900) had speculated upon earlier. From Searle's result for the total energy of a moving Heaviside ellipsoid [i.e., $E^e = (e^2\beta(1 + \frac{1}{2}\beta^2))(2R\sqrt{1 - \beta^2} \text{ arc sin } \beta)^{-1}$] Wien could almost deduce Weber's gravitational force between two particles one of which had an inertial mass M and the other a purely electromagnetic mass of $4/3(E_0^e/c^2)$ where E_0^e was the "energy of the resting ellipsoid," and this result was accurate to second order in v/c. Most likely Wien focused upon Searle's result for a Heaviside ellipsoid, instead of a sphere, because he extrapolated from a Lorentz contraction of bulk matter in its direction of motion to electrons in motion; this hunch turned out to be fortuitous.

In 1911 Wien was awarded the Nobel Prize in physics for his work on cavity radiation, particularly for the displacement law which, he wrote (1911), "exhausts the conclusions that can be drawn from pure thermodynamics with respect to radiation theory." In 1900 Wien was Professor of Physics at the University of Würzburg where he remained until 1920 when he went to the University of Munich. For further biographical information see Kangro (1976).

38. Quantitative studies of the rays emitted from the hot cathode of a vacuum tube can be dated from those of Julius Plücker (1858) and Johann Hittorf (1869). Eugen Goldstein (1876) dubbed these emanations "cathode rays." By 1897, wrote Thomson [(1897) – in the October issue of the *Philosophical Magazine*], "the most diverse opinions are held as to these rays." Chief among these opinions were: (1) these rays resulted from the gas in the vacuum tube having become ionized owing to contact with the cathode; (2) these rays were pulsations in the ether that were converted into light by transferring their energy to gas molecules in the vacuum tube; (3) these rays were negatively charged particles that had been emitted from the hot cathode, i.e., the emission hypothesis. The opinions (1) and (3) were held mostly by British physicists, and (2) was popular among the Germans with Goldstein, Hertz and Lenard among

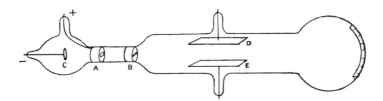

FIG.   J. J. Thomson's schematic of his apparatus for measuring the charge-to-mass ratio of the negatively charged particles that constitute cathode rays. The sense of the coordinate system is

y
$\llcorner$—x. (From *Philos. Mag.*)
z

them. Thomson supported the third hypothesis because, wrote Thomson (1897), "it has a great advantage over the aetherial theory since it is definite and its consequences can be predicted." The experiments of Thomson and Kaufmann were among those most frequently cited as having served to clarify the nature of cathode rays.

A description of Thomson's experimental method (see Fig.) suffices because of its similarity to other experiments on cathode rays; furthermore, Thomson's method included a means for deflecting cathode rays that would play an important role in the electron physics of 1907 – crossed electric and magnetic fields. Cathode rays are accelerated by the potential difference between the cathode C and the anode A, which is a metal plug. After passing through a slit in the anode, the cathode-ray beam is further collimated by a slit in the earthed plug B. The parallel-plate capacitor has plates D and E of length $l = 5$ cm and width 2 cm, and a plate spacing of 1.5 cm. The vacuum tube can be placed within Helmholtz coils which provide a uniform magnetic field normal to the plane of the paper, and acting only over the region between the capacitor plates. The beam's deflection is measured on a scale pasted to the end of the tube. Thomson used air, carbon dioxide and hydrogen at low pressures, and cathodes made of aluminium and platinum.

First Thomson demonstrated that cathode rays were deflected by an electrostatic field in the direction expected for negatively charged particles. Thus he disposed of the second "opinion" on the nature of cathode rays, which Hertz had supported owing in part to his erroneous data to the effect that cathode rays could not be deflected by an electrostatic field. Thomson next sought to determine "What are these particles? Are they atoms, or molecules, or matter in a still finer state of subdivision?" In order to reply to these questions, Thomson used only the magnetic field to measure the charge-to-mass ratio of one of the negatively charged particles that constituted a cathode-ray beam. In this experiment Thomson removed the capacitor from the tube. For a predetermined accelerating voltage, he measured the cathode ray's velocity by stopping the beam in a thermopile that was located between A and B. The N particles of mass m and velocity v expend their kinetic energy as heat W in the thermopile according to the equation

$$\tfrac{1}{2} N m v^2 = W. \tag{A}$$

Thomson then removed the thermopile and deflected the collimated beam with a uniform magnetic field oriented normal to the plane of the paper. The particles were deflected into a circular orbit whose radius ρ could be measured from the position of the phosphorescent spot on the tube's wall where the rays struck. In these runs the particles' orbits were restricted to the magnetic field region, and so the particles struck the top of the tube. From the force acting on a charged particle moving in a magnetic field Thomson obtained

$$m v^2 / \rho = e(v/c) B \tag{B}$$

or

$$m/e = B\rho/vc. \tag{C}$$

Since B, ρ, $Ne = Q$ (i.e., the amount of electricity in the beam) and v were known, then ε/m could be determined from the equation

$$\varepsilon/m = 2Wc/B^2\rho^2 Q. \tag{D}$$

Thomson found that v was of the order of $c/10$, and ε/m was of the order of 10^7. Furthermore, he ascertained that ε/m was independent of the nature of the gas in the tube (thereby eliminating the first "opinion" above, that had been proposed by Crookes), and of the material that constituted the cathode.

Contrary to most modern textbook descriptions of Thomson's measurement of ε/m, he described the application of crossed electric and magnetic fields as a "method of determining the values of m/ε and v [that] is much less laborious and probably more accurate than the former method [i.e., using only the magnetic field]; it cannot, however, be used over so wide a range of pressures." Within its domain of applicability, the crossed-fields method also offered an independent check on the cathode-ray velocities determined from a thermopile. Thomson applied the method of crossed fields as follows. First he determined the position of the undeviated cathode-ray beam from the phosphorescent spot on the end of the vacuum tube. Then he turned on the electrostatic field $\boldsymbol{E} = E\hat{j}$, in which a negatively charged particle's equation of motion was

$$m\,d^2y/dt^2 = -eE \tag{E}$$

or

$$d^2y/dx^2 = -eE/mv^2 \tag{F}$$

where Thomson set $v = v_x = $ constant because his data for these runs were comprised of only small deflections of the cathode-ray beam from the x-direction. From Eq. (F) the electrostatic field deflected a negatively charged particle through an angle θ, where

$$\tan \theta = -eEl/mv^2 \tag{G}$$

and l was the length of the parallel plate capacitor. If only the magnetic field $\boldsymbol{B} = B\hat{k}$ acted in the area between the capacitor plates, then the negatively charged particle's equation of motion was

$$m\,d^2y/dt^2 = e(v/c)B \tag{H}$$

or

$$d^2y/dx^2 = (e/m)(B/vc), \tag{I}$$

and again Thomson considered only small deflections from the x-direction. Then the magnetic field deflected the cathode ray through an angle ϕ where

$$\tan \phi = (e/m)(Bl/vc). \tag{J}$$

Thomson adjusted the electric and magnetic fields so that their deflections canceled each other, i.e., $\tan \theta = -\tan \phi$, and consequently

$$v/c = E/B \tag{K}$$

and

$$\varepsilon/m = (\tan \phi)(Ec/B^2l). \tag{L}$$

From these two methods for deflecting the negatively charged particles that constituted cathode rays, Thomson could conclude only that ε/m was of the order of 10^7.

Thomson (1897) went on to assume that the charge on the particles that constituted cathode rays was constant and was the same as the charge on the positive hydrogen ion from electrolysis. From electrolytic experiments the quantity ε/m_H (where m_H was the hydrogen ion's mass) had been measured to be of the order of 10^4. Therefore, the carrier of electricity in the cathode ray, and so too Lorentz's fundamental ion, was 10^3 times lighter than the hydrogen ion.

Thomson (1899) found that ε/m for the charged particles from the photoelectric effect was also of the order of 10^7. Then, Thomson measured the elementary charge e by using a method that became known as the Thomson-Townsend-Wilson method [see Holton (1978)]. Briefly, this method involved allowing the ions from the photoelectric effect to condense on water droplets. Then, with the aid of Stokes's law, Thomson compared the rate of fall of these droplets under gravity, and then under the influence of an external electrostatic field. As he had assumed in 1897, Thomson found that the elementary charge on these ions was of the order of 10^{-10} esu, which was the value for the charge of cathode rays and for the ionization produced by Röntgen rays. (Röntgen rays were produced when cathode rays struck the walls of the vacuum tube. They had been discovered in 1895 by Wilhelm Röntgen. By 1899 it was being established, principally by Thomson, that Röntgen rays were a form of light.)

In summary, in a series of experiments performed during the period 1897–1899, Thomson had clarified the nature of cathode rays, established the identity of the ions constituting cathode rays with those from the photoelectric effect, and demonstrated the connection between Röntgen rays and cathode rays. In addition, he performed one of the earliest measurements of the elementary charge e. In 1906 Thomson was awarded the Nobel prize "in recognition of the great merits of his theoretical and experimental investigations on the conduction of electricity by gases" (1906). For biographies of Thomson see Rayleigh (1942) and Heilbron (1976). (In 1936 Thomson's son, George P. Thomson, shared the Nobel Prize for demonstrating the wave nature of the electron.)

In 1905, Lenard received the Nobel Prize "for his work on cathode rays" (1905). The award presentation emphasized Lenard's having studied their formation, particularly by the photoelectric effect, and for demonstrating in 1894 that cathode rays could exist outside of a vacuum tube which he accomplished by allowing them to escape from the tube through an aluminum window – workers in the field came to refer to this window as a "Lenard window." Lenard emphasized (1905), that in addition to cathode rays another sort of radiation also escaped through the aluminum window which Röntgen went on to investigate in 1895. In Lenard's opinion Röntgen's discovery was "a good example of a lucky discovery." Lenard was deeply depressed that he had not made the discovery. Furthermore, he was bitter over not having been cited in Röntgen's seminal paper on X-rays because Röntgen may well have made his discovery with the vacuum tube that Lenard had given to him. In 1901 Röntgen was awarded the first Nobel Prize in physics. Nevertheless it was the discovery of X-rays that intensified interest in their source – cathode rays. At first Lenard, like his teacher Hertz, believed that cathode rays were ether vibrations. He became convinced of their corpuscular nature as a consequence of the experiments of Jean Perrin who showed (1895) that cathode rays were negatively charged particles by using a magnetic field to deflect them into a metal cylinder which then developed a negative charge; by Wien (1898) who, as Lenard's successor at Aachen, used a vacuum tube with a Lenard window to verify Perrin's results; and by Thomson (1897). The footnotes in Lenard's publications, e.g., [(1898), (1900)], and in his Nobel Prize address, reveal that he believed himself to have been in close competition with Thomson during the period 1897–1900. For biographical information on Lenard see Hermann (1973).

Kaufmann's contributions to the physics of cathode rays were mainly to confirm that they were constituted of negatively charged particles that emanated from the cathode. He accomplished this (1897a, b) by demonstrating that the deflections of a cathode-ray beam in an electric or magnetic field depended on only the accelerating potential. Despite the similarity of their experimental apparatus' and their data, Kaufmann [(1897a), submitted April 1897] did not consider himself on firm enough ground to discard the hypothesis that cathode rays were ether vibrations, and draw a hypothesis as bold as Thomson's. Kaufmann (1897a) concluded: "Therefore, I believe that I am justified to conclude that the hypothesis accepting that cathode rays are emitted particles is not sufficient for a satisfactory clarification of the relations observed by me." In April 1897 Kaufmann had read Thomson's brief note in the 11 March 1897 issue of *Nature* [55, 453 (1897)] and so he knew that Thomson had obtained similar empirical results; however, in a footnote to his (1897a) Kaufmann took the care to emphasize that Thomson's note in *Nature* had come to his attention only shortly before the conclusion of his own experiment. (In order to complete the sequence of Thomson's 1897 publications on cathode rays I add that on Friday evening of 30 April 1897 Thomson delivered a lecture at the Royal Institution entitled "Cathode Rays" [*The Electrician*, (21 May 1897), 104–109 (1897)]. Neither in the *Nature* note nor in the Friday evening discourse did Thomson describe the method of crossed fields.) Two further experiments on deflecting cathode rays in an electric or a magnetic field, and Thomson's (1897) served by 25 October 1897 to convince Kaufmann of the emission hypothesis.

Kaufmann (1897c) developed a numerical method for taking into account inhomogeneities in the applied magnetic field of the magnetizing coil in whose interior was the vacuum tube. Briefly, from Eq. (I)

$$y = \frac{\varepsilon}{mv} \int_0^{x_0} dx \int_0^x B \, dx, \tag{M}$$

where the quantity x_0 designated the distance along the x-axis traversed by the cathode rays, and x designated the portion of the path along the x-axis in the magnetic field B. For B as a function of x Kaufmann used an empirical formula that agreed with the results of mapping the magnetic field with a probe spool that was linked to a ballistic galvanometer.

Kaufmann's vacuum tube was similar to Thomson's (see Fig.). In the magnetic-deflection experiments, Kaufmann removed the capacitor plates and the magnetic field acted over the tube's entire length, beginning at the metal plug B. The velocity of a particle constituting the cathode rays could be determined from the accelerating potential P as

$$v = \sqrt{2\varepsilon P/m} \tag{N}$$

where P was measured in abvolts, and the maximum P was $12,100 \times 10^8$ abv or 12,100 V. Kaufmann combined Eqs. (M) and (N) and obtained

$$y = \sqrt{\frac{\varepsilon}{2mP}} \int_0^{x_0} dx \int_0^x B \, dx. \tag{O}$$

Kaufmann's data indicated that to 1.5% ε/m was constant and its average value was

$$\varepsilon/m = 1.77 \times 10^7 \tag{P}$$

in agreement, on the average, with Thomson's (1897) measurements.

In 1898 Kaufmann devised a method for improving his (1897c) determination of ε/m. In order to take into account any extraneous potential differences between the anode A and the end of the vacuum tube (see the figure from Thomson's apparatus), Kaufmann placed the vacuum tube from the plug B to the measuring scale inside a metal case that could be charged so as to provide an additional accelerating potential in this region. A photographic plate was cemented to the end of the metal case. In Kaufmann's arrangement the magnetic field extended over the entire metal case. With this apparatus Kaufmann revised his (1897c) measurement of ε/m to

$$\varepsilon/m = 1.86 \times 10^7.$$

For Kaufmann, however, the problem remained of accurately mapping the magnetic field. This was accomplished by Kaufmann's student S. Simon whose (1899) Inaugural Dissertation at Berlin was based on his design of a highly accurate magnetometer. Reverting to the method for magnetic deflections used in 1897 by Thomson and Kaufmann, Simon used Kaufmann's method for folding magnetic variations into deflection measurements in a magnetic field, i.e., Eq. (M), to obtain

$$\varepsilon/m = 1.865 \times 10^7, \tag{Q}$$

in agreement with his mentor's most recent measurement.

But basic problems remained in 1899, for as Thomson wrote (1899), "We have no means yet of knowing whether or not the mass of the negative ion is of electrical origin." Kaufmann (1901a, b) addressed himself to this problem.

By 1901 the name electron, coined in 1894 by G. Johnstone Stoney (1894), had been widely adopted for the negatively charged particle whose charge-to-mass ratio was of the order of 10^7 instead of labels such as cathode rays, ions, quanta and lightions, and was used interchangeably with the name β-ray.

For other discussions concerning topics in the history of the electron see Anderson (1966), Holton (1978), Kaufmann (1901a), Lenard (1905), Thomson (1906) and Whittaker (1973, vol. 1).

39. Assuming that only the electric field is present, we can derive Kaufmann's Eq. (1.113) as follows. We label the point a in Fig. 8 as $x = 0$. Then we have

$$\mu\, d^2y/dt^2 = eE \qquad 0 \leqslant x \leqslant x_1, \tag{A}$$

$$\mu\, d^2y/dt^2 = 0 \qquad x_1 < x \leqslant x_2, \tag{B}$$

$$v_x = \text{constant} \qquad 0 \leqslant x \leqslant x_2. \tag{C}$$

Since for small deflections $v_x = v$, then $d^2y/dt^2 = (d^2y/dx^2)v^2$ then Eq. (A) becomes

$$d^2y/dx^2 = eE/\mu v^2. \tag{D}$$

Integrating Eq. (D), with the proper boundary conditions, we obtain the result that

$$y = (eE/2\mu v^2)\, x(x - x_1). \tag{E}$$

The tangent to the electron's parabolic trajectory at $x = x_1$ is

$$\left(\frac{dy}{dx}\right)_{x = x_1} = \tan\theta = \frac{eE}{\mu v^2}\left(\frac{x_1}{2}\right). \tag{F}$$

The point $x = x_1/2 = h/2$ where h is the length of the capacitor plates (see Fig. 1.8). The electron's trajectory in the region $x_1 < x \leqslant x_2$ is linear, and so it strikes the y-axis at the point y_0, where

$$y_0 = (x_2 - x_1)\tan\theta = (eE/\mu v^2)(x_1/2)(x_2 - x_1). \tag{G}$$

Under experimental conditions the magnetic field is also operating, and Kaufmann emphasized that Eq. (G) had to be modified to take into account that it represents the projection of the electron's orbit on the x-y-plane. Consequently, Kaufmann replaced $x = x_1/2 = h/2$ with s_1 which is the projection of one-half of the electron's trajectory between the capacitor plates on the x-y-plane, and $(x_2 - x_1)$ (or x_2 in Kaufmann's Fig. 1.8) with s_2 which is the projection on the x-y-plane of the electron's trajectory from B to the photographic plate. From straightforward geometrical considerations based on Fig. 1.8, $s_1 = \rho \arcsin(h/2\rho)$ and $s_2 = 2\rho \arcsin(\sqrt{((x_2 - x_1)^2 + z_0^2)}/2\rho)$. It is important to note that when both fields are operating the projection on the x-y-plane of the electron's trajectory between the capacitor plates is not symmetric about the point of maximum curvature.

40. From Fig. 1.8 the result is [Kaufmann (1902a)]

$$\rho = \frac{z_0^2 + x_1^2 + x_1 x_2}{2z_0} + \frac{x_1^2 z_0}{4(z_0^2 + x_2^2 + x_1 x_2)}.$$

Kaufmann (1901c) wrote the second term incorrectly as

$$\frac{-x_1^2 z_0}{4z_0^2 + x_2^2 + x_1 x_2}.$$

41. Furthermore, Abraham pointed out the confusion in Kaufmann's terminology of true and apparent mass because taking the electron's total mass as electromagnetic, rendered the apparent mass "true" and the true mass "untrue."

Abraham had established himself as an expert in electrodynamics while a student of Planck's at Berlin. His Inaugural Dissertation (i.e., Ph. D. thesis) of 1897 focused on electrical oscillations in thin wires, and was subsequently useful for problems in wireless telegraphy. Abraham was Planck's assistant at Berlin until 1900 when he habilitated at Göttingen, i.e., he became a *Privatdozent*. Abraham's 1901 monograph in the *Enzyklopädie der Mathematischen Wissenschaften* was the first systematic exposition of vector analysis in the German scientific literature.

Despite the recognition of Abraham's ability as a physicist, his unusually sharp published criticisms brought down upon him the enmity of many influential members of the physics community. He was a *Privatdozent* at Göttingen during the years 1900–1909; during 1905 he was at Cambridge University and at the University of Illinois. From 1909 to 1915 he was a Professor at the Technische Hochschule in Milan. When asked how he got along with the Italian physicists, he was reported to have replied, "Excellently, since I am not yet fluent in the language" [Born and von Laue (1923)]. Owing to the war Abraham was forced to leave Milan in 1915. He returned to Germany where he participated in research on electromagnetic theory (with applications, for example, to wireless telegraphy) for the Telefunkengesellschaft. In 1919 he could obtain only a position as a substitute at Stuttgart. At last in 1921 he received the call to come to Aachen as a Professor of Theoretical Mechanics. However, in April 1922 in the course of the journey to Aachen he collapsed from what was diagnosed as a brain tumor and, after a painful interval, he died the following September.

Einstein was one of the recipients of Abraham's acid criticisms. However, Einstein was able to rise above the fray: "Physicists have a somewhat negative attitude toward my work on gravitation. It is once again Abraham who has demonstrated the most understanding. Indeed in '*Scientia*' he scolds vigorously all of relativity theory, but with understanding." [Letter of Einstein to M. Besso written towards the end of 1913, in Einstein (1972)]. In 1913 Einstein had strongly recommended Abraham to be his successor at the ETH. In a letter to Heinrich Zangger in Zurich, Einstein wrote that the ETH should proceed in electing his successor without him because "I have taken the side of the fearsome Abraham" (letter undated but probably written early in 1914 from Berlin). On 7 July 1915 Einstein wrote to Zangger that "I am still of the opinion that he [Abraham] would be the right man for Zurich."

Abraham's pungent criticisms were not the only factor that prevented him from obtaining a Professorship in Wilhelmian Germany. Aside from exceptional cases such as Einstein, being Jewish presented an almost insurmountable barrier for academic advancement. The shoddy treatment given Abraham after the war, however, may well have been rooted in these criticisms.

For further biographical information see Born and von Laue (1923) and Chapter 12, footnote 29.

42. So far as I know, Abraham was one of the first to emphasize the usefulness of the units in Eqs. (1.25)–(1.28) – which he referred to as "absolute Gaussian units," and which he wrote for the velocity of light in vacuum c, instead of the more commonly used V or $A = 1/c$. Abraham noted that this system of units showed clearly that the absolute velocity of the electron $v = c$ was a "critical velocity." Having read the papers of Thomson (1889), (1893), Searle (1897) and Wiechert (1900), Abraham was well aware of statements, based on the way factors of γ were occurring in forces and fields, to the effect that c could be the ultimate velocity.

43. Expansion of the retarded potentials for an accelerating electron reveals F^e to be the series

$$F^e = -\alpha_1 a + \alpha_2 \dot{a} + \cdots \text{(terms proportional to higher-order time}$$
$$\text{derivatives of } a) \tag{A}$$

where the term α_1 is dependent on the electron's structure, while α_2 is not. The quantity $\alpha_2 \dot{a}$ is the force exerted on the electron by its self-fields; it is the radiation-reaction force.

From Eq. (1.137),

$$F_{\text{ext}} = \alpha_1 a - \alpha_2 \dot{a}.$$

Depending on whether the transverse or longitudinal force components are taken, the α_1 becomes m_T or m_L. Consequently, in order to maintain an equivalent of Newton's second law in the electromagnetic world-picture, Abraham had to prove that $\alpha_2 \dot{a} \ll \alpha_1 a$. Abraham (1903) emphasized the evaluation of the relative sizes of the terms $\alpha_1 \dot{v}$ and $\alpha_2 \ddot{v}$ for the case of an electron acted on by a force oriented transverse to its trajectory, because, to a good approximation, this was the case in Kaufmann's experiment. For the motion under investigation $\dot{v} = v^2/R$ and $\ddot{v} = v^3/R^2$. The latter expression Abraham deduced in the following way. Since $\dot{v} \cdot v = 0$ then $\dot{v} = v\omega$, where $\omega = v/R$, and since $\ddot{v} \cdot \dot{v} = 0$, then $\ddot{v} = \dot{v}\omega$. Using Kaufmann's 1903 data Abraham estimated that

$\alpha_2\ddot{v}/\alpha_1\dot{v} \sim 10^{-9}$. Therefore, Abraham concluded that quasi-stationary motion was valid "in every practical case." For the sake of exactness $\alpha_2\ddot{v} = \frac{2}{3}(e^2/c^3)\ddot{v}\gamma^4$. Abraham's investigations in his (1903) and (1905) on the radiation-reaction force were quoted often, and have become the standard fare in today's texts on classical electrodynamics.

Circa 1905 the following dimensional analysis was also used for demonstrating that $\alpha_2\dot{a}$ could be neglected [Poincaré (1906)]. Since α_1 is dimensionally a mass, then it can be represented as

$$\alpha_1 = e^2/Rc^2 \tag{B}$$

and we can write α_2 as

$$\alpha_2 = e^2/c^3. \tag{C}$$

Then

$$\alpha_1 a \sim \frac{e^2}{Rc^2}\frac{d}{T^2}, \qquad \alpha_2\dot{a} = \frac{e^2}{c^3}\frac{d}{T^3} \tag{D}$$

where d is the distance over which the electron is accelerated from a velocity v_1 to a velocity v_2 during a time interval T. Thus, for $\alpha_2\dot{a} \ll \alpha_1 a$ then

$$R/cT \ll 1. \tag{E}$$

Hence Abraham's condition for truncating the series in Eq. (A) at the first term was that the time interval during which the electron underwent acceleration must have exceeded the time it took light to have traversed the electron's radius.

Abraham coined the term quasi-stationary motion because the electron behaved during this sort of motion like a quasi-stationary current. A current is quasi-stationary whose variation is much greater than the time it takes for light to traverse the length of the circuit under consideration. In the approximation of a quasi-stationary current the retardation of the fields is neglected and, for example, the magnetic energy of a wire is calculated as though its current were steady. Consequently the self-induction of a current-carrying wire is easily determined from the wire's magnetic field energy. Abraham (1905) emphasized that just as the self-induction was connected with the magnetic-field energy, the electromagnetic mass was linked with the momentum and energy of the electron's own electromagnetic fields. The approximation of quasi-stationary motion permitted Abraham to calculate the electromagnetic mass by assuming that the electron's self-fields were those for a charge moving with the uniform velocity v that happens to be the electron's instantaneous velocity; so the electron's momentum was the expression in Eq. (1.135).

Lorentz (1892a) had deduced the radiation-reaction force for the case of $v \ll c$. Von Laue (1909) deduced the radiation-reaction force by applying the relativistic transformations for force and acceleration to the radiation-reaction force exerted on an electron in its instantaneous rest system. In a footnote to the 1914 edition of his (1905), Abraham protested that he had precedence over von Laue because in the 1908 edition he had used the "theorem of relativity" to deduce the radiation-reaction force. In fact, the two derivations were identical and in a note "added in proof" von Laue (1909) mentioned Abraham's derivation. Most likely Abraham's derivation was not usually cited for the reason that he prefaced it with the statement that although the "theorem of relativity" was incorrect for use in problems concerning the electron's shape, it facilitated the mathematics when dealing with a point electron.

44. Abraham (1903) continued by demonstrating that rotational motion about the direction of motion of a spherical electron was also a distinctive motion, but that contributions to the electron's mass from its rotation were vanishingly small. Further detailed theoretical investigations of Schwarzschild (1903) and Herglotz (1903) supported Abraham's results on rotating electrons.

45. In his (1906) Kaufmann performed this calculation correctly. From Eq. (1.111) the total energy E^e of Abraham's electron is

$$E^e = \frac{e^2}{2R} \left\{ \frac{1}{\beta} \ln \left(\frac{1+\beta}{1-\beta} \right) - 1 \right\}. \tag{A}$$

To order β^4 Eq. (A) becomes

$$E^e = E_0^e \{1 + \tfrac{2}{3}\beta^2 + \tfrac{2}{5}\beta^4\}, \tag{B}$$

where the electron's "electrostatic energy" [Kaufmann (1906)] E_0^e is

$$E_0^e = e^2/2R. \tag{C}$$

Kaufmann employed Eq. (C) to rewrite Eq. (A) as

$$E^e - E_0^e = \tfrac{2}{3} E_0^e \beta^2 [1 + \tfrac{3}{5}\beta^2] \tag{D}$$

or

$$E^e - E_0^e = \mu_0^e (v^2/2) [1 + \tfrac{3}{5}\beta^2] \tag{E}$$

where $\mu_0^e = e^2/2Rc^2$.

For β small, the right-hand side of Eq. (E) approaches the kinetic energy from classical mechanics. Consequently, Kaufmann replaced $E^e - E_0^e$ with the kinetic energy that an electron obtains from an accelerating potential P (in absolute units), i.e., εP. Then Eq. (E) becomes

$$\varepsilon P = \mu_0^e (v^2/2) [1 + \tfrac{3}{5}\beta^2]. \tag{F}$$

In the β^2 limit an electron moving initially in the x-direction is deflected by a magnetic field in the z-direction to a position z [see Eq. (7.14)]

$$z = \frac{\varepsilon}{\mu_0^e v} \mathbb{B}(1 - \tfrac{2}{5}\beta^2) \tag{G}$$

(since to order β^2, $m_T = \mu_0^e [1 + \tfrac{2}{5}\beta^2]$). Kaufmann eliminated v between Eqs. (G) and (F) to obtain

$$\frac{\varepsilon}{\mu_0^e} = \frac{2z^2 P}{\mathbb{B}^2} [1 + \tfrac{1}{5}\beta^2]. \tag{H}$$

From Eq. (F), accurate to order β^2, we have

$$\varepsilon P = (\mu_0^e/2)\beta^2 c^2 \tag{I}$$

or

$$\beta^2 = \frac{2\varepsilon P}{\mu_0^e c^2} = 2\alpha \frac{P}{c^2} \tag{J}$$

where Kaufmann defined the quantity α to be

$$\alpha = 2Pz^2/\mathbb{B}. \tag{K}$$

Then Eq. (H) becomes

$$\frac{\varepsilon}{\mu_0^e} = \alpha\left[1 + \frac{2}{5}\alpha\frac{P}{c^2}\right]. \tag{L}$$

The quantity α was used by Simon to determine ε/μ_0 for cathode rays, i.e., $\alpha = 1.865 \times 10^7$. Kaufmann took Eq. (L), with its "correction term" $\frac{2}{5}\alpha P/c^2$, to be the appropriate extrapolation of Simon's results to slow β-rays. Substituting Simon's empirically determined α, and the corresponding accelerating potential $P = 8300\mathrm{V}$ Kaufmann obtained

$$\varepsilon/\mu_0^e = 1.878 \times 10^7 \tag{M}$$

for the appropriate value to compare with the ε/μ_0^e from Abraham's theory that had been obtained from empirically determined velocities.

Since on the left-hand-side of Eq. (L) the quantity ε/μ_0^e is in the absolute electromagnetic units, then abvolts must be used for the potential difference P (1 volt $= 10^8$ abvolt). Experimentalists, such as Kaufmann, often expressed potential differences in volts and/or abvolts. For calculations in the Gaussian (cgs) system, however, the volt must be converted to the statvolt (1 volt $= 1/300$ statvolt).

46. The transformations are:

$$x_r = \frac{\varepsilon}{\gamma}x'',$$

$$y_r = \varepsilon y'',$$

$$z_r = \varepsilon z'',$$

$$t_L = \gamma\varepsilon t'',$$

where Lorentz defined t_L as the "modified local time," and ε is an undetermined scale factor that Lorentz inserted so as to allow for the hypothesis of contraction. Lorentz deduced the two expressions for the velocity dependence of the bound ion's mass by comparing the forces and accelerations in S_r and the new fictitious double-primed reference system.

47. Lorentz (1904a) also criticized Planck's recent correct mathematical formulation of Stokes' ether theory. Lorentz considered it overburdened with hypotheses especially concerning the nature of the ether and consequently that it impeded progress toward a unification of electricity and gravitation.

He concluded by saying that Planck concurred. Earlier Planck had communicated his theory to Lorentz who had discussed it in (1899b). Lorentz (1899b) further emphasized that the local time was not a physical time. He showed that neglecting terms of order $(v/c)^2$, Hertz's equations for an ether in motion could be reduced to those "which would hold for an aether without motion." Lorentz's proof involved using a variant of the local time

$$t_L = t + \phi/c^2$$

where ϕ is the velocity potential of the ether in a Planck-Stokes ether theory. Thus, it is possible to formulate in the Planck-Stokes theory, which utilizes Hertz's equations, a theorem of corresponding states.

He concluded, "It is curious that in the two rival theories somewhat the same mathematical artifices may be used." In (1904b) Lorentz made similar assertions.

48. Brace, however, claimed that the absence of double refraction meant also the absence of any strain on the isotropic substances under investigation, and therefore his data were evidence against Lorentz's contraction. Larmor (1904) demonstrated that Brace's results could be brought into agreement with the theorem of corresponding states provided that the new coordinate transformations were employed which he had proposed in his (1897) – see footnote 50 below. Brace went on to use the apparatus from his (1904) experiment to perform several other high-precision ether-drift experiments – see footnote 23.

49. The entire footnote is: *Lorentz*, Amsterdam Zittingsverslag Akad. v. Wet. *12* (1904) (Amsterdam Proceedings, 1903–1904).

Holton (1973a) has pointed out an error in Whittaker's vol. 2 which is revealing of Whittaker's attitude towards relativity theory; namely, that Whittaker dated Lorentz's (1904c) as published in 1903. Volume 12 of the Amsterdam Proceedings contained the papers from 1903 and 1904. The reverse alphabetical order in the chapter on relativity theory in Whittaker's vol. 2 – "The Relativity Theory of Poincaré and Lorentz" – is probably indicative of Whittaker's regard for Poincaré. Other ahistoricisms in Whittaker's exposition of relativity are in my (1973).

For other discussions of Lorentz's classic (1904c) see Goldberg (1969), Holton (1973), Miller (1973, 1974) and Schaffner (1969).

50. Larmor (1897), (1900) extended Lorentz's proof from the *Versuch* of the covariance of the Maxwell-Lorentz equations to second order in v/c.

I do not discuss Larmor's transformations, which are equivalent mathematically to Lorentz's transformations of 1904, because in my opinion Larmor's work had an indirect effect, if any, on Lorentz's thinking toward the electron theory of 1904. For discussions of Larmor's work, see Schaffner (1972) and Kittel (1974).

Voigt (1887) had found that the wave equation in free space [e.g., Eq. (1.7)] was unchanged by the transformation

$$x' = x - vt, \tag{A}$$

$$y' = \frac{1}{\gamma} y, \tag{B}$$

$$z' = \frac{1}{\gamma} z, \tag{C}$$

$$t' = t - \frac{v}{c^2} x, \tag{D}$$

where the primed coordinates refer to a reference system in inertial motion relative to S. Voigt considered the primed reference system as merely a convenient one in which to solve problems concerning any sort of wave propagation in the free ether. The Eqs. (A)–(D) can be obtained from Lorentz's Eqs. (1.162)–(1.165) by setting $l = 1/\gamma$.

In *The Theory of Electrons* (1909) Lorentz wrote that not until recently had he become aware of Voigt's equations, which he correctly qualified were "equivalent" to the Lorentz transformations for considering radiation in the "*free* ether" (italics in original). Lorentz made a similar statement in a footnote to his (1904c) in the Teubner editions of *The Principle of Relativity*, which does not appear in the Dover reprint volume. Voigt's transformations could only have confused Lorentz in 1892, because of their inability to provide the plausibility argument required by Lorentz's contraction hypothesis.

51. For example, first transforming Eqs. (1.27) and (1.26) from S to S_r yields

$$\frac{\partial E_x}{\partial x_r} + \frac{\partial E_y}{\partial y_r} + \frac{\partial E_z}{\partial z_r} = 4\pi\rho, \tag{A}$$

$$\frac{\partial B_z}{\partial y_r} - \frac{\partial B_y}{\partial z_r} = \frac{1}{c}\left(\frac{\partial}{\partial t_r} - v\frac{\partial}{\partial x_r}\right)E_x + \frac{4\pi}{c}\rho(v + u_{x_r}). \tag{B}$$

Then from Eqs. (1.162)–(1.165) we have

$$\frac{\partial}{\partial x_r} = \gamma l\frac{\partial}{\partial x'} - \gamma l\frac{v}{c^2}\frac{\partial}{\partial t'}, \qquad \frac{\partial}{\partial y_r} = l\frac{\partial}{\partial y'}, \qquad \frac{\partial}{\partial z_r} = l\frac{\partial}{\partial z'}, \qquad \frac{\partial}{\partial t_r} = \frac{l}{\gamma}\frac{\partial}{\partial t'}.$$

Substituting this change of variables into Eqs. (A) and (B) gives

$$\gamma l\frac{\partial E_x}{\partial x'} + l\frac{\partial E_y}{\partial y'} + l\frac{\partial E_z}{\partial z'} = 4\pi\rho + \gamma l\frac{v}{c^2}\frac{\partial E_x}{\partial t'}, \tag{C}$$

$$l\frac{\partial B_z}{\partial y'} - l\frac{\partial B_y}{\partial z'} = \frac{l}{c\gamma}\frac{\partial E_x}{\partial t'} - \frac{v}{c}\gamma l\frac{\partial E_x}{\partial x'} + \frac{v^2}{c^4}\gamma l\frac{\partial E_x}{\partial t'} + \frac{4\pi}{c}\rho(v + u_{x_r}). \tag{D}$$

We obtain by elimination of $\partial E_x/\partial x'$ in (D) by means of (C),

$$l\frac{\partial}{\partial y'}\left(B_z - \frac{v}{c}E_y\right) - l\frac{\partial}{\partial z'}\left(B_y + \frac{v}{c}E_z\right) = \frac{l}{c\gamma}\frac{\partial E_x}{\partial t'} + \frac{4\pi}{c}\rho u_{x_r}. \tag{E}$$

Now (E) will be the same as its counterpart in S if

$$B'_{z'} = \frac{\gamma}{l^2}\left(B_z - \frac{v}{c}E_y\right), \qquad B'_{y'} = \frac{\gamma}{l^2}\left(B_y + \frac{v}{c}E_z\right), \qquad E'_{x'} = \frac{E_x}{l^2}.$$

Then

$$\frac{\partial B'_{x'}}{\partial y'} - \frac{\partial B'_{y'}}{\partial z'} = \frac{1}{c}\frac{\partial E'_{x'}}{\partial t'} + \frac{4\pi}{c}\rho'u'_{x'}$$

where

$$\rho'u'_{x'} = (\gamma/l^3)\rho u_{x_r}. \tag{F}$$

Since charge is conserved,

$$\int\rho\,dx_r\,dy_r\,dz_r = \int\rho'\,dx'\,dy'\,dz',$$

$$\int \rho J \, dx' \, dy' \, dz' = \int \rho' \, dx' \, dy' \, dz', \tag{G}$$

where Lorentz assumed that $\rho = \rho_r$, and J was the Jacobian for the transformation from S_r to S'. From (G) we have

$$\rho' = \rho J \tag{H}$$

and from Eqs. (1.162)–(1.165)

$$J = \frac{\partial(x_r, y_r, z_r)}{\partial(x', y', z')} = (\gamma l^3)^{-1}. \tag{I}$$

Therefore $\rho' = \rho/\gamma l^3$. Substituting (H) and (I) into (F) yields an equation for $u'_{x'}$:

$$u'_{x'} = \gamma^2 u_{x_r}.$$

52. This statement can be proved as follows. From equations (1.162) and (1.165)

$$dx' = \gamma l \, dx_r,$$

$$dt' = (l/\gamma) \, dt \left(1 - \gamma^2 \frac{v}{c^2} u_{x_r} \right) \tag{A}$$

where

$$u_{x_r} = dx_r/dt. \tag{B}$$

Then

$$u'_{x'} = dx'/dt' = \gamma^2 u_{x_r} \bigg/ \left(1 - \gamma^2 \frac{v}{c^2} u_{x_r} \right). \tag{C}$$

Consequently, if $u_{x_r} \ll 1$, (C) becomes

$$u'_{x'} = \gamma^2 u_{x_r}.$$

The equations for $u'_{y'}$ and $u'_{z'}$ follow in a similar manner. Lorentz then used Eqs. (1.162)–(1.165) to derive the transformation equations for the acceleration:

$$\frac{d^2 x'}{dt'^2} = \gamma^2 \frac{d^2 x_r}{dt^2} \frac{1}{(dt'/dt)} = \frac{\gamma^3}{l} \frac{d^2 x_r}{dt^2} \quad \text{(if } u_{x_r} \ll 1\text{)}.$$

Then by *cross multiplication*

$$a_{x_r} = \frac{l}{\gamma^3} a'_{x'}. \tag{D}$$

By a similar method the x and y components are found to be

$$a_{y_r} = \frac{l}{\gamma^2} a'_{y'}, \tag{E}$$

$$a_{z_r} = \frac{l}{\gamma^2} a'_{z'}. \tag{F}$$

It will be shown that these transformations are valid in S', where u_{x_r} is identically zero. It is important to note that the accelerations and forces are the same in S and S_r because these systems are related by a Galilean transformation.

53. Lorentz's result for the force per unit charge in Σ' was:

$$F_x = l^2\left[E'_{x'} + \left(\frac{u'}{c} \times B'\right)'_{x'} + \frac{v}{c^2}(u'_{y'} E'_{y'} + u'_{x'} E'_{x'})\right],$$

$$F_y = \frac{l^2}{\gamma}\left[E'_{y'} + \left(\frac{u'}{c} \times B'\right)_{y'} - \frac{vu'_x}{c^2}E'_{y'}\right],$$

$$F_z = \frac{l^2}{\gamma}\left[E'_{x'} + \left(\frac{u'}{c} \times B'\right)_{x'} - \frac{vu'_x}{c^2}E'_{z'}\right].$$

54. Equivalent states exist only if the forces in S and S' are related by

$$F(S) = (l^2, l^2/\gamma, l^2/\gamma)\,F(S'), \tag{A}$$

see the equations in footnote 53 written in S' where $u' = 0$; then

$$ma(S) = (l^2, l^2/\gamma, l^2/\gamma)\,ma(S'). \tag{B}$$

Write the transformation equations for the acceleration between S and S' as

$$a(S) = (l/\gamma^3, l/\gamma^2, l/\gamma^2)\,a(S'). \tag{C}$$

[See Eqs. (D)–(F) in footnote 52.] Comparing Eqs. (B) and (C) gives

$$m(S) = (\gamma^3 l, \gamma l, \gamma l)\,m(S'). \tag{D}$$

The result of comparing Eq. (D) with (1.185) is

$$d(\gamma lv)/dv = \gamma^3 l$$

which requires that l be a constant. But this constant must be unity in order that for $v = 0$, $l = 1$ and then the Eqs. (1.162)–(1.165) revert to the Galilean transformations.

55. Letter on deposit at the Algemeen Rijksarchief, The Hague, The Netherlands.

56. Abraham wrote that common to both theories were the hypotheses:

A. The "Maxwell-Hertz equations" were valid for charge-free space, and there was a reference system relative to which the velocity of light was $c = 3 \times 10^{10}$ cm/sec in every direction.

B. Electricity consisted of discrete positive and negative particles called electrons which moved about in an all-pervasive resting ether.

C. A charge in motion constituted a convection current.

D. Forces between charges were given by the Lorentz force.

In other words, both Abraham and Lorentz had based their theories of the electron on Lorentz's electromagnetic theory. Indigenous to Abraham's theory were the three hypotheses:

E. The electron was in equilibrium owing to the equality of its self-force and the external force "in the sense of the mechanics of rigid bodies."

F. The electron always maintained its shape. (This hypothesis, Abraham emphasized, was to be interpreted "as the equation of connection in the sense of Hertz's mechanics.")

G. The electron was a sphere of uniform surface or volume charge distribution.

Abraham listed Lorentz's additional special hypotheses as:

H. Bodies contract in their direction of motion.

In order to render hypothesis H "plausible," wrote Abraham, Lorentz proposed hypothesis I –

I. "Quasi-elastic forces" (i.e., molecular forces) transformed like electromagnetic forces.

K. Electrons contracted in their direction of motion, leading to Lorentz's predicting the electron's longitudinal and transverse masses.

In order to ensure the absence of double refraction in moving isotropic bodies, Lorentz's hypothesis K "is supplemented" with the hypothesis:

L. The mass of a molecule (containing many electrons) behaves like the electron's mass, and is therefore purely electromagnetic. We shall see that Abraham was not yet prepared to boldly extend results from the theory of electrons to molecules and ponderable matter.

In summary, Abraham's theory was composed of hypotheses A–G, and Lorentz's A–D, H–L. Therefore Lorentz could explain all first-order experiments, as well as Kaufmann's data and the second-order experiments of Michelson and Morley, Rayleigh and Brace, and Trouton and Noble. Abraham acknowledged that Lorentz's theory was broader but fatally flawed.

57. Wien (1904a) concluded from studying the Poynting vector for the radiation from an oscillating Heaviside ellipsoid that the case of $v = c$ had to be excluded as unphysical; furthermore, that no conclusions could be drawn for this case concerning the rigid electron. Abraham (1904b) responded in devastating detail, showing that the value of $v = c$ could be achieved for certain limiting cases concerning free rigid spherical electrons – for example, when their distance from the observer was much greater than their radius so that they could be considered as point particles. Wien's reply (1904b) revealed that he was rather stunned by the tenor of Abraham's criticisms. In summary Wien wrote: (1) Searle (1897) had noted that a point charge could not be regarded as the limiting case of a moving sphere; rather the electrostatic image of the moving sphere was a Heaviside ellipsoid. (2) After relating the great explanatory powers of Lorentz's (1904c) new theory of the deformable electron (leaving unsaid the narrowness of Abraham's theory), Wien emphasized that for $v > c$ "the quantity $\sqrt{1 - v^2/c^2}$ becomes imaginary," and this leads to unphysical results for the Heaviside ellipsoid. Wien concluded by writing that although a rigid electron could exceed the velocity of light, it was another matter whether this case had any physical meaning.

It is useful to survey research on hyperlight velocities because this topic was much pursued in German physics journals. Not unexpectantly, Heaviside [(1889), (1925, vols. I and III)] was the first to treat seriously the motion of a free electron for $v = c$. He asserted (1889) that the case of $v = c$ contained an unphysical singularity arising from treating the electron as having no structure. Furthermore, wrote Heaviside (1889) the electric field of a uniformly moving electron with $v > c$ remains mathematically real because the particle's disturbance is restricted to a cone of half-angle $\theta = \sin^{-1}(c/v)$ [see footnote 26, Eq. (D)] with the electron at the cone's apex. The similarity of Heaviside's restriction to the one in Cherenkov radiation, discovered in 1937, occurs owing to Heaviside's tacit assumption that somehow the ether could suppress disturbances exterior to the cone, thereby behaving like a medium with refractive index $N \neq 1$ for matter with velocity $v > c$. (Cherenkov radiation does not violate special relativity because it results from matter moving through a medium in which the velocity of light is c/N.)

As far as I know, Heaviside was also the only physicist, until relatively recent times, to emphasize the possibility of bringing a particle into the region $v < c$ from $v > c$.

Des Coudres (1900) elaborated upon Heaviside's (1889), and in the period 1904–1905 Sommerfeld wrote several lengthy papers on *Überlichtgeschwindigkeit* as it concerned Abraham's rigid electron. This sort of motion was treated mathematically by replacing $(v - c)$ with $(c - v)$ and then executing all necessary integrals, taking the real parts of field quantities wherever necessary. Sommerfeld found that if the electron had a volume distribution of charge then hyperlight velocity was physically possible, but it was not a force-free motion; rather an external force was necessary to balance the opposing forces from the electron's self-electromagnetic fields. On the other hand, hyperlight velocity was excluded for a surface-charged electron. It turned out that in the quasi-stationary approximation, the volume-charged electron moving at a hyperlight velocity developed a negative mass. Sommerfeld (1904c) stressed that Lorentz's new theory of the electron could not be fitted into his speculations: "As for velocity exceeding that of light [Lorentz's] hypothesis fails because in this case you can hardly speak of a 'Heaviside-hyperboloid'." By 1910 after having studied relativity theory Sommerfeld concluded that his earlier investigations on hyperlight velocities had been unfruitful [Sommerfeld, "*Autobiographical Skizze*" (c. 1950)]. For biographical information on Sommerfeld see Forman (1975).

In conclusion, an impetus to such investigations as Sommerfeld's may well have been Heaviside's 1897 retort to Thomson (1889) and Searle (1897) who doubted hyperlight velocities: "Don't be afraid of infinity" Heaviside (1925, vol. II).

58. Abraham (1904a) had nothing but praise for the "clear and unambiguous" manner in which Lorentz had proposed his hypotheses; he expressed the hope that other researchers in the domain of electron theory would follow Lorentz's example, instead of claiming that their investigations were "free of hypotheses." These sharp comments were undoubtedly aimed at Wien, with whom Abraham was in the midst of a bitter exchange that did not enhance his reputation in the German physics community.

59. Brace (1905a) called for further high-precision second- and third-order ether-drift experiments — for example, using the Kerr and Faraday effects to measure the velocity of light.

60. Letter on deposit at the Algemeen Rijksarchief, The Hague, The Netherlands.

61. The time rate of change of the electron's self-electromagnetic field energy can be written as

$$\frac{dE^e}{dt} = \frac{dE^e}{dv}\frac{dv}{dt}.$$

This represents the work done per unit time by the self-electromagnetic force. Therefore, $(1/v)\,dE^e/dt$ is the component of the self-electromagnetic force along the electron's direction of motion; consequently, $(1/v)\,dE^e/dv = m_L$ (with the assumption of quasi-stationary motion).

62. See Miller (1973) for an in-depth analysis of Poincaré's (1905a), (1906).

63. In the first letter Poincaré developed how he had corrected certain equations from Lorentz's (1904c), and he mentioned the problem with m_L. In the third letter Poincaré stated that he had resolved the problem with m_L and was in "perfect agreement" with Lorentz's "beautiful work" [see Miller (1979a) for an analysis of the three letters].

Instead of Lorentz's two-step procedure for transforming the derivatives of the space and time coordinates from S to Σ', Poincaré (1905a, 1906) proceeded thus,

$$\frac{\partial}{\partial x} = \gamma l \left(\frac{\partial}{\partial x'} - \frac{v}{c^2} \frac{\partial}{\partial t'} \right)$$

$$\frac{\partial}{\partial y} = l \frac{\partial}{\partial y'}$$

$$\frac{\partial}{\partial z} = l \frac{\partial}{\partial z'}$$

$$\frac{\partial}{\partial t} = \gamma l \left(\frac{\partial}{\partial t'} - \frac{v}{c^2} \frac{\partial}{\partial x'} \right).$$

In his (1909) book Lorentz used Poincaré's version of the Lorentz transformations to discuss the Maxwell-Lorentz equations accurate to all orders in v/c.

64. In a mathematical analysis of the Lorentz transformations that preceded his investigations of theories of the electron, Poincaré (1905a), (1906) proved that only if $l = 1$ could the subgroup of spatial rotations be connected with the Lorentz group. In other words, only if $l = 1$ could the Maxwell-Lorentz theory be covariant under Lorentz transformations whose velocity v was in an arbitrary direction.

65. Poincaré used the principle of relativity as a guide toward a theory of gravity whose force equation retained its form under a Lorentz transformation, and consequently according to which gravity propagated with the velocity of light.

66. According to Eq. (1.184), for $\mathbf{v} = v\hat{i}$

$$G^e = \frac{4}{3} E_0^e \gamma l \frac{v}{c^2}.$$

From Eq. (1.202) in S'

$$L^e = -l \frac{E_0^e}{\gamma}$$

A basic relation of Lagrangian mechanics Eq. (1.131) leads to

$$\frac{d}{dv} \left(-\frac{l}{\gamma} \right) = \frac{4}{3} \gamma l \frac{v}{c^2},$$

which is satisfied only by $l = \gamma^{-1/3}$.

67. Since S' is the spherical hollow deformable electron's rest system, then in S' the Maxwell stress tensor due to the electron's own coulomb field is

$$\ddot{\mathbf{T}}^{e'} = \frac{1}{4\pi}\left(\mathbf{E'E'} - \frac{1}{2}E'^2\,\ddot{\mathbf{I}}\right).$$

Since the electron's coulomb field $\mathbf{E}$ is everywhere normal to its surface

$$\ddot{\mathbf{T}}^{e'} = T^{e'}\,\hat{n}'\,\hat{n}'$$

where $T^{e'} = E^{e'2}/8\pi = e^2/8\pi R^4$, and so $T^{e'} = -p_0^c$ — that is, the total internal stress on the electron in S', due to its self-coulomb field and from the "forces due to the supplementary" forces, vanishes, thereby rendering the electron stable in its rest system.

68. The additional Lagrangian L^c contributes to the Lorentz-invariant action the term

$$J^c = \int p_0^c\,dV\,dt.$$

For $l = 1$, $dV\,dt$ is a Lorentz invariant, then so is p_0^c.

69. Klein and Needell (1977) have discovered that since 1905 Einstein had been a regular reviewer for the *Beiblätter zu den Annalen der Physik*, which was an important abstracting journal.

CHAPTER 2

EINSTEIN'S PHILOSOPHIC VIEWPOINT IN 1905

> The reciprocal relationship of epistemology and science is of noteworthy kind. They are dependent upon each other. Epistemology without contact with science becomes an empty scheme. Science without epistemology is — insofar as it is thinkable at all — primitive and muddled.
>
> A. Einstein (1946)

This reciprocal relationship was the leitmotif for Einstein's researches. Although the complex epistemology associated with his thinking did not emerge until after he formulated the general relativity theory in 1915,[1] the importance he ascribed to philosophic analysis of foundations is evident from the first part of the special relativity paper. This initial section is, in fact, nothing less than an epistemological analysis of the nature of space and time. It is essential therefore, to reconstruct Einstein's viewpoint in 1905, and examine how its interplay with his scientific research enabled him to break courageously with the prevailing currents of science and epistemology. By 1905 most physicists possessed all the relevant empirical data, and we can conjecture that Einstein was aware of many of them. Yet Einstein glimpsed these data and their relation to trends in epistemology in a new and deeper manner. This bold break permitted him to turn the contemporaneous programmatic intents of science sideways, in order to treat mechanics and electromagnetism on the same footing, instead of attempting to reduce one to the other.

In 1905 Einstein's epistemology emerges as concerned principally with determining what sorts of theories respond best to the pressing problems of contemporaneous physical theory. It was eclectic: it was, in part, Einstein's own version of Kant as viewed through the pages of, in particular, Hertz and Poincaré. Einstein's viewpoint then was also tempered by the works of Mach and Poincaré, from whom he learned both the value and pitfalls of reliance upon sense perceptions. Boltzmann and von Helmholtz encouraged Einstein to include in this epistemology his predilection for visual thinking.[2]

2.1. ASSESSING A SCIENTIFIC THEORY

A central problem for scientists has always been whether the empirical testability of a theory should be the sole criterion for its validity. Einstein (1946) considered testability as only one of two criteria by which we can

"criticize physical theories at all"; empirical testability he referred to as "external confirmation," and the other criterion as "inner perfection."
Einstein wrote (1946):

> The first point is obvious: the theory must not contradict empirical facts. However evident this demand may in the first place appear, its application turns out to be quite delicate.

Already in the 1905 relativity paper Einstein demonstrated just how delicate it was to apply this criterion because, for example, he avoided comparison of his prediction for the electron's transverse mass m_T with Kaufmann's earlier data from the period 1901–1903. It is reasonable to conjecture that at that time Einstein knew well that his prediction for m_T disagreed with those data. Although Kaufmann's data of late 1905 (which confirmed the theories of Abraham and Bucherer but not the Lorentz-Einstein theory, for by this time the predictions of Lorentz and Einstein for m_T were mathematically identical), was much discussed in the *Annalen*, and elsewhere, Einstein avoided responding to it until 1907 when he boldly wrote:

> In my opinion both theories [those of Abraham and Bucherer] have a rather small probability, because their fundamental assumptions concerning the mass of moving electrons are not explainable in terms of theoretical systems which embrace a greater complex of phenomena.

From this example — which has been explored in some depth by Gerald Holton (1973d) — we can see that Einstein was hesitant to permit experimental facts to decide the matter. He considered the theories of Abraham and Bucherer to be limited in scope, focusing principally on the electron, rather than second-order optical data as well. In fact, Kaufmann's data turned out to be invalid. The Kaufmann episode is a fascinating one in the history of science, and in Chapters 7 and 12 we shall analyze it further. Then we shall see just how "delicate" is the comparison between theory and experiment.

Einstein's second criterion — "inner perfection" — overlaps somewhat with the first one (1946):

> The second point of view is not concerned with the relation to the material of observation but with the premises of the theory itself, with what may briefly but vaguely be characterized as the "naturalness" or "logical simplicity" of the premises (of the basic concepts and of the relations between these which are taken as a basis). ... of the 'realm' of theories I need not speak here, inasmuch as we are confining ourselves to such theories whose object is the totality of all physical appearances [*die Gesamtheit der physikalischen Erscheinungen*].

Because of their restricted scope, Einstein did not consider the theories of Abraham and Bucherer to be in this class.

These two criteria can hardly be considered as a double-edged sword, and the vague manner in which he discusses them underscores this point (1946):

> The meager precision of the assertions contained in the last two paragraphs I shall not attempt to excuse by lack of sufficient printing space at my disposal, but confess herewith that I am not, without more ado (immediately), and perhaps not at all, capable to replace these hints by more precise definitions.

Thus we should perhaps say that the although essence of Einstein's scientific method was basically inarticulable, this method has nevertheless continued to prove its worth by its operative applicability in the hand of one who had a "sympathetic understanding of experience" [Einstein (1918b)].

2.2. ON THE MECHANICAL WORLD-PICTURE

> Enough of this. Newton forgive me; you found the only way which, in your age, was just about possible for a man of highest thought – and creative power.
>
> A. Einstein (1946)

Einstein's overall view of 19th-century attempts to base all of physical theory on mechanics, i.e., the mechanical world-picture, was that (1946):

> ... dogmatic rigidity prevailed in matters of principles: In the beginning (if there was such a thing) God created Newton's laws of motion together with the necessary masses and forces. That is all; everything beyond this follows from the development of appropriate mathematical methods of deduction.

His criticisms of the mechanical world-picture focused on its attempt to base "Maxwell's theory ... upon mechanics." He recalled (1946) that "It was Ernst Mach who, in his History of Mechanics, shook this dogmatic faith; this book exercised a profound influence upon me in this regard while I was a student."[3] The core of Mach's hard-hitting criticisms against the possibility of a mechanical world-picture was that: "*the simplest mechanical principles are of a very complicated character ... they can by no means be regarded as mathematically established truths but only as principles that not only admit of constant control by experience but actually require it*" (italics in original).[4] Consequently, the laws of mechanics could not be regarded as axioms because from time to time they may require revision – that is, one had to bear in mind that they were obtained from sense data, to which they must be ultimately reducible. Furthermore, Mach continued, "a real economy of thought cannot be attained by mechanical hypothesis." Although Einstein subsequently became disenchanted with Mach's epistemological viewpoint which emphasized empirical data, nevertheless he remained impressed with certain of Mach's personal characteristics (many of which were similar to his own): "I see Mach's greatness in his incorruptible skepticism and independence."

The narrowness of the curriculum at the ETH whetted Einstein's appetite for reading the classics of current physics. Philipp Frank, who was Einstein's successor at Prague and one of his biographers, has written [(1947), (1953)] that at home in the evenings Einstein "devoured the works of Helmholtz, Kirchhoff, Boltzmann, Maxwell and Hertz."

We may reasonably assume that at the ETH Einstein read Gustav Kirchhoff's influential *Lectures on Mechanics* (1876). Kirchhoff restricted the goal of mechanics to describing, in the simplest and most complete manner, motions taking place in nature;[5] Einstein could not have failed to notice that famous message. Einstein often referred to Kirchhoff's book in his scientific papers [e.g., Einstein (1905c)], and came to agree with the result of his foundational analysis that mechanics could not assume the role of the basic science.

Hertz too acknowledged his intellectual debt to Mach in the "Author's Preface" to his (1894), although he described his own goal as "tracing the phenomena of nature back to the principal laws of mechanics."[6] In fact, Hertz's long-term aim was nothing less than "to base the equations of motion of the ether upon the laws of mechanics." Joseph Sauter (1965), a colleague of Einstein's at the Patent Office in Bern, recalled that, sometime shortly after Einstein began working there, they discussed the possibility of formulating mechanical models for Maxwell's theory. Since he had recently published a paper on that topic, Sauter considered himself something of an expert. Sauter later recalled: "Immediately he said to me that models of Maxwell's theory did not interest him. He finished by saying, 'I am a heretic.'" Einstein told Sauter that while he was a student at the ETH he had stopped reading Hertz's book on mechanics "at the place where Hertz replaced each force with a connection." We therefore know that Einstein read at least Hertz's "Introduction," in which, as Robert S. Cohen has observed, "Hertz breaks cleanly with Mach" [Hertz (1894)]. Hertz attempted a reformulation of mechanics in which the action-at-a-distance force was replaced by a network of concealed masses connected to the actual physical system in such a way as to simulate the contiguous actions of an electromagnetic theory. This system of mechanics was based on axioms drawn from experience by "laws of the internal intuition" of the individual scientist, and is capable of yielding empirical predictions; these laws function, Hertz wrote, as "*a priori* judgments in Kant's sense." Hertz presented his reformulation of mechanics as a deductive system, in much the same style as Newton's *Principia*. Although Einstein disagreed with both Hertz's use of concealed masses and his long-term goal, nevertheless he could only have been impressed with Hertz's emphasis on the power of axiomatic presentation which predicated theory construction on axioms and intuition rather than empirical data. Einstein would take such a route in the 1905 relativity paper. On the whole, Einstein may well have appreciated the epistemological efforts by Mach, Kirchhoff, and Hertz to expose and strip away the unessential points from mechanics. As Max Jammer (1957) has written, for Mach, Kirchhoff, and Hertz, "it was a process of purification, of methodological clarification."

Einstein also studied Boltzmann at the ETH. In the preface to his widely-read *Lectures on Mechanics* (1897), Boltzmann had proclaimed his agreement "with Hertz's comments on the wealth of ideas in the relevant writings of Mach, even if I am by no means everywhere of the same opinion as Mach." Like Hertz, Boltzmann sought a mechanical foundation for Maxwell's equations. As Martin J. Klein (1970) has written, "For Boltzmann, mechanics was always the very core of physics," and, in Boltzmann's *Lectures on Maxwell's Theory* (1891), Einstein accordingly found a most un-Maxwellian version of Maxwell's theory.[7] But perhaps it was Boltzmann's (1897) "characterization" of his method for reformulating mechanics that most struck a responsive chord in Einstein:

> It is precisely the unclarities in the principles of mechanics that seem to me to derive from not starting at once with hypothetical mental pictures but trying to link up with experience from the outset.

Thus, as with Hertz, Einstein could follow a development of mechanics that stressed principles over empirical data; he had also found a view congenial to his predilection for visual thinking, which he had become increasingly aware of since his Aarau period of 1895–1896 [Holton (1973f)]. Boltzmann's notion of "mental pictures" was more unrestricted than Hertz's (1894) "images or symbols of external objects," which were actual conceptual frameworks [see Mach (1960)], such as Hertz's own version of mechanics.

While a student at the ETH, Einstein also read works by that giant of 19th-century science Hermann von Helmholtz. It is reasonable to assume that by 1905 Einstein would also have read, in addition to some of von Helmholtz's lectures on theoretical physics, the essay on "The Origins and Correct Interpretation of Our Sense Impressions" (1894a). That was one of von Helmholtz's many essays on the importance of memory pictures for creative thinking, for example:

> Thus the memory images of pure sense impressions can also be used as elements in combinations of ideas, where it is not necessary or even possible to describe those impressions in words and thus to grasp them conceptually...

> Indeed, the idea of a three-dimensional figure has no content other than the ideas of the series of visual images which can be obtained from it, including those which can be produced by cross-sectional cuts.

In this sense, we may rightly claim that the idea of the stereometric form of a material object plays the role of a concept formed on the basis of the combination of an extended series of sensuous intuition images. It is a concept, however, which, unlike a geometric construct, is not necessarily expressible in a verbal definition. It is held together or united only by the clear idea of the laws in accordance with which its perspective images follow on another.

The similarity of this passage with Einstein's description of thinking in the "Autobiographical Notes" (1946) is unmistakable. There, in reply to the question "What, precisely, is 'thinking'?" Einstein wrote:

When, at the reception of sense-impressions, memory-pictures emerge, this is not yet "thinking." And when such pictures form series, each member of which calls forth another, this too is not yet "thinking." When, however, a certain picture turns up in many such series, then — precisely through such return — it becomes an ordering element for such series, in that it connects series which in themselves are unconnected. Such an element becomes an instrument, a concept.

In his essay, "Physics and Reality" (1936), in which he offered a similar description of the dynamics of thinking, Einstein concisely defined thinking as "operations with concepts, i.e., the creation and use of definite functional relations between them [the concepts] and the coordination of sense experiences with these concepts." Furthermore, this process need not be expressible verbally; in fact, in the "Autobiographical Notes," Einstein wrote that thought is not only essentially "non verbal," but occurs in the unconscious, for how else could "we 'wonder' quite spontaneously about some experience"?[8] Einstein (1946) was as specific as he could be on the meaning of "wondering," which "seems to occur when an experience comes into conflict with a world of concepts which is already sufficiently fixed within us." Despite the wide use of visual thinking by scientists educated in German-speaking countries, no one before Einstein combined it so effectively with *Gedanken* experiments and with quasi-aesthetic notions. For example, Einstein began his 1905 papers on relativity and light quanta [as well as, in the (1907e), his first conjectures toward a general theory of relativity] by discussing *Gedanken* experiments exhibiting asymmetries that did not appear inherent in phenomena. He removed the asymmetries either by positing a new view on the structure of light (as in the paper on quanta) or by putting forth new axioms of great generality as in 1905 on special relativity, and in 1907 toward generalizing that theory.

Needless to say, I am not asserting that Einstein consciously took his notions of "concept" and of the dynamics of thinking from von Helmholtz. Rather, the relationship of Einstein to Hertz, von Helmholtz, Boltzmann, and Kirchhoff was best described by Frank, who wrote that from their works Einstein "learned how one builds up the mathematical framework and then with its help constructs the edifice of physics."[9] To which we add that Mach's "incorruptible skepticism" encouraged Einstein to begin questioning for himself whether a mechanical world-picture could succeed in deriving Maxwell's theory.

2.2.1. Assessment from the Viewpoint of "Confirmation by Experiment"

On assessing the mechanical world-picture through confirmation by experiment, Einstein (1946) wrote:

> If mechanics was to be maintained as the foundation of physics, Maxwell's equations had to be interpreted mechanically. This was zealously but fruitlessly attempted, while the equations were proving themselves fruitful in mounting degree.

Despite the fact that he had learned much from Hertz, Kirchhoff, and Boltzmann about building a theoretical system, to Einstein, among others, their efforts toward a mechanical basis for electromagnetism paled before the successes of Lorentz's electromagnetic theory.

Einstein may also have been convinced of the fruitlessness of attempting to deduce Maxwell's equations from mechanics by Henri Poincaré's critical essay (1900a) on the state of physics *circa* 1900, which appeared in German in the *Physikalische Zeitschrift* of 1900–1901. Poincaré not only identified the pressing problems of the time as the failure of the ether-drift experiments and the lack of explanation for Brownian motion and "the cause which produces the electric spark under the action of ultra-violet light," i.e., photoelectric effect, he also sketched his brilliant proof that any system satisfying the principle of least action admitted of an unlimited number of mechanical explanations. He delved further into this proof in his book *Science and Hypothesis* (1902), which Einstein read in the period 1902–1904.[10]

Einstein no doubt found this book provocative. The sheer sweep of the material presented in Poincaré's characteristically lucid style is dazzling; essays address the foundations of arithmetic and geometry, and a critique of mechanics emphasizes the importance of relative motion. All of this discussion was based upon a framework whose dynamics were explained more clearly than by Hertz or von Helmholtz — that is, the neo-Kantian framework emphasizing the role of those organizing principles for thinking which admit of the validity, for example, of non-Euclidean geometries (Chapter 1, footnote 28).

2.2.2. Assessment from the Viewpoint of "Inner Perfection"

Central to Einstein's assessment of the "inner perfection" of the mechanical world-picture was Mach's analysis of absolute space. Consistent with his epistemology, Mach considered that "space and time are well ordered sets of sensations"[11]; therefore, concerning absolute space, Mach wrote[12]:

> No one is competent to predicate things about absolute space and absolute motion; they are pure things of thought, pure mental constructs, that cannot be produced in experience. All of our principles of mechanics are, as we have shown in detail, experimental knowledge concerning the relative positions and motions of bodies.

Therefore the concepts of absolute space and absolute motion must be excluded from mechanics. Instead of referring motion to absolute space, Mach suggested referring all motion to the fixed stars (as in his well-known analysis of Newton's bucket experiment),[13] or perhaps to a "*medium*" filling all of space (i.e., an ether),[14] or to a mean velocity with respect to all the masses in the universe.[15]

In the "Autobiographical Notes," (1946) Einstein singled out Mach's first suggestion. We need only define, at a certain moment on the earth, one inertial reference system, and then refer inertial motions to it, thereby generating an infinity of inertial reference systems. Consequently, referring motion to the fixed stars could be circumvented, although this ultimate recourse for inertial motion must be kept in mind. Einstein found the emphasis on inertial reference systems "particularly offensive," and he illustrated the offensive consequence of Newton's principle of relativity (Corollary V from the *Principia*)[16] with a parable concerning inhabitants on a cloud-enshrouded planet. Their geometry would lead them to assume that space is isotropic, and yet their physical theories would have them attribute a preference or "absoluteness" to the vertical direction. This preference, Einstein wrote, was "precisely analogous" to that given inertial reference systems. Einstein's tendency to frame arguments such as Mach's into a quasi-aesthetic form, revealing asymmetries that should not be contained in laws of nature, was characteristic of his mode of thinking in the relativity paper. However, there, another basic asymmetry had to be removed before he could approach the one concerning the preferential status of inertial reference systems – that is, the notion of preferred reference systems in the ether having the mathematical properties of inertial reference systems though they were constrained to be at rest.

Einstein's parable is similar to one used in *Science and Hypothesis* (1902) by which Poincaré illustrated the usefulness of assuming that the earth rotates. In brief, "Space is symmetrical – yet the laws of motion would present no symmetry." Therefore, a Copernicus on the cloud-enshrouded planet would eventually emerge and assert that if we assume the earth turns, then asymmetries in the laws of nature could be explained if the earth were a noninertial reference system, rather than by physical space being nonisotropic. Poincaré went on to develop this parable in an effort to demonstrate the meaningfulness of only relative motions.[17] Beginning in 1907, however,

Einstein began to approach such problems as Poincaré's in a different way – by enlarging the special principle of relativity.

2.3. ON THE ELECTROMAGNETIC WORLD-PICTURE

In his discussion of the electromagnetic world-picture in the "Autobiographical Notes," Einstein commenced immediately to discuss the "inner perfection" of Lorentz's electromagnetic theory. By omitting the first criterion, he implied agreement with the current opinion that, of all existing electromagnetic theories, Lorentz's best explained the extant data. Poincaré, for example, had been explicit on the point in his (1900a) lecture, reprinted in German in the *Physikalische Zeitschrift*.

Einstein began his criticism of an electromagnetic world-picture by remarking that, "If one views this phase of the development of theory critically, one is struck by the dualism which lies in the fact that the material point in Newton's sense and the field as continuum are used as elementary concepts side by side." Although, for example, Poincaré's [(1900a), (1902)] criticism of Lorentz's theory emphasized its violation of the principle of action and reaction, Einstein was "struck by the dualism" of it: Why should both particle and field be considered as elementary? Einstein then mentioned a way to eliminate the blemish: To deduce the electron's mass entirely from its field energy, i.e., the electromagnetic world-picture. Abraham, Wien, Lorentz, and Poincaré did not advocate an electromagnetic world-picture for this reason, but for reductionistic purposes.

Einstein needed to have access to only the *Annalen der Physik* to be aware of research toward an electromagnetic world-picture. For in addition to Wien's paper of 1901 proposing this effort, the *Annalen* contained Abraham's major paper of 1903 detailing his field-theoretical description of a rigid spherical electron. In the relativity paper, Einstein used the nomenclature of Abraham's theory, i.e., the transverse and longitudinal masses. In the period 1901–1907 the *Annalen* was his main outlet for twenty scientific publications that contained citations to previous papers in this journal.

Einstein continued in the "Autobiographical notes":

> However, Maxwell's equations did not permit the derivations of the equilibrium of the electricity which constitutes a particle.

Thus, before 1905, Einstein was aware of Lorentz's 1904 theory of the electron, and about the fundamental defect that Abraham had pointed out in the 1904 *Physikalische Zeitschrift*. But this does not contradict Einstein's statement to his biographer Seelig in 1955 [see Born (1969)], for one can be aware of a paper without precisely knowing its contents. There exists documentation to the effect that, prior to writing the relativity paper, Einstein had not *seen* Lorentz's paper of 1904, and further evidence emerges from the relativity paper itself. In fact, scientists at major universities in Germany – for example, von Laue in Berlin and Kaufmann in Bonn could not obtain a copy (see Section 1.12.3), and as late as November 1904, the eminent electrodynamicist Emil Cohn (1904a) in Strassburg could obtain only the Dutch version. Most likely, Einstein learned of Lorentz's seminal paper of 1904 from Wien's (1904b) *Annalen* paper, or perhaps from Abraham's critical essay in the journal *Physikalische Zeitschrift*, or perhaps from Lorentz's well-circulated review paper (1904b).

Einstein continued in the "Autobiographical Notes":

> In any case one could believe that it would be possible by and by to find a new and secure foundation for all of physics upon the path which had been so successfully begun by Faraday and Maxwell...

Here he was perhaps not so much remembering the optimism of proponents of the electromagnetic world-picture, as recalling his own fascination with a field-theoretical description of nature. For, by 1905, Einstein (1946) knew that this goal could never be obtained using Lorentz's theory of the electron. The reason was that he had encountered a "second fundamental crisis ... the seriousness of which was suddenly recognized due to Max Planck's investigations into heat radiation (1900)." To this we must add that it was "suddenly recognized" by Einstein alone, because Planck's law for the black-body radiation spectrum was politely ignored from 1900 until 1905.[18]

2.4. CONTRA MECHANICAL AND ELECTROMAGNETIC WORLD-PICTURES

Max Planck [(1900a), (1901)] discovered that the continuous spectrum of the radiation emitted from a cavity within a heated substance could be described by the equation

$$\rho(v, T) = \frac{8\pi v^2}{c^3} \frac{hv}{e^{hv/kT} - 1} \tag{2.1}$$

where ρ is the energy density of the emitted radiation per frequency v at the cavity's absolute temperature T, and h is Planck's constant.[19] This spectrum was of great interest to scientists because of its independence from the cavity's structure; yet, despite intense efforts, the proper spectral distribution function had eluded them – until Planck's work. Owing to the model independence of ρ, Planck chose to base his derivation of it on the model of a solid from electromagnetic theory; namely, that the cavity's walls were composed of electrons on springs – oscillator electrons. To his consternation Planck found that deducing the radiation law from statistical mechanics necessitated supposing that the oscillator electrons exchanged energy with the radiation within the cavity only in discrete amounts hv, and this assumption of energy quantization violated the classical theory of heat radiation [cf. Kuhn (1978)]. Planck spent many years of his life seeking alternate derivations and interpretations in order to bring his radiation law back within the fold of classical theory.

For Einstein, Planck's law precipitated a "second fundamental crisis" in physical theory, because its violation ran deeper than classical radiation theory. Since mechanics permits any mode of vibration for the oscillator electrons, and, according to electromagnetic theory, radiation is emitted continuously, Planck's law violated both mechanics and electromagnetism, Einstein wrote in the "Autobiographical Notes": "all this was quite clear to me shortly after the appearance of Planck's fundamental work." His efforts at using conventional mechanics and electromagnetism to adapt the foundations of physical theory to this new law had failed: "It was as if the ground had been pulled out from under one, with no firm foundation to be seen anywhere, upon which one could have built." No wonder, then, that in 1902 he considered himself as a "heretic" regarding a mechanical basis for Maxwell's equations. Consequently, Einstein lost interest in finding a means of deducing Planck's law; he accepted it as valid and then sought what "general conclusions could be drawn from [it] concerning the structure of radiation and even more generally concerning the electromagnetic foundation of physics."

Leopold Infeld (1950), a colleague of Einstein's, has described a characteristic of Einstein's working style — "originality and obstinancy, the capability of traveling a lonely path for years." While keeping in mind problems concerning the crumbling foundations of physical theory, Einstein pursued a research problem not concerned with foundations. He investigated the nature of intermolecular forces in liquids in an effort to clarify further the phenomenon of capillarity.[20] His research tool was classical thermodynamics because it made no assumptions on the constitution of matter. Einstein (1919a) referred to such theories as "theories of principle"; their "merit ... is their logical perfection, and the security of their foundation (within their domain of applicability)." Einstein elaborated on this viewpoint in a letter to his old friend from the days of the "Olympia Academy" in Bern, Maurice Solovine, 24 April 1920: "[Thermodynamics] is nothing else than a systematic reply to the question: What form should the laws of nature assume in order that it be impossible to construct a perpetuum mobile?" [Einstein (1956)]. In order to investigate the form of the intermolecular force, Einstein assumed it to be analogous to the gravitational force. Early in the research, when a connection seemed imminent between phenomena in the small and in the large, Einstein wrote (14 April 1901) to his friend from the days at the ETH, Marcel Grossman [Seelig (1954)]: "It is a wonderful feeling to recognize the unity of a complex of phenomena which appear to direct physical observation as totally separate things." We may assume, therefore, that an important by-product of this research was to incline Einstein's thinking further toward theories embracing the widest possible phenomena.

Needless to say, Einstein found thermodynamics unsatisfactory for problems concerning the constitution of matter. Consequently, he turned to statistical mechanics, which he considered (1919a) as a "constructive theory." This sort of theory makes assumptions on the constitution of matter, and Einstein placed an equally high premium on it: "When we say we understand a group of natural phenomena, we mean that we have found a constructive theory which embraces them." In other words, constructive theories explain why phenomena occur; for example, according to Lorentz's theory the contraction of matter in motion occurs because of the dynamical interaction between its constituent electrons and the ether. On the other hand, a theory of principle is based on principles asserting the form that physical laws must assume in order to forbid certain phenomena. In a series of three papers, appearing in the period 1902–1904, Einstein rediscovered the statistical mechanics of Boltzmann and Gibbs, as well as discovering new insights concerning fluctuation phenomena.[21]

As M. J. Klein (1967) has written, fluctuation phenomena were always dismissed as unobservable. Using a particularly simple equation for calculating these phenomena, Einstein sought cases where they could be detected. The relevant equation in Einstein's (1904) is

$$\langle \varepsilon^2 \rangle = \langle E^2 \rangle - \langle E \rangle^2 = kT^2 \, d\langle E \rangle/dT, \qquad (2.2)$$

where T is the absolute temperature, $\langle \varepsilon^2 \rangle$ is the fluctuation in the average thermodynamic internal energy $\langle E \rangle$ of a system, and k is Boltzmann's constant. Einstein did not miss the point that this equation held for any kind of system. If $\langle \varepsilon^2 \rangle$ could be measured, then k could be calculated, thereby determining Avogadro's number and thus the dimensions of an atom (neither k nor Avogadro's number were known accurately in 1904). Using the Stefan-Boltzmann Law and Eq. (2.2), Einstein showed that cavity radiation exhibited measurable fluctuations.[22] He wrote to another charter member of the "Olympia Academy," Conrad Habicht on 14 April 1904 [Seelig (1954)]: "I have discovered in the simplest possible way the relationship between the size of the elementary units of matter and the wavelengths of radiation." Unlike his abortive attempt at relating the intermolecular and gravitational forces, here Einstein had successfully realized the "unity" of two seemingly disparate regimes of nature: the invisible microcosm and the observable spectrum of cavity radiation.

At this point it is difficult to imagine Einstein passing up the opportunity of inserting, for $\langle E \rangle$ into Eq. (2.2), the average energy of cavity radiation from Planck's law. After all, Einstein wrote that at the time he was not interested in deriving Planck's law, but rather in deducing its consequences. We know, in fact, that he made this calculation, for on 17 January 1952 he wrote to Max von Laue that "in 1905 I already knew that Maxwell's theory leads to false fluctuations of radiation pressure, and with it, to an incorrect Brownian motion of a mirror in a Planckian radiation cavity" [Holton (1973a)]. On 19

February 1955 Einstein wrote to his biographer Seelig that before 1905 he "had already previously found that Maxwell's theory did not account for the micro-structure of radiation and could therefore have no general validity" [Born, (1969)]. Thus, let us conjecture that Einstein substituted the average energy of black-body radiation in a cavity of volume V,

$$\langle E \rangle = V\rho(v, T)\,dv \tag{2.3}$$

into Eq. (2.2) yielding

$$\langle \varepsilon^2 \rangle = \left\{ \frac{c^3}{8\pi v^2} \rho^2 + hv\rho \right\} V\,dv. \tag{2.4}$$

Since the first term is proportional to $\langle E \rangle^2$, it can be deduced from Maxwell's theory, but the second term cannot, because it asserts that radiation possesses particulate, or discontinuous, properties. Therefore, as Einstein wrote to von Laue, Maxwell's theory did not describe completely the fluctuations in black-body radiation.

2.5. A NEW BEGINNING: THE TRIAD OF PAPERS IN VOL. 17

In the first of his three masterpieces on the nature of light in vol. 17 of the *Annalen der Physik* of 1905, Einstein neither used Planck's law, nor for that matter did he even mention the baffling wave-particle duality of light from Eq. (2.4). Using statistical thermodynamics, a constructive theory, Einstein deduced that the entropy of the type of radiation described by Wien's distribution law ($\rho^{\text{Wien}} \sim v^3 e^{-hv/kT}$) had the same form as the entropy of a gas of independent particles. He used this result to argue for the theoretical expedient, i.e., heuristic proposal, that light consists of quanta, each possessing an amount of energy $E = hv$. Einstein went on to show that, by means of this heuristic proposal, certain processes involving the transformation of light – for example, the photoelectric effect – could be more readily understood. Thus, Einstein went far beyond Planck's original assertion of quantizing only the energy of charged oscillators interacting with radiation.[23] It was not, therefore, an under-statement when Einstein described the light-quantum paper to his friend Habicht as "very revolutionary" [Seelig (1954)]; and certainly the well-established high-frequency limit of the politely ignored Planck radiation law, i.e., Wien's law, sufficed for this purpose. But I should like to propose another reason for Einstein's use of the Wien law: namely, that this law was sufficient for the purpose of Einstein's grand strategy in the three papers of vol. 17. In addition to suggesting solutions for the problems of photoelectric effect and Brownian motion, Einstein also sought to dem-onstrate in the first two papers the insufficiency of mechanics, electro-magnetism, and thermodynamics in volumes where fluctuation phenomena could not be overlooked; and then to propose, in the third paper of vol. 17, a theory of principle as a new beginning for problems concerning the constitution of matter.

In order to support this conjecture and to allow his strategy to unfold naturally, it is useful to review Einstein's stated intent in the light-quantum paper. Einstein (1905b) sought to remove the "profound formal distinction between the theoretical ideas which physicists have formed concerning gases and other ponderable bodies and the Maxwell theory of electromagnetic processes in so-called empty space." The central problem here, as Einstein expressed it in quasi-aesthetic terms, concerned the disturbing dualism that he had already observed in Lorentz's theory, i.e., between the discontinuous particle and continuous radiation field. In the light-quantum paper, Einstein chose to remove the tension with the heuristic proposal that, for explaining certain processes — "black-body radiation, fluorescence, the production of cathode rays by ultraviolet light, and other related phenomena connected with the emission or transformation of light," it was useful to assume that light possessed particulate properties; Wien's law sufficed for that purpose.

Two years later Einstein disclosed yet another theme in his light-quantum paper. In a 1907 paper, containing his first published criticisms of the mechanical and electromagnetic world-pictures, Einstein wrote (1907d) that his earlier investigations on the structure of light showed him that "the present-day electromagnetic world-picture [was] unsatisfactory." Einstein was referring here to the light-quantum paper of 1905 (but there was nothing in it concerning world-pictures), and to another paper published in 1906, which reveals the other theme in the light-quantum paper. Einstein wrote (1906b) that in the light-quantum paper he had shown "that Maxwell's theory, in connection with the electron theory, leads to consequences contradicting experiments on black-body radiation." Briefly, in the light-quantum paper, Einstein used kinetic theory for calculating the spectral distribution of an oscillator electron in thermal equilibrium with cavity radiation. But this spectral distribution led to an infinite amount of energy which was emitted from the cavity at each temperature, instead of the result expected from the Stefan-Boltzmann law.[24] Since this model of a solid was derived from the "electron theory" (i.e., Lorentz's theory), then Lorentz's theory in its current form was not only based on an unstable electron, but also failed to describe black-body radiation. In addition, Einstein (1905b) interpreted the predicted emission of an infinite amount of energy as meaning that "there can be no talk of a definite energy distribution between ether and matter." This interpretation could have been still another reason for Einstein, in the relativity paper, to declare Lorentz's ether to be "superfluous." Taking a safe approach based on Wien's law enabled Einstein in the light-quantum paper to avoid mentioning the paradoxical results of the wave-particle duality and yet criticize, albeit implicitly, the electromagnetic world-picture.

Perhaps even more subtle was that in the light-quantum paper, Einstein began criticizing the mechanical world-picture because he used notions from mechanics for calculating the oscillator electron's average energy; Einstein pursued this criticism further in the second paper (in vol. 17) where he again used a constructive theory. Of this paper on Brownian motion, Einstein wrote (1907d) that its results convinced him of the insufficiency of mechanics and of thermodynamics to account for the properties of all systems of matter in motion.[25] Moreover, Einstein continued, this state of affairs should be enough to convince everyone of the necessity for making fundamental changes in the basis of theoretical physics. Nevertheless [Einstein (1907d)], as long as one was not studying the instantaneous state of a system, in regions of space small enough so that fluctuation phenomena must be taken into account, then "the equations of mechanics and thermodynamics can be employed," and with this restriction in mind one could also make use of Maxwell's equations. This Einstein did in the third paper that he published in vol. 17 of the *Annalen*, the relativity paper. Here Einstein resolved the tension, or incompatibility, between the laws of mechanics and electromagnetism by proposing a principle of relativity applicable to both. The theory of relativity was necessarily a theory of principle. This was the only route remaining for one whose despair in 1900 of "discovering the true laws by means of constructive efforts" was reinforced by the limitations imposed upon theories of matter that he demonstrated in the light quantum and Brownian motion papers. For Einstein the theory of relativity was the phoenix rising from the ashes of physics in 1905; it was to be the new beginning. Looking back over the state of physical theory at that time, Einstein wrote (1923) that physicists were "out of [their] depth."

In summary, all three papers in vol. 17 of the *Annalen der Physik* (1905) contained overlapping themes: the paper on light quanta demonstrated the limits of Lorentz's theory and mechanics; the one on Brownian motion probed the limits of thermodynamics and mechanics; the one on relativity proposed a theory of principle, which specified what form the laws of physics should assume in order to be used again to investigate the structure of matter. The nature of radiation was the principal problem from which this triad of papers resulted.[26] Other physicists, such as Poincaré, also realized the importance of the problems Einstein addressed in these papers, but none of them realized their connectedness.

FOOTNOTES

1. For an analysis of the emergence of Einstein's complex viewpoint after 1915, see Holton (1973d).

2. This dimension of Einstein's thinking has been discussed by Holton (1973f).

3. Einstein's good friend, Michele Besso, whom Einstein immortalized in the relativity paper, introduced him to Mach's *Science of Mechanics* in 1897 [Seelig (1952)]. Michele Besso (1873–1955) attended the ETH from 1891 to 1895 where he studied engineering. Einstein and Besso met in 1896, shortly before Einstein entered the ETH,

at a musical soirée – like Einstein, Besso played the violin. In 1904 Einstein was instrumental in obtaining for Besso a position at the Patent Office in Bern. The two friends often walked home from work and discussed physics. When they were later separated Einstein thought of Besso as his best "sounding board" for new ideas. Their correspondence lasted over half a century and is as interesting as it is voluminous [Einstein (1972)].

In discussing Mach's influence on Einstein before 1905, it is apropos to cite only those portions of his *Science of Mechanics* that appeared in the edition of 1897 and the preceding one of 1888. For this purpose I use the notation EV = English version, and GV(1897) and GV(1888) for the German versions. The quotation cited in the text is from EV, p. 290 (italics in original); GV(1888), pp. 221–222; GV(1897), pp. 231–232. Mach further drove this point home in the final chapter of *Science of Mechanics*. For example, he wrote:

> Purely mechanical phenomena do not exist.... The view that makes mechanics the basis of the remaining branches of physics, and explains all physical phenomena by mechanical ideas, is in our judgment a prejudice.... The mechanical theory of nature is, undoubtedly, in an historical view, both intelligible and pardonable; and it may also, for a time, have been of much value. But, upon the whole, it is an artificial conception.

[EV, p. 599; GV(1888), p. 467; GV(1897), p. 486]. The effect of Mach on Einstein is discussed in Holton (1973e) and Blackmore (1972). For biographical information on Mach see Blackmore (1972) and Hiebert (1973).

4. Mach, *Mechanics*, EV, p. 599; GV(1888), p. 469; GV(1897), p. 489. Mach's analysis of mechanics was based upon an epistemology emphasizing sensationism. So for Mach the sole objective reality were sensations, such as colors, sounds, odors, pressures and heat. Mach referred to sensations as elements, claiming that our knowledge of the physical world was obtained from elements and from complexes (i.e., small groups) of elements. Theorizing begins from elements, and conversely statements in theories were physically meaningful only if they could be related directly to elements. For Mach the goal of science was merely an economical description of sense data: "Science itself therefore should be regarded as a minimal problem, consisting of the completest possible presentment of facts with the *least possible expenditure of thought*" [italics in original; EV, p. 586; GV(1888), p. 461; GV(1897), p. 480]. Mach bluntly repudiated the relativity theory in the Preface (dated 1913) to his posthumously published (1921). Einstein's (1922) reply on 22 April 1922 in Paris was, "what Mach has done is to make a catalogue, not a system. To the extent that Mach was a good mechanician he was a deplorable philosopher." For others of Mach's writings on the goal of science, see Mach (1888, 1897, 1960), (1910), (1921), (1943), (1959) and (1976).

5. For a discussion of Kirchhoff's view of mechanics, see Jammer (1957).

6. On Mach, Hertz wrote in the "Author's Preface" (1894): "In a general way I owe very much to Mach's splendid book on the *Development of Mechanics*." For some discussions of Hertz's mechanics, see Mach (1960), Jammer (1957), and Klein (1970).

7. Klein (1970) has written that "an unwary reader, glancing through the illustrations in Boltzmann's *Lectures*, might well wonder if he had mistakenly picked up a

book on the design of engineering mechanics." Although in Part II of his *Vorlesungen über die Prinzipe der Mechanik*, published in 1904, Boltzmann had shifted toward an electromagnetic description of nature [see Jammer (1961)]. Indeed, if Einstein had read Maxwell at all before 1905, it was very likely from Boltzmann's German translations liberally supplied with explanatory notes. In 1893 Maxwell's *Treatise* (1873) had been translated into German. Holton (1973c), for example, has found no evidence that Einstein ever read Maxwell's original papers, and so it is unclear as to what Frank meant by Einstein's having studied Maxwell.

8. Einstein's manner of thinking has been the subject of numerous analyses – for example, those of Holton (1973f), (1979). There is an analysis by that pioneering figure of Gestalt psychology, Max Wertheimer, who was a colleague of Einstein at Berlin. In his book *Productive Thinking* (1959), Wertheimer attempted to show that the dynamics of Einstein's thinking toward the special relativity theory could be explained by the guidelines of Gestalt psychology. This episode has been much misused by certain modern philosophers of science, because it centered upon the Michelson-Morley experiment. Placing this experiment in the proper historic context, Holton (1973d) has argued cogently that at most it indirectly influenced Einstein's thinking in 1905. Proper assessment of Wertheimer's Gestalt psychology has placed this episode in its proper light – that is, as a daring case study in Gestalt psychology, and nothing more [see Miller (1975b)].

9. Quoted from the original German-language edition of Frank's biography of Einstein. Holton (1973c) has pointed out several important omissions in the English-language edition.

10. Einstein read this book as part of the curriculum of the "Olympia Academy" in Bern [Seelig (1954)]. This was an informal study group composed of Einstein and his friends; among them were Conrad Habicht and Maurice Solovine. A sampling of other parts of the curriculum is: J. S. Mill's, *System of Logic*; David Hume's, *Treatise on Human Nature*; Plato's, *Dialogues*; works by Leibniz from which Einstein "was impressed by the sentence: "*Natura non facit saltus.*"

11. Mach, *Mechanics*, EV, p. 611; GV(1888), p. 437; GV(1897), p. 498.

12. *Ibid.*, EV, p. 280; GV(1888), p. 213; GV(1897), p. 223.

13. *Ibid.*, EV, pp. 277–279; GV(1888), pp. 211–213, 418–482; GV(1897), pp. 221–223, 233–234. I omit a detailed analysis of Mach's commentary on Newton's bucket experiment because it was more important for Einstein's thinking toward the general relativity theory, which is outside the scope of this book.

14. *Ibid.*, EV, p. 282; GV(1888), p. 215; GV(1897), p. 225.

15. *Ibid.*, EV, pp. 286–288; GV(1888), pp. 218–220; GV(1897), pp. 228–230.
In Chapter 3 we discuss the relevant passage from the 1905 paper. Mach's discussion in *Science of Mechanics* of referring motion to a mean velocity is therefore deserving of some commentary. According to Mach the motion of every mass m is affected by all of the other masses in the universe; however, Mach emphasized that the effect is small and can be neglected, and so we can assume that to a good approximation there is inertial motion. But motion relative to what? Mach offered the following suggestion. Uniform linear motion of a free body of mass μ can be described by asserting that the mean

velocity of the body with reference to the other masses in the universe remains constant; or in mathematical terms

$$\frac{d}{dt}\frac{\sum mr}{\sum m} = \text{constant},$$

where $r_1, r_2, \ldots$ are relative distances between μ and $m_1, m_2, \ldots$. Next, consider that two masses μ_1 and μ_2 interact by means of a central force $f(r)$, where r is their relative distance. In this case we must consider the mean momentum of the two-body system relative to the other masses in the universe. From Newton's third law the mean velocity of this two-body system with reference to the other masses in the universe remains constant:

$$\frac{d}{dt}\left[\mu_1 \frac{\sum mr}{\sum m} + \mu_2 \frac{\sum mr}{\sum m} \right] = \text{constant}.$$

Rotations of the two-body system can be referred to this mean velocity vector. Similar arguments obtain for n-body systems interacting through central forces.

Mach considered these meditations to be extremely important:

> ... we see that even in the simplest case, in which apparently we deal with the mutual interaction of *two* masses, the neglecting of the rest of the world is *impossible* ... We may indeed regard all masses as related to each other.

(italics in original; EV, pp. 227–228; GV(1888), p. 219; GV(1897), p. 229.) Einstein regarded as important to his thoughts toward a theory of gravitation Mach's observation that "we may consider all masses as related to each other" [Einstein (1918c)]. Einstein was careful to emphasize that what he referred to as the "Mach principle" was a generalization of Mach's ideas in *Science of Mechanics* — see the footnote on p. 241 of (1918c). For discussion of Mach's principle, see Pauli (1958), Weinberg (1972).

16. Newton's Corollary V states that the laws of mechanics are the same in all inertial reference systems.

17. Einstein (1921a) began an essentially technical lecture not with physics, but with a survey of "how are our customary ideas of space and time related to the character of our experiences?" For Poincaré, too, there was a relationship between prescientific and scientific knowledge, the latter being a development of the former. Not surprisingly, Einstein said:

> We now come to our concepts and judgments of space. It is essential here also to pay strict attention to the relation of experience to our concepts. It seems to me that Poincaré clearly recognized the truth in the account he gave in his book, "La Science et l'Hypothèse."

The description of the foundations of geometry that Einstein gave in this lecture was paralleled by Poincaré's in *Science and Hypothesis*. From our movements about the world of sensations, we discover the laws of geometry by noting relationships (i.e., displacements) between the positions of solid bodies relative to other solid bodies. Accordingly, we learn that only relative motions and positions have a meaning in our prescientific thinking and this should be the case in scientific theories also. However, even if one accepts uncritically the notions of absolute space and time, Einstein wrote in the "Autobiographical Notes," then one must also accept the "idea of action-at-a-distance which is unsuited to the ideas which one forms on the basis of the raw experiences of daily life." Mechanics was in this sense "primitive" because in "daily life" effects have dynamical causes — a fact we learn while discovering the laws of geometry. On the level of scientific theorizing, the "wonder" of the compass needle remained with the adult Einstein; for, at age four or five, Einstein was fascinated by the fact that the compass needle always points in the same direction, as if constrained to do so by some invisible force, and yet, from the world of perceptions, effects happen by contact. Perhaps here resides the root of Einstein's future preference for field-theoretical descriptions of nature. I have developed this theme in the predilection for field theories of many well-known scientists, e.g., Maxwell, Newton and Poincaré in my (1979b).

18. Klein (1962) writes in a detailed study of Planck's work:

> ... a revolutionary idea is not always recognized as such, not even by its propounder. Planck's concept of energy quanta went practically unrecognized in the literature of physics for over four years. His radiation formula was accepted as describing the experimental facts in a simple and adequate way, but the theory which he had proposed as a basis for this formula drew no attention until 1905.

19. Planck (1900a) deduced his radiation formula from thermodynamic considerations. In (1900b) he sought to place it on a firmer physical foundation, and so he used Boltzmann's probabilistic notion of entropy; in this paper the energy quanta first appeared. The (1901) paper is an elaboration of the first two. Planck's papers of 1900 are translated in ter Haar (1967). For discussions of Planck's researches, see Klein (1962), (1963), Jammer (1966), Hermann (1971), and Kuhn (1978).

20. Einstein's first published papers emerged from this research (1901), (1902a). These papers are discussed in Klein (1967) and Holton (1979).

21. Einstein's papers are (1902b), (1903), (1904). For discussion of these papers, see Klein (1967), (1975).
Indicative of his interest in constructive theories at this time, Einstein wrote to Besso in January of 1903 that he was embarking on "comprehensive studies in Elektronentheorie." During January of 1903 Abraham's (1903) may have appeared, or Einstein may have read Abraham's (1902). At any rate, with certainty I can conjecture that Einstein's use of the term Elektronentheorie in January of 1903 meant that at this time he was aware of the only substantial theoretical work on this subject — Abraham's.

22. For a detailed analysis of Einstein's adroit use of Eq. (2.2), see Klein (1964). Klein writes (1975) that although Einstein's concern in this series of papers was to "complete the mechanical basis of thermodynamics," even in the "first article ... there are suggestions that Einstein was conscious of the limitations of the mechanical world-picture and was eager to free his argument from some of its restrictions" (italics in original).

23. In retrospect, Einstein resurrected Planck's theory of black-body radiation twice over, once in 1905 (albeit indirectly) in his paper on light quanta, and then in (1907a) with work toward a quantum theory of specific heats and of solids. For a discussion of Einstein's work on specific heats, see Klein (1965) and Jammer (1966).

24. Briefly, according to classical radiation theory, the spectral function ρ is:

$$\rho = (8\pi v^2/c^3) kT.$$

Then the total energy density u of the radiation emitted from a cavity is:

$$u = \frac{8\pi}{c^3} kT \int_0^\infty v^2 \, dv = \infty,$$

instead of being proportional to T^4 as is required by the Stefan-Boltzmann law. In 1911 Paul Ehrenfest dubbed this result the "ultraviolet catastrophe" [see Klein (1970)].

25. See also Einstein (1946), where he wrote:

> The success of the theory of the Brownian motion showed again conclusively that classical mechanics always offered trustworthy results whenever it was applied to motions in which the higher time derivatives of velocity are negligibly small.

We have noted already in Chapter 1 that in electromagnetic theory this was known as the approximation of quasi-stationary motion.

26. This observation, first made by Holton (1973a) and Klein (1967), receives further support in this chapter through analysis of Einstein's (1906b), (1907d); I have discussed this point in my (1976), and it is elaborated upon in subsequent chapters.

3. *Zur Elektrodynamik bewegter Körper;*
von A. Einstein.

Daß die Elektrodynamik Maxwells — wie dieselbe gegenwärtig aufgefaßt zu werden pflegt — in ihrer Anwendung auf bewegte Körper zu Asymmetrien führt, welche den Phänomenen nicht anzuhaften scheinen, ist bekannt. Man denke z. B. an die elektrodynamische Wechselwirkung zwischen einem Magneten und einem Leiter. Das beobachtbare Phänomen hängt hier nur ab von der Relativbewegung von Leiter und Magnet, während nach der üblichen Auffassung die beiden Fälle, daß der eine oder der andere dieser Körper der bewegte sei, streng voneinander zu trennen sind. Bewegt sich nämlich der Magnet und ruht der Leiter, so entsteht in der Umgebung des Magneten ein elektrisches Feld von gewissem Energiewerte, welches an den Orten, wo sich Teile des Leiters befinden, einen Strom erzeugt. Ruht aber der Magnet und bewegt sich der Leiter, so entsteht in der Umgebung des Magneten kein elektrisches Feld, dagegen im Leiter eine elektromotorische Kraft, welcher an sich keine Energie entspricht, die aber — Gleichheit der Relativbewegung bei den beiden ins Auge gefaßten Fällen vorausgesetzt — zu elektrischen Strömen von derselben Größe und demselben Verlaufe Veranlassung gibt, wie im ersten Falle die elektrischen Kräfte.

Beispiele ähnlicher Art, sowie die mißlungenen Versuche, eine Bewegung der Erde relativ zum „Lichtmedium" zu konstatieren, führen zu der Vermutung, daß dem Begriffe der absoluten Ruhe nicht nur in der Mechanik, sondern auch in der Elektrodynamik keine Eigenschaften der Erscheinungen entsprechen, sondern daß vielmehr für alle Koordinatensysteme, für welche die mechanischen Gleichungen gelten, auch die gleichen elektrodynamischen und optischen Gesetze gelten, wie dies für die Größen erster Ordnung bereits erwiesen ist. Wir wollen diese Vermutung (deren Inhalt im folgenden „Prinzip der Relativität" genannt werden wird) zur Voraussetzung erheben und außerdem die mit ihm nur scheinbar unverträgliche

The first page of Einstein's "On the Electrodynamics of Moving Bodies" as it appears in the memorable Volume 17 of the *Annalen der Physik*. (From *Ann. d. Phys.*, reproduced with permission of the estate of Albert Einstein)

CHAPTER 3

AN ANALYSIS OF EINSTEIN'S INTRODUCTORY COMMENTS TO "ZUR ELEKTRODYNAMIK BEWEGTER KÖRPER"

> The phenomenon of the electromagnetic induction forced me to postulate the (special) relativity principle.
>
> A. Einstein [(1919), quoted in Holton, (1973f)]

In 1905, it was generally understood that terms such as "electrodynamics of moving bodies" referred to phenomena involving moving dielectric and magnetic media, i.e., macroscopic phenomena. For example, the title of Hertz's (1890b), "*Über die Grundgleichungen der Elektrodynamik für bewegte Körper.*" On the other hand, Abraham's major paper of 1903 on a theory of the *electron* was entitled "*Prinzipien der Dynamik des Elektrons.*" However despite its "title, which blended easily into the table" of contents of the *Annalen*, Einstein's "*Zur Elektrodynamik bewegter Körper*" (On the Electrodynamics of Moving Bodies) was not addressed solely to one or the other of these two topics. It was neither an explicit solution to problems in the electrodynamics of moving macroscopic media, nor a theory of the electron. To explore why Einstein chose this title, we must first examine the "experimental data" with which he opened the relativity paper.

3.1. EINSTEIN'S EXPERIMENTAL DATA

The content of Einstein's opening passages [1–25] is consistent with the style of physics papers both then and now[1]; yet Einstein developed a curious twist on the presentation of puzzling data by never discussing in detail (or referencing) any of the well-known "unsuccessful attempts to discover any motion of the earth relatively to the 'light medium'" (i.e., the ether-drift experiments). The only experiment described in some detail in the introductory section was also the sole example in the entire paper from the electrodynamics of moving bodies, as that class of problems was customarily understood. Furthermore, the case of magnet and conducting circuit in relative inertial motion was presented not as a puzzle concerning hard experimental data, but as a puzzle concerning certain "asymmetries" (presumably to be specified later) that arise from "Maxwell's electrodynamics – the way in which it is usually understood."

In 1919, Einstein recalled that the phenomenon of electromagnetic induction "forced [him] to postulate the (special) relativity principle." Why was Faraday's law so important to his thinking? Why does the discussion of electromagnetic induction not appear again until the second half of the paper? There Einstein claimed to have removed all asymmetries, and thus the problem of the "'seat' of electrodynamic electromotive forces (unipolar machines)" could be dismissed as "meaningless." The brusqueness, indeed the audacity, with which Einstein dismissed these problems belied that they were of great concern to many major contemporary electrodynamicists and engineers; in short, Faraday's law was not well understood in 1905. Einstein informed the scientific-engineering community that the root of misunderstandings resided not in problems concerning the constitution of magnets, but in kinematics. It appears, therefore, that in the opening paragraph Einstein posed a version of a perplexing problem (e.g., unipolar induction) and then developed a viewpoint so powerful as to render the problem meaningless. Then there is the second paragraph, which seems like a non sequitur; for how could Einstein claim that the data of magnet and conductor are related to those from the ether-drift experiments and mechanics?

Einstein did not claim that Faraday's induction law was incorrect, but rather that it was being misinterpreted. If it had been *generally* believed in the physics community "that Maxwell's electrodynamics – the way in which it is usually understood – when applied to moving bodies, leads to asymmetries ... ," then we might expect to find statements to that effect in contemporary elaborations on Maxwell's theory, notably by Hertz or Lorentz, and perhaps even a hint in Maxwell's own papers. That was not the case, however, especially for Faraday's induction law for the case of magnet and conducting loop in relative inertial motion. In 1831 Faraday reported that current would arise in a conductor if there was relative motion between the conductor and a magnet [Faraday (1965)].[2] Although not everyone agreed with Faraday's interpretation of the cause of the induced current, the elegance and simplicity of his experiments were appreciated by his contemporaries. Maxwell referred to Faraday's discovery again and again, stressing that the direction and magnitude of the induced current depended upon only the relative motion of magnet and conductor [Maxwell (1873)]. To substantiate this statement, Maxwell (1873) extended Faraday's law to moving conductors and proved its Galilean covariance.[3] Hertz (1890b) further extended Maxwell's proof to all the equations of his own version of Maxwell's theory. The Faraday induction law was also one of the fundamental equations of Lorentz's theory; Lorentz demonstrated in the *Versuch* that, to order v/c, electromagnetic induction was unaffected by the earth's motion through the ether, when this motion could be considered as inertial motion.

Einstein had studied the works of Hertz and Lorentz where Faraday's law was discussed, but they mentioned no asymmetries. Thus, what was the asymmetry that struck Einstein? Perhaps Einstein's clearest statement of his assessment of the Faraday induction law, as it applied to the case of magnet and conducting loop, may be found in an unpublished 1919 essay brought to light by Holton (1973f). In "Fundamental Ideas and Methods of Relativity Theory, Presented in their Development," Einstein wrote (italics in original):

> In the construction of special relativity theory, the following, [in the earlier part of this manuscript] not-yet-mentioned thought concerning the Faraday [experiment] on electromagnetic induction played for me a leading role.
>
> According to Faraday, during the relative motion of a magnet with respect to a conducting circuit, an electric current is induced in the latter. It is all the same whether the magnet is moved or the conductor; only the relative motion counts, according to the Maxwell-Lorentz theory. However, the theoretical interpretation of the phenomenon in these two cases is quite different . . . .
>
> The thought that one is dealing here with two fundamentally different cases was for me unbearable. The difference between these two cases could not be a real difference but rather, in my conviction, only a difference in the choice of the reference point. Judged from the magnet, there were certainly no electric fields, [whereas] judged from the conducting circuit there certainly was one. The existence of an electric field was therefore a relative one, depending on the state of motion of the coordinate system being used, and a kind of objective reality could be granted only to the *electric and magnetic* field together, quite apart from the state of relative motion of the observer or the coordinate system. The phenomenon of the electromagnetic induction forced me to postulate the (special) relativity principle. [Footnote:] The difficulty that had to be overcome was in the constancy of the velocity of light in vacuum which I had first thought I would have to give up. Only after groping for years did I notice that the difficulty rests on the arbitrariness of the kinematical fundamental concepts [presumably such concepts as simultaneity].

Clearly, what Einstein meant in 1905 by "Maxwell's electrodynamics" was Lorentz's version of it, or what, in 1907, he referred to using Abraham's (1903) terminology, as the "Maxwell-Lorentz theory." According to Einstein, the asymmetry in the Faraday induction law was made manifest by the clear definition of the nature of electricity in Lorentz's theory. Thus, even though the magnitude and direction of the induced current depended upon only the relative motion between magnet and conductor, the interpretation of the induction effect is fundamentally different for the case of moving magnet and resting conductor, and vice versa. In typical fashion, Einstein uses the somewhat subjective notion of a problem in science phrased in terms of a polarity, that is so "unbearable" that he is forced to resolve it by means of a very general principle.

3.2. THE CASE OF MAGNET AND CONDUCTOR

By using Faraday's induction law in the "way in which it is usually understood," Einstein next demonstrated in the relativity paper [7–13] that the

Maxwell-Lorentz theory "leads to asymmetries which do not appear to be inherent in the phenomena."

Consider the following *Gedanken* experiment which Einstein may well have had in mind. A conducting loop is situated between the pole faces of a permanent conducting magnet. The magnet begins to move with a uniform velocity v relative to the conducting loop (see Fig. 3.1); in terms used in Lorentz's theory of electromagnetism, the conducting loop is at rest in the ether. An observer S_1, at rest relative to the loop, notices that when the magnet is in motion a positive charge q in the wire moves upward. S_1 interprets the motion of q as due to the existence of an electric field generated by the moving magnet at the site of the conducting loop.

Present-day physics texts introduce Faraday's law by citing the second case discussed later by Einstein — that is, the conducting loop in motion relative to a resting magnet.[4] Although the idea that an electric field resulting from a moving magnet is the cause for current induced in a resting circuit may at first seem peculiar to a modern physicist, in Einstein's time a well-known text treated the example of magnet and conductor in relative motion much as Einstein did [Föppl (1894)]. Holton (1973c) has shown that Einstein read August Föppl's book during his stay at the ETH in 1896–1900. Consequently, it is safe to assume that Einstein, following Föppl's example, calculated the electric field E_1 that "arises in the neighborhood of the magnet" as follows: the differential form of Faraday's law is

$$\nabla \times E_1 = -\frac{1}{c}\frac{\partial B}{\partial t}, \tag{3.1}$$

where all derivatives are evaluated in the conductor's rest system. Expressing the magnet's field in terms of its vector potential A (i.e., $B = \nabla \times A$), Eq. (3.1) can be rewritten as

$$E_1 = -\frac{1}{c}\frac{\partial A}{\partial t} - \nabla\psi, \tag{3.2}$$

where the potential ψ is added for generality. An observer moving with the magnet measures no change in its magnetic field, and so the temporal variation of the vector potential A in the magnet's rest system vanishes, i.e., $dA/dt = 0$. The quantity dA/dt can be related to the variation of the vector field A measured by S_1 at rest in the ether by means of the convective derivative Eq. (1.10).[5] Since electromagnetic radiation is neglected then $\nabla \cdot A = 0$, and Eq. (1.10) becomes

$$dA/dt = \partial A/\partial t + B \times v + \nabla(A \cdot v). \tag{3.3}$$

The constancy of A in the magnet's rest system yields

$$-\partial A/\partial t = B \times v + \nabla(A \cdot v). \tag{3.4}$$

Then from Eqs. (3.4) and (3.2) the electric field at the site of the resting conductor due to the moving magnet is:

$$\boldsymbol{E}_1 = \boldsymbol{B} \times \frac{\boldsymbol{v}}{c} + \nabla[(\boldsymbol{A} \cdot \boldsymbol{v}/c) - \psi]. \tag{3.5}$$

According to Lorentz's theory (*Versuch*), the electromotive force (EMF) ε_1 measured by S_1 is:

$$\varepsilon_1 = \oint (\boldsymbol{F}_1/q) \cdot d\boldsymbol{l}, \tag{3.6}$$

where the force on the charge q due to the electric field $\boldsymbol{E}_1$ is $\boldsymbol{F}_1 = q\boldsymbol{E}_1$. Then from Eq. (3.5) we have

$$\varepsilon_1 = \oint \boldsymbol{B} \times (\boldsymbol{v}/c) \cdot d\boldsymbol{l} \tag{3.7}$$

since the contribution from the term $\nabla[(\boldsymbol{A} \cdot \boldsymbol{v}/c) - \psi]$ vanishes for a closed circuit. Thus, as Einstein wrote in the relativity paper, "there arises in the neighborhood of the [moving] magnet an electric field $[\boldsymbol{B} \times \boldsymbol{v}/c]$ with a certain definite energy." The direction of the field $\boldsymbol{B} \times \boldsymbol{v}/c$ is along the positive y-axis in Fig. 3.1. This result is consistent with the direction of motion of the positive charge q. From Eq. (3.7), the EMF is

$$\varepsilon_1 = (v/c)Ba, \tag{3.8}$$

and from Ohm's law the magnitude of the induced current is

$$i_1 = \varepsilon_1/R = vBa/Rc \tag{3.9}$$

where R is the resistance of the conducting loop, and the direction of flow for positive charges is clockwise.

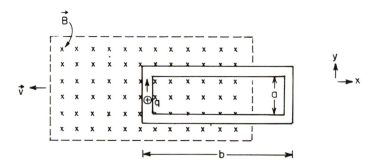

FIG. 3.1. The $\times$ indicates that the magnetic field $\boldsymbol{B}$ is into the plane of the paper. The field's source is a permanent conducting magnet which is moving with velocity $\boldsymbol{v}$ relative to the conducting loop. The loop is of length b and width a. The positive charge q moves upward as a result of the conducting magnet moving to the left with a velocity $\boldsymbol{v}$ relative to the resting conducting loop.

Einstein next turned to the converse case:

> But if the magnet is stationary and the conductor in motion, no electric
> field arises in the neighborhood of the magnet.

Consider now that the conducting loop is moving with a uniform linear
velocity $u = -v$ (in the direction of the positive x-axis) with respect to the
uniform magnetic field B (see Fig. 3.2); once again, in the terminology of
Lorentz's theory, the magnet is taken as at rest in the ether. An observer S_2, at
rest relative to the magnet, observes that a positive charge q in the wire loop
begins moving upward when the loop is in motion. S_2 interprets this effect as
resulting from the force F_2 exerted on the moving particle by the magnet's field
B (see Fig. 3.2), that is,

$$F_2 = q(u/c) \times B. \tag{3.10}$$

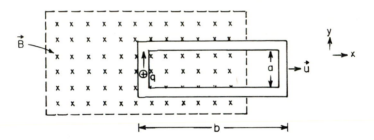

FIG. 3.2. The conducting loop moves with velocity $u = -v$ relative to the magnet.

This force is a portion of the Lorentz force. So according to S_2 "no electric field
arises in the neighborhood of the magnet"; rather, the motion of q in the
y-direction is due to the force F_2. From Eq. (3.10) we can see that the motion of
the charge q is clockwise, as it was in the first case (the magnet in motion and
the loop at absolute rest). This flow of charge in the wire loop implies the
existence of an electromotive force.

Einstein continued:

> In the conductor, however, we find an electromotive force, to which in
> itself there is no corresponding energy, but which gives rise—assuming
> equality of relative motion in the two cases discussed—to electric currents
> of the same path and intensity as those produced by the electric forces in
> the former case.

Unfortunately a clash of terminology occurs here. In 1905 the electromotive
force was defined as the "impressed force," which caused current to flow in a
conductor as it moved through a magnetic field [Föppl (1894)]; the

mathematical expression for this vector quantity was $u/c \times \boldsymbol{B}$.[6] So there is "no corresponding energy" to Einstein's electromotive force because this quantity could not, according to "Maxwell's electrodynamics ... in the way in which it is usually understood," be interpreted as an electric field.

From Eq. (3.10), the electromotive force (using modern terminology) is

$$\varepsilon_2 = \oint (\boldsymbol{u}/c) \times \boldsymbol{B} \cdot d\boldsymbol{l}. \tag{3.11}$$

Electrodynamicists referred to Eqs. (3.11) and (3.7) as Faraday's law; this result lends itself to an interpretation that the EMF arises because the circuit cuts the magnet's lines of force. Finally,

$$\varepsilon_2 = (u/c)\,Ba, \tag{3.12}$$

which is identical to the result in Eq. (3.8), since $u = v$.

Since $\varepsilon_1 = \varepsilon_2$, it follows that the current $i_2 \, (= \varepsilon_2/R)$ is identical with i_1. Thus, "assuming equality of relative motion in the two cases discussed [there arises] electric currents of the same path and intensity as those produced by electric forces in the former case." In the second case, however, there is no electric field; rather, there exists only the electromotive force "to which in itself there is no corresponding energy."

In summary, S_1 interprets the motion of charges in the wire loop as caused by an electric field due to the moving magnet. According to S_2, the electromotive force arises because of the force exerted on the charges in the moving wire: there is no electric field in the S_2 reference system. Yet, since only the relative motion between S_1 and S_2 seems to determine the magnitude and path of the observed current, it is curious that there is such a "sharp" distinction between how S_1 and S_2 explain why the charge q moves. Thus, as Einstein pointed out, in these cases there is an asymmetry which "does not appear to be inherent in the phenomena."

At least one of the "examples" Einstein referred to [19–25] concerned the problem of understanding unipolar induction, which he would assert (§6), he had solved. In his following mention of "the unsuccessful attempts to discover any motion of the earth relatively to the 'light medium'...," however, he jumped abruptly to a class of experiments hardly ever connected with fundamental problems of electromagnetic induction, namely those concerned with detecting the earth's motion relative to the ether ('light medium'). The "unsuccessful attempts to discover any motion of the earth relative to the 'light medium,' " some of which were listed in Chapter 1, imply that electrodynamical phenomena occur on the moving earth as if the earth were at rest. In other words (to the accuracy of these experiments), one could not discover by means of an electrodynamical experiment whether or not the earth was in motion with respect to the ether; consequently, to second order in v/c the velocity of light from terrestrial sources was the same as that measured by observers at rest in the ether. Thus, "to the concept of absolute rest there correspond no properties of the phenomena, neither in mechanics, nor in electrodynamics."

But how does the example of magnet and conductor and the "unsuccessful attempts" (presumably Einstein means to first and second order in v/c) to determine motion of the earth relative to the ether "lead to the conjecture that to the concept of absolute rest there correspond no properties of the phenomena ... ?"

The explanation can be found in the next passage, in which Einstein emphasizes a point of particular importance for him, namely, the "unsuccessful attempts," that were accurate "to quantities of the first order." Einstein returned to this point many years later in an interview on 4 February 1950 with R. S. Shankland (1963):

> He [Einstein] continued to say the experimental results which had influenced him most were the observations on stellar aberration and Fizeau's measurements on the speed of light in moving water. "They were enough," he said. I reminded him that Michelson and Morley had made a very accurate determination at Case in 1886 of the Fresnel dragging coefficient with greatly improved techniques and showed him their values as given in my paper. To this he nodded agreement, but when I added that it seemed to me that Fizeau's original result was only qualitative, he shook his pipe and smiled. "Oh it was better than that!"

What could Einstein have meant by "They were enough"? After all, by 1905 he was probably aware of most ether-drift experiments.[7] A quite plausible conjecture of why Einstein claimed that the first-order experiments "were enough" is that, to the best of Einstein's knowledge in 1905, they could be accounted for systematically by means of Lorentz's theorem of corresponding states as Lorentz presented it in his *Versuch*. In his (1907e) Einstein recalled how impressed he was by the mathematical statement of Lorentz's theorem of corresponding states — the hypothesis of the local-time coordinate:

> It required [i.e., that the key to a principle of relativity lay in an understanding of relative simultaneity, and hence in a redefinition of time] only the understanding that one should define as "time" what Lorentz introduced [in the *Versuch*] as an auxiliary quantity [*Hilfsgrösse*] and called the "local time."

3.3. ON THE ROOTS OF EINSTEIN'S VIEW OF ELECTROMAGNETIC INDUCTION

3.3.1. August Föppl and Electromagnetic Induction

Joseph Sauter (1965), a colleague of Einstein at the Patent Office, recalled that "Einstein was admitted to the post [at the Patent Office] without possessing an engineering diploma, but as a physicist *au courant* with Maxwell's theory." But Boltzmann's un-Maxwellian version of Maxwell's theory, or von Helmholtz's (1897) elegant methods for physical optics, or the papers of Hertz and Lorentz on the electrodynamics of moving bodies were hardly required reading for an assessor of patents. Recently, Holton (1973c) has brought to light "an almost forgotten teacher of Einstein"; namely, Föppl, the author of a widely-read text (1894) on the Hertz-Heaviside version of Maxwell's theory. From that text Einstein learned enough Maxwellian electromagnetism to enable him to obtain a position at the Patent Office in June 1902; Holton notes that Föppl's book was "just the kind of book an interested student would want if he were deprived of Maxwell's theory in course lectures."[8] The chapter of interest to us here is Chapter V: "*Die Elektrodynamik bewegter Leiter*" (The Electrodynamics of Moving Conductors), where Föppl analyzed the arrangement of magnet and conductor with which Einstein began the relativity paper.

Before launching into an analysis of the electrodynamics of moving bodies, Föppl was obliged to discuss relative and absolute motions, and what happened to the ether when bodies moved through it. Consistent with his high regard for Mach's views, Föppl considered space filled with ether because empty space was "not at all subject to experience."[9] By "experience" Föppl had in mind a concrete case — Maxwell's theory in which forces are propagated through an ether. For Föppl the ether played another role. Whereas in mechanics, reference systems fixed to the sun were at absolute rest, the ether could assume this role, and therefore absolute motion could be conceived of as motion relative to the ether. Since only relative motion was relevant for mechanics, continued Föppl, the same should be true for electromagnetism, and especially for phenomena where the laws of kinematics were used in conjunction with those of electromagnetism. But to Föppl this was not "a priori settled"; for example, one could not be certain whether the current arising in the circuit was the same for the cases of magnet in motion and resting electric circuit, and vice versa. He proceeded to discuss the case in which magnet and conductor were at relative rest, but in motion with respect to the ether. "Experiment," continued Föppl, without citing data, had not demonstrated any electromagnetic effects in either body. Therefore, the characteristics of the system of magnet and conductor were unchanged, and thus their state of relative rest was indistinguishable from one of absolute rest. This result led Föppl to conclude that only relative motion between magnet and conductor could produce electromagnetic effects.[10]

Nevertheless, Föppl was worried because, in his opinion, the necessity for such a careful investigation meant that perhaps the same conclusion could not be reached in every case.[11] In order to illustrate this point Föppl developed another thought experiment. Consider two charged "material points" moving in parallel paths at equal velocities relative to the ether, i.e., they are at relative rest. Since these charges constitute convection currents, then they are acted upon by each others' magnetic fields. Now, continued Föppl, if we remove all reference bodies so that, as understood within the context of Mach's analysis of mechanics, to a good approximation these charges could be considered to be inertial reference systems. But since the presence of magnetic fields distinguishes their state of relative rest from one of absolute rest, Föppl considered that the basic "axiom of kinematics" was violated here, i.e., that the reciprocal action of two bodies depends only on their relative motion. Föppl concluded that this example shows clearly the difficulties with which present-day electrodynamics must "struggle."

Having alerted the reader to these fundamental problems, Föppl returned to the simple but "significantly important" case of magnet and closed circuit. First, applying the Hertz-Heaviside theory required an analysis of how a moving body drags ether. The principal problem dealt with the transition layer between the moving and resting ether which Hertz had referred to as "surfaces of slip." Föppl avoided mathematical analysis of these surfaces by assuming that electromagnetic quantities varied continuously across them. Briefly, the boundary conditions were generally complicated because, for example, one had to decide how the electric field due to the moving magnet achieved its full value in the resting ether, whereas in the fully dragged portion of the ether it was zero.

Föppl considered first the case of a magnet undergoing motion relative to a conductor; such motion could be linear, rotational, or both. He deduced that the "electric force" (in Hertz's terminology) at the site of the resting conductor was the one in Eq. (3.5). Consistent with emphasis on only observable quantities (as was *de rigueur* in Mach's view and in Hertz's electrodynamics), Föppl stressed the observability of only the line integral of the electric force E_1. On the other hand, in his introductory remarks, Einstein referred to $B \times v/c$ as a field, because he wished to discuss its effect upon a charge q in the circuit, thereby throwing into relief the asymmetry between a field description in this case, and a description using forces in the inverse case.

For the inverse case – of a conductor moving with velocity $-v$ relative to the magnet – Föppl obtained the result in Eq. (3.10), and once again Faraday's law emerged for a closed circuit.

"But it is otherwise," Föppl cautioned, for an open circuit, because here the line integrals in the equations for the EMF are no longer over closed paths, and so the potential terms would contribute. Föppl gave unipolar induction as an example of electromagnetic induction in an open circuit. In Fig. 3.3, his example shows where a conducting wire makes sliding contacts with one pole and with the body of a cylindrical permanent conducting magnet that can rotate about its axis.[12] In 1832, Faraday had found that a current could be induced in the wire if there was relative motion between it and the magnet – for example, the same current was induced in the wire if it rotated counterclockwise relative to the resting magnet with a constant angular velocity ω, or the inverse case. Consider the case in which, when the magnet rotates, there is an

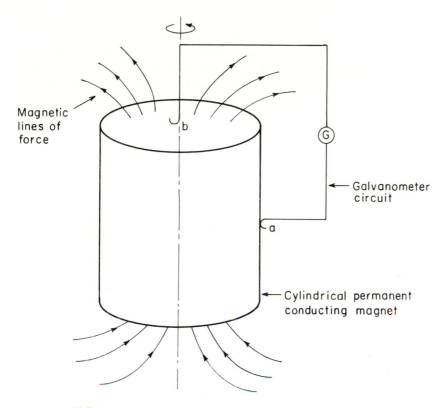

FIG. 3.3. The unipolar induction apparatus described by Föppl.

EMF in the wire of

$$\varepsilon_1 = \int_a^b E_1 \cdot dl = \frac{1}{c} \int_a^b \left(B \times \frac{v}{c} \right) \cdot dl + A \cdot \frac{v}{c} \bigg|_a^b - \psi \bigg|_a^b, \qquad (3.13)$$

where here $v = \omega \times r$. In order to maintain the view that Faraday's law had the same form for an open and a closed circuit, Föppl set $\psi = A \cdot v/c$ and interpreted it as the potential arising from a "simulated" electrical volume distribution of charge of amount $-(\omega/c)(B/2\pi)$, arising from the magnet's state of rotation.[13] Clearly this slight of hand could not be expected to save Faraday's law, and Föppl admitted as much. Rather, he had only demonstrated his faith in the general validity of Faraday's law in the form interpretable as line cutting, just as "Maxwell himself and most of his followers" believed in it. In one of his rare references to the works of others, Föppl cited J. J. Thomson's doubt about the validity of Faraday's law for open circuits. Thomson's detailed 1893 investigation of Faraday's law for open circuits led him to conclude that, in general, the electric force from Maxwell's theory included the term $A \cdot v/c$ but not the additional potential ψ. In fact, realizing the necessity of Faraday's law in differential form [i.e., Eq. (3.1)] for deducing inhomogeneous wave equations for light propagation, Thomson suggested methods for determining the validity of Faraday's law from the results of ether-drift experiments. If Einstein had purused Föppl's few references, he might have been struck by the connection between Faraday's law and the ether-drift experiments.[14]

On the form of the electric field due to a moving magnet, Föppl agreed with Thomson that further experiments were necessary because generally the contribution to the EMF of the term $A \cdot v/c$ was much smaller than that of $B \times v/c$.[15] There was also a major conceptual problem surrounding electromagnetic induction, and Föppl went on to illustrate it using the case of unipolar induction in Fig. 3.3 – that is, where is the site of the EMF? If one held that a magnet's lines of force rotated with it, then the seat of EMF was always in the circuit. The reasoning was as follows: when only the circuit rotated, it cut lines of force; when only the magnet rotated, its lines of force cut the wire circuit; when both rotated together in the same direction, there was no current induced because the circuit and the lines of force were at relative rest. This view was espoused by those who like Föppl professed relativism, because the induced current could be explained quite simply as resulting from relative motion of magnet and conductor, and no asymmetries confounded interpretation. Föppl's relativism demanded the physical existence of lines of force, and he mentioned a well-known reason for their existence; namely, removing a magnet removed also its lines of force, and this could be demonstrated easily with the use of iron filings liberally sprinkled on a piece of paper.

In view of the opening paragraph of the 1905 relativity paper, Einstein had been impressed by Föppl's frank confession of the problems surrounding electromagnetic induction. No other exposition of electromagnetic induction before Föppl's emphasized the interrelatedness of principles of mechanics and electromagnetism. In Lorentz's *Versuch*, Einstein could have read an exposition of electromagnetic induction *ab initio* free of any assumptions concerning ether drag, but problems created by a clash of principles in electromagnetic induction were nowhere discussed in the *Versuch*, nor were they in Föppl's 1894 book.

3.3.2. Unipolar Induction and the Interaction between Engineers and Scientists in German-Speaking Countries

Föppl's analysis of electromagnetic induction, steeped in the interplay between philosophy and physics, and emphasizing the *Gedanken* experiment, could only have made a lasting impression on Einstein. So we can assume that Föppl piqued Einstein's curiosity about the process of unipolar induction – a phenomenon that was much discussed in German physics and engineering journals to which Einstein had access at the ETH in Zurich, and then later at the Patent Office in Bern.[16] While one is examining those portions of unipolar induction that may have been of interest to Einstein, it is useful to remember that he originally considered engineering as a career, was raised in a household where electrical machinery was often discussed because his father was in the electrotechnical business, and his favorite uncle was an engineer; he attended a technologically oriented institute of higher education; and he worked in a Patent Office that often examined patents for electrical dynamos.[17]

The problem of unipolar induction had two related parts: whether Faraday's law could describe the EMF in open circuits, and whether lines of force rotate. By 1851, Faraday had convinced himself that lines of force participated totally in a magnet's translational motion, but not at all in its rotational motion [Faraday (1965), (1932)].[18] This viewpoint required two different explanations for the induced current in unipolar induction, depending upon whether the circuit or the magnet was moving. If only the circuit was in motion, then the current was interpreted as arising from line cutting. In the inverse case, according to Faraday, because of the magnet's turning through its own lines of force, charges of unequal sign appeared on the magnet's poles and equator; the resulting potential difference caused current to flow in the resting wire. Faraday explained that the absence of current when both magnet and wire rotated together in the same direction was due to currents, generated within the magnet, which neutralized the motion of charges in the wire.

The resolution of these problems of unipolar induction was important to geophysics (where the earth was taken as a spherical magnet with the atmosphere playing the role of the wire connecting points on the earth's surface[19]); to fundamental physical theory (where the question was whether Faraday's law could describe electromagnetic induction in open circuits[20]); to the electrical industry (where everyone knew that dynamos worked, there remained fundamental problems concerning their operation). For industry, the important point was that the unipolar dynamo was a true direct-current generator that required no expensive commutator assembly.[21] If only the armature rotated, then Faraday's law with its line-cutting interpretation could be used to calculate the EMF. But in 1895, unipolar machines were being designed which contained toothed rotating-field magnets, capable of inducing alternating current in the stationary armature. Engineers were unsure about how to calculate the reaction of the field magnet on the machine's frame, because they believed they needed to know whether the lines of force swept past the frame or oscillated in space.[22] The engineer C. L. Weber (1895) wrote in the *Elektrotechnische Zeitschrift* that whether lines of force rotate or not constitutes a "fundamental *Anschauung* of immediate importance and technical meaning, and not an academic moot point." Another German scientist Grotrian (1901) interested in applications wrote in the *Annalen der Physik* that "exact knowledge of the seat of EMF in unipolar induction is of importance in electrical technology for the building of unipolar machines."

The term *Anschauung* is important to understanding how problems of unipolar induction were approached within the German-speaking scientific-engineering community. In this context, *Anschauung* refers to the intuition through pictures constructed from previous visualizations of physical processes in the world of perceptions; it is something superior to viewing merely with the senses.[23] In this Kantian philosophical tradition, lines of force represented with iron filings, were assigned wide explanatory powers. German scientists and engineers saw lines of force everywhere, and discussed in highly visual terms the relative merits of Faraday's *Anschauung* versus Preston's *Anschauung* — the rotating-line viewpoint named for the British scientist S. Tolver Preston, who vigorously advocated it in the late 19th century [Preston, (1885a), (1885b), (1891)].[24] For example, Lecher's (1895) paper in the *Annalen der Physik*, reviewed Preston's *Anschauung* and also referred to Preston's (1885a) which emphasized that Faraday's *Anschauung* required two *different causes* for a single effect — namely, the current induced in the conductor. So as far as I know this is the only publication in which anyone before Einstein noticed a redundancy in interpretations of electromagnetic induction. Einstein viewed this redundancy as an asymmetry. If Einstein had read Lecher's paper (perhaps at the Patent Office as a result of a literature search concerning unipolar induction), he might have attempted to peruse the earlier paper of Preston, even though it was in English, because in the same footnote Lecher referred also to the discussion on unipolar induction in Föppl's book.

The close interaction between scientists and engineers in German-speaking countries can be attributed almost entirely to the influence of one man, the great German industrialist and scientist Werner von Siemens, who also designed and built the first unipolar dynamo for industrial purposes [von Siemens (1881)].[25] One of Siemens's philanthropic endeavors was to contribute toward establishing, in 1886 at the ETH, an institute for engineering and science; the director was Siemens's good friend and Einstein's bête noir, Heinrich Weber.

Circles are beginning to close, for Einstein attended the ETH where he learned much about applications of electromagnetism in machine design from Weber, and surely also about unipolar machines; perhaps, in addition, Einstein visited the massive Oerlikon works in Zurich where C. L. Weber was attempting to design alternating-current unipolar dynamos. Clearly, at the ETH Einstein had ample opportunity to observe engineers' attitudes [see also McCormmach, (1976)].[26]

3.3.3. The Approach of Hertz and Abraham to Unipolar Induction

On the highly theoretical side, Einstein had read Hertz's treatment of unipolar induction in the (1890b) paper on electrodynamics. Most likely Einstein read this paper in Hertz's reprint volume, *Electric Waves* (1892), where Hertz wrote that in his opinion "Maxwell's theory is Maxwell's system of equations," and these equations asserted nothing about the movement of lines of force. For Hertz, these lines symbolized only "special conditions of matter," and accordingly there was "no meaning" to the question, whether lines of force rotated. Hertz was convinced that the "laws of unipolar induction" could be "deduced" from his Galilean-covariant system of electrodynamics.

In 1904 the first of many descendants of Föppl's text on electromagnetism appeared – the version rewritten by Max Abraham (1904c). Abraham strove to bring Föppl's book up to date: in the introduction he discussed the advantages of Lorentz's version of the "Maxwell-Hertz theory," especially as applied to problems of light and wireless telegraphy. However, Abraham wrote, just as the study of mechanics should precede that of statistical mechanics, so should the study of Maxwell's theory come before Lorentz's. He promised to treat Lorentz's theory fully in a second volume.

Abraham rewrote Föppl's Chapter V entirely, emphasizing that whereas the "principle of relative motion" held exactly in mechanics, it was valid only to order $v/c = 10^{-4}$ in electrodynamics (where v is the earth's orbital velocity about the sun, 30 km/sec). The "principle of relative motion" was Abraham's terminology for Föppl's basic "axiom of kinematics." Abraham spelled it out in more detail, emphasizing that a uniform translation or rotation left unchanged the mutual electrodynamical phenomena between ponderable bodies. But this principle had empirical support in electrodynamics only to order $v/c = 10^{-4}$. Accordingly, to this order of accuracy, electromagnetic induction depended only on the relative velocity between circuit and magnet. Armed with this principle, Abraham developed electromagnetic induction from the case of circuit in motion and resting magnet, thereby avoiding problems of charge distributions appearing on the magnet – i.e., the source of the moving magnet's external electric field. Thus, for a closed circuit, Abraham obtained the Eq. (3.11), mentioning that this result could be due to the circuit's being cut by the magnet's lines of force. Without further ado, he treated the inverse case simply by replacing u with $-v$. Similarly, from the consistency of Hertz's theory of electromagnetism with the principle of relative motion, Abraham concluded that the EMF for unipolar induction with a permanent conducting magnet should also be calculable from Faraday's law, and was therefore independent of which portion of the system was in motion – that is, so long as the velocity $v = \omega r$ was much less than c.[27] As for lines of force, Abraham emphasized that they were merely a "geometrical illustration" of the magnet's field; his relativism was satisfied by the principle of relative motion. In summary, using this principle, Abraham could side-step Föppl's sleight of hand for saving Faraday's law, but at the expense of treating in detail only the case of circuit in motion.

Abraham concluded his updated version of Föppl's book by comparing the theories of Maxwell-Hertz and Lorentz, thereby providing a transition into his promised second volume. On the experimental side, while both theories were in agreement with the phenomenon of electromagnetic induction, Lorentz's theory agreed better with Eichenwald's 1903 data for the magnetic field set up by a moving dielectric.[28] As for the venerable principles of mechanics, unlike Lorentz's theory, the Maxwell-Hertz theory agreed with both the principle of action and reaction and the principle of relative motion. The Lorentz theory's violation of the principle of action and reaction was built into the theory and so did not entail disagreement with any empirical data. It was the principle of relative motion on which Abraham focused, presenting data which, in his opinion, decided between the theories of Maxwell-Hertz and Lorentz. Did Abraham discuss the interferometer experiment of Michelson and Morley? No. He discussed an experiment accurate to order v/c — Fizeau's experiment — which could not be explained by the Maxwell-Hertz theory, based as it was on a total ether drag. Therefore, Abraham concluded that Fizeau's data violated the principle of relative motion. In the second volume, Abraham demonstrated how Lorentz's theory, although not Galilean-covariant, could nevertheless explain Fizeau's experiment. Abraham's choice of Fizeau's experiment as critical for deciding between these two theories was probably predicated on the fact that both could explain the result of Michelson and Morley; that is, Hertz's theory explained the experiment *ab initio*, and Lorentz could offer the *ad hoc* hypothesis of contraction. Fizeau's experiment, on the other hand, offered a clear-cut case for a decision between the theories of Hertz and Lorentz.

Consistent with his support of an electromagnetic world-picture, Abraham's response to the competition between the laws of mechanics and electromagnetism was that perhaps we should be guided no longer by the principles of mechanics as they were interpreted within "*die alte Mechanik*," but should turn to data from electrical and optical phenomena. Furthermore, Abraham went on to emphasize that Lorentz's theory provided a better reply to certain epistemological problems raised by Föppl. Among such problems were: (1) Lorentz's all-pervasive resting ether could be considered as the sought-after "universal reference system," at absolute rest; (2) as for the three-dimensionality of space, which Föppl maintained because of its consistency with our sensations, Abraham believed that, unless the electromagnetic characteristics of space were clarified further, we would do well to refrain from conjecturing whether physical space was Euclidean or not. For the moment, Abraham suggested neglecting this problem.[29]

For Abraham, the Maxwell-Hertz theory could not even explain currently-known electromagnetic phenomena in the "narrowest sense." Abraham believed that an approach based on the electromagnetic world-picture was the correct one, and his volume concluded:

> The transformation of the fundamental concepts of geometry and mechanics for which the electron theory strives, is one of the greatest importance for all of science (*die gesamte Naturwissenschaft*).

Thus, on the theoretical side, Abraham's analysis may well have further underscored for Einstein the interlocking roles of principles of mechanics and electromagnetism, as well as those of geometry; on the experimental side, Abraham's emphasis on the far-reaching importance of data accurate to first order in v/c — in particular, the experiment of Fizeau — could have further stressed for Einstein the importance of this class of data. And we can well imagine that Abraham's (1903) brilliant use of mathematical symmetry arguments further impressed on Einstein an approach he used often in the relativity paper.

Abraham wrote the "Preface" to the second volume (1905) at Wiesbaden in March 1905, but by this time Einstein was already moving in a direction different from the one Abraham had proposed for confronting the pressing problems of the physics of 1905.

3.3.4. Lorentz's Approach to Electromagnetic Induction: A Symmetry

Lorentz's (1904a) monograph focused on the Maxwell-Hertz theory. Not surprisingly, his treatment of unipolar induction was similar to Abraham's. But Lorentz also introduced a new element — a notion of a certain sort of symmetry, first pointed out by Heaviside and Hertz, between electric and magnetic phenomena. Föppl had interpreted it as a purely geometrical operation, according to which a moving magnet (dielectric) generated an electric (magnetic) force $E = B \times v/c$ ($H = v/c \times D$);[30] and Hertz and Heaviside assumed the validity of Faraday's law. Thus, in the Heaviside-Hertz theory, interchanging D and E with B and H changed the equations for electric effects into those for magnetic effects (with the interchange also of dielectric and magnetic constants characterizing the material in question). Hertz (1890b) wrote that in 1888 Röntgen had, in fact, detected a magnetic field in the vicinity of a moving dielectric; however, the more precise 1903 experiment of Eichenwald disagreed with Hertz's prediction for the magnitude of the magnetic field, agreeing instead with Lorentz's prediction. Seeking to view unipolar induction within a wider framework, Lorentz wrote in the (1904a) monograph that, if Röntgen had used a permanent dielectric, then his experiment would have been the exact analog of unipolar induction.[31]

In a sequel monograph, "*Weiterbildung der Maxwellschen Theorie. Elektronentheorie.*" (Further Development of Maxwell's Theory. Theory of Electrons.), Lorentz discussed the source of the electric field due to a moving magnet. In Lorentz's theory there are three sorts of electrons (1904b): conduction electrons that constitute an electric current; positively and negatively charged electrons whose relative displacements produce dipole moments — these are polarization electrons; magnetization electrons whose rotation about an atom produced a magnetic moment M. Briefly, Lorentz developed the external electric field of a permanent conducting magnet in inertial motion as follows. He applied the modified Galilean transformations that contain the local time, i.e., Eqs. (1.76)–(1.79), to the time derivative of the vector potential from a magnetization electron. Lorentz found that the magnetization electrons contribute to the moving magnet's macroscopic electric field the macroscopic dipole moment

$$P = (v/c) \times M, \tag{3.14}$$

where a subscript r on M is unnecessary because in (1904b) Lorentz took $M_r = M$. The contribution of the conduction electrons to the electric field in S_r is $-\nabla_r \phi_r - (1/c)(\partial A_r / \partial t_L) + \nabla_r (A_r \cdot v/c)$ where A_r is the vector potential from a conduction electron which contributes a current ρu (see Chapter 3, footnote 10). Lorentz demonstrated that the term $\nabla_r (A_r \cdot v/c)$ contributes to the macroscopic electric field within the magnet the amount $-P$. Consequently, to order v/c, the total macroscopic effective force on the magnet's conduction electrons is of the same form as the Lorentz force on an individual free electron. The macroscopic force must vanish inside of the conducting magnet, which yields the moving magnet's external electric field to be

$$E = B \times v/c. \tag{3.15}$$

This example, Lorentz continued, illustrated how in certain phenomena dielectric and magnetization electrons had become intermingled. Although he provided no details, Lorentz claimed that unipolar induction with a rotating magnet could be further clarified by means of Eq. (3.14). Lorentz went on to emphasize that this result could also have been obtained without consideration of the dipole moment from the magnetization electrons simply by asserting that the magnet's interior is in an equilibrium state when the magnet is in inertial motion; consequently, from Eq. (1.84), $E_r = 0$ inside the magnet and $E = B \times v/c$ is the magnet's external electric field. In fact, continued Lorentz, this must be the case for it is consistent with Faraday's law for the EMF in a neighboring closed circuit at rest in the ether [see Eq. (3.7)]. On the other hand, wrote Lorentz (1904b), one could not in general ignore the term $\nabla_r (A_r \cdot v/c)$ because it was necessary for recovering Budde's (1880) result that agreed with observation [see Chapter 3, footnote 10].

A scientist struck by disturbing dualities could only have been shocked by the addition of another hypothesis for excluding a theoretically predicted effect of motion relative to the ether on electromagnetic phenomena, i.e., the compensation charge, and the profusion of postulated quantities necessary for discussing an effect (the current in electromagnetic induction) dependent upon only the relative velocity between magnet and conductor. On the other hand, a scientist fascinated by symmetries might well have been attracted to what Lorentz (1904a) described as the "duality between electric and magnetic phenomena." In his (1904b) Lorentz referred to this relationship as a "parallelism" because in his opinion it was weaker than the "duality" in the Hertz theory. Lorentz's reason was that in his theory moving magnetic (dielectric) matter generated a dielectric (magnetic) moment of $P = v/c \times M$

($M = P \times v/c$). The source of the difference between the two theories was that Lorentz's had an atomistic basis, which precluded symmetry among symbols representing directly observable quantities. Moreover, had Einstein read Lorentz's (1904b) it could not have escaped his attention that the local time was the mathematical source of Lorentz's "parallelism."

3.3.5. A Possible Scenario for Einstein's Thoughts on Electromagnetic Induction

Had Einstein read Lorentz's authoritative monographs (1904a, b) he would have been impressed with the local time coordinate that could account for first-order ether-drift experiments as Lorentz had shown in (1895), as well as for Lorentz's "parallelism" between electric and magnetic phenomena. If the local time was part of a coordinate transformation treating only inertial motion, then why not restrict considerations to inertial reference systems?

Furthermore, everyone seemed to agree on the validity of Faraday's law for the example of magnet and closed conductor in relative inertial motion; in inertial systems the principle of relative motion had its sharpest form, stating not only that all phenomena were dependent upon only the relative motion of ponderable bodies, but also that the laws of physical theory remained the same in these reference systems. Although Abraham did not emphasize this point in his book, Poincaré did so in *Science and Hypothesis* where he stated this principle as: "The movement of any system whatever ought to obey the same laws; whether it is referred to fixed axes or to the moveable axes which are implied in uniform motion in a straight line." Yet applying the principle of relative motion (within its domain of validity for electromagnetic theory) to this simple case of electromagnetic induction revealed to Einstein an astonishing situation regarding Lorentz's theory — that from the standpoint of an observer on the magnet, there is no displacement of real charges, as required by the principle of relative motion; for an observer on the circuit, there is no motion of the magnet's electrons other than their translation. Then how could the moving magnet be the source of the electric field that caused current to flow in the resting circuit? Föppl had avoided this problem by treating in detail only the effects due to a rotating magnet in which a real-volume charge distribution arises. Abraham had dodged this problem by considering first the case of conductor in motion, and then using the principle of relative motion to discuss the inverse case, all the while restricting himself to a closed conducting circuit. Abraham chose to ignore study of the source of the electric force arising from a permanent conducting magnet in inertial motion. Föppl, in particular, adroitly eluded a related problem which he already had the formalism to approach — that is, why current flows in a linear unipolar machine (Fig. 3.4); for here, too, relative to an observer on the magnet there is no displacement of real electricity, and yet current flows in the wire when there is relative motion. Faraday's explanation concerned line cutting, and Föppl could agree. Lorentz

(1904b) also avoided a detailed treatment of the linear unipolar dynamo even though he was in possession of an atomistically-based theory (Fig. 3.4).

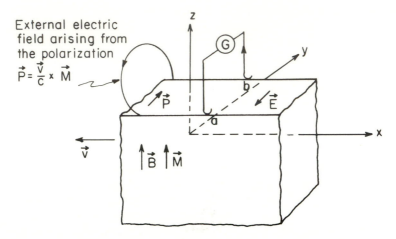

FIG. 3.4. A permanent conducting magnet moving in the direction of the $-x$-axis. The magnet's dimensions are taken to be much greater in the x-z-plane than in the y direction. The galvanometer circuit is stationary in the laboratory and makes sliding contacts with the moving magnet at a and b. The galvanometer indicates that current is flowing in the stationary wire from b to a. Lorentz's (1910) explanation for why current flows in the circuit is presented in Chapter 3, footnote 33.

Would Einstein at this time have decided to consider the class of problems concerning magnet and conductor in relative inertial motion? Let us assume that he had, and see what he found. We can imagine Einstein going one step beyond Lorentz – that is, to consider the process of electromagnetic induction from the microscopic viewpoint of the circuit, and in 1919 he wrote that this is what he did. Einstein asked himself: How would an observer (S_2) on the magnet in Fig. 3.1 explain the motion of a charge in the circuit moving with relative velocity $v = -u$? Quite simply, because S_2 could take Eq. (3.15) as an assured result since it could be deduced from either applying Lorentz's local time to the potential representation of the electric field, or from the theorem of corresponding states that concerns only the fields, i.e., Eq. (1.84). Then S_2 could set $v = -u$ in Eq. (3.15) and interpret the charge's motion as caused by an electric force $qu/c \times B$.

Thus, as Einstein wrote in the relativity paper, "in the conductor ... we find an electromotive force [$u/c \times B$] to which in itself there is no corresponding energy." What happened to the electric field observed by S_1 in Fig. 3.2? Besides being asymmetrical, as Einstein wrote in the relativity paper, the points of view of S_1 and S_2 could also have seemed to him unnecessarily complicated and redundant. In fact, Lorentz himself came to this conclusion in 1910 when he analyzed an example equivalent to a linear unipolar dynamo (1910).[33] Furthermore, Lorentz's theorem of corresponding states was based on the dynamical interaction of three sorts of electrons. To make matters more complicated, by 1905 Einstein had convinced himself that neither electrodynamics nor mechanics was sufficient to explore the structure of matter. No wonder, as Einstein recalled in 1919, the "phenomenon of the electromagnetic induction forced me to postulate the (special) relativity principle."

Consequently in the relativity paper, Einstein chose as a first approximation to focus on the problem of the interrelatedness of viewpoints, rather than on the source of the external electric field of a moving magnet – that is, he chose to be guided by a theory of principle. The principle had to be the one from mechanics which had been emphasized by Föppl, Abraham, and Poincaré and which was indispensable for a treatment of electromagnetic induction – the principle of relative motion.

In an illuminating recollection in 1952, as reported by R. S. Shankland (one that meshes with the unpublished essay of 1919), Einstein asserted [Shankland (1964)][34]:

> What led me more or less directly to the special theory of relativity was the conviction that the electromotive force $[u/c \times B]$ acting on a body in motion in a magnetic field was nothing else but an electric field.

Indeed, as Einstein wrote in the 1919 essay, his struggle over electromagnetic induction had convinced him that "the existence of an electric field was therefore a relative one, depending upon the state of motion of the coordinate system being used"; in §6 of the relativity paper, Einstein deduced this result using space and time transformations that incorporated a totally new concept in physical theory – the relativity of time.

Hence, beginning the relativity paper with the example of magnet and closed conducting circuit in relative inertial motion was one of Einstein's master strokes. He chose to view electromagnetic induction (i.e., "examples of this sort") as a clash of principles that occurred nowhere else. Einstein had learned well Föppl's lesson that electromagnetic induction involved both mechanics and electromagnetism. Yet while the principle of relative motion was exact in mechanics, when it was applied only to inertial systems in electromagnetism, Lorentz's principle led to "asymmetries" and was based upon a profusion of postulated quantities. Since electromagnetic induction did not, to order v/c, exhibit any effects other than the "reciprocal electrodynamic action of magnet and conductor," Einstein felt free first to "conjecture that to the concept of absolute rest there correspond no properties of the phenomena, neither in mechanics nor in electrodynamics," although that was expected for the laws of mechanics in inertial reference systems. Then, in the second-half of this long and dazzling sentence, Einstein focused on the optical ether-drift experiments, accurate to order v/c, according to which optical phenomena occurred on the moving earth as if the earth were at absolute rest. On the basis of these two observations Einstein could broaden his "conjecture" to the effect "for every reference system in which the laws of mechanics are valid, the laws of electrodynamics and optics [two disciplines which Lorentz's theory had supposedly unified] are also valid."

Considering the case of electromagnetic induction within the context of a theory of principle therefore enabled Einstein to cut directly to the heart of a basic problem of current science and technology – the tension or incompatibility of mechanics and electromagnetism. As a first step toward resolving that problem, Einstein did not attempt to reduce mechanics to an atomistically-based theory of electromagnetism, but, rather, proposed the "conjecture" that, to order v/c, both disciplines obeyed a principle of relative motion applied to inertial reference systems.

In these deliberations the first-order experiments "were enough" for Einstein because not only were they the ones stressed by Abraham and Lorentz, but in addition they could be explained by Lorentz's local time, which was also the root of Lorentz's "parallelism." Since the local time concerned viewpoints just as electromagnetic induction had, then possibly Lorentz's parallelism was not merely some sort of substitution symmetry, but concerned relating information between observers in two different inertial systems; Einstein returned to this point in §6 of the relativity paper. In a moment I shall discuss another "datum" that impressed Einstein with the fundamental importance of Lorentz's local time.

3.4. THE BASIS OF THE NEW VIEWPOINT

The tempo now quickens in the relativity paper. After conjecturing that "to the concept of absolute rest there correspond no properties of the phenomena, neither in mechanics, nor in electrodynamics," Einstein next extended this conjecture to cover empirical results accurate "to quantities of first order according to which for every reference system in which the laws of mechanics are valid, the laws of electrodynamics and optics are also valid." In taking a further, third step, Einstein then parted company with his contemporaries [26–30]:

> We will raise this conjecture (whose intent will from now on be referred to as the "Principle of Relativity") to the status of a postulate, and also introduce another postulate, which is only apparently irreconcilable with the former: light is always propagated in empty space with a definite velocity c which is independent of the state of motion of the emitting body.

It is a breathtaking sentence. In the first part of it, Einstein elevated to a *postulate*, or *axiom*, what previously was merely the conjecture that "for every reference system in which the laws of mechanics are valid, the laws of electrodynamics and optics are also valid." This postulate he called the "Principle of Relativity." It is a generalization of the principle of relative motion and refers here to the laws of electrodynamics, optics, and mechanics for such processes as occur *only* in inertial reference systems. Thus, Einstein raised the conjecture from the previous sentence, which was exact to order v/c, to a statement exact to all orders in v/c. But Einstein did not reveal how the asymmetry in electromagnetic induction was to be removed until the second half of the paper. Einstein chose to defer until §2 the exact statement of the principle of relativity [§2, 5-8]:

> The laws by which the states of physical systems undergo changes are independent of whether these changes of state are referred to one or the other of two coordinate systems moving relatively to each other in uniform translational motion.

Thus, Einstein boldly enlarged Newton's principle of relativity to the statement that neither mechanical nor electromagnetic nor optical experiments could reveal the motion of an inertial reference system.

According to Einstein's second postulate the velocity of light in empty space is a constant c in every inertial reference system and is independent "of the state of motion of the emitting body." This statement violated Lorentz's formulation of electrodynamics where it held axiomatically only in the ether-fixed reference system; this was perhaps one of the reasons for Einstein's emphasis that it was "only apparently irreconcilable" with the principle of relativity. Einstein provided no mention of experimental results supporting this postulate; in fact, evidence existed only to order $(v/c)^2$. Consequently this postulate was the one that drew the most criticism on both physical and epistemological grounds.

He continued [30-32]:

> These two postulates suffice in order to obtain a simple and consistent theory of the electrodynamics of moving bodies taking as a basis Maxwell's theory for bodies at rest.

Therefore, the two postulates do not lead *immediately* to a "simple and consistent theory of the electrodynamics of moving bodies"; rather, they lead merely to the "attainment " (i.e., by degrees) of such a theory. This theory, Einstein predicted, would be "based upon Maxwell's theory for bodies at rest." Electrodynamical processes would be considered from the standpoint of a reference system moving along with the charged body, i.e., the charged body's rest system. Then, by means of an as yet unspecified set of coordinate transformation equations that are a consequence of the two postulates, the Maxwell-Lorentz equations are transformed from the body's rest system to an inertial system in relative motion with respect to the charged body.

The axiomatic status of Einstein's two postulates of relativity placed them outside the scope of direct experimental observation; however, in the second half of the paper he deduced from them theorems making empirical predictions, e.g., the electron's transverse mass. Here Einstein's sympathy for the neo-Kantian viewpoint can be seen emerging, thereby permitting him to propose provisional axioms obtained somehow from "experiences." This approach was closer to the views of Hertz and Poincaré than to those of Mach, with the novel ingredient that data could mean also the results of the *Gedanken* experimenter.[35]

These considerations led Einstein next to the statement [32–36]:

> The introduction of a "luminiferous ether" will prove to be superfluous because the view here to be developed will introduce neither an "absolutely resting space" provided with special properties, nor associate a velocity-vector to a point of the empty space in which electromagnetic processes occur.

Thus, the two postulates lead not only to the "attainment" of a theory of the electrodynamics of moving bodies, but also to the result that a "luminiferous ether" was "superfluous," because Einstein's view of physical theory did not require a physical entity devoid of observable consequences. Here Einstein went beyond Mach.

Mach had never categorically rejected the ether nor declared it as superfluous, but rather, had adopted a wait-and-see attitude to questions on its reality. After all, the experiments that failed to detect the ether were accurate only to order $(v/c)^2$. Poincaré, on the other hand, believed that the ether was real because it established relationships among sensations, and this criterion of physical reality was central to his epistemology. He offered, in *Science and Hypothesis*, several other reasons for the necessity of an ether: to support electromagnetic disturbances in transit; to ensure the continuity of physical phenomena so that they may be described by differential equations instead of difference equations; if effects of the earth's motion on optical phenomena were found, then these motions would be relative to an ether instead of to an absolute space; by means of unknown mechanisms of compensation, the ether reacts back upon a moving emitter of unidirectional radiation in order to maintain the principle of action and reaction in Lorentz's theory. The last reason led Poincaré to suggest a mechanical experiment for detecting effects of the ether; therefore, even though he believed that optical experiments would be unaffected by the earth's motion, he was still able to write that "the ether is all but within our grasp." All but the last of Poincaré's reasons for the necessity of an ether were standard fare for 19th-century electrodynamicists. Yet any of the ethers contained in contemporaneous electromagnetic theories seemed to Einstein as encumbering physical theory. Most physicists agreed that Hertz's ether disagreed with optical phenomena; to Einstein, Lorentz's ether introduced asymmetries "not inherent in the phenomena." One of the asymmetries concerned the interpretation of electromagnetic induction. In addition, in the light-quantum paper, Einstein had demonstrated the superfluousness of Lorentz's ether for problems concerning black-body radiation.

By the "special properties" of an "absolutely resting space," Einstein could have meant the states of motion or of absolute rest of Hertz's or Lorentz's ether respectively; or perhaps also the "electromagnetic characteristics" of an ether-filled space could lead to speculations on the Euclideanness of physical space, which Abraham suggested were best left aside for the moment.

Einstein apparently made no mention of this point again until 1920 in his address at Leiden, "Relativity and the Ether":

> Mach did try to avoid the necessity of postulating an imperceptible real entity, by substituting in mechanics a mean velocity with respect to the totality of masses in the world for acceleration with respect to absolute space.

In Chapter 2 (footnote 15) I discussed just this point in Mach's *Science of Mechanics*. Thus, in 1905 Einstein declared as superfluous not only Lorentz's ether but also one of Mach's suggestions for avoiding the concept of absolute space – another point that doubtless did not escape Mach. Indeed, Einstein's *a priori* declarations of the postulates of relativity already indicated that he had gone beyond Mach.

Einstein carefully drew a distinction between "the view here to be developed" and the one to which he referred in the opening paragraph as "usually understood." Einstein probably did this to emphasize that he would not reformulate the Maxwell-Lorentz electrodynamics, but rather reinterpret it. For example, according to the view about to be developed, Faraday's induction law was not incorrect, but was misinterpreted.

3.5. FURTHER COMMENTARY ON THE ORIGINS OF EINSTEIN'S PRINCIPLE OF RELATIVITY

From 1900–1904 Einstein, as a result of studying Planck's radiation law, had discovered the limited applicability of mechanics and electrodynamics. On this period he wrote in the "Autobiographical Notes":

> By and by I despaired of the possibility of discovering the true laws by means of constructive efforts based on known facts. The longer and the more despairingly I tried, the more I came to the conviction that only the discovery of a universal formal principle could lead us to assured results. The example I saw before me was thermodynamics. The general principle was there given in the theorem: the laws of nature are such that it is impossible to construct a *perpetuum mobile* (of the first and second kind).

Although he may have already realized the limits of classical thermodynamics, its thrust as a theory of principle offered a guide toward a general principle embracing every branch of physical theory.

Eintein continued in his "Autobiographical Notes": "How, then, could such a universal principle be found?" In reply he described an important consideration toward its discovery. Was the reply couched in the form of a discussion of the unsuccessful first- and second-order experiments to determine the earth's motion relative to ether, e.g., as Lorentz had done in the (1895) *Versuch*? No. Or was this general principle obtained from an elegant epistemological discussion of these experiments, as Poincaré had done in *Science and Hypothesis*? No. Instead we find the description of a *Gedanken* experiment which Einstein thought of at age sixteen while attending the Kanton Schule at Aarau:

> After ten years of reflection such a principle resulted from a paradox upon which I had already hit at the age of sixteen: If I pursue a beam of light with the velocity c (velocity of light in a vacuum), I should observe such a beam of light as a spatially oscillatory electromagnetic field at rest. However, there seems to be no such thing, whether on the basis of experience or according to Maxwell's equations. From the very beginning it appeared to me intuitively clear that, indeed from the standpoint of such an observer, everything would have to happen according to the same laws as for an observer who, relative to the earth, was at rest. For how, otherwise, should the first observer know, i.e., be able to determine, that he is in a state of fast uniform motion?
>
> One sees that in this paradox the germ of the special relativity theory is already contained. Today everyone knows, of course, that all attempts to clarify this paradox satisfactorily were condemned to failure as long as the axiom of the absolute character of time, viz., of simultaneity, unrecognizedly was anchored in the unconscious. Clearly to recognize this axiom and its arbitrary character really implies already the solution of the problem. The type of critical reasoning which was required for the discovery of this central point was decisively furthered, in my case, especially by the reading of David Hume's and Ernst Mach's philosophical writings.

To explore why this *Gedanken* experiment posed a paradox, it is necessary to fill in Einstein's line of argument. Consider an observer on a moving platform who is trying to catch a point on an electromagnetic wave (labeled A) whose source is at rest in the ether. Consider also that this observer is performing optical experiments in which light rays — whose sources were at rest in the ether — move parallel to A. As the observer approaches the point on A, different results are obtained for the optical experiments; for example, interference fringes shift as light rays in the observer's apparatus require more time to interfere with one another. Finally, when the observer catches up with the point on A, a standing electromagnetic wave is observed, and light rays no longer reflect from the mirrors at rest on the moving platform. According to

Maxwell's equations written relative to the ether, there is no such entity as a standing electromagnetic wave. However, ether-based theories of electromagnetism did not preclude an inertial observer's catching up with a light wave and riding beside it at zero relative velocity [see Eq. (1.13)]. Yet, to order $(v/c)^2$, the velocity of light on the moving earth equaled that predicted relative to an ether-fixed reference system by ether-based theories. At sixteen, Einstein must have known of some, or perhaps all, of the famous ether-drift experiments, thus accounting for the comment: "However, there seems to be no such thing ... on the basis of experience ...." As for the latter portion of Einstein's sentence ("or according to Maxwell's equations"), we know that by the period 1895–1896 Einstein was already somewhat familiar with Maxwell's equations and had encountered Hertz's works. In 1894 or 1895, he had sent an essay to his uncle Caesar Koch entitled "*Über die Untersuchung des Ätherzustandes im magnetischen Felde*," which proposed a method for detecting elastic deformations of the ether by sending light rays into the vicinity of a current-carrying wire.

According to Maxwell's equations expressed in a reference system fixed in the ether, the velocity of light is a constant and is independent of the emitter's motion. Although Einstein recalled that it was "intuitively clear" that observers in the ether and on the moving platform should have the "same laws" of physics, this intuition does not explain his later strategy. The validity of a principle of relative motion was intuitively clear also to Poincaré who adopted a different method for dealing with the results of the ether-drift experiments. Furthermore, the principle Einstein discussed here obtains for both inertial and noninertial motions. Consequently, the paradox for Einstein was that electromagnetic theory predicted occurrences which to order $(v/c)^2$ could not be observed.

Lorentz and Poincaré were also unknowingly working toward resolution of this paradox, one order of magnitude at a time. In the *Versuch*, Lorentz used the local time to explain systematically why, to first order in v/c, the velocity of light on the moving earth had the same value as that measured by an observer in the ether. Then he explained the Michelson-Morley experiment, accurate to $(v/c)^2$, by postulating, in an *ad hoc* manner, the hypothesis of contraction. Poincaré's famous criticism in 1900 of this hypothesis – "An explanation was necessary, and was forthcoming; they always are; hypotheses are what we lack least" – may have reinforced Einstein's notion that the first-order experiments "were enough."

Whereas in 1895 it was "intuitively clear" to Einstein that a principle of relative motion should apply to the paradox, by 1905 his considerations on electromagnetic induction made this point abundantly clear. As Holton (1973e) has pointed out, the paradox and the problem of electromagnetic induction are related because they both concern how different observers view electromagnetic phenomena. Perhaps Einstein's reason for emphasizing the *Gedanken* experiment from Aarau was that the observer on the moving platform could perform all possible ether-drift experiments and any positive result would contradict the principle of relative motion; but to order $(v/c)^2$ no positive results had been obtained. Having in mind the principles of thermodynamics, Einstein realized that he could resolve this paradox in one stroke by promoting a principle of relative motion into an axiom, thereby reducing the *Gedanken* experiment to a mere fairy tale. Support for this bold move came from his work on electromagnetic induction, in particular, his cutting through problems of the structure of matter to view this process as an arena for a clash of principles. Thus, Einstein courageously moved counter to the prevailing currents of theoretical physics by resolving problems in a Gordian manner; namely, by formulating a view in which certain problems became nonexistent for phenomena that occurred in inertial reference systems. Einstein's two axioms of relativity theory do not explain the failure of the ether-drift experiments or, equivalently, why the measured velocity of light always turns out to be c, and why one cannot catch up with a light wave. Rather, it is axiomatic that the space of every inertial reference system is homogeneous and isotropic for the propagation of light. Einstein makes use of this property of space many times in the relativity paper.

In the passage where Einstein discussed the *Gedanken* experiment from Aarau, he acknowledged that Mach and Hume had given him a clue to the "type of critical reasoning" required for its clarification. Mach's "incorruptible skepticism" gave Einstein the support to question current methods of theoretical physics, and it is reasonable to conjecture that Hume's conclusion that exact knowledge could not be induced directly from sense experiences prevented Einstein from lapsing into the "dogmatic slumber" characteristic of the other physicists around 1905.[36] Although Einstein was not in full agreement with Kant's reply to Hume's message, he moved in a similar direction. Einstein elevated two principles of physics above the realm of sense-perceptions to Kantian-like categories, or concepts, which could function as organizing principles for the sense-data of the laboratory or *Gedanken* experimenter. In Einstein's neo-Kantian framework, however, the concepts or categories were not unalterable. It is appropriate to postpone until the next chapter how Einstein was driven to bring into play new notions of simultaneity and of time. Suffice it to say here that once again the local time appeared, because the paradox involved relating electric and magnetic fields between different points of view.

Thus far in the works of others, e.g., Abraham and Poincaré, only a principle of relative motion has been mentioned, but not a principle of relativity. Why might Einstein have used this name instead of the principle of relative motion? It turns out that the term "relativity of position" had been used in analyzing the foundations of geometry, in particular, concerning the possibility of testing geometry by empirical spatial measurements. In one such analysis, Bertrand Russell (1897) claimed the equivalence of the "homogeneity of space" (i.e., unless acted on by physical forces a body's shape remained unchanged when the body was moved from one place to another) with the "relativity of position" (i.e., the equivalence of all points in space). Although Einstein probably had not read Russell's monograph before 1905, he had read a portion of Poincaré's well-known criticism of it in *Science and Hypothesis*.[37] In this analysis, Poincaré demonstrated, as was his opinion, the meaninglessness of testing the nature of physical space. For, according to Poincaré, experiments related the positions of bodies to one another, and did not discuss the relations of bodies and space; thus, data could be interpreted according to any geometry, though it was more convenient to use Euclidean geometry. Because a discussion of the relationship of bodies to one another brings into play the laws of rigid-body mechanics, Poincaré extended the notion of the relativity of space from geometry into mechanics, where he referred to it as the "law of relativity." This law asserts that the state of bodies at any time depends only upon their relative positions and relative velocities at some initial time. The version of the law of relativity for inertial systems was Poincaré's principle of relative motion, and he was emphatic as to its validity. "The contrary hypothesis is singularly repugnant to the mind,"[38] wrote Poincaré, because this principle was a development of prescientific knowledge, having its basis in our early notion of the meaningfulness of relative motions only. Poincaré considered the relativity of position, the law of relativity, and the principle of relative motion to be conventions.

In the concluding chapter of *Science and Hypothesis*, Poincaré reviewed the history of electrodynamics, emphasizing certain "remarkable" phenomena such as unipolar induction. He wrote of a "principle of relativity," setting the stage for mention of the principle as follows. Although he praised Lorentz's theory for being "attractive," he was still hesitant to throw his full support behind it, because the theory "seemed" to contain absolute velocities, thereby violating the "relativity of space," a term he used interchangeably with the law of relativity. The context of Poincaré's discussion of this point might have reminded Einstein of a similar one in Föppl's book, for Poincaré analyzed a *Gedanken* experiment involving two charged conductors of like charge, in linear motion, but at relative rest. Since they constitute convection currents, then the conductors should attract one another. If the attraction could be measured, then their "absolute velocity" could be determined. Poincaré continued: " 'No!' replied the partisans of Lorentz. 'What we could measure in that way is not their absolute velocity, but their relative velocity with respect to the ether, so that the principle of relativity is safe.' " This was, of course, one of the standard reasons given for an ether, but Poincaré used it also in support of extending the principle of relative motion from mechanics into "physical theory" – that is, into electromagnetic theory, with particular reference to Lorentz's theory. But consistent with his emphasis on empirical data Poincaré was forced to postpone discussion of this point because recent measurements by Victor Crémieu disagreed with Rowland's determination of the convection current thereby making questionable this key concept of Lorentz's theory, and in fact most other electromagnetic theories. I have discussed the Crémieu episode in my (1973). By 1903 it turned out that Crémieu's measurements were incorrect.

Although Einstein may have disagreed with the liberties that Poincaré took with the principle of relativity, nevertheless Poincaré's extension of the principle of relative motion to cover electrodynamical phenomena, without discussing reductionistic possibilities, could have impressed Einstein enough by 1905 to adopt Poincaré's nomenclature for one of the postulates of the new "view."

This section would not be complete without mentioning that, unknown to Einstein in 1905, after the appearance of Lorentz's theory of the electron in May 1904, Poincaré had become a partisan of Lorentz's theory. He referred to Lorentz's theorem of corresponding states as a principle of relativity for the physical sciences that included only the relative motion of ponderable bodies (Section 1.14).[39] Although Poincaré's principle of relativity is stated in a manner similar to Einstein's, the difference in content is sharp. The critical difference is that Poincaré's principle admits the existence of an ether, and so considers the velocity of light to be exactly c only when it is measured in coordinate systems at rest in the ether. In inertial reference systems, the velocity of light is c and is independent of the emitter's motion as a result of certain compensatory effects such as the mathematical local time and the hypothesis of an unobservable contraction. Consequently, Poincaré's extension of the principle of relative motion into the dynamics of electrons was a statement whose mathematical basis resided in electromagnetic theory, and not in mechanics; it was in agreement with Abraham's view to reinterpret principles of the old mechanics. Furthermore, Poincaré considered his principle of relativity as neither an axiom nor a convention, because it lacked sufficient empirical support (e.g., additional ether-drift experiments were necessary).

Three weeks before Einstein submitted the relativity paper for publication, a brief version of Poincaré's reformulation of Lorentz's theory of the electron appeared in the *Comptes rendus*. "*Sur la dynamique de l'électron*" shows that, more than anyone else, Poincaré came closest to rendering electrodynamics consistent, but not to a relativity theory.

3.6. FURTHER ASYMMETRIES

In the first sentence of the relativity paper Einstein used the plural form asymmetries. The case of magnet and conductor revealed one asymmetry. In his first invited lecture at Salzburg (1909b), Einstein discussed another: Lorentz's ether had contained a preferred set of reference systems relative to which the velocity of light was exactly c and independent of the emitter's motion. Einstein's reasoning (1909b) was probably similar to what he had used in 1905: Lorentz's theory postulates a preferred role to a reference system K at rest in the ether over an inertial system K', where K and K' are related by a coordinate transformation similar to the Galilean one; so could we not consider the case in which the roles of K and K' are interchanged? Einstein considered it as "unnatural" to distinguish between K and K'. "Thus it follows," continued Einstein, "that one can obtain a satisfactory theory only by rejecting the ether hypothesis."[40] Accordingly, rejection of Lorentz's ether (or, as in 1905, regarding it as superfluous) resolved an asymmetry that Einstein considered to be built artificially into Lorentz's electrodynamics. Since K and K' were now equivalent inertial systems, Einstein regarded them as conforming to a version of the principle of relative motion. However, according to the principle of relative motion from mechanics, an observer in K' measures the velocity of light V emitted from a source moving with velocity v relative to K' as $V = c + v$; but, from ether-drift experiments accurate to order $(v/c)^2$, $V = c$, not $c + v$. Lorentz's attempts to remove this disagreement were constructive: the hypothesis of the local-time coordinate and the hypothesis of contraction. Both hypotheses were based upon other underlying assumptions concerning how molecular forces were altered as a result of the dynamical interactions of electrons with the ether. On the other hand, Einstein's axiomatic approach declared it to be a foregone conclusion that $V = c$ always, and furthermore was independent of the emitter's motion. Hence, besides removing another asymmetry not "inherent in the phenomena," this axiom is part of the Gordian means for resolving the paradox from Aarau. In §5 of the relativity paper, Einstein derived a new law of velocities which has the limiting case of $V = c$ resulting from the relativity of simultaneity.

Einstein went on, at Salzburg in 1909, to say that the principle of relativity and the principle of the constancy of the velocity of light led to the "so-called relativity theory." Einstein would have preferred to call his theory exactly the opposite: "*Invariantentheorie*." In fact, it was Planck who happened upon "*Relativtheorie*" when in (1906b) he was groping for a name to distinguish the Lorentz-Einstein theory from Abraham's; Planck referred to Abraham's theory as "*Kugeltheorie*" (sphere theory).[41]

Despite these new elements in his theory, Einstein often emphasized its continuity with Newton's mechanics. For example, we find in the opening section of his paper "Foundations of the General Theory of Relativity" (1916) the statement that "the special theory of relativity does not depart from classical mechanics through the postulate of relativity, but through the postulate of the constancy of light." In 1905 the accepted course of action was to abandon mechanics altogether in the quest for the unified field theory promised by an electromagnetic world-picture.[42]

3.7. FURTHER HEURISTICS: ANOTHER ROLE FOR THE PRINCIPLE OF RELATIVITY

For Einstein the principle of relativity, in addition to its role as a restrictive principle, played still another role as a heuristic principle. He first discussed it in this sense in a reply to Paul Ehrenfest, who had misunderstood Einstein's intent in the relativity paper. Ehrenfest (1907) wrote, that the solution to problems concerning the stability of the deformable electron (such as the one raised by Abraham) should be obtainable from the "Lorentz relativity-electrodynamics" by pure deduction, thus implying that Einstein was able to frame Lorentz's electrodynamics as a closed logical system. To which Einstein (1907c) replied:

> The principle of relativity or, more precisely, the principle of relativity together with the principle of the constancy of the velocity of light, is not to be interpreted as a "closed system," not really as a system at all, but rather merely as a heuristic principle which considered by itself, contains only statements about rigid bodies, clocks and light signals. Anything beyond that that the theory of relativity supplies is in the connections it requires between laws that would otherwise appear to be independent of one another.

In subsequent years Einstein made similar statements. For example, in the chapter "The Heuristic Value of the Theory of Relativity," in his book *Relativity, the Special and General Theory* (1917a), Einstein wrote:

> General laws of nature are co-variant with respect to Lorentz transformations.

> This is a definite mathematical condition that the theory of relativity demands of a natural law, and in virtue of this, the theory becomes a valuable heuristic aid in the search for general laws of nature. If a general law of nature were to be found which did not satisfy this condition, then at least one of the two fundamental assumptions of the theory would have been disproved.

Hence, a statement of physics (expressed mathematically) can be considered as a general law of nature if it maintains its form under the Lorentz transformation. In this sense the principle of relativity represents an aid to theorizing — it is a heuristic principle.[43]

Einstein concluded the introductory section to his (1905d) with the statement:

> The theory to be developed is based — like all electrodynamics — on the kinematics of the rigid body, since the assertions of any such theory concern the relationships between rigid bodies (coordinate systems), clocks, and electromagnetic processes. Insufficient consideration of this circumstance is the root of the difficulties with which the electrodynamics of moving bodies presently has to contend.

Thus, while Abraham, Lorentz, and Poincaré were concerned with such esoteric notions of the dynamics of electrons in 1905, as group theory and the radiation-reaction force, Einstein focused his attention on *kinematics*. Connecting the case of magnet and conductor with optical phenomena had convinced Einstein of the necessity to reconsider basic concepts of mechanics within a theory of principle. This is a surprising conclusion to a surprising beginning.

FOOTNOTES

1. The asterisk in this passage and the corresponding footnote were added subsequently by Arnold Sommerfeld to the reprint of Einstein's paper in the collection of related essays originally issued by Teubner. The footnote says: "The preceding memoir by Lorentz was not at this time known to the author." The "memoir of Lorentz" is Lorentz's (1904c). What Einstein in 1905 may have been aware of in Lorentz's (1904c) has already been discussed. There were only four notes by Einstein in the 1905 relativity paper and no references. Among recent survey discussions of Einstein's introductory section are Holton (1973e) and Miller (1977c).

2. See Williams (1971) for a discussion of Faraday's researches.

3. Needless to say, this result was not exact.

4. E.g., Weidener and Sells (1975); Purcell (1975); Resnick and Halliday (1978). Sometimes the case of magnet in motion is considered qualitatively in order to illustrate Lenz's law (see Resnick and Halliday). In their purple-covered edition of 1967, Halliday and Resnick appended a brief but illuminating discussion of the case of magnet in motion relative to an observer at rest on a closed conducting loop.

5. Föppl (1894) discussed the convective derivative at length; its presentation in more modern vector notation is in the first of many descendents of Föppl's book, rewritten by Abraham [Abraham (1904c)]; see also Cullwick (1959).

6. For discussion of the notion of impressed forces in the Hertz-Heaviside version of Maxwell's electromagnetic theory, see Whittaker (1973, vol. 1). A modern discussion of impressed forces can be found in the most recent descendent of Föppl's book, the one of Becker and Sauter (1964, vol. 1), who write that the quantity ε_2 "is usually defined (somewhat unfortunately) as the" EMF.

7. See Sec. 1.15.

8. As Louis Kollros, a classmate of Einstein's at the ETH, recalled (1956); "we waited in vain for an exposition of Maxwell's theory." Einstein strove to fill this gap by reading independently "the works of Helmholtz, Maxwell, Hertz, Boltzmann and Lorentz." Kollros continued, "He neglected his courses knowing that at examination time he could study the notes taken concisely by his friend Grossmann."

9. Holton (1973c) discusses the influence of Mach on Föppl. Of particular importance to Föppl was Mach's insistence upon a foundational analysis, and, in fact, Föppl considered that he worked in the spirit of Kirchhoff, Hertz and Mach. Thus, writes Holton, we "see some evidences of the kind of approach to physics which would appeal to a young reader with the kind of background, or lack of background, of Einstein in the late 1890's." On the other hand, Föppl's adherence to Mach's

philosophy led him to be skeptical "on the work of Einstein and Minkowski" [letter of Föppl to Mach, 11 January 1910, in Blackmore (1972)]. Continued Föppl: "I can clearly see that provisionally the question is merely one of hypothesis, whose confirmation by experience, that is, of the movement of perceivable bodies, is completely lacking at least for the moment."

10. As far as I know this experiment was never actually performed. It was, however, one of a group of examples much discussed at that time. Among other examples were: (1) whether there is any electrodynamic action between a charge and a current-carrying circuit at relative rest but moving through the ether; (2) same problem, except replace the circuit with a conducting magnet. That in the first case there should be no effect due to absolute motion was first explained by E. Budde (1880) in defense of Clausius' electromagnetic theory. According to Budde a charge of electrostatic origin is induced on the circuit's surface of just the correct amount to cancel electrodynamic forces due to absolute motion. Budde extended this result to the second case. FitzGerald (1882) deduced this result for both cases from Maxwell's theory. FitzGerald began this important paper (he also included here the convection current into Maxwell's theory) by asserting his feeling that "as it is very unlikely that anything depends on absolute motion, the motion here spoken of must be with respect to something, and this something can hardly be any other thing than what is known as the ether in space." This reason for an ether is similar to that of Föppl and Budde, and in fact was standard fare for the 19th-century electrodynamicists.

In summary, the absence of an effect in these two cases due to motion relative to the ether, and not to their relative motion, resulted from cancellations among terms containing apparently unobservable velocities, i.e., velocities relative to the ether, thus ensuring that physical effects arise due only to the relative motion among ponderable bodies.

The method of postulating terms for the purpose of cancelling effects arising from motion relative to the ether that could not be observed empirically is clear from Lorentz's (1895) derivation of Budde's compensation charge. To order v/c the electric fields in S and S_r written in terms of potentials differed in form by the term $\nabla_r(A_r \cdot v/c)$ — i.e., E_r should be of the same form as E in Eq. (A) of Chapter 1, footnote 24 instead of $E_r = -\nabla_r\phi_r - (1/c)(\partial A_r/\partial t_L) + \nabla_r(A_r \cdot v/c)$ where $\phi_r = (1/c)\int(\rho/r_r)$ $\cdot dx_r\, dy_r\, dz_r$ and $A_r = (1/c)\int(\rho u/r_r)\, dx_r\, dy_r\, dz_r$, ρu is the current from an ion moving relative to S_r, and ϕ_r and A_r are evaluated at the time $t_L - r_r/c$. According to the extra term in E_r a current-carrying circuit at rest on the earth's surface exerts a force on a neighboring charge q at relative rest on the earth of the amount $F'_r = q\nabla_r(A_r \cdot v/c)$. The force F'_r arises from the charge's motion relative to the ether through the circuit's magnet field. Since F'_r is derivable from a potential then, continued Lorentz, it agrees with Eq. (H) in Chapter 1, footnote 24 if $\phi = -(A_r \cdot v/c)/(1 - v^2/c^2)$. Considering now that all electromagnetic quantities are replaced by their averaged values that denote the effects of bulk matter, then from Eq. (1.59) the potential ϕ has its origin in a volume charge distribution of $\rho'_r = -v \cdot J_r/(c^2 - v^2)$, where J_r is the value of ρu averaged over all the ions in the circuit, i.e., J_r is the conduction current in the circuit. But, to order v/c, no effect of F'_r had been discerned empirically. Consequently, like Budde, Lorentz postulated that F'_r acted also on the circuit itself producing a volume charge density $\rho''_r = (v \cdot J_r)/(c^2 - v^2)$ (the compensation charge), that is distributed over the circuit to produce a force $F''_r = -F'_r$. The postulate of the compensation charge, invented for the purpose of saving the theorem of corresponding states, required Lorentz to assume that, contrary to what we expect for a current–carrying circuit at rest, i.e., that the macroscopic charge density vanished, in S_r the macroscopic charge density is not zero over parts of the circuit. Over the entire circuit, however, $\int \rho''_r \, dx_r \, dy_r \, dz_r = 0$ because ρ''_r is a static charge distribution.

Perhaps to avoid using the coordinate transformations that were postulated only for electrodynamical problems, and not for the theorem of corresponding states, in (1904b) Lorentz set directly $\rho''_r = (\boldsymbol{v} \cdot \boldsymbol{J})/c^2$, which is the source of the potential $\boldsymbol{A}_r \cdot \boldsymbol{v}/c$.

Until (1904b) Lorentz did not emphasize the representation of the electric field in terms of potentials for what I conjecture is the following principal reason: neither ϕ_r nor A_r entered into the theorem of corresponding states which included neither bodies with net charge nor conduction currents. In fact, the extra term in the electric field in S_r was related to the extra term in Gauss' law that violated the covariance of the Maxwell-Lorentz equations, unless $\boldsymbol{u} = \boldsymbol{0}$ (see Chapter 1, footnote 27). On the other hand, Lorentz required this extra term in the electric field in S_r for postulating the compensation charge, which enabled him to explain the failure of ether-drift experiments that concerned bulk matter with conduction currents or permanent magnetization, which the 1895 theorem of corresponding states did not encompass. At the end of his (1904b) Lorentz discussed how unsatisfactory was this procedure.

Lorentz's lectures (1910–1912) contained the most straightforward derivation of the compensation charge which, of course, was implicit in the (1895) and (1904b) derivations. In S, $\boldsymbol{\nabla} \times \boldsymbol{E} = -(1/c)(\partial \boldsymbol{B}/\partial t) = (\boldsymbol{v}/c \cdot \boldsymbol{\nabla})\boldsymbol{B} = \boldsymbol{\nabla} \times (\boldsymbol{B} \times \boldsymbol{v}/c)$; then $\boldsymbol{E} = \boldsymbol{B} \times \boldsymbol{v}/c$, and consequently in S the Lorentz force on q vanished. The compensation charge on the circuit is $\rho = \boldsymbol{\nabla} \cdot \boldsymbol{E}/(4\pi) = (\boldsymbol{J} \cdot \boldsymbol{v})/c^2$. Although the (1910–1912) derivation was more "relativistic" since it was not performed from the viewpoint of a comoving observer, nevertheless Lorentz asserted a similarity between the dynamical reasons for the compensation charge and the polarization $\boldsymbol{P}$ that arose in a moving permanent conducting magnet (see Chapter 3, footnote 33).

In Lorentz's theory, to order v/c, the comoving observer in S_r cannot measure the non-zero charge distribution in parts of the circuit because it annuls the force of the circuit on the neighboring charge at relative rest. In Minkowski's theory the surprising result that whereas for a comoving observer a current carrying circuit has no average charge density anywhere, while for a *laboratory* observer parts of the circuit contain a net charge distribution, results from the relativity of time [see Pauli (1958) and Rosser (1964)].

11. FitzGerald (1882) was also worried. He wrote that "although it is unlikely that anything depends on absolute motion" cases such as those in footnote 10 above "should be completely investigated." Lorentz in the *Versuch* (1895) was able to demonstrate that to first order in v/c electromagnetic induction phenomena depend only upon the relative motion between the two different parts of the system — for example, magnet and conductor, or two induction coils; Des Coudres (1889) had demonstrated this empirically for the latter arrangement.

12. See Miller (1977b) for a discussion of the history of effects due to rotating magnets, in particular unipolar induction. Weber (1841) was the first to use the term unipolar induction.

13. The proof is as follows: According to Föppl the electric field $\boldsymbol{E}$ for electromagnetic induction should be written as $\boldsymbol{E} = \boldsymbol{E}' + \boldsymbol{E}''$ where $\boldsymbol{E}' = \boldsymbol{B} \times \boldsymbol{v}/c + \boldsymbol{\nabla}(\boldsymbol{A} \cdot \boldsymbol{v}/c)$ and $\boldsymbol{E}'' = -\boldsymbol{\nabla}\psi = -\boldsymbol{\nabla}(\boldsymbol{A} \cdot \boldsymbol{v}/c)$. Incidentally, writing $\boldsymbol{E}$ in terms of $\boldsymbol{E}'$

and E'', Föppl used the theorem that any vector field could be decomposed into a solenoidal part E' and an irrotational part E'' — that is, $\nabla \cdot E' = 0$ and $\nabla \times E'' = \mathbf{0}$. For the case of unipolar induction in Fig. 3.3, $E' = \mathbf{0}$, and so $\nabla \cdot E = \nabla \cdot E'' = -2\omega B/c = 4\pi\rho$.

14. I add here, however, that Einstein knew little English until 1913 when he began taking lessons. See the letter from Einstein to Besso (Zurich, the end of 1913) in Einstein (1972).

15. Föppl could have cited here a well-known result of von Helmholtz (1875): The potential measured between the ends of a wire rotated about an axis parallel to a uniform magnetic field was within three per cent of the value predicted by Faraday's law.

16. Wiedemann's comprehensive 1885 review of unipolar induction lists upwards of thirty papers for the period 1860–1885; most of them appeared in the *Annalen*. The next comprehensive review was in 1895 by E. Lecher who cited eight papers in the period 1885–1895, and seven of them were published in the *Annalen*. For further citations, see Miller (1977b).

17. In a report dated 17 December 1907, Einstein refused the patent application for a dynamo because, he wrote, "The patent is described incorrectly, inaccurately, and unclearly" [Flückiger (1974)].

18. It is good to remember that most of Faraday's key experiments on electromagnetic induction concerned relative rotatory motion between magnet and conductor.

19. In 1832 Faraday (1965, vol. 1) hypothesized that atmospheric effects such as the Aurora Borealis and Australis could also be explained in this manner by considering the earth to rotate through its own magnetic lines of force.
The external and internal electric fields and total charge distributions for a spherical conductor rotating about a diameter parallel to an externally imposed uniform magnetic field are the same as for a rotating spherical conducting magnet. These problems had been solved by Jochmann (1864) using Weber's action-at-a-distance electromagnetic theory and by Larmor (1884) using Maxwell's theory, and they provided models for investigating atmospheric effects. See also Hertz's (1880) detailed calculations on these problems in his Inaugural Dissertation, Berlin, 15 March 1880. See Miller (1977b) for discussion of these calculations.

20. E.g., in addition to the works of Föppl (1894), Thomson (1893), those listed in Wiedemann (1885), Larmor (1884), Lecher (1895), see also Larmor (1895).

21. The unipolar dynamo is a low voltage, high current machine. For a discussion of the state of the art of unipolar dynamos in the late 19th century, see Uppenborn (1885), Crocker and Parmly (1894).

22. For descriptions of these machines, see Weber (1895), Arnold (1895) and Miller (1977b). Arnold supported the rotating-line viewpoint. Weber criticized Arnold's supporting statements in the light of Lecher's (1895).

23. I have previously discussed the importance of the notion of *Anschauung* in my (1977b) and (1978a). See also Arnheim (1971).

24. One of Preston's reasons for supporting the rotating line viewpoint was its consistency with Ampère's theory of magnetism.

25. Unlike his counterpart in America, Edison, Siemens cultivated contacts among the academic community. For example, Siemens (1881) wrote that his "good friend Kirchhoff" had assisted him in designing this unipolar dynamo; and one of Siemens' sons married von Helmholtz's daughter. Some of Siemens' contributions to the electrical industry are discussed in my (1977b). Biographical information can be found in Georg Siemens (1957) and Siemens (1893).

26. In fact, the Kaiser Wilhelm Institute, where Einstein became the first Director of its section for research in physics in 1913, was originally conceived of by Walther Nernst to be sister institute to the *Physikalische-Technische Reichsanstalt*, Charlottenburg, founded by von Siemens and headed originally by von Helmholtz.

27. Abraham took the circuit *ab* (see Fig. 3.3) to be completed within the permanent conducting magnet.

28. Röntgen (1888) detected a magnetic field due to a dielectric disc spinning between the plates of a capacitor that served as the source of a static electric field. Eichenwald (1903) rotated the disc and capacitor together. According to Hertz's theory Eichenwald should have observed no magnetic field because the current due to the charges on the capacitor was cancelled by the current due to the true charges in the dielectric (the source of the electric polarization D). In Lorentz's electron-based theory, however, only the polarization charges (the source of macroscopic dipole moment P) are carried in motion, and the predicted magnetic field in this case is independent of the dielectric's constant, in agreement with Eichenwald's result. Eichenwald also demonstrated empirically that Hertz's prediction for the magnetic field disagreed with a more accurate version of Röntgen's experiment. In summary, Lorentz found that the current curl($D \times v/c$) should be curl($P \times v/c$); he referred to these terms as the "Röntgen current" [Lorentz (1904a)]. The theories of Hertz and Lorentz were in agreement for effects due to conductors, but not for nonconductors.

29. Whereas Föppl (1894) had deemed the point of the Euclideaness of physical space worthy only of a footnote, Abraham developed it in the text itself.

30. This symmetry was first emphasized by Heaviside (1894, vol. 1) and Hertz (1884). Föppl (1891) wrote on the duality principle's usefulness for the design of dynamos; this paper was cited in the second volume of Boltzmann's (1891). Consequently, it is technically possible that Einstein's interest in electrical machinery caused him to have read Föppl's (1891).

31. Lorentz had written similarly elsewhere on unipolar induction (1904d).

32. Lorentz (1904b) succeeded in extending his electromagnetic theory to bulk matter that was dielectric or magnetic. The ultimately successful electromagnetic theory of bulk matter was Minkowski's (1908a) because it included the Lorentz contraction and Einstein's relativity of simultaneity. I shall not discuss in any detail research on the electromagnetic properties of bulk matter since it did not serve to illuminate the foundations of relativity theory during 1905–1911. See Pauli (1958) for an exposition of the theories of Lorentz and Minkowski concerning bulk matter.

33. Lorentz's (1910) approach to the linear unipolar induction dynamo is an interesting mixture of Einstein's relativity theory and Lorentz's theory of electrons. By 1910 Lorentz had further developed his electromagnetic theory of moving bulk matter by explicitly including transformation equations for the dipole moment that results from magnetization and conduction electrons. Lorentz's (1910) explanation for the linear unipolar induction phenomenon in Fig. 3.4 is thus: Since to an observer in S there is no displacement of conduction electrons in addition to the magnet's inertial motion, then the macroscopic dielectric displacement vanishes in S, i.e., $D = 0$ and $E = -4\pi P$. To order v/c, P is the quantity in Eq. (3.14). Since the magnet is infinitely long, then $4\pi M = B$, and the electric field at the site of the resting circuit in Fig. 3.4 is $E = B \times v/c$ which yields the correct EMF in the circuit. (Note that this electric field is totally electrostatic in origin and consequently would not cause current to flow in a neighboring closed circuit.) Lorentz (1910) wrote of the "impossibility to distinguish sharply between polarization and magnetization electrons." Consistent with the dynamical nature of his theory Lorentz (1910–1912) interpreted the dipole moment P as not arising "merely from the transformation formulae" for "we should keep in mind that it is also a consequence of the particular property of the physical forces determining the motions of charges which produce within the molecules a magnetic moment" (1910–1912). For, according to Lorentz, the atom's magnetic moment results from the circulation about the atom's nucleus of a magnetization electron. A translation of the atom results in a distortion of the electron's orbit accompanied by the electron's lingering in one half of the orbit a little longer than the other: "As a consequence, the mean electric moment of the moving atom differs from zero." In Minkowski's relativistically covariant theory the polarization P arises from the relativistic transformation equations for the electric and magnetic field quantities. The charge distribution that is obtained from P is interpreted as being apparent – it is a charge distribution whose effects can be measured only by an observer at rest in the laboratory [see Born (1910a), Pauli (1958)]. Becker's (1932) is the earliest fully relativistic treatment of the linear unipolar machine. See also Rosser (1964) and Miller (1977b).

34. Einstein characteristically added [Shankland (1964)]:

> There is, of course, no logical way leading to the establishment of a theory but only groping constructive attempts controlled by careful consideration of factual knowledge.

35. Thus, perhaps it was to Einstein's theory of relativity that Mach was referring to in the "Preface" to the seventh edition of the *Science of Mechanics*, written 5 February 1912:

> At the end of the last century my disquisitions on mechanics fared well as a rule; it may have been felt that the empiricocritical side of this science was the most neglected. But now the Kantian traditions have gained power once more, and again we have the demand for an *a priori* foundation of mechanics.

36. By 1737 Hume had concluded that exact scientific laws could not be induced from sense-perceptions. For example, notions of cause and effect arise from our belief or expectation in the recurrence of patterns previously learned through sense

experience. Kant's response to Hume's skepticism was boldly to raise causality to a category to which we are irresolutely driven by the *a priori* intuitions of Euclidean space and an absolute time. In effect, Einstein reversed Kant's "Transcendental Analysis" in the *Critique of Pure Reason* by obtaining the new notions of space and time from categorical statements.

Seelig (1952) writes that while at Aarau, Einstein did not participate in any of the numerous beer parties because he took seriously Bismarck's advice that "beer makes one dumb and lazy." Instead, continued Seelig, Einstein became "intoxicated on Kant's *Critique of Pure Reason*." Max Talmey, a medical student who dined weekly with the Einstein family, introduced the thirteen year old Albert to Kant's writings. Talmey recalled that "Kant's works, incomprehensible to ordinary mortals, seemed clear to him." Einstein [(1949), (1944)] discussed his views of Hume and Kant.

37. Pages 265–269 of Poincaré's (1899) book review of Russell's (1897) constitute pages 75–79 of Chapter V, "Experiment and Geometry," of Poincaré's *Science and Hypothesis.*

38. Einstein (1917a) used similar terms in discussing a principle of relativity: "... we should retain the principle of relativity, which appeals so convincingly to the intellect because it is so natural and simple."

39. Not surprisingly, Poincaré (1904) considered this expansion of the principle of relative motion as "irresistibly imposed upon our good sense."

40. On Lorentz's theory distinguishing between K and K', Einstein (1920) wrote in a subjective tone so similar to the one in his unpublished 1919 essay:

> Such an asymmetry of the theoretical structure, to which there is no corresponding asymmetry in the system of empirical facts, is intolerable to the theorist .... The most obvious line to adopt in the face of this situation seemed to be the following: – There is no such thing as the ether.

41. The term *Invariantentheorie* appealed to Arnold Sommerfeld whose teacher, Felix Klein, had emphasized the geometrical interpretation of transformation groups, i.e., the Erlanger Program. After reviewing why in four-dimensional space-time geometry the Lorentz transformations were the proper group for electrodynamics, in (1948) Sommerfeld wrote: "Relativity theory is accordingly an *Invariantentheorie* of the Lorentz group. The name relativity theory was an unfortunate choice: the relativity of space and time is not the essential thing, which is the independence of laws of nature from the viewpoint of the observer."

42. Einstein could well have been influenced somewhat by the epistemological aspects of the researches of Emil Cohn and of the energeticists.
(i) *Emil Cohn's Electrodynamics*
For describing moving dielectric and magnetic media Cohn [(1900), (1902)] used a combination of Hertz's and Lorentz's theories suitably modified so that to every order in v/c the field equations were covariant under the transformation Eqs. (1.76)–(1.79), and a version of Eqs. (1.84)–(1.85). Cohn speculated on neither the nature of the ether, nor the nature of electricity (his theory was not based upon an atomistic conception of electricity), nor did he attempt to reduce the laws of electromagnetism to those of mechanics. He sought to formulate a theory with a firm empirical base. Cohn

emphasized that although his theory was provisional, perhaps at some future time the ether could be utilized as a "heuristic concept" to aid in its completion. However, with a note of caution he added that "such a metaphorical term [ether] should not acquire an importance relative to the theory in question." In summary, Cohn (1904b) asserted that his guiding theme was "scientific economy (*wissenschaftliche Ökonomie*)." Nevertheless, Cohn did consider space to be filled with an all-pervasive dielectric substance.

Cohn's theory was criticized by Lorentz (1904b) and in great detail by Wien (1904a) who demonstrated that according to Cohn the velocity of light depended on the source's motion even relative to an observer at rest in the ether. The crux of the matter was that Cohn's theory did not distinguish between matter and field owing to the theory's lack of an atomistic foundation; consequently, for example, the theory often failed to discuss processes in vacuum [e.g., see Lorentz (1904b), Abraham (1905)]. Cohn's defense (1904a, b) of his theory focused upon appropiate alterations of the equations and epistemological arguments. For example, Cohn emphasized that his field equations for moving objects contained only the true space and time coordinates; on the other hand, Lorentz's were written relative to a "distorted" reference system with space and time coordinates that required "ideal comoving measuring rods [and] clocks" to distinguish them from the true space and time coordinates. Thus, Cohn continued, "No conceivable experiment could distinguish between the two systems of explanation" (i.e., Cohn's and Lorentz's for moving bulk matter that could be magnetic or dielectric). Lorentz's set of space and time coordinates that Cohn addressed in his (1904a) were the ones from Lorentz's (1904c), to which Cohn had access to only the Dutch version.

Einstein could well have been intrigued by Cohn's [(1900), (1902)] assertions on the superfluousness of an ether, as well as Cohn's Machian presuppositions. Had Einstein read Cohn's (1904a, b) he would have been struck by Cohn's epistemological analysis of Lorentz's space and time transformations of 1904, especially Cohn's suggestion of using ideal measuring rods and clocks for investigating the difference between Lorentz's ideal and real coordinates. In addition, critiques of Cohn by Lorentz, Wien and Gans (1905) may have underscored for Einstein the limits of an inductive approach toward theory construction.

In his (1907e) Einstein wrote that he thought highly of Cohn's "competent works," although he refrained from discussing them owing to their inapplicability to the review paper's subject matter. Evidently by 1913 Cohn accepted the special relativity theory because Einstein's (1915b) referred to Cohn's (1913) as "an excellent presentation of the subject matter" (i.e., special relativity theory); the unusually critical Pauli (1958) referred to Cohn's (1913) book as a useful reference. See Hirosige (1966) for a survey of Cohn's electrodynamics.

(ii) *Energeticism*

According to the energetic world-picture the laws of thermodynamics and the principle of least action should be taken as a description of physical phenomena; thus, assumptions on the constitution of matter were unnecessary. Like the positivists, the energeticists were adamant in their demand that physical theories contain no metaphysical quantities, and so both groups were anti-atomistic. On this point Einstein (1946) took them to task: "The antipathy of these scholars [Ostwald, Mach] towards atomic theory can indubitably be traced back to their positivistic philosophical attitude. This is an interesting example of the fact that even scholars of audacious spirit and fine instinct can be obstructed in the interpretation of facts by philosophical prejudices." Einstein recalled that his theory of Brownian motion "convinced the skeptics ... of the reality of atoms." The most important portion of the energeticist's program for Einstein's thinking toward the special relativity theory was that since energy is the only form of reality, then there is no need for an ether to transport it. For an early and famous diatribe against the energeticists see Planck (1896). See Merz (1965, vol. 2) for further discussion of the energetic world-picture and also Nye (1972).

43. Other examples of Einstein's use of heuristics are: The heuristic proposal of (1905) that light has a granular structure.

In (1907e) Einstein emphasized the "heuristic value" of extending the principle of relativity to include accelerating reference systems as a "positive step" toward a theoretical treatment of such problems. Einstein's style of argument for this extension was similar to the one he had used so effectively two years earlier – a quasi-aesthetic argument (Chapter 12, footnote 23).

CHAPTER 4

SIMULTANEITY AND TIME

The qualitative problem of simultaneity is made to depend upon the quantitative problem of time.

H. Poincaré (1898)

We have to take into account that all our judgments in which time plays a part are always judgements of simultaneous events.

A. Einstein (1905d)

When he spoke in Zurich in 1909 for the first time on special relativity, it was neither at the university nor at the ETH, but in a room of the Carpenters' Union at a town restaurant. For writing he had only a small blackboard on which he drew a horizontal line: it was a space of one dimension that he was going to relate to his new notion of time. He began by saying, "Consider at each point of this straight line a clock – that is, infinitely many clocks." After having developed his theory for more than an hour, he suddenly stopped and excused himself for having spoken for so long. He inquired, "How late is it, because I have no clock?"

L. Kollros (1956)

4.1. SOME PRE-EINSTEINIAN ANALYSES OF TIME AND SIMULTANEITY

4.1.1. Newton, Kant, and Mach

To Isaac Newton (1687) time was "absolute, true ... and from its own nature flows equably without relation to anything external." By 1781 Immanuel Kant had elevated the time of Newton to knowledge that we possessed before all else – an *a priori* intuition. As Kant wrote in his monumental *Critique of Pure Reason*, "Time is not an empirical concept that has been derived from experience." Little more than a century later the Newtonian conception of absolute time was sharply criticized by Ernst Mach (1889). Mach considered Newton's notion of absolute time as an "idle metaphysical conception" since it was unavailable to our sense-perceptions (italics in original):

We arrive at the idea of time ... by the connection of that which is contained in the province of our *memory* with that which is contained in the province of sense-perception.

Mach's reliance on sense-perception meant that if an observer at a point A flashed a beam of light to an observer at a point B, then the observer at B could conclude that the time at which he receives the light ray is coincident with its time of emission from A. However, by the late 17th century it was known that the speed of light was enormous, but not infinite.

4.1.2. Poincaré

By cleanly separating the problems of defining simultaneity from those for time, Poincaré (1898) analyzed the notion of time in a manner more exact than Mach's. Poincaré's epistemology was rooted in sense-perceptions, but his deep regard for the efficacy of mathematical analysis of foundations enabled him, in contrast with Mach, to demarcate clearly between the two problems.

In "Measurement of Time" (1898) Poincaré defined two "facts" as being "simultaneous" when "the order of their succession" could be interchanged. He quickly pointed out, however, that this definition was inapplicable to "two physical facts which happen far from one another, and that, in what concerns them, we no longer even understand what this reversibility would be; besides succession itself must first be defined." Thus in analyses of two widely separated physical facts, the notions of causality, simultaneity, and time became intermingled. Consequently, Poincaré considered this sort of simultaneity (which I refer to as distant simultaneity) as more important than the simultaneity of two facts which "analysis cannot separate without mutilating them," i.e., local simultaneity. For Mach, this distinction was unnecessary because he pushed Hume's analysis of causality onto a path that sidestepped Kant, and relegated the notion of cause and effect to the mere facilitating of an economy of thought; by this reasoning the law of causality did not find any meaningful mathematical expression in the laws of physical theory.

On the other hand, Poincaré's (1898) analysis focused on exposing certain vicious circles in commonly held notions of "what is understood by simultaneity or antecedence." For example, the vicious circle inherent in defining cause and effect in terms of a time sequence: fact A is the cause of fact B if A occurred before B; but A must have occurred before B if A was the cause of B. So Poincaré was led to state — "We say now, post hoc, ergo propter hoc; now, propter hoc, ergo post hoc; shall we escape from this vicious circle?" Yes, replied Poincaré, but only through recourse to sense-perceptions, and this escape was not foolproof; especially so because a single effect could have a multiplicity of causes. But by making assumptions based on "convenience and simplicity," a scientist could deal with these problems. Mach agreed, particularly regarding the convenience and simplicity problem — e.g., his analysis of why to a good approximation we could disentangle multiple causes and focus on two body systems (Chapter 2, footnote 15); Poincaré, however, had something else in mind concerning "convenience and simplicity."

Poincaré (1898) discussed the "definition implicitly supposed" by scientists of simultaneity. For example, Poincaré wrote, astronomers such as Römer, measured the velocity of light as follows (italics in original):

He has begun by *supposing* that light has a constant velocity, and in particular that its velocity is the same in all directions. This is a postulate without which no measurement of this velocity could be attempted. This postulate could never be verified directly by experiment; it might be contradicted by it if the results of different measurements were not concordant. We should think ourselves fortunate that this contradiction has not happened and that the slight discordances which may happen can be readily explained.

The postulate, at all events, resembling the principle of sufficient reason, has been accepted by everybody.

In the context of the ether-based theories of light of the late 19th and early 20th centuries, this is a rather paradoxical statement because, according to Poincaré, central to the astronomer's measurement of the velocity of light is his postulating the isotropy of space for the propagation of light. As for Poincaré as researcher in electromagnetic theory, it was an altogether different story because, like all ether-based theories, Lorentz's theory is predicated on the exact isotropy of space only for observers at rest in the ether. The theorem of corresponding states systematically explained the failure of first-order experiments to detect direction-dependent effects, and Michelson and Morley's second-order results were explained by additional postulates asserting that the space of inertial reference systems appeared to be isotropic owing to additional compensating effects such as the Lorentz contraction. Poincaré undoubtedly also applied this analysis to Fizeau's ingenious 1849 measurement of the velocity of light from a terrestrial source (Fig. 4.1): A ray of light is emitted from A, then after traveling over a known distance $\overline{AB}$ it is reflected from a mirror at B back to A. At A there is a clock, and Fizeau measured the time of the round-trip journey by interposing a rotating toothed wheel in front of A. I shall refer to this type of experiment as a two-way one-clock measurement.

FIG. 4.1. Although this two-way one-clock experiment requires no new definition of simultaneity, it is based on assuming the isotropy and homogeneity of space for light propagation.

Assume that light left the source at time t_A and returned at time t'_A. According to Fizeau the velocity of light c is

$$c = 2\overline{AB}/(t'_A - t_A). \tag{4.1}$$

Fizeau made two tacit assumptions: (1) equality of the to-and-fro light velocities; (2) isotropy of space for light propagation. Fizeau could not have measured the one-way velocity of light by placing a clock at B which was synchronized with the one at A. For consider (Fig. 4.2) that $c_\rightarrow$ is the velocity of

FIG. 4.2.   This one-way two-clock experiment is flawed because it contains the vicious circle of defining time in terms of velocity.

the light emitted from A at the time t_A, and then received at B at the time t_B and registered on a clock placed at B; thus the velocity $c_\rightarrow$ is

$$c_\rightarrow = \overline{AB}/(t_B - t_A). \tag{4.2}$$

But measurement of $t_B - t_A$ requires a knowledge of $c_\rightarrow$. As Poincaré wrote (1898): "Such a velocity could not be measured without *measuring* a time" (italics in original). Poincaré concluded that "the qualitative problem of simultaneity is made to depend upon the quantitative problem of the measurement of time." He emphasized this point with two examples: (1) For accurate distance measurement, such as determining latitudes and longitudes, a clock is carried that has been set for the time in Paris. (2) In sending a telegraph message between Paris and Berlin, the finite velocity of light is neglected and so cause and effect are considered as locally simultaneous; continued Poincaré, "to be rigorous, a little correction would still have to be made, because it would be well within the errors of observation," although in practice it was disregarded.

   In summary, for Poincaré, clocks spatially separated could be synchronized (1) by transporting a duplicate clock in synchrony with one over to the other or (2) by telegraphing the time and considering the velocity of light to be infinite. Trapped by his sensationism, Poincaré wrote: "We have not a direct intuition of simultaneity, nor of the equality of two durations." The most convenient way to solve the quantitative problem of time measurement was, thus, to use a definition of physical time that permitted the "laws of physics, mechanics and astronomy" to be expressed in a convenient and simple form — the time from the Galilean transformations.

   There was another definition of time available: Lorentz's hypothesis of the local time coordinate. In his lectures of 1899, Poincaré (1901) demonstrated that the difference between Lorentz's local time t_L, and the true time t could be neglected because for two clocks placed 1 km apart on the earth's surface, $t_L - t = vd/c^2 = \frac{1}{3} \times 10^{-9}$ sec, where Poincaré used $v = 30$ km/sec (i.e., the earth's orbital velocity about the sun). Then Poincaré (1904) analyzed how two observers at relative rest, but in inertial motion relative to the ether, could synchronize their clocks by exchanging light signals. Observers resting in the ether could synchronize their clocks in this manner to read the true time, but continued Poincaré, the clocks of the inertial observers would register the local time. However, since the difference between t_L and t_r $(= t)$ was so small, "it matters little since we have no means of perceiving it [the local time]."

4.2. THE NATURE OF TIME AND THE GEDANKEN EXPERIMENT FROM AARAU

> As for the influence of Mach on my thinking, it has certainly been very great. I remember very well how, during my early years as a student, you [Besso] directed my attention to his treatise on mechanics and to his theory of heat, and how these two works made a deep impression on me. Frankly, however, I cannot clearly see to what extent they affected my own work. So far as I can recall, David Hume had a greater direct influence on me; I read him at Bern in the company of Conrad Habicht and Solovine.
>
> A. Einstein to M. Besso, 6 January 1948, in Einstein (1972)

Two of Einstein's major 1907 papers (1907d, e) contained discussions of the state of physical theory as he had perceived it in 1905. In his (1907e) Einstein described the tension between mechanics and electromagnetism: The mathematical statement of the "principle of relativity" from mechanics (the Galilean transformations) did not leave unchanged the equations of Lorentz's electrodynamics. Although certain "fundamental assumptions" enabled Lorentz to explain every optical experiment that was accurate to order v/c, Lorentz's modified Galilean transformations did not leave unchanged the laws of mechanics.[1] Einstein continued, the failure of the Michelson-Morley experiment, accurate to second order in v/c, forced Lorentz to introduce an "ad hoc hypothesis" in order to "rescue" the theory: "however there emerges in an astonishing manner the fact that it was only necessary to fix satisfactorily the concept of time in order to overcome the difficulties pointed out above." It required only the understanding that one should define as "time" what Lorentz introduced as an auxiliary quantity, and called the "local time." That is, the transformation rules for the laws of mechanics and of electromagnetism depended upon two different notions of time, one physical the other mathematical, despite Lorentz's goal of unification. Thus whereas current physics considered a tension to exist between mechanics and electromagnetism that was rooted in the inability of mechanics to describe the velocity of light, Einstein delved deeper and found that current physics rendered mechanics and electromagnetism incompatible. If we accept the veracity of Einstein's recollection of the *Gedanken* experiment from his Aarau period, we can also conjecture that his (1907e) recollections illuminate his technical route toward the special relativity theory. My reasons in support of this statement are as follows: Einstein considered the *Gedanken* experiment as posing a paradox in which "the germ of the special relativity theory is already contained" — (i) It was "intuitively clear" that a principle of relativity had to be maintained; (ii) yet according to ether-based theories of light a moving observer should be able to detect an ether drift. Clearly, (i) and (ii) are mutually contradictory. From Lorentz's modified transformation Eqs. (1.76)–(1.79), that served as the basis for the theorem of corresponding states, Einstein could have derived the result (see footnote 1, Eq. (G))

$$w_{x_r} = \frac{w_x - v}{1 - (v/c^2)w_x}.$$ (4.3)

If w_x is the velocity of a point on a light wave, i.e., $w_x = c$, then $w_{x_r} = c$, instead of the expected result from Newton's mechanics, $w_{x_r} = c - v$. Unlike Lorentz's approximate theorem of corresponding states, Newton's exact principle of relativity asserts the equivalence of S and S_r; accordingly from the inverse of Eq. (4.3) Einstein would have obtained $w_x = c$ for the case where w_{x_r} was a point on a light wave observed from S, i.e., $w_{x_r} = c$. Thus Einstein realized that, to first order in v/c, Eq. (4.3) produced a result that agreed with the intuition of the *Gedanken* experimenter (see also Section 3.6).[2] Since Lorentz's modified transformations differed from the Galilean transformation only in the local time coordinate, then might the local time be the "time?" But this step required Einstein's asserting that the times in inertial reference systems were different. Yet the absoluteness of time had always been accepted. Why? Furthermore, the *Gedanken* experimenter's intuition demanded an examination of the relation between Newton's principle of relativity and Lorentz's theorem of corresponding states; after all, Lorentz's modified transformation equations of 1895 were Newtonian in their spatial coordinates, and their time coordinate had been invented for use in electromagnetic theory. But imposing a Newtonian unity upon these transformation equations necessitated rejecting Lorentz's ether, and with it the dynamical interpretations of an enormously successful and, for the most part, satisfying theory. Clearly, the notion of time was the central point about which a reconsideration of physical theory had to turn: to his surprise Einstein had found that the notion of time was both the central point and the Achilles heel of the current electrodynamics of moving bodies. Einstein's analysis of electromagnetic induction resulted in clarifying the relation between the observers in Lorentz's transformations of 1895. For aid in analyzing the nature of time he (1946) turned to the "critical reasoning [in] David Hume's and Ernst Mach's philosophical writings." Their analyses of the limits imposed by our sense-perceptions on our notions of causality, and of time, enabled Einstein to realize that the high value of the velocity of light, compared with the other velocities we encounter daily, had prevented our appreciating that "the absolute character of time, viz., of simultaneity, unrecognizedly was anchored in the unconscious." Poincaré (1898) extended the analyses of Hume and Mach to simultaneity, and Einstein read his conclusion in *Science and Hypothesis* (1902):

> Not only have we no direct intuition of the equality of two periods, but we have not even direct intuition of the simultaneity of two events occurring in two different places. I have explained this in an article entitled "Mesure du Temps."

On the basis of our analysis thus far we learn that the writings of Hume, Mach, and Poincaré persuaded Einstein to be skeptical of inducing exact laws of nature from empirical data. Consequently, we may conclude that Einstein's referring in his (1907e) to Lorentz's local time as a "concept" is in harmony with his later use of the term — an element that could bring order to the data of the *Gedanken* experimenter. Hence, Einstein's foundational analysis of physical theory went far beyond science as it is normally conceived; from an analysis of electromagnetic induction into an analysis of sensations, and then into an analysis of thinking itself. He concluded that the customary sensation-based notions of time and simultaneity resulted in a physics burdened with asymmetries, unobservable quantities and ad hoc hypotheses. Guidance from the neo-Kantian view that was predicated upon the usefulness of organizing principles such as the second law of thermodynamics, allowed Einstein to rise above the current state of physical theory: he proceeded to enlarge Newton's principle of relativity to include Lorentz's theory, and then he raised to axioms this principle and the one governing the velocity of light in Lorentz's ether-based system S. In this manner he discovered that it would be better to begin anew with examination of the "kinematics of a rigid body [and of the] relationships between rigid bodies (coordinate systems), clocks and electromagnetic processes." He did just that in Part I; and in §1 he rediscovered some of Poincaré's arguments from "The Measure of Time" (1898), in addition to offering a notion of geometry radically different from that of Poincaré's in *Science and Hypothesis*. Thus ended what Einstein recalled in the unpublished 1919 essay as the "groping for years" (quoted on p. 145).

4.3. EINSTEIN'S NOTION OF A PHYSICAL GEOMETRY

I. KINEMATICAL PART

§1. *Definition of Simultaneity*

> Let us consider a coordinate system in which the equations of Newtonian mechanics hold. For precision of demonstration and to distinguish this coordinate system verbally from others which will be introduced later, we call it the "resting system."

In §1 [§1, 1–6] Einstein analyzed the notion of time in an inertial reference system (the "resting system"), which he distinguished only "verbally" from similar systems. The usefulness of this artificial device soon becomes clear.

> If a material point is at rest relatively to this coordinate system, its
> position can be defined relative to it by rigid measuring rods employing
> the methods of Euclidean geometry, and can be expressed in Cartesian
> coordinates.

This [§1, 7–9] is the prelude to the core of Einstein's results in Part I. The
simplicity of the above passage, so typical of many to follow in this paper,
could be as misleading now as it was then; for Einstein asserted that the
elimination of unobservable quantities, and ultimately asymmetries too, from
physical theory required an examination of how distance and time were
measured — an examination of kinematics. No wonder that most readers
skipped over Part I with its discussion on the nature of space and time, in order
to reach the more familiar material in the electrodynamical §§9 and 10.

Banesh Hoffmann, a collaborator of Einstein's at the Institute for Advanced
Study, gives a description of a reader's possible response to Einstein's line of
thinking in the §1 that captures well what may also have taken place in 1905:

> Watch closely. It will be worth the effort. But be forewarned. As we
> follow the gist of Einstein's argument we shall find ourselves nodding in
> agreement, and later almost nodding in sleep, so obvious and un-
> important will it seem. There will come a stage at which we shall barely be
> able to stifle a yawn. Beware. We shall by then have committed ourselves
> and it will be too late to avoid the jolt; for the beauty of Einstein's
> argument lies in its seeming innocence.

"Seeming innocence," indeed, for Einstein began first by defining the concept
of an inertial reference system — one in which "Newtonian mechanics hold."
(In §3 he deigned to describe the inertial reference system in even more
detail — "three rigid material lines, perpendicular to one another, and originating from a point.") In §1 Einstein further took the reader by the hand and
explained that the position of an object relative to the "resting system" was
determined with a rigid measuring rod whose markings conformed with
Euclidean geometry. What could be simpler than this physical procedure? Yet
here Einstein set up a kinematics which he applied to electrons whose distance
from the origin of an inertial reference system could be determined only by the
measuring rod of a *Gedanken* experimenter: Einstein's measurement procedure was possible in-principle.[3]

Consistent with his view that the "theory here to be developed" was a theory
of principle, Einstein introduced the rigid measuring rod as an irreducible
object. But the rigid measuring rod of the theory of special relativity, as
Einstein wrote in "Geometry and Experience" (1921b), did not find its "exact
correspondence in the real world" — for example, variations due to temperature gradients and to external forces. Nevertheless, Einstein considered the

concept of the rigid measuring rod as admissible because it could be associated unambiguously with the "practically-rigid bodies" of the world of sense-perceptions.

In his lecture, "Fundamental Ideas and Problems of the Theory of Relativity," (1923) Einstein referred to the "postulate" that "concepts and distinctions are only admissible to the extent that observable facts can be assigned to them without ambiguity (stipulation that concepts and distinctions should have meaning)," as the "stipulation of meaning." He went on to emphasize that "this postulate, pertaining to epistemology, proves to be of fundamental importance." Although in 1905 Einstein may not have had a name for this postulate, nevertheless he used it well: (1) It enabled him to give an in-principle physical meaning to the notion of distance from an inertial reference system composed of practically-rigid rods. (2) It permitted him to begin bringing to fruition the dream of such philosopher-scientists as Bernhard Riemann and von Helmholtz to give meaning to the characteristics of physical space. Abraham (1904c) had considered this problem best left aside pending further clarification of electromagnetic theory; Poincaré took this problem to be meaningless since the objects of Euclidean geometry and those of the world of sense-perceptions had no correspondence. Although in "Geometry and Experience" (1921b) Einstein agreed with Poincaré's viewpoint *sub specie aeternitatus,*[4] nevertheless, heeding Poincaré's (1902) analyses of geometry, classical mechanics, and electromagnetic theory may well have provided Einstein with a clue to formulating his stipulation of meaning. This stipulation led Einstein to the notion of a "practical geometry" as distinct from a "purely axiomatic geometry" which made no assertions about the world of sensations. Einstein (1923) wrote that he attached "special importance" to the practical geometry because without it he would have "been unable to formulate the theory of relativity."

However, even though Einstein felt that the "'system of coordinates' concept as well as of the motion of matter relative" to a reference system satisfied the stipulation of meaning, the concept of an inertial system did not, and he considered this to be the "logical weakness" of the theory of special relativity. In 1923 he recalled that it would have been "logically more correct" to have applied the stipulation of meaning to the laws governing the constitution of matter and then to have deduced the properties of measuring rods and clocks; but in 1905 he had proceeded in the reverse manner: he had discussed simultaneity and time using irreducible measuring rods and then, in Part II, he had applied the new notion of time to the electrons in an electrodynamics mathematically equivalent to Lorentz's.[5]

4.4. SIMULTANEITY AND TIME

The Cartesian coordinates of a moving particle are functions of time that are determined through the solutions of equations of motion. Einstein [§1, 10–13]

stressed the importance of a clear understanding of "time," without which this
description "has no physical meaning." Here Einstein almost certainly had in
mind the root of the current tension and incompatibility that he perceived
between mechanics and electromagnetism — the notion of time which he then
went on to link with the notion of simultaneity [§1, 13–17]:

> We have to take into account that all our judgments in which time plays a
> part are always judgments of *simultaneous events*. If, for instance, I say,
> "That train arrives here at 7 o'clock," I mean something like this: "The
> pointing of the small hand of my watch to 7 and the arrival of the train are
> simultaneous events."*

Einstein's "simultaneous events" were the arrival of the train at his position on
the platform and his noting its time of arrival. These two events occurred so
close in time that their temporal order could be arbitrarily reversed, but as
Einstein wrote in the footnote to this passage:

> * We shall not here discuss the inexactitude which lurks in the concept
> of simultaneity of two events at (approximately) the same place, which
> must be removed through introducing an abstract concept.

Thus, in the remainder of the relativity paper Einstein used the term event to
mean an occurrence at a particular position, measured relative to an inertial
reference system by a rigid measuring rod, and whose time was registered by a
clock at that point. Once again Einstein has carefully used an in-principle
physical procedure to define a notion that was unclear in the sensationist-based
ether-physics — simultaneity.[6] By the impreciseness in the "concept of simultaneity of two events at approximately the same place" Einstein could only
have meant that two events occurred so close in space to one another that their
succession could be interchanged. He realized that the time of an event could be
defined properly through examining the connection in time of two events
occurring at different places far enough apart so that their temporal order
could not be arbitrarily interchanged, i.e., distant simultaneity. Therefore, to
Einstein the problem of simultaneity was linked with the problems of the causal
order of events, and of their times. On the other hand for Poincaré the
difference between local and distant simultaneity could ultimately be neglected, because he considered the problem of simultaneity to be sensation-based
and therefore qualitative; the problem of defining time was quantitative
because it could be resolved by means of the Galilean transformation. For
Einstein simultaneity and time were quantitative problems.

   Thus far Einstein has given physical meaning to the concept of an event. In
order to remove ambiguities inherent in current notions of "what we
understand by time," Einstein [§1, 18–32] next suggested two procedures that
could be carried out in-principle: (1) "Substituting 'the position of the small

hand of my watch' for 'time.'" But this procedure suffices to define time in the watch's proximity only, and not for the "times of events" distant. (2) A clock is located at the origin of an inertial reference system (designate the origin as A). At time $t = 0$ a train begins a journey along the x-axis. Upon reaching its destination at B the train emits a light signal that is received at A. Since the velocity of light c and the elapsed time are known by the observer at A, then the event's location B can be determined. Although there is nothing wrong with this method of coordinating a single clock with distant events, it is impractical because it depends on the distance between B and the master clock at A. In summary: method (1) emphasized the arbitrariness of the times at A and B; method (2) demonstrated the inconvenience of recording events using a one-way signal process.

To relate the clocks at A and B in a "practical" way, Einstein proposed a *Gedanken* experiment [§1, 33–46]. A light ray is emitted from A at time t_A; it is received at B at time t_B; then it is reflected back to A where it is received at time t'_A; Einstein's *Gedanken* experiment is a two-way two-clock experiment (see Fig. 4.3).

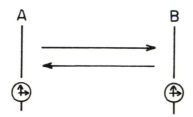

FIG. 4.3. Einstein boldly applied the two axioms of relativity to this *Gedanken* experiment and declared the definition of simultaneity to be identical with that of clock synchronization. We note that without the two axioms of relativity, placing clocks at A and B removes neither the vicious circle inherent in the experiment in Fig. 3.1, nor the necessity for Fizeau's assumptions, nor Poincaré's claim that the definition of simultaneity is a qualitative problem.

In Fizeau's one-clock two-way experiment, any values for the to-and-fro velocities of light sufficed so long as the sum of their inverses was $2/c$: suppose a clock were placed at B, then from Eqs. (4.1) and (4.2), and the equation for the velocity of the reflected ray, i.e.,

$$c_\leftarrow = \overline{AB}/(t'_A - t_B) \tag{4.4}$$

we obtain

$$1/c_\rightarrow + 1/c_\leftarrow = 2/c. \tag{4.5}$$

But, as Poincaré stressed, there is a vicious circle in a one-way two-clock experiment. This is clear from Eq. (4.5) which is one equation for two unknowns. Consequently, Einstein proposed "further definitions" in order to define "a common 'time' for A and B":

$$t_B - t_A = t'_A - t_B, \qquad (4.6)$$

or, equivalently,

$$c_{\leftarrow} = c_{\rightarrow} = c. \qquad (4.7)$$

But Einstein need not have referred to Eqs. (4.6) or (4.7) as definitions, because they are consequences of the two postulates of the relativity theory; equivalently, the isotropy and homogeneity of the space of inertial reference systems for the propagation of light are consequences of the two postulates of relativity. If space were not *isotropic* for light propagation then certain directions could be preferential; if space were not *homogeneous* for light propagation then certain points could be preferential.[7]

In summary, Einstein (1917a) considered the isotropy of space as a "very powerful argument in favor of the principle of relativity." Poincaré (1898), on the other hand, wrote concerning the isotropy of space for light propagation, that "this postulate could never be verified directly by experiment." Lorentz's 1904 theory ensured the isotropy of space to all orders in v/c by means of several interlocking hypotheses which were based on the dynamics of electrons interacting with the ether. Like Poincaré, Einstein realized the contradictory conceptions of light propagation held by electrodynamics and astronomy. The two postulates of relativity theory enabled Einstein to eliminate the asymmetrical treatment of light propagation by two branches of physics which he took axiomatically to be on equal footing. This is a good example of Einstein's ability to dissect a physics problem into subproblems of ascending degrees of complexity: in 1905 he replaced the complex problem of the dynamics of the electron with another problem that was then soluble — a consistent description of the "kinematics of the rigid body," which was a prerequisite to solving the dynamics of the electron. Even though the inertial system — composed of rigid rods and an associated isotropic and homogeneous space — was a logical inconsistency, nevertheless its use as a concept enabled Einstein to connect empirical data in a new way; this, after all, was the role he assigned to concepts. In his (1907e) Einstein found that widening the principle of relativity to include the noninertial reference system, led him to reject the light axiom[8] and by 1913 he discovered that higher-level theories, discussing the problem of gravitation and noninertial reference systems, paid the price for increased unification of a greater "poverty of concepts."

4.5. THE PROPERTIES OF EINSTEIN'S METHOD FOR CLOCK SYNCHRONIZATION

From Eq. (4.5) follows the central result of Part I [§1, 47–66]:

$$t_B = \tfrac{1}{2}(t_A + t'_A). \qquad (4.8)$$

According to Eq. (4.8), the time at which the light is received at *B* is one-half the sum of the time at which the light signal was emitted from *A* (t_A) and the time at which the light signal reflected from *B* was received at *A* (t'_A). Thus, the time of the clock at *B* should be set as midway between the times of t_A and t'_A. The definition of synchronization Eq. (4.8) (i.e., of distant simultaneity, or more exactly of time) is independent of the distance between *A* and *B*, thereby avoiding a pitfall of an earlier method of clock synchronization proposed by Einstein.

Einstein assumed that the "definition of synchronization" is free of "contradictions" and stated that it is reciprocal and transitive. Its reciprocity follows directly from the second postulate, and its transitivity follows from Eq. (4.6).[9] Consequently from reasoning based on applying the postulates of the relativity theory to "certain (imaginary) physical experiments," Einstein overcame the problems inherent in the "connect in time series of events occurring at different places, or — what comes to the same thing — to evaluate the times of events occurring at places remote from the watch." The two-way, two-clock "imaginary" or *Gedanken* experiment enabled Einstein to give an in-principle operational definition of "simultaneous," or "synchronous," and "time." Thus Einstein shifted the problem of defining the distant simultaneity of two events (or, as Poincaré wrote, facts) to that of defining the synchronization of two distant clocks. He succeeded in defining the time of an event as the time registered on a clock located at the event's spatial position.[10] Furthermore, the synchronization of a clock synchronized according to Eq. (4.6) with a specified clock at relative rest is permanent as long as the two clocks remain at relative rest.

By Eq. [§1.2] being in "agreement with experience" Einstein could well have meant Fizeau's measurement of the velocity of light which utilized it. Although, since Fizeau's experiment used only one clock, no new notion of simultaneity was required. Nevertheless Einstein's and Fizeau's equations are only mathematically equivalent because in the relativity theory *c* is a universal constant, and the quantity time is redefined. Einstein seemed to remind the reader that by postulation the special relativity theory avoided unknown quantities in a Fizeau experiment, such as $c_\rightarrow$ and $c_\leftarrow$.

In his conclusion to this section, Einstein emphasized that thus far he had discussed only the time in the "resting system." In the next section, Einstein discussed the consequences of the two postulates when synchronized clocks in two inertial systems were compared.[11]

FOOTNOTES

1. The Galilean transformation Eqs. (1.37)–(1.40) yield

$$dx_r = dx - v \, dt \qquad \qquad \text{(A)}$$

or since $dt = dt_r$, then

$$w_{x_r} = w_x - v \qquad \text{(B)}$$

where $w_{x_r} = dx_r/dt$ and $w_x = dx/dt$. Eq. (B) is the Newtonian law for the addition of velocities. For constant linear relative motion between S and S_r we obtain from Eq. (B)

$$a_{x_r} = a_x \qquad \text{(C)}$$

and so

$$F_{S_r} = F_S. \qquad \text{(D)}$$

Consequently, Newton's laws of motion have the same form in S and S_r.

From Lorentz's Eq. (1.79)

$$dt_L = dt - (v/c^2)\,dx \qquad \text{(E)}$$

or

$$dt_L = dt(1 - (v/c^2)w_x). \qquad \text{(F)}$$

Division of Eq. (A) by Eq. (F) yields

$$w_{x_r} = \frac{w_x - v}{1 - (v/c^2)w_x}, \qquad \text{(G)}$$

where $w_{x_r} = dx_r/dt_L$ and Einstein could have interpreted w_{x_r} as the velocity of a point relative to S_r. Since Eq. (G) is not the Newtonian addition rule for velocity, then using Lorentz's modified transformations, Eq. (D) is invalid in S_r.

2. As one of the many bonuses offered by his point of view, in §5 Einstein demonstrated the validity of Eq. (4.3) to all orders in v/c.

3. As Holton (1973d) has written: "This was the kind of operationalist message which, for most of his readers, overshadowed all other philosophical aspects in Einstein's paper. His work was enthusiastically embraced by the groups who saw themselves as philosophical heirs of Mach, the Vienna Circle of neopositivists and its predecessors and related followers providing a tremendous boost for the philosophy that had initially helped to nurture it."

Defining a concept by means of a measurement was later developed by two philosophical "heirs" Bridgman (1949) and Reichenbach (1958) who referred to it as an "operational definition" and a "coordinative definition," respectively. Their terminology is to be understood relative to a view of science that was not Einstein's — empiricism. See Einstein's (1949) criticisms of Bridgman and Reichenbach.

4. In Miller (1979b) I have compared the views of Einstein and Poincaré on the nature of the geometrical axioms. Briefly, in Poincaré's view, if an experiment contradicted an assertion of Euclidean geometry then he would have changed the laws of physics governing that phenomenon. Einstein often expressed his admiration of Poincaré's foundational analysis of geometry, even though he disagreed with Poincaré's conclusions [see Einstein (1917a), (1921a)].

5. Owing to the unambiguous connections between the measuring rods of special relativity and complexes of sense-perceptions, Einstein placed the special relativity theory on a low level in what he referred to (1936) as the "stratification of the scientific system." The general theory of relativity and a theory providing a unified description of

the gravitational and electromagnetic fields, i.e., a unified field theory, fit into his description of higher-level theories which are systems "of the greatest conceivable unity, and of the greatest poverty of concepts of the logical foundation, which are still compatible with the observation made by our senses." For example, in the general theory of relativity physical reality is attributed to the metric tensor $g_{\mu\nu}$ that describes the gravitational field instead of to the differential distance between two space-time points. A goal of the higher-level unified field theory is to remove the duality inherent in the special and general theories where field and particle were introduced on an equal footing; in the unified theory the elementary particles, and thus the measuring rods too, would be deduced from field equations. This theory, however, eluded Einstein.

6. This definition of event is analogous to the one in Hermann Minkowski's (1908b) geometrical version of the Lorentz transformations, which defines an event as the intersection of two world lines (i.e., trajectories in space-time) – for example, the intersection at 7:00 of the trajectories of myself at rest on the train platform and the moving train.

7. The abstracting journal *Jahrbuch über die Fortschritte der Mathematik*, described Einstein's "theory of relativity" to have been based upon three postulates ("*Voraussetzungen*"): The two principles that Einstein proposed (i.e., the ones concerning relativity and the propagation of light), and "the definition of the common time for two moving clocks according to which the 'time' taken by the light to proceed from one clock to the other should be equal to the time of the return jouney [*Jahrb. Fort. Math.*, 36, 920–921 (1905), and the date of publication was listed as 1908 which is reasonable because the term relativity theory was coined by Planck (1906b)]." Kaufmann, on the other hand, interpreted Einstein's generalization of Lorentz's theory of the electron to have been based on one principle – the principle of relative motion (see Section 7.4.1).

8. It is apropos to discuss Einstein's initial attempt at a general theory of relativity in Chapter 12.

9. Assume that a clock at A is synchronous with clocks at B and C. From the definition of synchronization we have:

$$t_B - t_A = t'_A - t_B, \tag{A}$$

$$t_C - t_A = t'_A - t_C. \tag{B}$$

Eqs. (A) and (B) are consistent only if $t_B = t_C$. Therefore the clocks at B and C are synchronous. Reichenbach (1969) proved that attempting to spread the simultaneity of clocks in the neighborhood of A, i.e., arbitrarily close to A, led to a definition of the synchronization of two widely separated clocks that was symmetrical but not transitive.

10. Einstein (1921a) addressed the criticism that "without justification" the velocity of light played a critical role in the special relativity theory. He replied that it was "immaterial" what sort of process defined time, and he chose the velocity of light because about it "we know something certain" from electromagnetic theory. Needless to say, by 1905 "something certain" was known also about the velocity of sound. Besides the complication that the velocity of sound depended on the characteristics of the medium through which it was propagated, Einstein considered the interaction of light with matter as the most pressing problem confronting the physics of 1905. As he

emphasized often (1907e), (1923), "the special theory of relativity is an adaptation of physical principles to Maxwell-Lorentz electrodynamics" (1923). Chapter 7 discusses criticisms of the second axiom.

11. Many recent philosophical analyses of special relativity theory focus upon: (1) other methods to synchronize clocks that avoid assumptions on the one-way velocity of light, e.g., infinitely slow clock transport; (2) the consequences of definitions of clock synchronization other than Einstein's Eq. (4.8). For discussions of (1) see Bowman (1978), Bridgman (1962), Ellis and Bowman (1967), Grünbaum (1973) and Salmon (1975). As first emphasized by the logical-empiricist Reichenbach [(1949), (1958)] the relation

$$t_B = t_A + \varepsilon(t_{A'} - t_A)$$

where $0 < \varepsilon < 1$ provides definitions of clock synchronization that do not violate causality. The definition consistent with the special relativity theory is $\varepsilon = \frac{1}{2}$, other values of ε lead to different one-way light velocities. Certain philosophers, e.g., Reichenbach (1958) and Grünbaum (1973), hold that empirical data cannot decide the value of ε; however, the choice $\varepsilon = \frac{1}{2}$ leads to descriptive simplicity. Winnie's (1970) analysis explores the Reichenbach-Grünbaum thesis of the "conventionality of simultaneity" through the use of ε-dependent Lorentz transformations. He concludes that different choices of ε result in "kinematically equivalent versions of special relativity." Giannoni (1978) continued Winnie's analysis and developed an ε-dependent "generally covariant" formulation of mechanics and electrodynamics. Beyond the pedagogic sense, what sort of further understanding of the relativity theory this analysis can yield is yet to be seen. For example, it mutilates the symmetries that Einstein believed were intrinsic to the description of physical processes.

CHAPTER 5

LENGTH AND TIME ARE RELATIVE QUANTITIES

> [In order to measure the Lorentz] contraction we should have to operate with two perpendicular rods, and with two mutually interfering pencils of light, allowing the one to travel to and fro along the first rod, and the other along the second rod. But in this way we should come back once more to the Michelson experiment...

H. A. Lorentz (1895)

> The special relativity theory resulted in appreciable advances [among them was that] it enforced the need for clarification of the fundamental concepts in epistemological terms.

A. Einstein (1923)

A noncircular and unambiguous method for measuring length using light rays depended on Einstein's extending to every inertial reference system the axiom concerning the propagation of light relative to ether-fixed reference systems. The relativity of time and simultaneity was a startling consequence of Einstein's having combined axiomatics with an epistemological analysis of the measurement of time and length. Consequently, Einstein may well have believed it necessary to prepare the reader of 1905 for the consequences of an axiomatic approach to electrodynamics via kinematics, by presenting in §2 an illustrative example of how an observer in one inertial reference system notes how observers in another inertial system check the synchronization of their clocks.

Einstein began §2 by restating the two postulates that he had proposed in the introductory section [§2, 1–12]:

§2. *On the Relativity of Lengths and Times*

The following considerations are based on the principle of relativity and on the principle of the constancy of the velocity of light. We define these two principles thus —

1. The laws by which the states of physical systems undergo changes are independent of whether these changes of state are referred to one or the other of two coordinate systems moving relatively to each other in uniform translational motion.

2. Any ray of light moves in the "resting" coordinate system with the definite velocity c, which is independent of whether the ray was emitted by a resting or by a moving body. Consequently,

$$\text{velocity} = \frac{\text{light path}}{\text{time interval}}$$

where time interval is to be understood in the sense of the definition in §1.

[In 1905 Einstein used the terms postulate and principle interchangeably.]

Having in §1 broken the vicious circle in clock synchronization, Einstein could define the velocity of light as:

$$\text{"velocity} = \frac{\text{light path}}{\text{time interval}}\text{"} \qquad (5.1)$$

where the light path need not be closed.

In fact, Einstein (1907e) rewrote Equation (5.1) as

$$c = r/(t_B - t_A) \qquad (5.2)$$

where r is the distance between A and B and he prefaced Eq. (5.2) with the statement that c was a "universal constant." Einstein continued in (1907e): "It is by no means natural to expect that the hypothesis stated above, which we call the principle of the constancy of the velocity of light, should be actually satisfied in nature, yet — at least for a coordinate system of a certain state of motion — it is made likely by the confirmations which Lorentz's theory, that is based on the assumption of an absolutely resting ether, have obtained by experiment." Einstein's two footnotes to this passage stated that by "Lorentz's theory" he meant the one in the (1895), and in the second footnote he wrote: "In particular the fact that this theory [Lorentz's (1895)] of the dragging coefficient (Fizeau's experiment) is in agreement with the available data." Therefore, for Einstein, Lorentz's (1895) explanation of Fizeau's 1851 experiment furnished sufficient support for the "principle of the constancy of the velocity of light."

The universality of the second principle is perhaps clouded by Einstein's use in 1905 of the term "resting system": yet any inertial system can be the "resting system." Throughout the 1905 paper, Einstein referred to a particular inertial reference system K as the resting system in order to emphasize how, in a certain sense, the relativity theory was a generalization of Lorentz's electromagnetic theory in which K was the ether-fixed reference system. As Einstein (1907e) wrote, the special relativity theory "originally arose from the coalescence of H. A. Lorentz's theory and the principle of relativity [from mechanics]."

Einstein continued in the relativity paper [§2, 13–58] with an epistemological analysis, based upon in-principle operational definitions, of the concepts of simultaneity and length as these quantities referred to observers in different inertial reference systems.

Using Einstein's notation from his §3, consider two inertial reference systems k and K, equipped with measuring rods and clocks that are synchronized according to the procedure in §1. Suppose that initially k and K are relatively at rest and are superposed, i.e., their x-, y-, and z-axes are aligned. Observers in k and K measure the length of a rod lying along their common x-axes by comparing it with identical measuring rods at relative rest, i.e., by a congruence measurement. Let this length be l. Then k is accelerated to a velocity v relative to K, and Einstein assumes that the rod's characteristics are unaffected by the sudden acceleration.

Einstein next presented two in-principle operational procedures, or *Gedanken* experiments, in which we could "imagine [the moving rod's] length to be determined." (1) When an observer resting in k determined the rod's length by a congruence measurement, "according to the principle of relativity," the result had to be l. The magnitude of any discrepancy would be a measure of violation of this principle. Einstein defined the length l as "the length of the rod in the moving system."[1] (2) An observer at rest in K also measured the rod's length. "At a definite time," i.e., instantaneously, the observer in K marked off on K's x-axis the positions of the moving rod's end points A and B. Einstein referred to this length as r_{AB} — "the length of the (moving) rod in the resting system." According to the "two principles" of Einstein's theory the lengths l and r_{AB} differ, although, as Einstein emphasized, "current kinematics tacitly assumes [them to be] precisely equal." The crux of the matter was that the high velocity of light led scientists to assume the equivalence of "these two operations" and so the moving rod in "geometrical respects could be perfectly represented by *the same* body *at rest* in a definite position." Here Einstein carefully, albeit implicity, differentiated between two different notions of a body's shape — the geometrical and what he referred to as (1907e) as the "kinematical." The observer in k measures the geometrical shape, and the K-observer the kinematical shape; "current kinematics tacitly" assumed their equality.

"We imagine further," wrote Einstein in the relativity paper, that the clocks at A and B in k are always synchronized with those in K. Einstein then offered the following *Gedanken* experiment. "Imagine" that two observers in k at A and B check their clocks' synchronization, using the two-way, two-clock procedure from §1, and assume also that an observer in K monitors this operation with clocks that are synchronized with those at A and B in k. In consequence, the observer in K notes three events E_1, E_2 and E_3 on K's clocks: E_1 — a ray of light leaves A when A coincides with the position of a clock in K at time t_A; E_2 — after traveling the distance r_{AB} (relative to K) the ray is received at B when B coincides with a clock in K at time t_B; E_3 — the ray returns to A after traveling the distance r_{AB} when A coincides with the clock in K at time t'_A. (Einstein's footnote emphasized that the times discussed were always those in K.) Einstein's mathematical analysis relates E_1 with E_2, and E_3 with E_4 as follows:

$$c(t_B - t_A) = r_{AB} + v(t_B - t_A), \tag{5.3}$$

$$c(t'_A - t_B) = r_{AB} - v(t'_A - t_B), \tag{5.4}$$

and he used the second postulate for calculating the left-hand sides of Eqs. (5.3) and (5.4). From Eqs. (5.3) and (5.4) follow Einstein's results:

$$t_B - t_A = r_{AB}/(c - v), \tag{5.5}$$

$$t'_A - t_B = r_{AB}/(c + v). \tag{5.6}$$

Thus, if the clocks at A and B on the moving rod remain in synchrony with those in K, they are not synchronous to observers in k. Consequently, wrote Einstein, "we can attribute no *absolute* meaning to the concept of simultaneity" – simultaneity is a relative concept. Einstein continued by emphasizing a consequence of the relativity of simultaneity: two spatially separated events that are simultaneous in K are not simultaneous in k, which "is in motion relatively to" K.

If, for the purpose of defining time, terms of order v/c are neglected, i.e., c is considered to be much larger than every other velocity as the ether theorists assumed, then Eqs. (5.5) and (5.6) become equal and the clocks at A and B remain in synchrony.

As far as I know Einstein's demonstration in §2 of the relativity of simultaneity is rarely analyzed. Most likely the reason is that the relativistic transformation equations for the space and time coordinates provide the quickest means for deducing the properties of time and simultaneity in moving reference systems. But Einstein's purpose in §2 was essentially epistemological and he did not require excess mathematical baggage to make his point – that is, by referring all measured quantities to the "resting system" K, whose analogue in the ether-based theories satisfied a principle of relativity and a light axiom, Einstein analyzed the meaning of the time of two moving clocks A and B by exploring how observers in another inertial reference system measure the time for a to-and-fro light exchange between A and B.

But what about the relativity of lengths which Einstein signaled in the title to this §2? Perhaps Einstein wished the reader to deduce a rough version of this result, thereby also allowing the reader to be convinced that it followed from the relativity of simultaneity. For example, at this point in the relativity paper, the reader could have deduced the relativity of length as follows. The observer in k at A registers the time interval for light to traverse to-and-fro the rod of length l in k, $\Delta t' = t'_A - t_A$ as

$$\Delta t' = 2l/c. \tag{5.7}$$

From Eqs. (5.5) and (5.6) the observer in K registers the time interval as

$$\Delta t = 2r_{AB}/c(1 - v^2/c^2). \tag{5.8}$$

Comparing these two time intervals yields

$$\frac{\Delta t'}{\Delta t} = \frac{l(1 - v^2/c^2)}{r_{AB}}.$$

(5.9)

We cannot progress any further toward an exact result, however, without knowing how r_{AB} and l are related — that is, we need the transformation equations relating the space and time coordinates in k and K. Perhaps Einstein wanted the reader to realize from Eq. (5.9) that in a geometry where lengths were measured using light rays and where clocks were synchronized by exchanging light signals — and thus simultaneity was a relative notion (i.e., $\Delta t' \neq \Delta t$) — then r_{AB} could not equal l. Or perhaps Einstein was inviting the reader to substitute into Eq. (5.9) a mathematically equivalent result from Lorentz's electrodynamics — that is, $r_{AB} = l\sqrt{1 - v^2/c^2}$, the Lorentz contraction, thereby obtaining the result from Einstein's §4 for the relativity of time.

The notion of measuring lengths using light rays had been proposed before Einstein by A. A. Michelson (1881) and (1887), Lorentz [(1892b), (1895)] and Poincaré [(1905a), (1906)]. Lorentz and Poincaré had proposed, and then rejected, two methods for measuring the Lorentz contraction: (1) a congruence measurement, but the measuring rod would likewise undergo a contraction; (2) a light geometry using two transverse rods of equal length, but this was just another Michelson-Morley experiment. Einstein realized that unambiguous length measurement of moving objects using light rays had to be taken in combination with properly synchronized clocks. To this we add Einstein's insistence that the problem of determining changes of length should for the present be treated as a kinematical problem.

In summary, the far-reaching results of §2 were that there were no such notions as the true time or the true length of an object; rather these were relative concepts: for example, the length of the rod was either l or r_{AB}, depending upon the rod's motion relative to an inertial observer.

FOOTNOTE

1. Einstein (1921b) elaborated on the relativistic notion of congruence as follows: We make two "tracts," i.e., marks, on a practically-rigid body. If the tracts on B were congruent with those on another practically-rigid body A, when A and B were at relative rest, they would be congruent whenever A and B were brought together.

CHAPTER 6

THE RELATIVISTIC TRANSFORMATIONS

> The new feature of [the 1905 relativity theory] was the realization of the fact that the bearing of the Lorentz transformation transcended its connection with Maxwell's equations and was concerned with the nature of space and time in general. A further new result was that the "Lorentz invariance" is a general condition for any physical theory. This was for me of particular importance because I had already previously found that Maxwell's theory did not account for the micro-structure of radiation and could therefore have no general validity.
>
> A. Einstein to C. Seelig, 19 February 1955, quoted from Born (1969)

Whereas Lorentz and Poincaré considered the Lorentz transformations as a separate postulate necessary for deriving the covariance of electromagnetic theory, in this section Einstein deduced these transformations from two axioms which concerned "the nature of space and time in general."

6.1. EINSTEIN'S DERIVATION OF THE RELATIVISTIC TRANSFORMATION EQUATIONS

Einstein here [§3, 1–31] presents striking imagery – the reader of 1905 had to "imagine" two inertial reference systems (k and K), constructed from three mutually perpendicular rigid rods originating from a common point (their origin); the rods were of grandiose dimensions; they glided through space at velocities approaching c; they were equipped with sets of measuring rods and networks of synchronized clocks. Our exposure to modern art, and to such futuristic conceptions as the massive star ships that traverse current science fiction, have prepared us to imagine easily Einstein's mechanical models of inertial reference systems.

Einstein's detailing the in-principle operational procedures already introduced in §1, for establishing unambiguous physical meaning to the space and time coordinates in k and K, underscores his opinion that the physics of 1905 was so unclearly formulated that an introduction to a foundational analysis was worth repeating. The spatial coordinates relative to k (K) were determined by rigid measuring rods at rest in k (K). Einstein designated the spatial coordinates in k and K as (ξ, η, ζ) and (x, y, z), respectively; complete specification of an event required determination of its temporal coordinate by means of a properly synchronized clock at the event's spatial position – for example, a position on the train platform as the locomotive passes by.

If [§3, 32–33], for example, the transformation equations relating (x, y, z, t) with (ξ, η, ζ, τ) were quadratic then a point located by a single set of coordinates relative to k could correspond to two sets of coordinates relative to K; another possibility was that the velocity of light could depend upon the orientation of the emitter. But since the axioms of relativity theory "attribute to space and time" the "properties of homogeneity," then every point in the space of inertial reference systems must be equivalent.

Consistent with the internal structure that Einstein imposed upon the 1905 exposition of his theory, he derived the relativistic space and time transformations from the definition of simultaneity in §1, Eq. (4.8) — that is, from the definition of clock synchronization in an inertial reference system. He proceeded [§3, 34–71] in two steps by relating the sets of space and time coordinates in k (ξ, η, ζ, τ) with those in K (x, y, z, t) through an auxiliary set of space and time coordinates in k $(x', y'(=y), z'(=z), t'(=t))$, whose spatial portion x', y, z transforms according to the Galilean transformations; but every time coordinate is relativistic. Einstein's intermediate Galilean system bears some similarity with Lorentz's 1895 transformation Eqs. (1.76)–(1.79).

Consider the operation of synchronizing two clocks situated on the ξ-axis of k. One clock is at k's origin A and the other is at the point B a distance x' from A. A light ray emitted from A at time τ_0, is received at B at time τ_1, and then reflected back to A where it is received at time τ_2. According to the definition of clock synchronization in k

$$\tau_1 = \tfrac{1}{2}(\tau_0 + \tau_2). \tag{6.1}$$

For Einstein Eq. (6.1) expressed more than a mathematical relation, each symbol held physical content through an in-principle operational definition of time — "τ is nothing else than the collection of the data of clocks at rest in the system k, which have been synchronized according to the rule given in §1." Toward deducing the relativistic space and time transformations, Einstein next evaluated the functional dependence of τ_0, τ_1, and τ_2 on x', y, z and t. After a time t_1 an observer in K notices that a light ray emitted from A at an earlier time t arrives at B. At this point in the derivation Einstein avoided discussing a contraction of length because x' is a Galilean coordinate. Einstein could not deduce this contraction until after he derived the space and time transformation equations.

The light ray leaves the origin of k, i.e., from A, at time τ_0, as registered by the clock at A in k, and at the time t when A was at the position of an observer on K's x-axis; this is event E_1 whose time is

$$\tau_0 = \tau_0(0, 0, 0, t); \tag{6.2}$$

in what follows the times τ in k are functions of (x', y, z, t), i.e., $\tau = \tau(x', y, z, t)$; so in Eq. (6.2) the three zeros indicate that $x' = y = z = 0$.

The next event E_2 in k is the arrival of the light ray at B which is a distance x' from A; this time in k is τ_1. Einstein used the second postulate in K in order to calculate the time of arrival as registered by an observer in K at the place where E_2 occurred in k; so the light ray traversed the distance x between A and B in a time $t_1 = x/c$. Since $x = x' + vt_1$, then

$$t_1 = x'/(c - v). \tag{6.3}$$

Accordingly, the time interval between emission and reception of the light ray was measured in K as

$$t + t_1 = t + x'/(c - v), \tag{6.4}$$

and so τ_1 is

$$\tau_1 = \tau_1(x', 0, 0, 0, t + x'/(c - v)). \tag{6.5}$$

The event E_3 is the reception of the light ray at A after a time τ_2. According to K the time t_2 for the ray reflected from B to arrive at A is

$$t_2 = x'/(c + v), \tag{6.6}$$

and the round trip time is thus

$$t + t_1 + t_2 = t + x'/(c - v) + x'/(c + v). \tag{6.7}$$

Consequently, the time τ_2 becomes

$$\tau_2 = \tau_2(0, 0, 0, t + x'/(c - v) + x'/(c + v)). \tag{6.8}$$

The velocity of light in Eqs. (6.5) and (6.8) is not c because x' is a Galilean coordinate. Einstein removed this coordinate shortly; but at this point its purpose was to avoid discussing an effect not yet deduced – the length contraction.

Using Eqs. (6.1), (6.2), (6.5) and (6.8), the definition of clock synchronization in k becomes

$$\tau\left(x', 0, 0, t + \frac{x'}{c - v}\right) = \frac{1}{2}\left\{\tau(0, 0, 0, t) + \tau\left(0, 0, 0, t + \frac{x'}{c - v} + \frac{x'}{c + v}\right)\right\}. \tag{6.9}$$

Toward obtaining differential equations whose solutions provide the functional dependence of τ on (x', y, z, t), Einstein took x' to be infinitesimal and expanded both sides of Eq. (6.9) as a series in x'. Neglecting terms higher than first order the result is

$$\frac{\partial \tau}{\partial x'} + \frac{v}{c^2 - v^2}\frac{\partial \tau}{\partial t} = 0. \tag{6.10}$$

Eq. (6.10) gives only the dependence of τ on x' and t. Einstein next obtained differential equations for τ as functions of y and z by means of the following variation on the *Gedanken* experiments for the synchronization of clocks positioned along the ξ-axis of k. The problem is to synchronize the clock A with another at a point Q which is on k's η-axis, at a distance $y' = y$ above A. The y-axes of k and K remain parallel to one another and at time t_1 are a distance $d = vt_1$ apart. Observers in k at A and Q begin the synchronization procedure when k and K are superposed. At time τ_A the observer at A sends a ray of light that is received at Q at time τ_Q and then reflected back to A where it is received at time τ'_A. The clocks are synchronized if

$$\tau_Q = \tfrac{1}{2}(\tau_A + \tau'_A). \tag{6.11}$$

Since this procedure takes place along k's η-axis, then $x' = z = 0$. By analyzing the event in the way just discussed, the observer in K determines that the light ray takes a time t_1 to travel from A to Q. During the time t_1, the y-axes of k and K become separated by a distance $d = vt_1$. With respect to the observer in K the light ray connecting A with Q has traveled a distance $r = ct_1$ which is the hypotenuse of a right triangle. Consequently, the observer in K measures the distance y traversed by the light ray connecting A to Q as:

$$y = \sqrt{r^2 - d^2} \tag{6.12}$$

or

$$y = \sqrt{c^2 - v^2}\, t_1. \tag{6.13}$$

Thus, the velocity of light along the y-axis of k as measured in K is $\sqrt{c^2 - v^2}$. The functional dependence of τ_A on x', y, z, t is

$$\tau_A = \tau(0, 0, 0, t). \tag{6.14}$$

From Eq. (6.13) we may write τ_Q as

$$\tau_Q = \tau(vy/\sqrt{c^2 - v^2}, y, 0, t + y/\sqrt{c^2 - v^2}), \tag{6.15}$$

and τ'_A as

$$\tau'_A = \tau(2vy/\sqrt{c^2 - v^2}, 0, 0, t + 2y/\sqrt{c^2 - v^2}). \tag{6.16}$$

Thus Eq. (6.11) can be written as

$$\tau\left(vy/\sqrt{c^2 - v^2}, y, 0, t + \frac{y}{\sqrt{c^2 - v^2}}\right)$$
$$= \frac{1}{2}\left[\tau(0, 0, 0, t) + \tau\left(2vy/\sqrt{c^2 - v^2}, 0, 0, t + \frac{2y}{\sqrt{c^2 - v^2}}\right)\right]. \tag{6.17}$$

The result of an expansion of Eq. (6.17) for y considered as infinitesimal is

$$\partial\tau/\partial y = 0. \tag{6.18}$$

Identical considerations obtain for the synchronization of clocks along the z-axis of k. The result is

$$\partial \tau / \partial z = 0. \tag{6.19}$$

Hence, τ is independent of y and z. Einstein could next deduce a form for τ consistent with Eqs. (6.10), (6.18), and (6.19)

$$\tau = a(v)\left[t - \frac{v}{c^2 - v^2}x'\right], \tag{6.20}$$

with $a(v)$ an arbitrary function of v which Einstein determined later. Since Eq. (6.10) contains partial derivatives of τ we may add an arbitrary constant to Eq. (6.20); but this means only that the light ray was emitted from some point other than k's origin, and so Einstein omitted it.

Einstein next introduced relativistic spatial coordinates by demanding that the second postulate be satisfied in k — that is, that a ray of light emitted at $\tau = 0$ propagated along the direction of increasing ξ in k according to the equation

$$\xi = c\tau. \tag{6.21}$$

Substituting for τ from Eq. (6.20) yields

$$\xi = ac\left[t - \frac{v}{c^2 - v^2}x'\right]. \tag{6.22}$$

The Galilean coordinate x' could be written in terms of the time t in the "resting system" K by means of the equation

$$x' = x - vt. \tag{6.23}$$

Relative to K the light ray emitted in k travels a distance $x = ct$ (since x is a relativistic coordinate); consequently from Eq. (6.23)

$$t = x'/(c - v). \tag{6.24}$$

Substituting Eq. (6.24) into Eq. (6.22) gives

$$\xi = a\frac{c^2}{c^2 - v^2}x', \tag{6.25}$$

which shows that the relativistic coordinate ξ and Galilean coordinate x' are related by a scale factor. Upon eliminating x', Eq. (6.25) becomes

$$\xi = \frac{a}{1 - v^2/c^2}(x - vt). \tag{6.26}$$

Similar considerations for η and ζ yield, i.e., $\eta = c\tau$ and $\zeta = c\tau$,

$$\eta = (a/\sqrt{1 - v^2/c^2})y, \tag{6.27}$$

$$\zeta = (a/\sqrt{1 - v^2/c^2})z. \tag{6.28}$$

Then, without prior warning Einstein replaced $a(v)$ with $\phi(v)\sqrt{1 - v^2/c^2}$ to obtain

$$\tau = \phi(v)\gamma(t - (v/c^2)x), \tag{6.29}$$

$$\xi = \phi(v)\gamma(x - vt), \tag{6.30}$$

$$\eta = \phi(v)y, \tag{6.31}$$

$$\zeta = \phi(v)z. \tag{6.32}$$

But why did Einstein make this replacement? It seems as if he knew beforehand the correct form of the set of relativistic transformations (in a moment he proves that $\phi = 1$). For the sake of continuity discussion of this point is postponed to the commentary.

Einstein next [§3, 72–85] demonstrated the consistency of Eqs. (6.29)–(6.32) with the second postulate. Previously he had invoked this postulate in k through Eq. (6.21), with similar equations for η and ζ. Assume that at $t = \tau = 0$ the origins of k and K coincide. At this time a spherical light pulse is emitted from the origin of K. A point x, y, z, t on the wave front satisfies the equation

$$x^2 + y^2 + z^2 = c^2t^2. \tag{6.33}$$

Using Eqs. (6.29)–(6.32) to replace x, y, z, t with ξ, η, ζ, τ yields

$$\xi^2 + \eta^2 + \zeta^2 = c^2\tau^2. \tag{6.34}$$

This wave is also spherical in k and also travels with velocity c with respect to k. Thus Eqs. (6.29)–(6.32) satisfy the two postulates of the relativity theory, and "after ten years of reflection" the paradox from the Aarau period was resolved. The form invariance of Eq. (6.33) was the first proof in the relativity paper that the two postulates of relativity were "only apparently irreconcilable." Furthermore, we can discern in the first paragraph of the passage under discussion from the 1905 paper, an effect on Einstein of Lorentz's *Versuch*. For Einstein's rest system K could be taken in analogy with one of Lorentz's reference systems fixed in the ether, in which the velocity of light was always c, and was independent of the emitter's motion.

However, any Φ in Eqs. (6.29)–(6.32) satisfies the second postulate [§3, 86–112]. In order to determine the value of Φ that is consistent with the principle of relativity, Einstein investigated the length of a moving rod. He used a two-step procedure:

(1) A third relativistic system of coordinates K′ moves relative to k with a velocity $-v$. If all three coordinate origins coincide at $t = t' = \tau = 0$ (where primed coordinates refer to K′), then measurements in the three systems can be related by means of appropriate forms of Eqs. (6.29)–(6.32). Transforming from k to K′ yields:

$$t' = \phi(-v)\gamma(\tau + (v/c^2)\xi), \tag{6.35}$$

$$x' = \phi(-v)\gamma(\xi + v\tau), \tag{6.36}$$

$$y' = \phi(-v)\eta, \tag{6.37}$$

$$z' = \phi(-v)\zeta, \tag{6.38}$$

(in an unfortunate shift of notation Einstein used x', y', z' to mean relativistic and not Galilean coordinates, as had designated x', y', z' earlier in §3) while transforming from k to K gives

$$\tau + (v/c^2)\xi = \phi(v)t/\gamma, \tag{6.39}$$

$$\xi + v\tau = \phi(v)x/\gamma, \tag{6.40}$$

$$\eta = \phi(v)y, \tag{6.41}$$

$$\zeta = \phi(v)z. \tag{6.42}$$

Upon substituting Eqs. (6.39)–(6.42) into (6.35)–(6.38) we obtain

$$t' = \phi(v)\phi(-v)t, \tag{6.43}$$

$$x' = \phi(v)\phi(-v)x, \tag{6.44}$$

$$y' = \phi(v)\phi(-v)y, \tag{6.45}$$

$$z' = \phi(v)\phi(-v)z. \tag{6.46}$$

Since Eqs. (6.44)–(6.46) are independent of t, then K and K′ must be the same system; therefore,

$$\phi(v)\phi(-v) = 1. \tag{6.47}$$

The transformation from K to K′ is the identity transformation.

Next Einstein investigated the relationship between $\Phi(v)$ and $\Phi(-v)$. Consider that in k there is a rod of length l (as measured by observers at rest in k) oriented along the y-axis. In k its coordinates are $\eta_1 = 0$ and $\eta_2 = l$. Thus, the length of the rod in k is $\eta_2 - \eta_1 = l$. From Eq. (6.31) the length of the rod in K is

$$y_2 - y_1 = l/\phi(v). \tag{6.48}$$

According to Eq. (6.48) the rod's length measured by an observer in K depends only upon the magnitude of the relative velocity. Since the rod's dimension parallel to its direction of motion depends upon only the magnitude of v, then it is reasonable to assume "from reasons of symmetry" that this holds also for dimensions normal to the rod's motion. In other words, if the rod's dimensions were a function of its direction of motion, then the space of inertial reference systems could not be isotropic, i.e., inertial space would have lacked a basic symmetry property; therefore, "from reasons of symmetry"

$$\phi(v) = \phi(-v). \tag{6.49}$$

Eqs. (6.49) and (6.47) yield

$$\phi^2(v) = 1 \tag{6.50}$$

or

$$\phi(v) = \pm 1. \tag{6.51}$$

In subsequent publications Einstein wrote simply that the negative solution need not be considered [e.g., (1907e)]. Hence Eqs. (6.29)–(6.32) become[1]:

$$\tau = \gamma(t - vx/c^2), \tag{6.52}$$

$$\xi = \gamma(x - vt), \tag{6.53}$$

$$\eta = y, \tag{6.54}$$

$$\zeta = z. \tag{6.55}$$

To paraphrase a statement made earlier in §3 by Einstein: Eqs. (6.52)–(6.55) represent how the collection of the data of rods and clocks in k are transformed to K. All of the symbols in these equations have a physical meaning. Since all inertial reference systems are equivalent the relativistic transformation equations possess the property that their inverse can be obtained simply by interchanging v with $-v$ and the Greek symbols with the Roman symbols – this property is referred to as reciprocity.

6.2. COMMENTARY TO EINSTEIN'S §3

6.2.1. On Einstein's Derivation of the Relativistic Transformations

How did Einstein know that he had to make the further substitution $a = \phi(v)\sqrt{1 - v^2/c^2}$ in order to arrive at those space and time transformations in agreement with the postulates of relativity theory? From Lorentz's *Versuch* Einstein knew that the spatial coordinate x very likely required a transformation other than the Galilean one (recall Lorentz's use of two sets of transformations in 1895); from Einstein's own testimony in (1907e), he had realized that a version of the local time should be the physical time. Lorentz's plausibility argument in the *Versuch* for the hypothesis of contraction could have been useful to Einstein because it was based partly on the spatial transformation $x' = \gamma x_r$. But there was no such exact mathematical comparison for Einstein's time coordinate. Although Lorentz's (1904c) version of the local time was discussed in the scientific literature by Cohn (1904a) – see Chapter 3, footnote 42, there is no extant documentation that Einstein was aware of Cohn's (1904a) previous to writing the relativity paper.

Rewriting Eq. (6.20) as

$$\tau = a\gamma^2\left(t - \frac{v}{c^2}x\right) \tag{6.56}$$

and inverting it along with Eqs. (6.26), (6.27) and (6.28) gives

$$t = \frac{1}{a}\left(\tau + \frac{v}{c^2}\xi\right), \tag{6.57}$$

$$x = \frac{1}{a}(\xi + v\tau), \tag{6.58}$$

$$y = \frac{1}{a}\sqrt{1 - v^2/c^2}\,\eta, \tag{6.59}$$

$$z = \frac{1}{a}\sqrt{1 - v^2/c^2}\,\zeta. \tag{6.60}$$

Except for the function a, these equations are the same as Voigt's (see Chapter 1, footnote 50)[2]. It is straightforward to prove that choices other than $a = \sqrt{1 - v^2/c^2}$ serve to transform Eq. (6.33) into Eq. (6.34). So from the *Versuch* Einstein knew that $a = \sqrt{1 - v^2/c^2}$, i.e., $\phi = 1$. But bearing in mind the undefined quantities lurking in Lorentz's theory, Einstein may well have based his own first proof on the one in his (1917a) for comparing the lengths of measuring rods between inertial reference systems; that proof used certain notions not yet developed until §4 of the relativity paper: Assume that along the x-axes of K and k there are measuring rods of unit length and define the length of the unit measuring rod in K as measured by an observer in k as $l_k^{\rm K}$. The principle of relativity asserts that $l_{\rm K}^{\rm k} = l_k^{\rm K}$ (see also §4, 18–19). If simultaneously in K two observers mark the endpoints of the rod in k as it moves by them, then from Eq. (6.26) we have

$$l_k^{\rm K} = a/(1 - v^2/c^2) \tag{6.61}$$

since $t = 0$, and $l_{\rm K}^{\rm K} = 1$. Consider now that simultaneously in k two observers measure the length of the rod in K. From Eq. (6.58) we have

$$l_{\rm K}^{\rm k} = 1/a \tag{6.62}$$

where $l_k^{\rm k} = \varDelta\xi = 1$. Since $l_k^{\rm K} = l_{\rm K}^{\rm k}$, then we have from Eqs. (6.61) and (6.62) that

$$a = \sqrt{1 - v^2/c^2}, \tag{6.63}$$

as usual the negative solution is rejected. Thus, Eqs. (6.26), (6.27), (6.28) and (6.56) become Eqs. (6.52)–(6.55). An additional bonus from this method of proof is that the function $\phi(v)$ does not appear. If $l_k^{\rm K}$ were not a function of the relative velocity's magnitude only, then the moving rod's length, when measured in another reference system, could depend upon the rod's orientation in violation of the principle of relativity. Since this method of derivation was inconsistent with the relativity paper's internal structure, Einstein had to replace $a(v)$ by $\phi\sqrt{1 - v^2/c^2}$ and then prove that $\phi(v) = 1$. I conjecture that what Einstein had in mind in 1905 by "from reasons of symmetry" was the condition that $l_K^k = l_k^{K}$.[3]

In summary, on the basis of this chapter it seems as if Einstein knew beforehand the spatial portion of the relativistic transformations and an approximate version of the correct time coordinate. Using the spatial portion as a guide he deduced the time coordinate from demanding that the velocity of light remain invariant. It is difficult to imagine that Einstein first derived the relativistic transformations by the method described in the 1905 paper; in fact, he never used this method again.[4]

6.2.2 A Comparison Between the Space and Time Transformations of Einstein and Lorentz

Although the mathematical form of Einstein's transformation equations is the same as Lorentz's, the equations differ in physical content. In Einstein's transformation equations every space and time coordinate has a physical interpretation, whereas Lorentz and Poincaré interpreted the relativistic transformation equations merely as mathematical coordinate transformations. For Einstein the relativistic transformations related experimental data obtained on the rods and clocks of observers in two inertial reference systems. Einstein's rest system K could be compared with a preferred reference system at rest in the ether; thus for Lorentz only the coordinates x, y, z, t had physical meaning. On the other hand, Einstein's reference system k played the role of Lorentz's 1904 reference system Σ' and in less exact forms Lorentz's mathematical coordinate systems Q' (1892a), R' (1892b), S'' (1895) or the modified Galilean transformations in (1895). For example, in the (1895) and (1904c), Lorentz's plausibility argument for the Lorentz contraction hypothesis involved cross-multiplying the quantities $\sqrt{1 - v^2/c^2}$ in the spatial portions of the S'' and S' transformations, respectively. But since in special relativity K and k were equivalent, then Einstein could move between k and K by changing v to $-v$, and interchanging Greek and Roman letters. For Lorentz in 1904 this interchange had no physical interpretation because the system K was fixed in the ether. In 1905 Poincaré attributed only a mathematical interpretation to the reciprocity property of the Lorentz transformations — that is, reciprocity corresponded to a rotation of K and k by $180°$ about their common y-axes.

Chapter 1 outlined Lorentz's circuitous route from 1892 to the space and time transformations of 1904. His several derivations focused upon transforming the equations of electromagnetism, written in S to an inertial reference system S_r, and then to a fictitious reference system having all the properties of a system fixed in the ether; in the fictitious reference system the optics and electrodynamics of moving bodies were reduced to problems concerning bodies at rest. Chapter 1 also discussed how Lorentz's thinking was often finely tuned to recent experimental data—for example, it was Kaufmann's

(1901–1903) data that urged Lorentz's return to the transformations of 1899. Then in 1904 Lorentz presented the space and time transformations as one of the several postulates of a theory of the electron. In summary, Einstein saw in Lorentz's *Versuch* (1895) three different coordinate systems (S'', S_r, S) related in two different ways, i.e., Eqs. (1.63)–(1.66) and (1.76)–(1.79), and postulated for use in two different types of problems (S'' as an aid in electrodynamical problems, but not in optical problems). Furthermore, Lorentz's plausibility argument for the hypothesis of contraction used the transformation for electrodynamical problems, even though the Michelson-Morley experiment concerned the optics of moving bodies. The physical interpretation of Lorentz's electron theory was rendered further unclear when Poincaré symmetrized Lorentz's space and time transformations from 1904 by eliminating S_r, because Poincaré related the imaginary system Σ' to the ether-fixed system S, with which inertial observers could not communicate (at least to second order in v/c). But for purposes of solving problems in the electrodynamics and optics of moving bodies, the laboratory reference system was taken as S, thereby tacitly assuming that unknown velocities relative to the ether could be discarded since they were negligible compared to the velocity of light.

In *The Theory of Electrons* (1909), Lorentz further discussed his ether-based version of the Lorentz transformations. There he referred to the coordinates in $S(x, y, z, t)$ and $S_r(x_r, y_r, z_r, t_r(= t))$ as the "true" coordinates and "true" time; those in $S'(x', y', z', t')$ as the "effective coordinates" and the "effective time." Then, as if in response to Einstein's recollection in (1907e) (that what Lorentz called "an auxiliary mathematical quantity" was in fact the physical time), Lorentz wrote in the 1916 edition of *The Theory of Electrons*:

> If I had to write the last chapter now, I should certainly have given a more prominent place to Einstein's theory of relativity by which the theory of electromagnetic phenomena in moving systems gains a simplicity that I had not been able to attain. The chief cause of my failure was my clinging to the idea that the variable t only can be considered as the true time and that my local time t' must be regarded as no more than an auxiliary mathematical quantity. In Einstein's theory, on the contrary, t' plays the same part as t . . . .

In summary, Lorentz and Poincaré considered the Lorentz transformations as a mathematical device for deducing a principle of relativity for Lorentz's theory of the electron. As Einstein emphasized in the quotation at the start of Chapter 6, he considered the relativistic transformations to have a much deeper connotation.

FOOTNOTES

1. The relativistic transformations for the relative velocity taken in an arbitrary direction are:

$$\xi = r + \frac{1}{c^2}(\gamma - 1)(r \cdot v)v - \gamma vt,$$

$$\tau = \gamma\left(t - \frac{1}{c^2}(r \cdot v)\right).$$

These equations remove any dependence of the relativistic transformations on the choice of coordinate axes. They first appeared, although not in vector form, in Herglotz (1911).

2. Thus, incorrect is the footnote added to the *Physikalische Zeitschrift* reprint of Voigt's (1887): "...except for the employment of the irrelevant factor $[\sqrt{1 - v^2/c^2}$, Voigt's transformations are] identical to. the Lorentz transformation from the year 1904." The editors of the *Physikalische Zeitschrift* reprinted Voigt's (1887) paper to remind their readers, on the occasion of the birthday celebration of the principle of relativity, of Voigt's early work. Since Voigt's paper was received for reprinting on 8 October 1915, it is unclear what is meant by the birthday celebration of the principle of relativity. Einstein's special relativity paper of 1905 was received 30 June, and published on 26 September.

3. Lorentz (1904b) referred to Φ as l and proved that it was unity from the equations for the electron's longitudinal and transverse masses, in conjunction with the condition that l becomes unity in the limit of $v \ll c$. Poincaré (1906) was not satisfied with Lorentz's proof, and proved the unity of l in two different ways: from group theory and from the transformation properties of the equations for quasi-stationary motion.

4. For example, Einstein (1917a) deduced the relativistic transformations by calculating the linear transformation that changed Eq. (6.33) into (6.34); Sommerfeld proposed this method in his footnote to Einstein's §3.

Von Laue (1911a) derived the Lorentz transformations by calculating the space-time linear transformations that left the d'Alembertian invariant, *and* was also consistent with the principle of relativity.

W. von Ignatowsky (1910b) was one of the first to investigate the possibility of deducing the Lorentz transformations without invoking the second postulate. His assumptions were: (1) Einstein's principle of relativity; (2) the isotropy and homogeneity of space, which implies linearity of the transformation equations; (3) reciprocity. Von Ignatowsky could identify the "universal space-time constant" in the resulting transformations as the velocity of light c only by recourse to an electrodynamical argument whose validity, he claimed, was independent of the principle of relativity: to a resting observer the equipotential surface for the convection potential of a moving point charge is a Heaviside ellipsoid. His other misunderstandings of Einstein's theory of relativity are discussed in Chapter 7, for he went on in (1910b) to deduce from the Lorentz transformations that disturbances could propagate with hyperlight velocities.

P. Frank and H. Rothe (1911) pointed out that von Ignatowsky had made several additional "tacit" assumptions—for example, (1) the coordinate transformations constitute a linear homogeneous group of transformations; (2) the length contraction should depend on only the magnitude of the relative velocity, which entered through considerations of the Heaviside ellipsoid. Frank and Rothe's more general derivation was based upon only two explicit assumptions: (1) finite-linear-homogeneous transformation equations could be constructed from one-parameter continuous groups for infinitesimal transformations, i.e., Lie groups; this had been one of Poincaré's (1906) techniques for investigating the Lorentz transformations [see Miller (1973)]. The relative velocity was the single parameter. (2) The length contraction should depend on only the magnitude of the relative velocity. Like von Ignatowsky, Frank and Rothe were unable to identify the invariant velocity in the resulting Lorentz transformations as the velocity of light c. As Pauli (1958) has written: "From the group-theoretical assumption it is only possible to derive the general form of the transformation formulae, but not their physical content." In summary, the importance of the investigations of von Ignatowsky and of Frank and Rothe was that they were the first to attempt dissecting out the assumptions made in deducing space-time transformations. In a sequel publication, Frank and Rothe (1912) asserted that their derivation of the Lorentz transformations was "even simpler than the classical derivation from the invariance of the quadratic form $x^2 - c^2 t^2$." Thus the method of derivation, referred to in Sommerfeld's footnote to Einstein's §3 and commonplace in textbooks today, was by 1912 already "classic."

Robertson (1949) demonstrated how the experiments of Michelson and Morley (1887), Kennedy and Thorndike (1932) and Ives and Stillwell (1938) served to specify completely the coefficients of a linear transformation of space and time coordinates in a four-dimensional space; the result is Lorentz's transformations.

For discussions of the consequences of space-time transformations other than Lorentz's see Reichenbach (1969) and Winnie (1970). Resnick (1968) gives a careful elementary discussion emphasizing that both postulates of the theory of relativity are required to deduce the Lorentz transformations.

CHAPTER 7

THE RELATIVITY OF LENGTH AND TIME

> *A priori* it is quite clear that we must learn something about the physical behaviour of measuring-rods and clocks from the equations of transformation, for the magnitudes x, y, z, t are nothing more nor less than the results of measurements obtainable by means of measuring-rods and clocks. If we had based our considerations on the Galilei transformation we should not have obtained a contraction of the rod as a consequence of its motion.
>
> A. Einstein (1917a)

Straightaway [§4, 1–2] Einstein emphasized the "physical meaning" of the results to follow: "Physical Meaning of the Equations Obtained Concerning Moving Rigid Bodies and Moving Clocks."

7.1. MOVING BODIES: EINSTEIN'S LENGTH CONTRACTION

> [If] one should define as 'time' what Lorentz introduced as an auxiliary quantity, and called the local time [then] the hypothesis of H. A. Lorentz and Fitzgerald appears as a secondary consequence of the theory.
>
> A. Einstein (1907e)

Einstein quickly demonstrated that a body having a spherical shape relative to an observer in its rest system k, Eq. [§4.1], appeared to be an ellipsoid of revolution to an observer in K, Eq. [§4.2] [§4, 3–13]. Although the sphere's dimensions normal to its direction of motion remain unchanged, its dimension along the common direction of the x-axes of k and K "appears shortened in the ratio $1 : \sqrt{1 - v^2/c^2}$, i.e., the greater the value of v, the greater the shortening"; consequently, length is a relative quantity.[1] The terseness of Einstein's derivation of the contraction of moving bodies was perhaps meant to demonstrate not only the power of his axiomatic method vis-à-vis Lorentz's in 1895 which required interlocking hypotheses and fictitious reference systems, but also to emphasize that the Lorentz contraction is a "secondary consequence of the theory." In addition, Einstein's derivation set the stage for §8 where he discusses the transformation properties of the energy of a light pulse.

As v approaches c the contraction in the direction of motion becomes more pronounced (from Eq. [§4.3] the semi-minor axis is $R\sqrt{1 - v^2/c^2}$). The ellipsoid gets flatter until when $v = c$ it appears to an observer in K as a plane perpendicular to the ξ-axis; nevertheless, to an observer in k the object remains a sphere as required by the principle of relativity. For $v > c$ the ellipsoid's semi-minor axis becomes imaginary, and so, as Einstein wrote, "our deliberations become meaningless."[2] (This was also Wien's (1904b) reason for c being the ultimate velocity, and from his readings in the *Annalen* Einstein could have been aware of speculations on hyperlight velocities, particularly the controversy between Abraham (1904b) and Wien (1904b).) Perhaps Einstein did not mention here that unphysical results ensue also for $v = c$ because this was obvious from the relativistic transformations of §3. This result may have provided further evidence for the impossibility of catching up with a point on a light wave, thereby viewing it as a standing wave. In (1917a) Einstein again demonstrated from the length contraction that the velocity of light in vacuum is the ultimate velocity, and he added (1917a) —

> Of course this feature of the velocity c as a limiting velocity also follows clearly from the equations of the Lorentz transformation, for these become meaningless if we choose values of v greater than c.

Einstein's statement supports my assertion above that certain results concerning $v \geqslant c$ were contained already in §3 of the 1905 relativity paper.[3] Einstein concluded his analysis of the shape of a moving body by asserting the reciprocity of length contraction.

Einstein's footnote to the passage under discussion ("That is, a body possessing spherical form when examined at rest.") refers to what in (1907e) he defined as the "geometrical shape"; he went on to define the shape measured by the observer in K as the "kinematical shape." Einstein's definitions served to emphasize that his length contraction differed fundamentally from Lorentz's which asserted that something happened to the rod in S_r, but compensating effects prevented the inertial observer from ascertaining changes in shape. Einstein's (1907e) derivation of the relativistic contraction exposed more clearly than the one in 1905, the relation between the relativities of length and time: In k there is a rod of length $\xi_2 - \xi_1 = l_0$. Two observers in K simultaneously, i.e., $t_1 = t_2$, mark on K's axis the rod's length as $x_2 - x_1 = l$. From Eq. (6.53) Einstein obtained $l = l_0\sqrt{1 - v^2/c^2}$.[4]

7.2. MOVING CLOCKS: THE RELATIVITY OF TIME

Consider [§4, 20–48] that K's x-axis is lined with synchronized clocks, and at k's origin there is one clock. If the origins of k and K coincided at $t = \tau = 0$, then at a later time t the origin of k has moved a distance $x = vt$ relative to K's origin, and [see Eq. (6.52)] the clock at k's origin registers a time observed from K of

$$\tau = t\sqrt{1 - v^2/c^2} \qquad (7.1)$$

where the quantity τ is the elapsed time on the clock at k's origin, and t is the time on a clock located at $x = vt$ on K's x-axis; consequently, the slowing down of a moving clock, i.e., time dilation, is measured by comparing one clock with at least two others. In order to emphasize the physical content of time dilation, Einstein rewrote the result as

$$\tau = t - (1 - \sqrt{1 - v^2/c^2})t, \qquad (7.2)$$

so the time interval in k is thus slower than the one in K by an amount

$$(1 - \sqrt{1 - v^2/c^2}). \qquad (7.3)$$

Expanding the square root in Eq. (7.3), and retaining terms up to second order in v/c, gives $\frac{1}{2}(v^2/c^2)t$. Thus, like length time too was a relative quantity which depended on only the magnitude of the relative velocity between two inertial reference systems. The relativity of time was a new result of Einstein's theory, and Einstein proceeded to investigate it further.

Suppose that separated clocks at A and B on K's x-axis are properly synchronized. Let the clock at A be instantaneously accelerated to the velocity v and transported over to B where it is instantaneously decelerated to rest.[5] A "peculiar consequence" of the prediction of the relativity of time is that the clocked moved from A to B is not synchronous with the one that remained at rest at position B: it is slower than the one at B by an amount of time $(1 - \sqrt{1 - v^2/c^2})$. This difference disposed of Poincaré's 1898 suggestion for synchronizing distant clocks by transporting a clock synchronous with a master clock. Einstein then took this "peculiar consequence" one step further by investigating the case in which the points A and B coincided, and their synchronized clocks C_1 and C_2 were side by side in K. Clock C_1 remains at rest but C_2 undergoes a round trip journey along the sides of a polygon. We may assume that C_2 moves in a succession of inertial reference systems $k_1, \ldots, k_n$, and may thus neglect the abrupt accelerations at each corner of the polygonal path when C_2 is transferred from one inertial system to another. Since each inertial system moves with velocity v relative to K, then Einstein claimed that when C_2 returned to K it lagged behind C_1 by the amount $(1 - \sqrt{1 - v^2/c^2})$, and he assumed the validity of this result when the polygon was replaced by a continuous curve. As an example, Einstein predicted that a clock placed at the equator ran slower than a clock placed at one of the poles. His extension to the case where C_2's trajectory was a continuous curve was unwarranted in 1905, but perhaps he considered that this case could always be treated as the limiting case of a many-sided polygon.[6] (Sommerfeld's footnote to this passage asserts that a pendulum-clock had to be excluded from the considerations of §4 because its oscillations are determined by gravity and hence cannot be discussed kinematically. See Appendix.)

Einstein's (1907e) discussion of time dilation did not contain any "peculiar consequences" because he proposed a real experiment, and not a *Gedanken* experiment, to verify the time dilation phenomenon. In 1907 Einstein went far beyond his intent in 1905 of basing relativity theory upon clocks and measuring rods taken as irreducible structures; now he defined a clock as any periodic process — for example, an atomic oscillator emitting a frequency v_0 as measured in k. From Eq. (7.1) an observer in K measures the frequency

$$v = v_0 \sqrt{1 - v^2/c^2}, \tag{7.4}$$

and Einstein wrote that "the equation [7.4] admits of a very interesting application" since Stark (1906) had observed a Doppler shift in the radiation emitted by canal rays. The "interesting application" of Eq. (7.4) was that it predicted a Doppler effect when the observer's line of sight was normal to the emitter's direction of motion, i.e., a transverse Doppler effect.[7] Einstein asserted that observation of this effect vindicated also time dilation because "we can consider the ion to be a clock." Lorentz's theory did not predict a transverse Doppler effect, although Einstein made no mention of this point (see Chapter 10).

Einstein's prediction was tested successfully in 1937 by two American physicists H. E. Ives and G. R. Stilwell, although Ives maintained that their research verified the "Larmor-Lorentz" theory (1938).[8]

7.3. ON THE RECIPROCITY OF RELATIVISTIC PHENOMENA

> The behaviour of measuring rods and clocks in translational motion, when viewed superficially, gives rise to a *remarkable paradox* which on closer examination, however, vanishes (italics in original).
>
> H. A. Lorentz (1910–1912)

In his lectures at Leiden University (1910–1912) Lorentz analyzed two paradoxes rooted in inferring that the reciprocity of the relativistic contraction and time dilation were achieved merely by interchanging Greek and Roman letters, and replacing v with $-v$. After all, according to the relativity theory all effects were relative. Lorentz first discussed the length contraction.

Consider that the reference systems K and k are equipped with identical meter sticks of length l and are positioned on the systems' x-axes; Lorentz referred to these rods in k and K as 1 and 2, respectively. Rod 1, wrote Lorentz, "appears" to the K-observer to be smaller than the rod 2; rod 2 appears to the k-observer as smaller than rod 1. "But how," continued Lorentz, "can the rod 1 be shorter than and, at the same time, longer than rod 2?" There is no paradox here, rather, Lorentz emphasized that there is a "flagrant contradiction" which could be removed by recognizing that the K and k observers "avail themselves of *different criteria* for measuring lengths" (italics in original). The K observer's measurement is made by marking simultaneously the endpoints of k's rod, but that operation is not simultaneous for the k observer. Thus, concluded Lorentz, resolution of the "contradiction" or "paradox" resulted from recognizing the "important role played therein by their criteria of simultaneity." Lorentz emphasized that Einstein's contraction was not merely an "appearance" but could "actually be observed" in photographs taken from K or k, just as "one photographs the dilatation of a heated body."[9]

"The behaviour of moving clocks," continued Lorentz at Leiden, "gives rise to a similar paradox." Equation (7.1) describes the result of comparing one clock in k with two clocks in K; in order to display this result for any time difference we write

$$\Delta \tau = \Delta t \sqrt{1 - v^2/c^2}. \tag{7.5}$$

According to Lorentz a paradox occurs when we invoke the reciprocity of the relativistic transformation equations in order to achieve the inverse measurement by replacing v with $-v$ and interchanging Greek and Roman letters, which yields:

$$\Delta t = \Delta \tau \sqrt{1 - v^2/c^2}. \tag{7.6}$$

From Eq. (7.5)

$$\Delta \tau < \Delta t, \tag{7.7}$$

and from Eq. (7.6)

$$\Delta \tau > \Delta t. \tag{7.8}$$

But, stressed Lorentz, Eqs. (7.5) and (7.6) "express entirely different things" because Eq. (7.5) describes the result of comparing one clock in k with two spatially separated clocks in K, and Eq. (7.6) expresses the inverse — that is, the measurement of time dilation is not symmetric. Consequently, although time dilation is reciprocal insofar as a clock in k (K) runs slower by the same rate when read by two spatially separated clocks K (k), the measurement process is unsymmetric. The reciprocity of time dilation leads to paradoxical results when "viewed superficially," as Lorentz wrote (1910–1912) — that is, when the reciprocity is demonstrated without taking proper cognizance of the physical content of Eqs. (7.5) and (7.6).

7.4. COMMENTARY ON EINSTEIN'S §4

In addition [to the theories of Abraham, Lorentz and Bucherer] there is to be mentioned a recent publication by Mr. A. Einstein on the theory of electrodynamics which leads to results which are formally identical with those of Lorentz's theory.

W. Kaufmann (30 November 1905)

7.4.1. Kaufmann's 1905–1906 Assessment of Einstein

To the best of my knowledge Kaufmann was the first to mention in print Einstein's relativity paper. Kaufmann's (30 November 1905) note contained some of the data that he had promised Lorentz in his letter of 16 July 1904. Kaufmann's data agreed well with Abraham's and Bucherer's predictions for m_T, but not the one of Lorentz and Einstein. I discuss Kaufmann's experiments of late 1905 in the next section and their impact on Einstein in Chapter 12. Of importance here is how Einstein's research was received as being "formally identical" with Lorentz's and, even more to the point, as a useful generalization of Lorentz's.

Kaufmann explicated this equivalence in a longer paper that had been sent from Bonn on 1 January 1906, and received at the *Annalen* on 3 January 1906. Although the "Lorentz-Einstein fundamental assumption" (i.e., the principle of relativity) fared even worse under the more refined 1905 data, Kaufmann's critical comparison of their views on the measurement of length favored Einstein's. [In fact, this was Kaufmann's third refinement in two months because in a letter dated 4 November 1905 to Sommerfeld, Kaufmann reported not-yet-finalized results that supported the electron theories of Abraham and of Bucherer and "definitely eliminated" Lorentz's (on deposit at the American Philosophical Society).] For Kaufmann pointed out the following serious shortcoming in Lorentz's theory: Lorentz's hypothesis of contraction, in conjunction with certain hypotheses concerning how intermolecular forces transform, served not only to explain the negative results of second-order optical experiments, but also to predict the deformed electron's transverse mass; however, Lorentz's theory could not produce an unambiguous result for the length of a moving rod. Kaufmann's reason was that the moving rod's length depended upon a quantity that had hitherto resisted empirical determination – the "absolute velocity or the velocity relative to the light ether." In order to demonstrate this point Kaufmann proposed the following "*Gedankenexperiment*." The length of a rod resting on the earth and oriented parallel to the earth's velocity v relative to the fixed stars was

$$l = l_0 \sqrt{1 - v^2/c^2} \tag{7.9}$$

with l_0 as the rod's "primitive length." But, continued Kaufmann, "we have no right to assume that the fixed stars, to which we refer terrestrial motion, are actually at absolute rest in the ether." Kaufmann supposed that if they had a velocity v_0 relative to the ether, then the moving rod's length, as measured on the earth, was less than l; rather it could be

$$l' = l_0 \sqrt{1 - (v_0 + v)^2/c^2}. \tag{7.10}$$

To second-order accuracy in the velocities, the length change, which Kaufmann called δl, was a function of the unknown velocity v_0, i.e.,

$$\delta l = l' - l = -\frac{1}{2} l_0 \frac{v^2}{c^2} - l_0 \frac{v_0}{c} \frac{v}{c}. \qquad (7.11)$$

At this point it was unnecessary for Kaufmann to mention Lorentz's 1895 demonstration that the hypothesis of contraction could not be directly observed, or equivalently that terrestrial observers could never determine the true length l_0.

"It is indeed remarkable," continued Kaufmann, that from entirely different hypotheses Einstein obtained results which in their "accessible observational consequences" agreed with Lorentz's, but Einstein avoided difficulties of an "epistemological sort," i.e., the inclusion of unknown velocities. Kaufmann viewed Einstein's success as based upon his raising to a postulate the "principle of relative motion" which he placed at the "apex of all of physics (*die gesamte Physik*)." Then, applying this postulate to the propagation of light, Einstein deduced "a new definition of time and of the concept of 'simultaneity' for two spatially separated points, where Einstein's relationship between the time in two inertial reference systems was identical with Lorentz's local time," i.e., Einstein had derived the Lorentz transformations. Kaufmann went on to emphasize that whereas Lorentz obtained his observationally accessible results through approximations (e.g., that of quasi-stationary motion), Einstein's kinematics of the rigid body, and his electrodynamics, were exact for inertial systems.

Clearly Kaufmann had studied the kinematical part of Einstein's paper carefully, but it was the electrodynamical portion that stuck him as being most important. Although in this part Einstein discussed only a point electron, the identity mathematically of Einstein's and Lorentz's results for the electron's transverse mass was conclusive evidence for Kaufmann that Einstein was merely generalizing Lorentz's theory of the deformable electron. Einstein (1905d) had deduced the electron's m_T and m_L by relativistic transformations between the electron's instantaneous rest system and the laboratory inertial system of the electron's acceleration and of the external electromagnetic fields acting on the electron. An inappropriate choice for the definition of force led Einstein in 1905 to a prediction for m_T [Eq. (12.24)] that differed from Lorentz's by the factor $\sqrt{1 - v^2/c^2}$. Kaufmann was the first to note in print that their predictions should have been the same (1906).

In summary, Einstein was considered to have replaced Lorentz's assumptions of how electrons interact with the ether with purely phenomenological assumptions concerning how clocks were synchronized using light signals, i.e., Einstein had improved Lorentz's theory of the electron. At this juncture Kaufmann mentioned that the difference in their views on length and time merited the designations "Lorentz's theory" and "Einstein's theory." Before long there was only a "Lorentz-Einstein theory" and a "Lorentz-Einstein principle of relativity"; in fact, Kaufmann (1905) had already referred to the "Lorentz-Einstein" theory of the electron.

7.4.2. Kaufmann's 1905 Data

> I anticipate right here that the ... measurement results are not compatable with the
> Lorentz-Einstein fundamental assumption.

W. Kaufmann (1906)

Kaufmann's 1906 opus was his most detailed description thus far of his measurements and data analysis. Let us review his 1906 "general theory of the orbit curve." Immediately Kaufmann wrote that he would consider only "deflections infinitely small relative to the apparatus' dimensions." Since in practice this was not the case he had to reduce the "actual observations to infinitely small deviations," which were directly proportional to the applied electric and magnetic fields and could be easily calculated from the Lorentz force equation. For infinitely small deflections in the magnetic field, the electron's radius of curvature in the x-z-plane ρ is large and so

$$1/\rho = d^2z/dx^2, \tag{7.12}$$

where the exact equation is $1/\rho = d^2z/dx^2/(1 + (dz/dx)^2)$. Consequently

$$d^2z/dx^2 = (e/m_T vc)B. \tag{7.13}$$

The reduced magnetic deflection is

$$z' = (e/m_T vc)\mathbb{B}. \tag{7.14}$$

Kaufmann calculated the quantity $\mathbb{B}$ through a double integration of the function describing how the magnetic field varied with x over the apparatus.

Similarly Kaufmann calculated the reduced electric deflection to be (where $d^2y/dt^2 = v_x^2\, d^2y/dx^2$),

$$y' = (e/m_T v^2)\mathbb{E} \tag{7.15}$$

and the quantity $\mathbb{E}$ is the result of a double integration over a function that describes the empirically determined fringing field of the capacitor which acts on an electron when it enters and leaves the region between the capacitor plates. Kaufmann used more exact equations than previously for calculating y' and z' from the empirical curve coordinates. In order to obtain Eq. (7.15) Kaufmann replaced the quantity v_x with the electron's total velocity v. Kaufmann proved that this approximation was valid to one part in 300 because the largest deflection with only the electric field operating was 0.17 cm, and consequently $v_y \ll v_x$. As he had done in his papers of 1901 and 1903, Kaufmann noted that the electric deflection could not be treated independently of the magnetic deflection. His more exact 1906 expression for the reduced coordinate y' in terms of the empirical curve coordinates (y_0, z_0) can be deduced from the results in footnotes 39 and 40 of Chapter 1 and the equation $y' = y_0 h x_2/2s_1 s_2$ which, strictly speaking, is valid only for a homogeneous electric field, as $y' = y_0(1 - 0.145z_0^2)$. The reduced coordinate z' is $z' = z_0(1 - 0.160z_0^2)$.

Guided by Runge's (1903) suggestion to apply the method of least squares to a linear relationship between y' and z', Kaufmann used different curve constants than he had previously. He rewrote Eqs. (1.155) and (1.156) as

$$z' = \frac{1}{R_1} \frac{1}{\beta\psi}, \tag{7.16}$$

$$y' = R_2 \frac{1}{\beta^2\psi}, \tag{7.17}$$

where

$$\frac{1}{R_1} = \frac{\varepsilon}{\mu_0^e} \frac{\mathbb{B}}{c} \tag{7.18}$$

and

$$R_2 = \frac{\varepsilon}{\mu_0^e} \frac{\mathbb{E}}{c}. \tag{7.19}$$

Equation (7.17) could also be expressed as

$$y' = \frac{R_1 R_2}{\beta} z'. \tag{7.20}$$

Kaufmann's choice of curve constants R_1 and R_2 was suggested by Runge (1903) in order that the linear relationship between y' and z' [Eq. (7.20)] depended on only the applied electric and magnetic fields. Kaufmann listed the quantities $\psi(\beta)$ for the electron theories under consideration as

$$\text{(Abraham)} \quad \psi(\beta) = \frac{3}{4\beta^2} \left(\frac{1 + \beta^2}{2\beta} \ln\left(\frac{1 + \beta}{1 - \beta}\right) - 1 \right), \tag{7.21}$$

$$\text{(Lorentz)} \quad \psi(\beta) = (1 - \beta^2)^{-1/2}, \tag{7.22}$$

$$\text{(Bucherer)} \quad \psi(\beta) = (1 - \beta^2)^{-3/2}, \tag{7.23}$$

where most German physicists referred to the theory of the constant volume electron as Bucherer's theory. In Kaufmann's notation the equation for the reduced curve in the Lorentz-Einstein theory is

$$y'^2 = Cz'^2 + D^2 z'^4 \tag{7.24}$$

where

$$C = \mathbb{E}/\mathbb{B} \tag{7.25}$$

and

$$D = (cC/\mathbb{B})(\mu_0^e/\varepsilon). \tag{7.26}$$

Kaufmann executed his data analysis in four steps:

(1) A least squares fit for R_1 and R_2 using the method suggested by Runge (1903) which Kaufmann developed as follows. He set $u = 1/\beta\phi$ and $1/\beta^2\phi = v = f(u)$. Consequently $y' = R_2 v = R_2 f(R_1 z')$. Then, for each inserted value of z', he calculated from closely spaced values for β closely spaced values for u and v, which he used to seek the values for R_1 and R_2 that minimized the square of the quantity $y' - R_2 f(R_1 z')$ summed over the set of data z' obtained from a given photographic plate. The least squares fit for C and D was easier because Eq. (7.20) did not contain the Lorentz-electron's velocity. Kaufmann found that all three theories fit the reduced curve excellently. The values of R_1, R_2, C and D that he determined from a least squares fit, Kaufmann referred to as "curve constants."

(2) Comparison of the curve constants with the values of R_1, R_2, C and D calculated from $\mathbb{E}$, $\mathbb{B}$ and ε/μ_0^e from cathode rays, which he referred to as "apparatus constants." Quite rightly Kaufmann noted that there was no universal agreement on ε/μ_0^e for cathode rays. Consequently he adroitly eliminated ε/μ_0^e from the comparison between the curve and apparatus constants by comparing instead the product of R_1 and R_2; the constant C in Eq. (7.24) is independent of ε/μ_0^e. Kaufmann found that the discrepancies between the curve and apparatus constants were

$$
\begin{array}{ll}
\text{Abraham} & -3.5\%, \\
\text{Lorentz} & -10.4\%, \\
\text{Bucherer} & -2.5\%.
\end{array} \tag{7.27}
$$

The Lorentz-Einstein theory was a poor third.

(3) Assuming that the observation errors could be distributed over $\mathbb{E}$ and $\mathbb{B}$, Kaufmann hoped to minimize their effect on ε/μ_0^e by calculating ε/μ_0^e from the "geometrical mean" of $\mathbb{E}$ and $\mathbb{B}$, and the apparatus constants. From Eqs. (7.18) and (7.19) Kaufmann obtained for the theories of Abraham and of Bucherer that $\varepsilon/\mu_0^e = c\sqrt{R_2/\mathbb{B}\mathbb{E}R_1}$, and from Eqs. (7.25) and (7.26) for the Lorentz theory that $\varepsilon/\mu_0^e = (cC/D)\sqrt{C/\mathbb{B}\mathbb{E}}$. He compared these values of ε/μ_0^e with the mean value of ε/μ_0^e from the three electron theories in the limit β^2, which was

$$
\varepsilon/\mu_0^e = 1.878 \times 10^7, \tag{7.28}
$$

where in the limit of $v = 0$, the value of ε/μ_0^e in Eq. (7.28) reduced to Simon's. [We recall — see note 45 in Chapter 1 — that in his (1903) and (1905) Kaufmann had incorrectly calculated the quantity in Eq. (7.28).] Kaufmann found that the three electron theories produced values of ε/μ_0^e that deviated from Eq. (7.28) by the following amounts:

$$
\begin{array}{ll}
\text{Abraham} & -2.9\%, \\
\text{Lorentz} & -11.6\%, \\
\text{Bucherer} & -3.7\%.
\end{array} \tag{7.29}
$$

Once again the Lorentz-Einstein theory was a poor third.

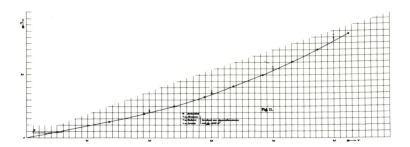

FIG. 7.1. Kaufmann's "reduced curve" from his paper (1906). The symbol ⊙ designates pairs of values (y', z') that Kaufmann obtained from the empirical curve coordinates (y_0, z_0). The points × = Abraham, ○ = Bucherer and + = Lorentz, Kaufmann "calculated from apparatus constants." (From *Ann. Phys.*)

(4) Kaufmann's delivered his *coup de grâce* to the Lorentz theory by using the apparatus constants and convenient values of z', (i.e., $z' = 0.1, 0.2, 0.3, 0.4, 0.5$) to calculate values of y' for each of the electron theories, and then to compare these calculated coordinates with the reduced curve. His results are in Fig. 7.1. From the empirical measurements (⊙ = *beobachtet*), Kaufmann's accuracy for y' was $\pm 1.7\%$ about the central value. Lorentz's values deviated most from the reduced curve. Kaufmann did, however, caution that these results were contingent on the correctness of Simon's value for ε/μ_0^e. Then he went on to emphasize that $R_1 \cdot R_2$ and C were independent of ε/μ_0^e and he left unsaid that Lorentz's theory had failed this comparison too [see Eq. (7.27)]. Kaufmann considered that his data weighed decisively against the Lorentz-Einstein fundamental assumption, and he minced no words about it — "*to base all of physics, including electrodynamics and optics, upon the principle of relative motion at present would be considered as a failure* (italics in original)." Consequently although the deformable electron theory, as amended by Einstein, explained the absence of effects due to the ether, it was invalid; thus, wrote Kaufmann, experiments seeking to exhibit the "*absolute resting ether*" should be continued (italics in original). In Kaufmann's opinion the only alternative for saving "Einstein's theory" was to modify Maxwell's equations for a resting body in such a way that, upon transformation from the electron's rest system to another inertial reference system, the electrons transverse mass would agree with the existing data; however, Kaufmann preferred not to take this drastic "step." The next step, continued Kaufmann, was to perform further electron-deflection experiments in order to decide between the theories of Abraham and Bucherer.

As we have come to expect of Kaufmann, a "*Nachtrag*" to his (1906) was received, 20 April 1906, at the *Annalen*. Fortunately, Kaufmann wished to add only that he had forgotten to mention Cohn's electrodynamics. In a manner consistent with his earlier assessment of the theories of Lorentz and Einstein, Kaufmann wrote: "Cohn and Lorentz achieve as an *end result* the full independence of all observable phenomena from a uniform translation ... Einstein, on the other hand, places the goal of the above theories as a *postulate*, at the apex, and achieves the same system of equations through pure mathematics" (italics in original).

7.4.3. Abraham and Kaufmann versus Planck: 1906

> Recently H. A. Lorentz, and in a still more general form A. Einstein, introduce the principle of relativity.
>
> M. Planck (1906a)

> Gentlemen: Every physicist specializing in the newest branch of electrodynamics, the mechanics of the electron, has awaited with great excitement the results of Mr. W. Kaufmann's 1906 research. He has performed delicate measurements of the electric and magnetic deflections of β-rays from radium ... in my opinion [Kaufmann's data are] not a definitive verification of [Abraham's theory] and a refutation of [Lorentz's].
>
> M. Planck (1906b)

Planck continued in his address (1906a) of 23 March 1906 before the German Physical Society by noting that the principle of relativity applied to both mechanics and electrodynamics, thereby serving as a "grand simplification for every problem of the electrodynamics of moving bodies; therefore, the question of its admissibility deserves to be placed in the foreground of every theoretical researcher in these areas." On Kaufmann's disconfirming data, at this time Planck could say only that Kaufmann's method of data analysis was "not quite simple," and that perhaps with some elaborations the principle of relativity could be made compatible with these data. Planck considered it particularly noteworthy that Einstein's derivation of the electron's mass did not depend on assuming any particular shape for the electron. He proposed to explore further the "fundamental equations of the mechanics" of a moving point charge in order to determine the equations that would replace the "customary Newtonian" formalism. This investigation was the first step in Planck's program to apply the principle of relativity to mechanics and electrodynamics, and then to examine the resulting formalism for errors and inconsistencies. (Planck's results are discussed in Section 12.2.)

On 19 September 1906, at the 78th *Naturforscherversammlung* at Stuttgart, Planck (1906b) scrutinized Kaufmann's data analysis. Without assuming infinitely small deflections, Planck integrated the electron's equations of motion from the Lagrange formalism to obtain curve coordinates (y_0, z_0). Using Kaufmann's values for z_0 as input, he found that although the values of y_0 calculated from Abraham's theory agreed better with the observed values of

y_0 than the ones from Lorentz's theory, the differences between the calculated values of y_0 from the two theories were less than the deviation of Lorentz's theory from the observed values of y_0. Planck concluded that a comparison of the calculated and the empirical values of y_0 "is in my opinion not a definitive verification of [Abraham's theory] and a refutation [of Lorentz's]." Planck went on to emphasize the important role in his analysis of the value chosen for ε/μ_0^e, i.e., either Simon's or Kaufmann's value calculated from the various electron theories to order β^2 accuracy.

Secondly, Planck turned to Kaufmann's coordinates for "infinitely small deviation," i.e., y' and z'. Using the pair of coordinates (y', z') calculated in Kaufmann's (1906) that were closest to the origin of the reduced curve (precisely: 0.1350, 0.0246), Planck deduced from Eq. (7.20) that $\beta = 1.034$, and he emphasized that this result was independent of ε/μ_0^e. But $\beta > 1$ was at variance with the two electron theories, and Planck asserted that Kaufmann's "theoretical interpretation of measured quantities" was unclear. Planck, however, could offer no suggestions as to errors in Kaufmann's experimental method or data analysis.

Kaufmann spoke first in the ensuing discussion. To begin with he circulated a drawn curve (probably the reduced curve in Fig. 7.1), "as well as the five original plates on which you notice two curve branches symmetrical as if from a draftsman." Since, continued Kaufmann, for every pair of points other than the one closest to the origin, his calculations agreed with Planck's then Kaufmann claimed that he had committed no calculational errors. Consequently, it followed from the "observational statistics that neither Lorentz's nor Abraham's theory agrees" with the data. Kaufmann, however, emphasized the "strong" deviation between Lorentz's theory and empirical data ("10–12%") vis-à-vis Abraham's theory ("3–5%") which was beyond his experimental accuracy. Nevertheless, Kaufmann was willing to entertain the possibility of small errors of observation that could combine in such a way as to increase the deviations between theory and experiment.

Planck replied that unknown errors could conspire to bring Lorentz's theory into better agreement with the empirical data. Thus Planck concluded that "from these bare data [i.e., y_0 and z_0] the fact that the deviation of one theory is smaller would not follow a preference for it."

Bucherer took exception to Planck's comment that his theory of the electron was insufficiently developed to be included in Planck's analysis, and then went on to give an uninvited lecture on his new version of Maxwellian electromagnetism.[10]

On Kaufmann's behalf, Runge noted the possibility of large observational errors near the origin of the empirical curve.

Planck replied first to Runge. Since $y' \ll z'$, then in Eq. (7.17) small variations in y' resulted in even larger β's. Consequently, Planck concluded, the part of the reduced curve closest to the origin had to be eliminated from analysis. Planck dismissed Bucherer's long-winded comment with the inquiry whether Bucherer had a Lagrangian formulation for his new theory, to which Bucherer replied that he did not. Exit Bucherer, but not for long, as we shall see.

Abraham's satirical wit colored the first sentence of his contribution to the discussion session which concerned the factor of two disagreement between the predictions of Abraham and Lorentz-Einstein for m_T: "When you look at the numbers you conclude from them that the deviations of the Lorentz theory are at least twice as big as those of my mine, so you may say that the sphere theory represents the deflection of β-rays twice as well as the *Relativtheorie*." A great deal of laughter was reported to have followed this remark. Abraham went on to astutely emphasize a basic difference between the views of Lorentz and Einstein concerning the physics of electrons. Abraham reminded everyone that Lorentz's theory of the electron conflicted with the aim of an electromagnetic world-picture because in his opinion it required nonelectromagnetic forces to prevent the deformable electron from exploding. "According to the *Relativtheorie*," Abraham continued using Planck's name for the new theory, Kaufmann's data could not "be understood as purely electromagnetic processes, but as phenomena for which electrodynamics is not sufficient for clarification." Planck replied to Abraham that although the rigid electron led to a purely electrical theory of matter, it was only a postulate; furthermore, the Abraham and Lorentz-Einstein theories were based on incompatible postulates. Planck continued by emphasizing that he was "sympathetic" toward the theory of Lorentz-Einstein, but that experiment must decide between the competing theories of the electron.

At this point in the proceedings a tense atmosphere prevailed. In addition to the participants mentioned thus far, R. Gans pointed out the inconclusiveness of the experiments of Michelson-Morley and Trouton-Noble concerning which hypothesis of contraction was the correct one. Sommerfeld attempted to defuse this situation with some conciliatory remarks. He felt that Planck's "pessimistic attitude" was somewhat premature owing to the extraordinary difficulty of Kaufmann's experiment; perhaps there was some unknown source of error lurking. We can imagine him next turning to the contingent led by Abraham and Kaufmann before he said that perhaps scientists under 40 preferred the "electrodynamic postulate" and those over 40 the "mechanical-relativistic postulate. I prefer the electrodynamic one." Laughter followed undoubtedly because, besides the fact that Sommerfeld was nearly 40, two of the chief proponents of the electromagnetic world-picture were absent — Lorentz and Poincaré who were well over 40. Kaufmann pressed on, however, claiming that the "principle of relativity" was not as valuable as Planck had contended because it referred only to inertial reference systems. Planck's reply to Abraham and Kaufmann was as follows. As to the limitations of the principle of relativity, Planck concurred and, in fact, went even further; he pointed out that truly inertial motion could not be detected. But such criticism missed the point, which Planck stressed was: "The claim is [based on] the fact that what is not observable in mechanics [absolute motion] is likewise not observable in electrodynamics." This was exactly Einstein's message, although Planck, like most others, did not distinguish between the principles of relativity of Einstein and of Lorentz-Poincaré, thus Einstein's results were interpreted as a generalization of Lorentz's theory of the electron, and so the goal of an electromagnetic world-picture seemed to come that much closer. Einstein's relativity paper was considered as important mostly for the wrong reasons; the fact that he continued to sit in the Patent Office until 1909 certainly belies any notion of a scientific revolution in 1905.

In summary, as of September 1906 Planck could find nothing wrong with Kaufmann's experiment. Planck's correspondence reveals that he had been strongly affected by the seemingly insurmountable criticisms of Abraham and Kaufmann, and that by letter Bucherer had elaborated on his own principle of relativity. For example, on 7 July 1907 Planck wrote to Einstein that he was happy that "for the present" Einstein did not share "Mr. B's opinion," because "so long as the supporters of the principle of relativity constitute as tiny a band as they do at present, it is doubly important that they agree with each other."

We shall see in Chapter 12 that Einstein did not share Planck's concern over Kaufmann's data. In fact, although Einstein was aware of Kaufmann's 1906 data, he did not reply to them until his review paper (1907e), that he prepared in November of 1907 and which appeared in early 1908.

On the other hand, Einstein responded immediately to a basic misunderstanding of the premises of his first paper on electrodynamics that appeared in the 1907 note of a contemporary Paul Ehrenfest. For the most part during 1907–1911, Einstein left the defense of his view to Ehrenfest, who became a good friend. But first Einstein had to set Ehrenfest straight regarding the thrust of Ehrenfest's own note (1907).

7.4.4. The 1907 Einstein-Ehrenfest Exchange

> The Lorentz relativity-electrodynamics in the formulation which Mr. Einstein has published it, is commonly regarded as a closed system.
>
> P. Ehrenfest (1907)

Ehrenfest inquired about the following problem: In 1903 Abraham had proved that a rigid ellipsoidal uniform charge distribution could move in stable force-free translational motion only if its velocity were parallel to its major axis.[11] But, wrote Ehrenfest, "according to Mr. Einstein [Lorentz's electron] when in uniform translation experiences a Lorentz contraction"; therefore, Abraham's question became, could a deformable electron be set into a uniform force-free translation in any direction even though the electron's major axis was normal to its direction of motion? Ehrenfest contended that if that sort of motion could not be deduced from the principle of relativity, then a new hypothesis must be introduced, or else the principle of relativity would be violated.

In reply Einstein (1907c) emphasized first that "the principle of relativity or, more precisely, the principle of relativity together with the principle of the constancy of the velocity of light" did not constitute a "closed system." For example, Einstein wrote that applying the space-time transformations to Maxwell's equations could only result in the transformation formulae for the electromagnetic field quantities; similarly the laws for fast moving point electrons could be deduced from those for slow moving point electrons, as he had done in §10 of the relativity paper. These heuristic uses of the principle of relativity, wrote Einstein, "permit certain laws to be traced back to one another (like the second law of thermodynamics)."

Determining the laws of motion of electrons with extension, continued Einstein, required proceeding along an electrodynamical route, guided by the principle of relativity—for example, to avoid "vagueness" would require assumptions concerning the electron's charge distribution, and Einstein proposed that the charge be distributed uniformly over a rigid frame with the proper forces to ensure the frame's stability. Then, within the "framework" of a "dynamics of rigid bodies," the problem of the electron's motion could be solved in a "deductive manner" without arbitrariness. Although Einstein did not cite Poincaré's [(1905), (1906)] results, some of which Kaufmann (1906) had mentioned, he may have had them in mind when he wrote of a rigid-body theory that avoids vagueness. Kaufmann (1906) had described Poincaré's pressure as an "unknown universal exterior pressure," and this could not have satisfied Einstein who already mistrusted speculations on the constitution of matter.[12] Although in Einstein's view, the goal had not been met of developing a relativistic kinematics of a "rigid body (i.e., one not deformable through exterior forces)," he had himself recently derived the "kinetic energy of an ideal body in parallel translation that is not interacting with other bodies." This result appeared later in an important paper "The Consequences of the Principle of Relativity for the Inertia of Energy" (1907d).

7.4.5. Einstein on Rigid Bodies and Hyperlight Velocities

> Using hyperlight velocities we could telegraph into the past.
>
> A. Einstein [Quoted by Sommerfeld in the discussion session to von Ignatowski (1910b)]

Einstein (1907d) extended the result from his fourth published paper of 1905 (the mass-energy equivalence) to a charged rigid body that persisted in inertial motion, even though it was acted on by external electromagnetic forces. But lest the unwary believe that the "goal" was being approached of a "dynamics of parallel translation," Einstein reminded his readers of the simplifying assumptions that permitted execution of the calculation (Section 12.5.3), and of the following "drastic difficulty" with the notion of the rigid body. He proposed the "simple special case" of a rigid rod of length l (as measured in its rest system k) moving along the z-axis of K with a relative velocity v. Along the rod's axis let equal and opposite impulses be exerted simultaneously with respect to k on the two ends of the moving rod. Relative to K, however, the impulses are not simultaneous; rather, the impulse striking A at time t_A acts first, and then at a later time t_B the impulse at B acts, where

$$t_B = t_A + \gamma l v/c^2. \tag{7.30}$$

Consequently an observer in K should notice a change in the rod's velocity after the impulse at A acts, because that impulse is not instantaneously compensated for by the impulse applied at B. Therefore the impulse at A does work on the rod, and the energy of the rod should increase during the interval $\Delta t = t_B - t_A$. But to the observer in K the rod's velocity remains unchanged, violating the principle of the conservation of energy. Einstein made two proposals for removing this difficulty: (1) There arises instantaneously in the rod a straining force that propagates with an infinite velocity, thereby compensating for the force at A; Einstein went on in the 1907 paper to demonstrate that this proposal violated the principle of relativity. (2) There arises in the rod an "unknown *Qualität*" that propagates with a "finite velocity" along the rod causing the acceleration required for compensating the force at A. But Einstein could not speculate further along this line of thought because, he wrote, "we are still far from possessing a dynamics of parallel translation of rigid bodies."

Einstein continued in the 1907 paper. According to classical mechanics $t_B = t_A$, i.e., the impulse at A is propagated instantaneously to B. Einstein next demonstrated that "any instantaneous propagation of an effect," or propagation at ultralight velocity was "incompatible with the *Relativitätstheorie*" (by this time Einstein had adopted Planck's name for his theory): Consider that relative to the x-axis of K a long "material strip (*Materialstreifen*)" moves with velocity $-v$. An observer A at $x = 0$ sends a signal along the strip to an observer B at $x = l$; the signal's velocity relative to the strip is w, and the strip could have been excited by A's striking it. Using a result from §5 of the relativity paper, the signal's velocity relative to K is

$$u = \frac{w - v}{1 - wv/c^2}. \tag{7.31}$$

The time interval Δt for the signal to traverse the distance l, on the x-axis is, $\Delta t = l/u$, or from Eq. (7.31)

$$\Delta t = \left[\frac{1 - wv/c^2}{w - v} \right] l. \tag{7.32}$$

If $w > c$, then there is a range of velocities v for which $\Delta t < 0$ — that is, the signal is received at B before it was sent. "In my opinion," concluded Einstein, "if [Eq. (7.32)] is regarded as pure logic then it contains no contradictions; however, it absolutely clashes with the character of our total experience, and in this manner is proven the impossibility of the hypothesis, $w > c$." In summary, hyperlight velocities would lead to reversal of causality, but this phenomenon has never been observed. In the (1907e) Einstein repeated this argument almost verbatim in order to further illustrate the relativistic addition formula for velocity. He probably felt obliged to take this route for proving the incompatibility of ultralight velocities with the relativity theory, because he wished to discuss how materials transmit pulses, that is, to prove the incompatibility of the notion of the rigid body with the relativity theory.[13]

Further developments concerning the notions of a rigid body and of hyperlight velocities resulted from Hermann Minkowski's geometrical formulation of the theory of electrons.

7.4.6. Hermann Minkowski

> Oh, that Einstein, always cutting lectures – I really would not have believed him capable of it.
>
> H. Minkowski (*circa* 1908)

Although by 1907, at age 43, Hermann Minkowski was assured lasting fame as a mathematician, his wider reknown among physicists and philosophers resulted directly from the linking of his name with that of his erstwhile former student at the ETH, Einstein. But that was not yet the case during Minkowski's productive period on electron theory, 1907–1908, for he considered Einstein's theory to be a generalization of Lorentz's. This section focuses upon the kinematical content of Minkowski's work.

Max Born (1969) recalled that Minkowski became interested in the electrodynamics of moving bodies at Göttingen in 1904, and then in 1905 conducted a seminar in which "we studied papers by Hertz, FitzGerald, Larmor, Poincaré and others, but also got an inkling of Minkowski's own ideas which were published only two years later." One of the papers they studied of Poincaré's may well have been Poincaré's (1905a), at any rate by 1907 Minkowski had learned Poincaré's methods and wrote that his goal was to exhibit those symmetries in the equations of physics which "had not occurred to any of the previously mentioned authors, not even to Poincaré himself" (the others were Einstein, Lorentz and Planck).

Minkowski's keen geometric intuition had been evident already at age 17 from his prize essay, to the Académie des Sciences, on a geometric analysis of the theory of quadratic forms. This penchant, coupled with an interest in the theory of invariants, prepared him to appreciate the fine points in the penultimate analysis of electron physics by the master of these mathematical fields, Poincaré. From Minkowski's papers of 1907–1908 it is clear that he was struck by Poincaré's discovery of the group property of Lorentz's transformations, and by Poincaré's proposal that the Lorentz invariance of the quantity $c^2t^2 - x^2 - y^2 - z^2$ could be utilized for seeking other quantities that transformed under the Lorentz group, as did (x, y, z, ict).

By 1907 Minkowski had moved toward a conception of the ether that permitted him to appreciate better than Poincaré how the close relationship between mathematics and physics was further revealed by Einstein's 1905 notion of relativity.

In his 1907 lecture, "*Das Relativitätsprinzip*" (1907), Minkowski explained that the mathematician's familiarity with certain necessary concepts from the theory of invariants placed him in a particularly good position to understand new ideas in physics, whereas the physicists were currently making their way painfully through a "primeval forest of obscurities."

Minkowski set himself the task of continuing Poincaré's exploitation of the theory of invariants in a four-dimensional vector space, into which he inserted the genial idea of a geometrical interpretation of Lorentz's transformations, replete with Einstein's relativity of time. Minkowski's results of 1907–1908 were no mere feats of mathematics, for they led to the formulation of such new disciplines as a relativistic theory of elasticity and a Lorentz-covariant theory of the electrodynamics of moving bulk matter, and to a deeper understanding of the foundations of relativity theory. In order to avail himself of the tools from the theory of invariants, Minkowski used space and time coordinates with numerical subscripts – that is, $x_1 = x$, $x_2 = y$, $x_3 = z$, $x_4 = ict$, and referred to the quantity $(x_1, x_2, x_3, x_4) = (\boldsymbol{x}, x_4)$ as a "space-time vector of the first kind" with a Lorentz-invariant magnitude $x_1^2 + x_2^2 + x_3^2 + x_4^2$.

Minkowski obtained the Lorentz transformation properties of the space-time vectors of the first kind from Lorentz's transformations, written in the numerically subscripted form:

$$x_1 = \gamma(x_1' - i(v/c)x_4'), \tag{7.33}$$

$$x_2 = x_2', \tag{7.34}$$

$$x_3 = x_3', \tag{7.35}$$

$$x_4 = \gamma(x_4' + i(v/c)x_1'), \tag{7.36}$$

where the primed and unprimed coordinates refer to k and K, respectively. Minkowski could then write the Lorentz transformations of these quantities in the same way that vector quantities were transformed in a three-dimensional vector space

$$x_\mu = \sum_{v=1}^{4} a_{\mu v} x_v', \qquad (\mu = 1, 2, 3, 4) \tag{7.37}$$

where $a_{\mu v}$ are the coefficients for a Lorentz transformation which may be displayed in matrix form as

$$[a_{\mu\nu}] = \begin{pmatrix} \gamma & 0 & 0 & -i\gamma v/c \\ 0 & 1 & 0 & 0 \\ 0 & 0 & 1 & 0 \\ i\gamma v/c & 0 & 0 & \gamma \end{pmatrix}. \tag{7.38}$$

The invariant normalization condition for the space-time vector of the first kind follows from Eq. (7.37) and the orthogonality of the transformation matrix $[a_{\mu\nu}]$. Some other space-time vectors of the first kind that Minkowski discussed were those whose transformation properties had been commented upon in 1905 by Poincaré—for example, $(A, i\phi)$, $(\rho v/c, i\rho)$ and $(F, iF\cdot v/c)$.

Minkowski's style of exposition bore the stamp of the elegant mathematician that he was, and consequently his mathematical formalism was the subject of two long 1909 review papers by Sommerfeld [(1910a), (1910b)] directed to physicists (in which Sommerfeld coined the term "four-vector (*Viervektor*)").

If in Newtonian physics an equation has the proper transformation properties under a Galilean transformation, then it is a candidate to be a law of nature. Minkowski's transformation formalism, based upon Eq. (7.37), offered the means for testing whether the laws of electron physics had the proper Lorentz transformation properties. For example, Minkowski was able to display the Maxwell-Lorentz equations in a form exhibiting their Lorentz "invariance (or better covariance)."[14, 15]

Minkowski referred to this result as the "theorem of relativity" and his reason for attributing it to Lorentz was probably that from 1892–1904 Lorentz had been attempting to find reference systems that simulated those fixed in the ether. Minkowski referred to the assertion that the as-yet-undiscovered laws governing phenomena in ponderable bodies could also be Lorentz covariant, as the "postulate of relativity," which he also credited to Lorentz. In Minkowski's view Einstein's contribution was the "principle of relativity" which asserted the expected covariance as a "definite relationship between real observed quantities for moving bodies." Furthermore, Minkowski continued, Einstein had shown that the postulate of relativity, with its attendant hypothesis of contraction, was not "artificial" but rather resulted from a "new concept of time" which the "phenomena had forced upon" Einstein.

It was in this empiricistic vein that Minkowski began his famous semi-popular lecture "Space and Time" (1908b): "The views of space and time which I wish to lay before you have sprung from the soil of experimental physics, and therein lies their strength. They are radical." Minkowski's theme was that mathematics could plumb the depths of the empirical data in which were rooted the work of Einstein and Poincaré. Classical physics, Minkowski continued, was covariant under two separate groups of transformations: (1) the group of rotations in three-dimensional space that left unchanged the quantity $x^2 + y^2 + z^2$, and this group had only a geometrical interpretation; (2) the group of Galilean transformations — he referred to this group as G_∞ because it asserted an infinite velocity for light. But empirical data, continued Minkowski, in particular the Michelson-Morley experiment, had forced Lorentz to propose both the theorem and the postulate of relativity, leading to the covariance of electrodynamics under the group of Lorentz transformations G_c. Therefore G_c connected the two separate covariances from classical mechanics through the constancy of the large but not infinite velocity of light. "To establish this connection" Minkowski proposed "to consider the graphical representation of [see Fig. 7.2]

$$c^2 t^2 - x^2 - y^2 - z^2 = 1.\text{"}^{16} \tag{7.39}$$

Minkowski proclaimed that it was the quantity c that caused "three-dimensional geometry [to become] a chapter in four-dimensional physics [and] space and time are to fade away into the shadows, and only a world in itself will exist." According to Minkowski the four-dimensional space-time is filled with "world-points," i.e., values of x, y, z, t, each of which is "substance" (which could be ponderable matter), and each trajectory of which is a "world-line."

Minkowski's analysis of the invalidity of the concept of a rigid body turned upon the incompatibility of the groups G_∞ and G_c. The rigid body belonged to G_∞, but Michelson and Morley's experiment demonstrated that G_c was the proper group for the optics of moving bodies. Lorentz's contraction, continued Minkowski, "lies in this very invariance in optics for the group G_c," and this hypothesis "becomes much more intelligible" from the space-time diagram:

> This hypothesis sounds extremely fanciful. For the contraction is not to be thought of as a consequence of resistances in the ether, or anything of that kind, but simply as a gift from above, — as an accompanying circumstance of the circumstance of motion.

Thus Minkowski distinguished between Lorentz's and Einstein's contractions, the latter being a kinematical phenomenon; however, Minkowski's exhibition of this phenomenon using electrons rather than measuring rods, displayed his incomplete understanding of the differences between their theories. For example, Minkowski's view of Einstein's contribution to the new geometrical space-time structure was "that the time of one electron is just as good as that of the other, that is to say, that t and t' are to be treated identically." Therefore, even though Minkowski recognized an important difference between the views of Lorentz and Einstein concerning length contraction, he nevertheless concluded "Space and Time" as follows:

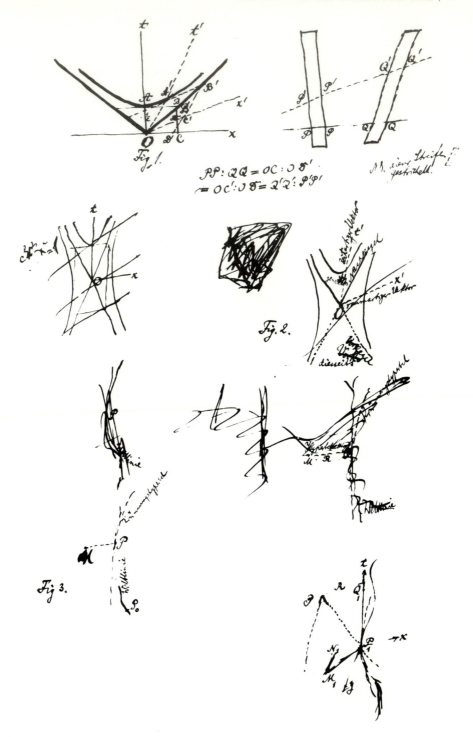

$$PP : QQ = OC : O\,\delta'$$
$$= OC' : O\,\delta = Q'Q' : P'P'$$

The validity without exception of the [postulate of relativity], I like to think, is the true nucleus of an electromagnetic world-picture [*Weltbildes*] which, discovered by Lorentz, and further revealed by Einstein, now lies open in the full light of day.

In summary, for Minkowski Einstein's research prepared the way for completion of the electromagnetic world-picture within a four-dimensional space-time framework in which Lorentz's ether was de-materialized and replaced by a "world" comprised of space-time points. During 1909–1911 Max von Laue used Minkowski's formalism to unfold further some new wonders of the special relativity theory.

Starting in 1912 Einstein used Minkowski's work as a guide toward generalizing the special relativity theory. Einstein (1916) began his major paper on the general relativity theory thus:

> The generalization of the relativity theory was facilitated through the form that Minkowski had given to the special relativity theory. He was the first mathematician to clearly perceive the formal equivalence of the space and time coordinates; this made possible the construction of the [general] theory.

7.4.7. Max Born's Attempt to Complete the Electromagnetic World-Picture

Upon Minkowski's untimely death in 1909, his assistant at Göttingen, Max Born, attempted to complete Minkowski's version of the electromagnetic world-picture. Born recalled (1969) having discovered Einstein's relativity paper early in 1908 while in Breslau, and then "combining Einstein's ideas with Minkowski's mathematical methods ... found a new, direct way to calculate the electromagnetic self-energy (mass) of the electron ...". Born sent off a draft to Minkowski who was sufficiently impressed with Born's work to invite him to become his assistant. A direct result of their short-lived collaboration was

---

FIG. 7.2. Minkowski's sketches of space-time diagrams for his address "Space and Time" (1908b). His figure numbers refer to the address' text which was published posthumously by Teubner. A version of the left-hand diagram in Minkowski's figure 1 is in footnote 16 of this chapter, and the one on the right Minkowski used for his geometrical derivation of the Lorentz contraction. Minkowski's figure 2 further explains the four-dimensional space-time representation (*Vorkegel* = front cone, *Nachkegel* = back cone, *raumartiger Vektor* = space-like vector, *zeitartiger Vektor* = time-like vector, *diesseits* [0] = before [0], and *jenseits* [0] = after [0]). Minkowski used his figure 3 to illustrate the orthogonality of the velocity and acceleration vectors in four-dimensional space-time (*Weltlinie* = World-line and *Krümmungshyp.* = hyperboloid of curvature). Minkowski's figure in the lower right-hand corner became figure 4 in the published text. He used this unnumbered figure for a geometrical illustration of the Liénard-Wiechart potentials. For a detailed discussion of Minkowski's thinking toward the space-time diagrams see Galison (1979). (Courtesy of Niedersächsische Staats- u. Universitätsbibliothek Göttingen, *Math. Arch.* **60**: 2, fol. 24/25.)

Born's *Habilitationsschrift* at Göttingen which used Minkowski's space-time formalism for developing a kinematics of rigid bodies. Straightaway Born (1909a) asserted his goal: "*The practical value of the new definition of rigidity must prove its worth in the dynamics of the electron*" (italics in original). As Einstein had written in 1907, this research effort required more than electrodynamical assumptions for ensuring the moving electron's rigidity. Born explained that whereas classical mechanics defined a body as being rigid if the distance r_{ij} between any two points within the body remained unchanged,

$$r_{ij}^2 = (x_i - x_j)^2 + (y_i - y_j)^2 + (z_i - z_j)^2, \qquad (7.40)$$

this condition was not Lorentz invariant, and so the classical rigid body was excluded by the relativity principle. Born's strategy was to define the rigid body through a set of differential equations describing the motion of a body that maintained its shape in its resting system, but which was compressible to inertial observers. He then combined these differential equations with the equations of electrodynamics. To this end Born replaced Eq. (7.40) by two Lorentz-invariant conditions of rigidity: (1) the constancy for every inertial observer of the line element $c^2(dt)^2 - (dx)^2 - (dy)^2 - (dz)^2$; (2) small variations from an equilibrium configuration in the body's rest system should be independent of the "proper time," which Minkowski had defined as the time in the body's rest system. Born prophetically added that the second condition "would become of importance in adapting the relativity principle to elasticity theory."[17] There followed a mathematical *tour de force* in which Born integrated the mechanical equations of motion in four-dimensional space-time for the special case of a body undergoing an arbitrarily large but constant acceleration; he referred to this sort of motion as hyperbolic since the body's trajectory in four-dimensional space-time was a hyperbola (an electron undergoing hyperbolic motion does not radiate). In order to obtain "Lorentz's formula" for m_T Born had to make certain electrodynamical assumptions for the electron's self-electromagnetic fields in the case of hyperbolic motion: (1) The resultant force from the electron's self-fields was cancelled by the external fields. (2) Since a "cathode or Becquerel ray" would persist in uniform accelerated motion in an external electric field without imposing additional external forces, then the self-force normal to the electron's linear trajectory must also vanish. This assumption required Born to assume that the electron was a uniform spherical charge distribution in its rest system. These two assumptions enabled him to deduce the same equation for m_T that Lorentz had deduced from quasi-stationary motion. So Born considered that he had already accomplished more than the only other author who had discussed the problem of accelerations and "without clarifying the state of affairs." His reference was to Section 18 of Einstein's (1907e) entitled, "Space and Time in a Uniformly Accelerated Reference System"; Born's *faux pas* was pointed out to him in September 1909 when he presented a summary version of his results to the *Naturforscherversammlung* at Salzburg at which Einstein presented his first invited paper. Born (1910b) spoke of his work as a major step toward completing Lorentz's theory of the electron from which Einstein in 1905 had removed a "formal objection": that the longitudinal mass had different values, depending on whether it was calculated from the electron's energy or momentum. According to Born, Einstein had succeeded by assuming the validity of Newton's law of motion for "infinitely slow motion of a mass point" and then used the "principle of relativity to infer the law of motion for an arbitrarily moving mass point." Very likely Born gave Einstein, rather than Poincaré, the credit for removing the inconsistencies in the longitudinal mass because Poincaré's derivation of Newton's law was invalid for arbitrary motion. Although Born could say that Einstein "had almost worked out a complete electrodynamic foundation for the electron's apparent mass," Born claimed that he had gone further by extending Minkowski's formalism to an electron of finite extension, and that his theory applied to arbitrarily large accelerations, whereas Lorentz's was restricted to quasi-stationary motion.

In the following discussion Sommerfeld complimented Born on the elegance of his approach, but pointed out its insufficiency because "the principle of relativity can discuss only uniform motion." Born's misunderstanding of special relativity was rooted in his belief that it provided a complete description of the electron's accelerated motion from a single reference system moving along with it, i.e., the electron's rest system, which in Born's case was a noninertial system. Evidently Born was at a loss for words and could only reply weakly, "But not every hypothesis that we make on the acceleration satisfies the principle of relativity." To which Sommerfeld retorted, "Of course not!"

7.4.8. Ehrenfest's Paradox

Having been interested earlier in how the "Lorentz relativity" treated problems of rigidity, Born's (1909a, b) piqued Ehrenfest's curiosity over the notion of a relativistic rigid-body. In a short note (1909) Ehrenfest proposed a definition of a relativistic rigid-body that was equivalent to Born's: every element of a moving relativistic rigid-body exhibits to an observer at rest the Lorentz contraction determined by the body's instantaneous velocity relative to the resting observer. Born's definition of a rigid body, noted Ehrenfest, was not based on measurements made by observers at rest in the laboratory, but on "Minkowski's method of measurement" which concerns a "continuum of infinitesimal observers" that move along with the points of a moving body and discern that each infinitesimal neighborhood of the rigid body remains undeformed. Although both definitions of the relativistic rigid-body agreed with the foundations of relativity theory, continued Ehrenfest, these definitions led to contradictions. For the purpose of illustration Ehrenfest applied his definition to uniform rotation about a fixed axis. Consider a solid cylinder of radius R and height H (as measured when it is at rest), set into motion with a constant angular velocity ω about a fixed axis normal to the cylinder's endfaces. Let R' be its radius, as measured by an observer at relative rest. Since elements along the periphery of an endface move with the linear velocity $R'\omega$ tangent to the direction of motion, they must undergo a Lorentz contraction; therefore, $2\pi R' < 2\pi R$. But any radius of an endface is directed normally to the direction of rotation; therefore, $R' = R$. Consequently, concluded Ehrenfest, the radius of each endface obeys two mutually contradictory conditions, and this contradiction has become known as Ehrenfest's paradox.

Since according to relativity theory a rigid body could not be set into rotation about a fixed axis, then it had only three of the expected six degrees of freedom. Gustav Herglotz (1910) and also Fritz Noether (1910) had independently reached this latter result by direct examination of Born's theory.[18] Herglotz (1910) wrote that it was of interest to determine whether "this new rigid body in the system of the electromagnetic world-picture" had six degrees of freedom as had the "customary rigid body in the system of the mechanical world-picture."

7.4.9. Abraham, Born and Planck Respond to Ehrenfest

Born (1910b) replied to Ehrenfest and Herglotz claiming that the rigid body of a relativistic kinematics bore no "actual analogy to the rigid body of the customary mechanics"; supporting this point Born referred to Ehrenfest's (1909) and to a 1909 discussion with Einstein at Salzburg during which, implied Born, they had independently discovered the Ehrenfest paradox. Born went on to explain that any "assumption on the kinematical constitution of the electron is purely hypothetical" and that his assumptions enabled him to deduce an electron mass in agreement with current data. Consequently, Born presently was interested in only the predictions of his theory and not in the merely expedient intermediate steps. Whether his rigid electron could rotate about a fixed axis was not presently important for a dynamics of the electron. Furthermore, continued Born, if we considered a macroscopic body not to be a continuum, but as comprised of an enormous number of rigid electrons, then the body could be set into rotation about a fixed axis because its microscopic electrons could then move in trajectories that were closed orbits – that is, the macroscopic body behaved as an elastic body composed of rigid electrons.

It was Planck who best perceived the direction ideas would take concerning the nature of deformed moving bodies. In reply to Ehrenfest he wrote (1910a): "The task of determining the deformation of any accelerating body is, in the relativity theory as in the customary mechanics, essentially a problem of elasticity." Planck's point was that if one knew how to calculate the stresses on a body, then from the relativity theory one could determine the deformation that would result from changing its velocity from zero to a value v. Planck pointed out a distinction between the "apparent" Lorentz contraction of the cylinder's rim once the cylinder was rotating, and the deformation of the rim due to its having been accelerated from rest to a linear velocity $v = \omega R$. In a note added in proof Planck gave his opinion of Born's recent reply to Ehrenfest of which he had just become aware; he doubted whether Born's (1910b) restriction of the rigidity criterion to a single electron could clear matters concerning Ehrenfest's objections.

Abraham (1910) seized on Ehrenfest's paradox as another criticism of the relativity theory. He wrote that he was waiting to see if the "relativity theorists" could succeed in giving a "satisfactory definition of rigidity or length and thus clarify the logical permissibility of the space and time concepts of Einstein and Minkowski." Furthermore, Abraham deprecated Planck's attempt to dodge Ehrenfest's objection by replacing the rigid body with an elastic body since no one had as yet proposed a relativistic theory of elasticity.

7.4.10. Von Ignatowski versus Born, Ehrenfest and Sommerfeld

With the entrance of Waldemar von Ignatowski (1910a, b) the controversy concerning the nature of a rigid body took another turn which serves to illuminate the misunderstandings of the foundations of relativity theory circa 1910, despite the existing mathematical superstructure of vectors and tensors in a four-dimensional space. In a note added in proof to his earliest examination of Born's rigid-body theory, von Ignatowski (1910a) wrote that he had just become aware of Ehrenfest's 1909 paper and disagreed with its conclusion: "In general by measurement we determine the true shape and dimensions of a rigid body if the body is resting. Measurements on a moving body yield only an apparent value."[19] At the *Naturforscherversammlung* Königsberg, September 1910, von Ignatowsky elaborated upon an interpretation of Einstein's principle of relativity which can best be described as one containing the best of both possible worlds. He (1910b) accepted c as a "universal space-time constant" which served as the basis for Minkowski's assertions that: (1) The length contraction was not, as von Ignatowski quoted from Minkowski, a "gift from above" but a "consequence of the motion"; (2) from the "concept of synchronized clocks and synchronized measurement" we could show that "substantial points" in space-time could not move with hyperlight velocities. On the other hand, von Ignatowski took a nonkinematical position in order to support what he claimed was a direct result of using clocks synchronized with light signals: although substantial points could not move with hyperlight velocities "an effect of a force or volume dilatation, and so forth, propagates with the velocity $V [> c]$." His derivation of this result, from the Lorentz transformations, is worth reviewing because Sommerfeld and Born, among others, were unable to address it directly; moreover, the episode with von Ignatowski shows how Minkowski's space-time formalism served at times to confuse matters of interpretation.

Assume, wrote von Ignatowski, that two observers A' and B' in the inertial system k moving with velocity v relative to K, had a rigid rod of length $x'_2 - x'_1 = l$; these observers were also equipped with synchronized clocks.

Then A' and B' decided that at time t' they would raise their ends of the rod the same distance above the x'-axis; the Lorentz transformation for time Eq. (6.52) yields a nonzero time difference in K

$$t_2 - t_1 = (v/c^2)(x_2 - x_1). \qquad (7.41)$$

Von Ignatowski's interpretation of this result went beyond the relativity of simultaneity, because from Eq. (7.41) he deduced what he interpreted as the velocity of a disturbance traveling from A' to B',

$$V = (x_2 - x_1)/(t_2 - t_1) = c^2/v, \qquad (7.42)$$

and consequently, wrote von Ignatowski, "the existence" of a hyperlight velocity was an "immediate consequence of the existence of synchronized clocks" in k and K. Since, he continued, $\Delta t > 0$, then to an observer in K the A' end of the rod rose before the B' end and the observer in K noted that owing to the Lorentz contraction the moving rod was deformed in the direction of its motion, and this bend ("*Knick*") moved with the velocity $V = c^2/v$.

Another of von Ignatowski's arguments supporting the result $V > c$ was based on Einstein's 1907 illustration of the velocity addition theorem that used a material strip (Section 7.4.5). He took Einstein's observers A and B to be A' and B', Einstein's w to be V, and Einstein's time interval Δt to be the time interval in k. So, according to von Ignatowski, if $V = c^2/v$, then $\Delta t' = 0$ and not $\Delta t' < 0$. The error in this argument was von Ignatowski's conveniently redefining Einstein's w to be the rate at which a disturbance traveled in the bar owing to the K-observer's measuring that A' picked up his end of the bar before B', i.e., due to the relativity of simultaneity, whereas Einstein's w was a signal velocity and so it could not exceed c. Sommerfeld apparently misunderstood von Ignatowski's redefinition of Einstein's quantities because his comment focused on the fact that relativity permitted hyperlight velocities, e.g., the phase velocity in anomalous dispersion, but these velocities carried no information. (Three years previously Sommerfeld (1907) had to lecture Wien on this point, and acting in an editorial capacity Wien had voiced his concern to von Ignatowski (1910a).) Sommerfeld went on to mention two personal communications from Einstein concerning hyperlight velocities: (1) Using "hyperlight velocities we could telegraph into the past" – Einstein very likely had referred to a variant of his 1907 result from the velocity addition equation; (2) although the point of intersection of two crossed rulers could travel faster than light as the angle at the point of intersection was decreased, this point could not be used to transmit information.

Von Ignatowski's reply further demonstrated his dynamical interpretation of the contraction of moving bodies: V was not a signal velocity, and, furthermore, it could in-principle be measured by arranging for the moving bent rod to collide with two hooks; consequently, continued von Ignatowski, a hyperlight velocity emerged from a theory of the rigid body. Sommerfeld responded only to the latter point by stating his belief that "the concept of a rigid body must be modified since the reaction in it cannot propagate with a hyperlight velocity." Evidently von Ignatowski's redefinition of Einstein's kinematical quantities merited no comment from Sommerfeld.

In reply to Sommerfeld, von Ignatowski emphasized that much of his work was based upon Born's rigid-body theory. To which Born replied that, since the relativity theory forbade hyperlight velocities to carry information (he referred here to Einstein's comment on telegraphing into the past), then "perhaps we must give up the concept of the rigid body," and move toward developing a relativistic theory of elasticity.[20] He then went on to demonstrate how Minkowski's theory discussed hyperlight velocities. Nevertheless, von Ignatowski decided to push his point by repeating *his version* of Einstein's 1907 proof that one could not telegraph into the past. No further comments were recorded.

It was to von Ignatowski to whom Ehrenfest (1910) responded in what Klein (1970) has referred to as "crushing detail." Ehrenfest elaborated on his original 1909 presentation with the following *Gedanken* experiment. Consider a disk with regularly spaced markings on its surface, e.g., equally spaced circles concentric about the disk's center. Let these marks be recorded on a piece of tracing paper by an observer at rest relative to the disk. Now rotate the disk at constant angular velocity ω about an axis through its center, and assume that the observer does not rotate with the disk. If the observer instantaneously registered the rotating disk's markings on another piece of tracing paper, he would find upon comparison with the other piece of paper that the radial lengths are the same, but the circumference measured before and during rotation differ. Ehrenfest concluded: "How could one imagine a tracing construction with such contradictory properties?"

In a paper on a relativistic theory of elasticity, von Ignatowski (1911a) conceded to Ehrenfest's criticisms; he had become convinced, in agreement with Planck, that the deformation of a moving body was a problem in elasticity.

Perhaps displeased with the way in which Born and Sommerfeld replied to von Ignatowski, Ehrenfest (1911) responded again to him in a short sharply worded note about which Klein has written, Ehrenfest later admitted "that he was not proud of the way he had disposed of Ignatowski." Besides mentioning Wien's "warning" [referenced by von Ignatowski (1910a)], Ehrenfest gave a detailed demonstration, that used Minkowski's space-time diagram, for illustrating Einstein's "catchphrase 'telegraphy into the past,'" and this demonstration could well have been aimed also at Born.

7.4.11. Varičak and Einstein

> [Einstein's] contraction is, so to speak, only a psychological and not a physical fact.
>
> V. Varičak (1911)

[Varičak's note] must not remain unanswered because of the confusion that it
could bring about.

A. Einstein (1911b)

Ehrenfest's first reply to von Ignatowski had provoked a note by V. Varičak
(1911) who focused upon the epistemological problem of the reality status of
the contraction of moving bodies. From Prague Einstein wrote to Ehrenfest in
St. Petersburg on 12 April 1911, inviting him to comment on Varičak [Klein
(1970)]. From a footnote added to Ehrenfest's second rebuttal of von
Ignatowski, it appears that by this time Ehrenfest had already been in
communication with Varičak; in fact, he had already set Varičak straight on
the meaning of Einstein's contraction. But owing to the seriousness of the
situation Einstein (1911) decided to enter the controversy surrounding the
length contraction "because of the confusion" that Varičak's note could
provoke.

Varičak (1911) considered that Ehrenfest's paradox could be "clarified"
from Lorentz's viewpoint which treated the contraction of a moving body as an
"objective" occurrence:

> According to Lorentz every element of the periphery is shortened
> independent of the observer, while the radial element remains unchanged.

Einstein's contraction, continued Varičak, was "only an apparent subjective
phenomenon produced by the manner in which our clocks are regulated and
lengths measured." Consequently, Varičak concluded: "It seems to me that the
tracing constructions must be identical" because, according to Lorentz, a
moving body underwent a real contraction, whereas Ehrenfest's paradox
resulted from measuring length using synchronized clocks.

In order to clarify this "opinion" Varičak analyzed the length measurement
of a rod in inertial motion. Consider that at one end of the rod there is a mirror
and at the other a light source, and that an observer in K measures the rod's
length in k through measuring the time required for a to-and-fro light exchange
on the moving rod. The resulting length is less than the rod's length measured
in k by congruence. As for the observer in K, wrote Varičak:

> But he remains conscious of the fact that the contraction is, so to speak,
> only a psychological and not a physical fact.

Varičak's reason was that the K-observer did not make a congruence
measurement; if he had, and here Varičak referred specifically to an Ehrenfest
Gedanken experiment with tracing papers, "I believe that both tracings will be
identical" – that is, tracings made by the observer in K both with and without
the rod at relative rest. Einstein's kinematical interpretation of relativistic

effects was not lost on Varičak who interpreted Einstein's length contraction as an apparent effect because nothing happened to the moving rod; furthermore, the occurrence of Einstein's contraction depended on an observer's being present in order to measure lengths using light signals, or equivalently, using properly synchronized clocks. On the other hand, according to Varičak, Lorentz's contraction was an objective phenomenon for it occurred whether an observer was present or not. Varičak pushed this difference further by writing that, whereas Lorentz could explain the results of such ether-drift experiments as that of Rayleigh and Brace, "we would not at all have approached this problem using Einstein's relativity principle."

Near the beginning of his paper Varičak wrote that the "radical difference between the views of Lorentz and Einstein" had been emphasized strongly by Gilbert H. Lewis and Richard C. Tolman, both of M.I.T. With certainty we can conjecture that Varičak gleaned many of his ideas from their paper of 1909, as well as gaining further support from a recent review paper of Jakob Laub which he also cited in the same footnote.

(i) *Lewis and Tolman*

> ... the physical condition of the electron obviously does not depend upon the state of mind of the observer.
>
> Lewis and Tolman (1909)

E. T. Bell (1945) observed that many American scientists appreciated the power of the axiomatic method. We have seen that most of the important European scientists, e.g., Lorentz, Planck and Poincaré, preferred a constructive approach where all hypotheses had a firm empirical base; they preferred from the outset to deal only with dynamical theories. Bell's observation rings true for Lewis and Tolman (1909) who interpreted Einstein's research thus[21]: Whereas Lorentz had explained every existing ether-drift experiment, his theory "did not, however, prevent determination of absolute motion by other analogous experiments which have not yet been tried. Einstein went one step further ... he concludes that similar attempts will also fail"; consequently Einstein proposed the two principles of relativity which "generalize a number of experimental facts, and are inconsistent with none." They cited, in particular, Bucherer's (1908a) experiment on the deflection of β-rays (see Chapter 12) which confirmed Einstein's prediction for the electron's transverse mass. Lewis and Tolman considered Einstein's principle of relativity as comprising two "laws"[22] — (1) absolute motion could not be detected in inertial reference systems; (2) the constancy of light velocity. Therefore, Einstein's principle of relativity was "radically different" from Lorentz's.

It was the light axiom with which Lewis and Tolman were most uncomfortable, for it "constitutes the really remarkable feature of the principle of relativity, and forces us to the strange conclusions which we are about to deduce" — the relativity of length and time, which they interpreted as a "scientific fiction ... in a certain sense psychological." For example, they continued, in Lorentz's theory a moving electron undergoes a "real distortion" owing to its interaction with the ether, whereas in Einstein's theory an electron viewed by inertial observers moving in different directions, each one "naively considering himself to be at rest, will appear shortened in a different direction by a different amount ... but the physical condition of the electron obviously does not depend upon the state of mind of the observer." Thus they chose to interpret the relativity of length and time as changes in the units of length and time, and not in the quantities themselves. They considered Einstein's principle of relativity to be a temporary expedient toward developing a new mechanics which included accelerated reference systems, and could be applied to "complex systems," i.e., presumably they meant atoms.

(ii) *Laub*

> [In Lorentz's view] observed changes in a moving body are of an objective sort.
>
> J. Laub (1910)

Although by 1910 Jakob Laub had collaborated with Einstein on two papers, a poor choice of words in his otherwise authoritative review, "The Experimental Foundation of the Principle of Relativity" (1910), caused it to be cited by Varičak in support of his view concerning Einstein's contraction. In the review paper Laub carefully distinguished the axiomatic statements of Einstein's theory from the constructive ones of Lorentz, explaining that in Lorentz's theory an inertial observer at rest relative to a light source could detect the apparatus' motion — that is, as Lewis and Tolman (1909) wrote, this observer could not naively consider himself to be at rest. But then Laub wrote that according to Lorentz "observed changes in a moving body are of an objective sort."

(iii) *Einstein replies to Varičak.* In reply to Varičak, and perhaps to set straight his friend Laub, Einstein wrote in (1911b)

> The question of whether the Lorentz contraction is real or not is misleading. It is not 'real' insofar as it does not exist for an observer moving with the object. This is indeed what Ehrenfest has made clear in a very nice way.

Einstein divided his rebuttal to Varičak into two parts:

> (a) From Einstein's point of view the contraction phenomenon is a subjective or psychological one and is therefore not experimentally observable.

(b) From Einstein's point of view the contraction phenomenon is a
result of our methods of measurement, i.e., the use of synchronized clocks.

Einstein replied to (a) with a short description, similar to the one from the
(1907e) paper, of an experimental method for detecting the contraction of a
moving rod.

Einstein disposed of (b) with the following *Gedanken* experiment, which did
not require the use of synchronized clocks. Consider two rods of equal rest
lengths $\overline{A'B'} = \overline{A''B''} = l_0$, and moving in opposite directions with any
velocities less than that of light, where $A'(A'')$ and $B'(B'')$ are the bars'
endpoints. When A' and A'' overlap then so does B' and B'', and if at this
instant two observers in K mark the spatial positions of these two events at A^*
and B^*, respectively, then the length A^*B^* is less than l_0.[23] Consequently
Einstein disposed of his critics without becoming involved in the rigid body
controversy.

7.4.12. Aftermath of the Rigid-Body Controversy: Mostly Planck, Poincaré and Lorentz

> Planck is blocked by some undoubtedly false preconceptions [regarding Einstein's
> efforts toward a generalized relativity theory]. . . .
> Poincaré was in general simply antagonistic [presumably toward all of Einstein's
> work in relativity]. . . .
>
> A. Einstein (November 1911)
>
> I shall have to seek out Lorentz in order to discuss fundamental matters.
>
> A. Einstein (late 1913)

At the 83rd *Naturforscherversammlung* at Karlsruhe in September 1911,
Sommerfeld (1911) could say that although he had been asked to speak on the
relativity theory, he decided otherwise because it was already in "the safe
possession of the physicist." In his opinion, the frontier problems concerned
the meaning of Planck's energy quantum and Einstein's light quantum,
Einstein having proposed the latter notion in the "same memorable year 1905,
even before presenting the principle of relativity." The nature of light was also
the theme for that first summit meeting of physics, the Solvay Congress of 30
October to 3 November 1911 at Brussells. Consequently by 1911, owing mainly
to Ehrenfest's fundamental studies, capped off by Einstein's reply to Varičak,
Einstein's theory had been demarcated from Lorentz's.

But not every major physicist accepted Einstein's theory with its consequences for space and time, and neither was Einstein's theory as yet
completely understood. For example, as Holton (1973c) has written, "it was
somehow Einstein's tragic fate to have the contribution he most cared about
rejected by the very men whose approval and understanding he would have
most gladly had" – among them were Planck, Poincaré and Lorentz.

Planck made two great discoveries in his lifetime: the energy quantum and Einstein. Although he was one of the principle of relativity's earliest supporters, he never accepted one of the special relativity theory's principal consequences – the superfluousness of Lorentz's ether – for example, Planck's reliance on the ether in his attempts to reconcile his theory of radiation with classical electromagnetic theory [see Kuhn (1978)]; nor could Planck accept the relativity theory as a theory of principle.[24] Like Abraham, Planck had been impressed with the unifying power of the Lagrangian formalism. Yet for Planck the formulation of a generalized dynamics went far deeper than for Abraham. Using the Lagrangian in the principle of least action was progress toward a unification of the world-picture because the action quantity was Lorentz invariant. This route squared with Planck's own neo-Kantian attempt to reach a world-picture based on unchanging organizing principles that hinted at the objective reality beyond the appearances. Planck made his philosophical position abundantly clear in a polemic with Mach, which was conducted in the pages of the *Physikalische Zeitschrift* [(Planck (1909), (1910b)), such was the importance of philosophy to the German-speaking physicist]. Their differences of opinion on points epistemological and ontological were irreconcilable. Einstein may well have been somewhat pleased with Planck's stance, since it was congenial to his own growing movement away from his youthful interest in Mach's philosophy of science. On the other hand, Einstein must have been chagrinned by Planck's excessive reliance on empirical data, and his reluctance to part with Lorentz's ether. For example, as I shall explore in Chapter 12, in 1908 many principal physicists were pleased that Kaufmann's disconfirming data of 1906 could now be discounted in favor of the more exact data from others that supported the Lorentz-Einstein theory. Planck persisted in seeking further empirical support. On the other hand, Einstein had never doubted the outcome in the first place.

Although there were disagreements on fundamental matters, Planck was immensely impressed by the depth and breadth of Einstein's ability as a physicist. In the summer of 1913 Planck personally offered to Einstein, then a professor at his old school the ETH, the Directorship of the physics research branch of the newly-formed Kaiser Wilhelm Institute, a professorship at the University of Berlin, and informed Einstein that he would be elected to the highly prestigious Royal Prussian Academy of Science. In late 1913, before departing for Berlin, Einstein wrote to his confidante and former coworker at the Patent Office, Michele Besso:

> In the meantime I shall have to seek out Lorentz in order to discuss fundamental matters. He exhibits interest, and so does Langevin. Laue is not accessible when it concerns matters of principle, also not Planck, Sommerfeld is. The free and impartial view is hardly very characteristic of the German adult (blinkers!).[25]

At that time what Einstein considered as fundamental matters were deep indeed, for he was immersed in the titanic struggle that culminated in the general relativity theory of 1915, Einstein's second *Annus Mirabilis*.

Poincaré's epistemology demanded Lorentz's ether and a constructive view of theoretical physics. Although by 1908 Poincaré considered Lorentz's prediction for m_T to have been vindicated empirically, nevertheless the prediction of his own Lorentz-covariant theory of gravity gave too small a value for the advance of Mercury's perihelion and so he never considered the principle of relativity to be a convention [Poincaré (1912a)]. At the Solvay Congress [Langevin (1912)] there were no recorded discussions on the relativity theory, but behind the scenes there were many heated debates on this topic. We know this because in November 1911 Einstein wrote to Heinrich Zangger at Zurich [Hoffmann (1972)]: "Poincaré was in general simply antagonistic (against the theory of relativity) and, for all his acuity showed little understanding of the situation." Despite the outcome of their conversations, Einstein had impressed Poincaré who, shortly after Solvay, wrote a letter supporting Einstein for a position at the ETH [Seelig (1954)]: "Einstein is one of the most original thinkers I have ever met .... . Since he seeks in all directions, one must ... expect the majority of the paths on which he embarks to be blind alleys"; no doubt to Poincaré special relativity was one of them. Poincaré (1912c) said that "everything happens as if time were a fourth dimension of space" and so "some physicists want to adopt a new convention" for simultaneity. But, he continued, "those of us who are not of this opinion can legitimately retain the old one."

In (1912b) after briefly mentioning Einstein's work on cavity radiation, Poincaré went on to discuss "Lorentz's principle of relativity." Poincaré neither cited in print Einstein's special relativity theory nor did he ever associate Einstein's name with Lorentz's theory of the electron.

Of Lorentz, Einstein wrote many years later [de Haas-Lorentz (1957)], "for me personally he meant more than all the others I have met on my life's journey." Since 1909 Lorentz had been one of Einstein's principal scientific correspondants. In late 1913 Lorentz was one of Einstein's sources of inspiration in his research toward the general relativity theory. But for Lorentz, "Einstein's theory" of electrodynamics was another "form" of his own theory of the electron, as Lorentz wrote in *The Theory of Electrons* (1909). According to Lorentz's theory of electrons two inertial observers "would be alike in all respects," and so it could be concluded that there was no reason for asserting "that neither of them is in possession of the 'true' times or of the 'true' lengths." In Lorentz's view Einstein "laid particular stress" on this point, and based his theory on "what he calls the principle of relativity." Therefore, continued Lorentz:

> Einstein simply postulates what we have deduced, with some difficulty
> and not altogether satisfactorily, from the fundamental equations of the

electromagnetic field. By doing so, he may certainly take credit for making us see in the negative result of experiments like those of Michelson, Rayleigh and Brace, not a fortuitous compensation of opposing effects, but the manifestation of a general and fundamental principle.

But despite the "fascinating boldness" of this principle, Lorentz preferred the form in which he had "presented the theory," and he interpreted Einstein's relativity of time and length within the context of his own dynamical ether-based framework. For example, since the observer in k, wrote Lorentz, was "unconscious" of his motion through the ether, the observer in k measured lengths according to the effective coordinates (ξ, η, ζ), and consequently took "for a sphere what really is an ellipsoid." Assuming that the forces on k's clocks transform properly when these clocks read the time τ, and so run $(1 - v^2/c^2)^{-1/2}$ times slower than K's clocks, Lorentz went on in *The Theory of Electrons* (1909) to deduce Einstein's second postulate from the hypotheses of the Lorentz contraction and the local time.

Despite the deep understanding of Einstein's kinematics which he displayed in his Leiden lectures (1910–1912), Lorentz continued to pursue his opinion that the two theories were equivalent both mathematically and in their empirical predictions; consequently, the choice between them "and especially the question of true time can be handed over to the theory of knowledge." Lorentz continued at Leiden by pointing out that although "the modern physicists, as Einstein and Minkowki, speak no longer about the ether at all," something was necessary to transmit electromagnetic fields and energy.[26]

At his Göttingen Wolfskehl lectures, Lorentz (1910) said: "To discuss Einstein's principle of relativity here in Göttingen where Minkowski taught seems to me to be a particularly welcome task." But after discussing Minkowski's relativization of Newton's second law, Lorentz defined the "Minkowski mass" appearing in this law as $\mu_0^e = 2e^2/(3Rc^2)$ – the electron's electromagnetic mass, instead of the mechanical mass from Einstein's special theory of relativity.

In 1922 at the California Institute of Technology, Lorentz again spoke on the theme of the complete equivalence between his theory and Einstein's; in particular, he said "my notion of time is so definite that I clearly cannot distinguish in my picture what is simultaneous and what is not."

Both Lorentz and Planck (and needless to say Abraham too) continued to maintain the observational equivalence of the two theories despite the 1908 suggestion of Einstein and Laub (1908a) of a decisive experiment to distinguish between them; namely, the repetition by H. A. Wilson of his 1904 experiment in which the dielectric sandwiched between the plates of a capacitor was replaced by a magnetic dielectric.[27] Laub repeated their prediction in his review paper (1910), and in 1913 the experiment was performed with results supporting Einstein and Laub. In addition, there was Einstein's 1907 prediction for a transverse Doppler effect.

On the basis of those existing data Laub (1910) was confident enough to believe in the complete vindication of the prediction of Lorentz and Einstein for m_T (although Lorentz and Poincaré accepted Bucherer's data, some others awaited further data); but Laub preferred Einstein's theory, and it was on this note that he concluded his review paper, "thus indeed becomes displayed, the fact that between mathematics and its application there stands an internal harmonious connection."

7.4.13. The Relativity of Time: 1908–1913

(i) *Max von Laue's Das Relativitätsprinzip* (1911a)

> ...more numerous are those who disagree with [the special relativity theory's] intellectual content; to whom, in particular, the relativity of time with its often paradoxical consequences appears unacceptable.
>
> M. von Laue (1911a)

Max von Laue wrote, in the first textbook on the special relativity theory, *Das Relativitätsprinzip* (1911a), that "a definite distinction between the developed theory of Lorentz and the theory of Einstein is not possible", although von Laue neither mentioned the ambiguity in Lorentz's notion of at rest in an inertial system cited in print by Drude (1900) and Kaufmann (1906), nor did von Laue discuss the suggested experiment of Einstein and Laub (1908a) which could distinguish between the theories of Einstein and Lorentz.

In a section surveying experimental data (based upon Laub's (1910) paper) von Laue wrote that although the electron-deflection experiments of Bucherer were not yet decisive, "the author of this book places great value on the fact that there exists no empirical evidence against the theory"; indeed, throughout the book von Laue left no doubt as to his optimism concerning the future of the relativity theory. In fact, von Laue's book was more than just a well-organized exposition of the relativity theory; it was a treatise in the purest sense of the term because totally new were his discussions of the influence of elastic tension on the momentum and energy densities, as well as of the transformation properties of tensions, and this material, directly influenced Herglotz (1911) to develop a relativistic theory of elasticity.

Von Laue tells us in his "*Vorwort*" of a fundamental problem with the relativity theory; namely, that whereas many researchers were not yet satisfied with the relativity theory's empirical base, "more numerous are those who disagree with its intellectual content ... in particular, the relativity of time with its often paradoxical consequences appears unacceptable." Therefore, some physicists took exception to Sommerfeld's declaration of 1911 that the special relativity theory was already so well established that it was no longer on the frontiers of physics. For example, J. D. van der Waals Jr. (1912), of The

Netherlands found Sommerfeld to be rather presumptuous because there was as yet no causal explanation for relativistic effects such as the variation of mass and length with velocity. Although by this time (1911) criticism of these effects was indeed outside the mainstream, nevertheless the need for a causal, i.e., dynamical, explanation for effects underlay why many physicists considered the relativity of time to have paradoxical consequences. Although a change in the spatial dimensions of a body was acceptable to everyone, a relativity of time was not.

(ii) *Abraham.*   Max Abraham in the third edition dated 1908 of his (1905) wrote on the relativistic time dilation: "As Einstein himself comments, this result leads to highly peculiar consequences" – that is, two clocks originally synchronous register a time difference upon being reunited after one of them has made a round trip journey. In Abraham's viewpoint the time on a clock should be independent of the clock's motion. So instead of Einstein's result for time dilation, i.e., Eq. (7.1), Abraham grouped the factor $\sqrt{1 - v^2/c^2}$ with the velocity of light, i.e.,

$$c' = c\sqrt{1 - v^2/c^2}, \tag{7.43}$$

where, using a terrestrial source, c' is the measured velocity of light on the earth, and c is the velocity of light measured in a reference system on the fixed stars, or to a good approximation on the sun. In analogy with Einstein's result [Eq. (7.1)], Abraham wrote Eq. (7.43) as

$$c - c' = c[1 - \sqrt{1 - v^2/c^2}], \tag{7.44}$$

and this equation could also explain the Michelson-Morley experiment because Eqs. (7.43) and (7.44) could be expected for an ether-based theory of electromagnetism containing a "postulate of relativity" that made assertions only on the relativity of length, and not time. Thus, for example, Abraham deduced the distance measured in K(d) traversed by a light ray in k(d') as follows: Since $d' = c'T'$, then from Eq. (7.43) and the assertion that $T' = T$ we obtain

$$d' = d\sqrt{1 - v^2/c^2}, \tag{7.45}$$

where $d = cT$, and thereby Abraham could deduce the length contraction from assuming the absoluteness of time, and not of the velocity of light in vacuum. But, continued Abraham, Lorentz had succeeded in explaining the null results of ether-drift experiments at the expense of including "absolute velocities" of moving electrons. Consequently wrote Abraham, of Lorentz's "theorem of relativity," which had been elaborated upon by "Poincaré, Einstein, Planck and Minkowski," did not eliminate absolute motion. Thus, whichever way we choose to view electrodynamics, either time dilation or the nonconstancy of the velocity of light "must depend upon the system's absolute motion." In other words, time dilation had to be interpreted as having a dynamical cause.

(iii) *Lewis and Tolman* (1909).   In 1909 Lewis and Tolman (1909) dismissed as "untenable" that the velocity of light could depend upon its source's motion, and yet they were dissatisfied with Einstein's manner of expressing this result because it led to "certain curious conclusions as to the comparative readings of clocks" which conclusions they relegated to a long footnote.

(iv) *Lorentz* (1910–1912).   Lorentz (1910–1912) analyzed what von Laue referred to as the "intellectual content" surrounding the relativity of time, demonstrating that Einstein's formalism contained no inconsistencies (see Section 7.3); nevertheless Lorentz preferred his own view of the absoluteness of time and simultaneity.

(v) *Paul Langevin.*   On 10 April 1911, one month before von Laue's preface was written, at the Philosophy Congress at Bologna, Paul Langevin added a new twist to Einstein's peculiar consequence by replacing the clocks at *A* and *B* with human observers. Like Minkowski's lecture, "Space and Time" (1908b), Langevin's popularization of relativity theory for philosophers, "*L'évolution de l'espace et du temps*" (1911), had an immediate impact which we can gauge from the comments of one of the philosophers present. Henri Bergson (1922) wrote:

> ...it was Langevin's address to the Congress of Bologna on 10 April 1911 that first drew our attention to Einstein's ideas. We are aware of what all those interested in the theory of relativity owe to the works and teachings of Langevin.

After developing how Einstein's relativity theory removed the tension between mechanics and electromagnetism, Langevin discussed certain of the theory's consequences by using a few "concrete examples" well chosen to stimulate the audience. For example, a space traveler travels a distance *L*, as measured by someone at rest on the earth, in a straight line to a star in one year, and then abruptly turns around and returns on the same line. Langevin asserted that for an appropriate velocity (relative to the earth) less than that of light, one could demonstrate that for the space traveler having aged two years, two hundred years had elapsed on the earth. He discussed the "asymmetry" between the two measurements of the duration of separation, by imagining that the voyager and the earth exchanged light signals. According to the §7 of the relativity paper an inertial observer moving away from a source of radiation that emitted signals with a frequency v_0 in its rest system, would receive the signals at the rate [see Eq. (10.16)]

$$v_1 = v_0 \sqrt{\frac{1 - v/c}{1 + v/c}} \tag{7.46}$$

where v is the observer's velocity relative to the source. Inversely an inertial observer moving linearly toward the source would receive signals at the increased rate

$$v_2 = v_0 \sqrt{\frac{1 + v/c}{1 - v/c}}. \qquad (7.47)$$

During the outward journey, explained Langevin, the voyager and earth observer would receive each others' signals at the rate v_1. At the turn around point the voyager would instantaneously register a change in received signal frequency from v_1 to v_2, whereas the earth observer had to wait a time L/c to receive signals at the rate v_2 and thereby realize that the voyager had changed direction. Therefore, the voyager would receive an equal number of signals at the rates v_1 and v_2, but the earth observer would receive more signals of the sort v_1 than v_2.[28] Langevin attributed the cause of the asymmetry between the reference systems of the earth and the voyager to be the "acceleration . . . which only the voyager has undergone" at the journey's midpoint. Acceleration had a particularly important meaning for Langevin because "every change of velocity, every acceleration has an absolute meaning"; for whereas Einstein's theory rendered an ether superfluous for the analysis of inertial motions, "we must not conclude from that, as we have done sometimes, that the ether is nonexistant and inaccessible to experiment." Langevin supported the point thus: An accelerating charged body is the source of electromagnetic waves that "propagate in *le milieu*," and conversely every electromagnetic wave has as its source an accelerated charged particle; consequently, "we have thus grasped the ether [or *le milieu*] through the intermediary of accelerations [because] the ether reveals its reality as the vehicle, supporting the energy transported by these waves." Langevin presented the concrete example of asymmetric aging to illustrate the "absolute nature of acceleration."

Although Langevin did not discuss further his view of a relativity theory containing an ether, we can conjecture that the root of his decision for including an ether was his dissatisfaction with a kinematical account of the relativity of time; rather, according to Langevin, this effect could be causally explained by the spacecraft ("*boulet de Jules Verne*") interacting with the ether, and the interaction manifested itself in the spacecraft's acceleration.

(vi) *Emil Wiechert.*   In September 1911 Emil Wiechert independently proposed the example of asymmetric aging within his own framework of relativity theory. Wiechert took heed of a comment by Lorentz at the 1910 Wolfskehl lectures in Göttingen: "If the relativity principle had general validity in nature you would still not be in a position to ascertain if the reference system you happen to use is the unique one. You would then arrive at the same results as if, following Einstein and Minkowski, you would deny the existence of an ether and a true time and thus take all reference systems as equivalent. Which of these two viewpoints you will follow [i.e., the ones of Lorentz and of Einstein-Minkowski] may remain to each individual to choose." Wiechert's attempt to accomplish the program of Einstein and of his late "Göttingen colleague and friend" Minkowski was to propose a nonisotropic four-dimensional space-time in which he designated the properties in different directions with the term "*Schreitung*." Consequently, Wiechert could assert that the space traveler and observer were affected by different *Schreitungen*, and this caused their asymmetrical aging.

(vii) *von Laue and the clock paradox*

> Moreover, in my opinion, such problems as whether the ether or absolute time exists can be omitted without damage from physical considerations ... I am not asserting that these problems are uninteresting, on the contrary they seem to me of great philosophical meaning. But precisely for this reason they should remain reserved for treatment with the methods of philosophy.
>
> M. von Laue (1912a)

In a critique of Wiechert, von Laue (1912a) wrote that of all the paradoxes associated with the relativity theory, the most perplexing was that a "clock's rate should depend upon its state of motion." He considered Langevin's recent "highly readable and charming lecture" to have explained this phenomenon adequately. Most likely von Laue agreed with Langevin's explanation for asymmetric aging because it was independent of an ether affecting inertial motion, but that was not the case for Wiechert's explanation. Von Laue's principal criticisms of Wiechert were: (1) Unless demanded by empirical data, "the entire discussion of the existence of the ether and of absolute time can be banished from physical discussions"; such problems should "remain reserved for treatment with the methods of philosophy." (2) There is an analogy between the isotropy of three-dimensional space and the geometrical law that two arbitrarily placed points determine a straight line and thus also a direction that is equivalent to every other direction; this analogy holds also for the four-dimensional space-time which should be isotropic.[29] In von Laue's opinion the asymmetric aging could be demonstrated directly from the Minkowski space-time diagram (von Laue's *Gedanken* experiment used two groups of physicists).

In a subsequent exposition of relativity theory written for philosophers, von Laue (1913) returned to the problem of asymmetric aging, and he further explicated what was the paradoxical state of affairs; namely, how could "purely kinematical considerations" distinguish between a moving clock and one at rest? This has become known as the clock paradox, and when the clocks are replaced by human observers it is called the twin paradox. In the (1912a) discussion von Laue had mentioned one critic claiming that Einstein's peculiar result revealed a "hidden error" in the relativity theory. Whereas von Laue's 1911 deduction of the asymmetric aging from Minkowki's formalism only indirectly addressed this point, in 1913 he focused upon the particular example of a linear outward-and-return jouney and obtained Einstein's Eq. (7.1). But in 1913 von Laue added another ingredient[30]: while B remained in one inertial system, A made the outward and return journeys in "two separate inertial systems," and this *accounted* for the asymmetric aging. To the best of my knowledge von Laue was the first to carefully explain that the "obvious objection" concerning the clock's undergoing acceleration during the change of inertial system could be avoided because "we can make the times of uniform motion arbitrarily greater than those of acceleration;" therefore, a clock always registers the proper time in its rest system, and time dilation depends on only the clock's instantaneous velocity. Von Laue went on to suggest that the dependence of a clock's time on its motion could be tested experimentally if we accept that "any periodic system is a clock" – for example, radiation from canal rays, and here he certainly had in mind Einstein's 1906 proposal for detecting the transverse Doppler effect.

Using two different inertial systems for A, von Laue avoided Langevin's claim that the asymmetric aging was caused by the instantaneous acceleration of A's rest system at the journey's mid-point.[31] Von Laue's vociferous avoidance of an ether set him in a group referred to by Abraham (1914a) as "*die radikalen Relativisten.*" Detailed analysis of von Laue's 1913 resolution of the clock paradox reveals that the fallacy in this paradox could be exposed in a one-way journey discussed first by Einstein in 1905, and then elaborated upon by Lorentz in his Leiden lectures (1910–1912). Consider that two clocks C_1 and C_2 are placed side by side and are synchronous in K. With von Laue we assume that the acceleration times are much less than any other time entering this demonstration. Event E_1 takes place when C_2 is placed into k and k is accelerated to velocity v; this occurs at the space-time points

$$x_1 = 0, \qquad x'_1 = 0,$$

$$t_1 = 0, \qquad t'_1 = 0,$$

where primed coordinates refer to k. Event E_2 occurs when, after traveling a distance l along K's x-axis, the clock A is removed from k and placed at rest in K at $x = l$; the space-time coordinates of E_2 are

$$x_2 = l, \qquad x'_2 = 0,$$

$$t_2 = T, \qquad t'_1 = T',$$

where T' and T are related by Eq. (7.1). Event E_3 occurs when, after being at rest for a time α in K, C_2 is placed in the inertial reference system K' which is traveling with velocity $-v$ relative to K; the coordinates of E_3 are:

$$x_3 = l, \qquad x''_3 = 0,$$

$$t_3 = T + \alpha, \qquad t''_3 = T' + \alpha,$$

where double-primed coordinates refer to K'. Event E_4 occurs when C_2 and C_1 are reunited in K; the space-time points for this event are

$$x_4 = 0, \qquad x_4'' = 0,$$

$$t_4 = 2T + \alpha, \quad T_4'' = T''.$$

The total time elapsed for C_2 on the round trip journey is $T' + T'' + \alpha$. Since

$$T' = T'' = T\sqrt{1 - v^2/c^2}$$

then

$$T' + T'' + \alpha = 2T' + \alpha$$

or

$$2(T - T') = 2T(1 - \sqrt{1 - v^2/c^2})$$

and, as Einstein predicted, C_2 runs slower than C_1 by the amount $1 - \sqrt{1 - v^2/c^2}$, that is, $T' < T$. But since according to the special relativity theory all motion is relative, could we not consider the clock C_2 as at rest, with C_1 moving in the opposite sense? The result of this journey is $T' > T$. On the other hand, we have just shown that $T' < T$. This so-called clock paradox, however, arises from the same misuse of the relativistic transformations that Lorentz emphasized in his Leiden lectures. Thus, the asymmetry of time dilation for a one-way journey reveals the fallacy of the clock paradox, and so also of the twin paradox where the clocks C_1 and C_2 are replaced by people. Furthermore, as von Laue wrote (1913), "this difference in the information of the two clocks remains 'absolute' in every reference system" – that is, the quantity $T - T'$ is Lorentz invariant because it is the difference between the proper times of two clocks at relative rest, and therefore the sign of this difference cannot be arbitrarily changed. Alternatively one could say that, if C_2 remained in a single reference system, the symmetry between the rest systems of C_1 and C_2 was broken by C_2's instantaneous acceleration at the turn around point.

(viii) *Einstein.* So poorly circulated was Einstein's address to naturalists (1911c) that his statement on the application of time dilation to life forms went virtually unnoticed[32]:

> If we placed a living organism in a box ... one could arrange that the organism, after any arbitrary lengthy flight, could be returned to its original spot in a scarcely altered condition, while corresponding organisms which had remained in their original positions had already long since given way to new generations. For the moving organism the lengthy time of the journey was a mere instant, provided the motion took place with approximately the speed of light.

But Einstein's guiding epistemology based on the belief that science is a development of prescientific thinking, could well have caused his dissatisfaction with only accounting for time dilation; in 1918 he used the principle of equivalence from the general relativity theory for proposing a dynamical explanation for the clock paradox [Einstein (1918a)].[33] Einstein's reasoning based on the general relativity theory, did not satisfy everyone. By 1921 the consequences of the special relativity theory for the nature of time were being so heatedly debated that S. Arrhenius' presentation on the occasion of Einstein's Nobel Prize contained the following statement [Einstein (1923)]: "It will be no secret that the famous philosopher Bergson in Paris has challenged (special relativity)." The criticisms of Bergson and his followers, most notably Herbert Dingle, focus on the relativity of time, and can be responded to in an elementary way. The essence of the matter, however, is that they seek within the special relativity theory a dynamical reason for relativistic effects; Einstein and Langevin also worried about this point (von Laue avoided it by using two reference systems for the space traveler), and Einstein began to resolve it within the higher level general relativity theory. But the special relativity theory has no logical inconsistencies and thereby provides a completely accurate *account* of the asymmetrical aging process. That was all that Einstein had intended in 1905.[34]

FOOTNOTES

1. Had Einstein not restricted himself to $t = 0$, he would have obtained

$$\gamma^2(x - vt)^2 + y^2 + z^2 = R^2,$$

which is the equation for a moving ellipsoid of semi-minor axis $R\sqrt{1 - v^2/c^2}$ and semi-major axis R, which is centered at $x = vt$, $y = z = 0$. Poincaré (1906) performed this general analysis for deducing the volume of a moving electron for any value of l. His ultimate goal was deriving the correct transformation equation for the electron's charge density.

2. See Grünbaum (1973) for a philosophical analysis of relativity theory requiring Einstein's having assumed *a priori* that the velocity of light in vacuum is the ultimate velocity. There is no extant archival material to support Grünbaum's claim. For my responses, based upon the material in this chapter, see Miller [(1975a), (1977a)].

3. In §10 Einstein proved that bodies with mass could not be accelerated to the velocity of light.

4. In order to demonstrate explicitly the connection between the length contraction and the relativity of simultaneity, Einstein could have proceeded as follows. Consider that in k there is a rigid rod at rest on k's ξ-axis. An observer in k measures by congruence the length of this rod as $\xi_2 - \xi_1 = l_0$. In order to relate l_0 and l (the length measured by K-observers) consider comparing the lengths in k and K by means of the transformation equation inverse to Eq. (6.53),

$$x_2 - x_1 = \gamma[(\xi_2 - \xi_1) + v(\tau_2 - \tau_1)]. \tag{A}$$

The length measurement in K involves two spatially separated observers on the x-axis simultaneously marking the positions of the ends of the moving rod. Consequently, the length measurement in K involves two events: (1) $E_1(x_1, 0, 0, t_1)$ and (2) $E_2(x_2, 0, 0, t_1)$ — that is, E_1 and E_2 are simultaneous events in K. The transformation equation inverse to Eq. (6.52) relates these two events between K and k as follows:

$$t_2 - t_1 = \gamma[(\tau_2 - \tau_1) + (v/c^2)(\xi_2 - \xi_1)], \tag{B}$$

and since $t_2 = t_1$, then

$$\tau_2 - \tau_1 = -(v/c^2)(\xi_2 - \xi_1), \tag{C}$$

and consequently in k the two spatially separated events in K are not simultaneous, i.e., τ_1 occurred before τ_2. Substitution of Eq. (C) into Eq. (A) yields

$$l = l_0\sqrt{1 - v^2/c^2}, \tag{D}$$

and this method of derivation displays the intimate relationship between the relativity of time and of length.

5. Einstein's assertion that the time of a clock depends on only the clock's velocity and not on its acceleration was first addressed by von Laue (1913): acceleration can be neglected if the time interval during which it occurs is much smaller than any other time entering the problem [see Møller (1972) for further discussions]. Experiments completed by 1957 indicate that the lifetime of a π^0-meson in its rest system is unaffected by the severe deceleration it suffers when brought to rest, thereby offering a consistency check on Einstein's prediction for time dilation. See Marder (1971) for a summary of elementary-particle experiments related to time dilation, and also footnote 8 below. Clock mechanisms affected by acceleration can be devised. For example, a chronometer whose mechanism is an electric circuit completed in a plastic hemispheric bowl by a metal ball rolling back and forth touching a wire running along the bowl's periphery and another wire at the bowl's bottom. Acceleration affects the ball's frequency of oscillation.

6. Einstein's 1905 argument concerning the behavior of the clock moving on a continuously closed curve is valid if the curve is a geodesic in Riemannian space-time. [See McCrea (1957).]

7. Einstein already had proposed this experiment in his (1907b). Stark's 1906 experiment lacked sufficient accuracy for detecting the transverse Doppler shift in the spectral lines emitted by the moving canal rays (hydrogen ions).

8. Six years before Ives, Kennedy and Thorndike (1932) had repeated the Michelson-Morley (1887) experiment using an interferrometer with arms of unequal length. The absence of a fringe shift could not be explained by Lorentz's contraction, but could be explained by Einstein's Eq. (7.14), i.e., by time dilation. Ives asserted that the importance of time dilation demanded a positive experiment; furthermore, the more general hypothesis of contraction could explain the null result of Kennedy and Thorndike. Ives' experiment was ingenious: he avoided the difficulties of an exact transverse observation of the radiation from canal rays by measuring the arithmetic mean of the wavelengths of radiation emitted in the forward and backward directions relative to the laboratory; he compared this result to the wavelength of an undisplaced spectral line [see French (1966) and Rosser (1964) for further discussions of Ives' data].

Ives' principal criticisms of Einstein's relativity theory were its second postulate and the absence of an ether – in short, Einstein's theory did not *explain* why effects such as length contraction occurred. Ives based his version of the Lorentz-Larmor theory on the following hypotheses (1937b); the existence of a "fixed ether," which could be identified as "the seat of the pattern of radiant energy received from the fixed stars"; it contained two sets of measuring rods and clocks, and one set was "unaffected by transport"; Lorentz's contraction; the hypothesis that the velocity of light was exactly c only relative to the observers at rest in the ether who use the set of rods and clocks that were unaffected by transport. From these hypotheses Ives succeeded in deducing a result for the optical Doppler effect observed in an inertial reference system. However, the deduction required the following artificial limiting process in order to remove unmeasurable quantities. According to Ives the time it took light to travel a distance d between two points a and b in an inertial reference system traveling at velocity v relative to the ether was determined by a clock that is moved from a to b at a velocity W, and the distance d was measured with a rod that moved past a and b at a velocity Y, where W and Y were velocities relative to the inertial reference system. In the limit that W and Y are reduced to zero, i.e., infinitely slow transport of the clock and rod, terms containing the unmeasurable quantity v vanished, and Ives obtained the exact equation for the optical Doppler effect – namely, the result from Einstein's special relativity theory of 1905. Among the criticisms against Ives' theory was the difficulty to define "fixed stars," and just this criticism happened to have been levelled against Lorentz's theory 32 years before Ives' deliberations appeared in print by Kaufmann (1906) and then later by Silberstein (1914). Ives remained a vigorous anti-relativist to the end of this life.

Jones (1939) used the special relativity theory for discussing the Ives-Stilwell experiment; he was critical of their not using a separate source for obtaining the center of gravity of the observed spectrum. Independently of Jones, Otting (1939) in Munich repeated the Ives-Stilwell experiment using a separate source as a spectral reference, thereby providing an accurate check on Ives' results. In view of the times in Germany, it was meritorious that Otting used only the relativistic transformation law for wavelength and made no mention of an ether-based theory of electromagnetism; needless to say he did not mention Einstein either. Otting's research was part of his dissertation, and he gave his forwarding address as Wünsdorf, *Panzertruppenschule*.

To this day the Ives-Stilwell experiment is the only positive proof for Einstein's prediction of time dilation. The various experiments utilizing elementary particles involve a vicious circle because their data analysis depends on special relativity; consequently these experiments test only the *consistency* of the special theory of relativity. For example, the interpretation of the data from the Rossi-Hall experiment (1941) required the Bethe-Bloch equation.

9. However, almost half a century later James Terrell (1959) showed that the contraction of a moving body in the direction of its motion could not be seen on a photograph. Using the relativistic transformations Terrell predicted that a more mind astounding phenomenon than either Lorentz or Einstein had supposed because the difference in times for light to arrive at the camera in K from different points on the solid body in k, would cause the body in k to be photographed rotated but not contracted; in fact, depending upon the body's state of motion, an observer in K might see its back first.

Terrell (1959) considers the case where the object subtends a small solid angle at the observer's camera, and proves that stereoscopic viewing would correct for the rotation and reveal a distortion, which is not a Lorentz contraction. For example, a solid object moving directly toward an observer with velocity v will be seen in three dimensions to be elongated along its direction of motion by the amount $\sqrt{(c+v)/(c-v)}$. For further discussion of Terrell's paper see Weisskopf (1960). Weinstein (1960) arrived independently at conclusions similar to Terrell's but less general.

10. In fact, continued Bucherer, he had thought a great deal about his theory, and was disturbed over its disagreement with the "theory of dispersion," i.e., the Rayleigh-Brace experiments. Yet whatever problems his own theory faced, Kaufmann's experiments did not constitute the sole criterion for deciding the "dynamics of the electron." There were, in Bucherer's opinion, certain other key theoretical considerations: (1) Lorentz's electron theory required nonelectromagnetic forces; (2) "Einstein's relativity theory" violated Gauss' law. Bucherer's second criterion was incorrect and was indicative of his confusion over the role and use of the space and time transformations. Bucherer continued by describing a modified version of Maxwellian electromagnetism that, as he wrote in a subsequent elaboration (1908b), was a "phenomenological method of calculating electromagnetic effects which should harmonize with all facts of observation, leaving it to further endeavors to find a physical interpretation of this method." This electromagnetic theory of magnetic poles interacting with electric charges *ab initio* contained only relative velocities, a principle of relativity and the principle of action and reaction. It predicted that electrons moving at angles other than $90°$ to the magnetic field direction should experience a force other than the one predicted in Lorentz's theory—that is,

$$e\frac{\boldsymbol{v}}{c} \times \boldsymbol{B}\left[1 - \frac{v^2}{c^2}(\hat{v} \cdot \hat{B})^2\right]^{-1},$$

instead of $e\boldsymbol{v}/c \times \boldsymbol{B}$. Bucherer promised to test this prediction in 1908 using β-rays. The result of this experiment and its important by-product are discussed in Section 12.4.4.

11. Ehrenfest (1906) had accepted Kaufmann's (1906) data as disconfirming Lorentz's theory of the electron and agreed with Kaufmann's opinion that the data could not decide between the electron theories of Abraham and of Bucherer-Langevin. Ehrenfest went on to use a theorem of Liapounof to prove the instability of the Bucherer-Langevin electron. Consequently, neither had Ehrenfest read the papers Bucherer [(1904), (1905)] and Langevin (1905) nor Poincaré's (1905a) which had been also referenced by Kaufmann, and where Poincaré mentioned his proof that only the deformable electron of Bucherer-Langevin is stable under exclusively electromagnetic forces. Clearly, this was Ehrenfest's first foray into electron physics and, considering that he was then in Vienna, it was not a well-researched effort.

12. Einstein could also have been thinking of Sommerfeld's (1904–1905) results that a rigid spherical electron with a uniform volume (surface) distribution of charge could (not) be accelerated beyond the velocity of light in vacuum.

13. He could have deduced the restriction against hyperlight velocities by analyzing how an observer in k measures the elapsed time for a signal of velocity w (relative to K) to traverse a distance l in K. From the relativistic transformation equation for time

$$\Delta t' = \gamma(\Delta t - (v/c^2)\Delta x)$$

since $\Delta t = l/w$ and $\Delta x = l$ then

$$\Delta t' = (l/w)(1 - wv/c^2).$$

If $w > c$, then there is a v such that $\Delta t' < 0$. Here the signal is not transmitted through a material substance.

14. Minkowski (1908a) associated the electromagnetic field with the double subscripted quantities $F_{12} = B_z$, $F_{23} = B_x$, $F_{13} = -B_y$, $F_{14} = +iE_x$, $F_{24} = +iE_y$, $F_{34} = +iE_z$, where $F_{\mu v} = -F_{v\mu}$. Then he could write the Maxwell-Lorentz Eqs. (1.26) and (1.27) as

$$\sum_{v=1}^{4} \frac{\partial F_{\mu v}}{\partial x_v} = \frac{4\pi}{c} S_\mu \qquad (\mu = 1, 2, 3, 4) \tag{A}$$

where $S_\mu = (\rho w/c, i\rho)$, and the Maxwell-Lorentz Eqs. (1.25) and (1.28) could be written as

$$\frac{\partial F_{\mu v}}{\partial x_\lambda} + \frac{\partial F_{\lambda \mu}}{\partial x_v} + \frac{\partial F_{v\lambda}}{\partial x_\mu} = 0 \tag{B}$$

where λ, μ, v can take the values 1, 2, 3, 4. The Eqs. (A) and (B) demonstrate the Lorentz covariance of the Maxwell-Lorentz equations. The "complete symmetry" in the Maxwell-Lorentz equations was for Minkowski the symmetry missed even by Poincaré.

For historical discussions of Minkowski's researches see Hilbert (1909–1910), Holton (1973d), Miller (1973), Reid (1970), Pyenson (1977) and Galison's (1979) detailed analysis.

15. "Space and Time" (1908b) focused on the space-time diagrams which Minkowski had only mentioned in the two previous expositions of his viewpoint (1907), (1908a). More than any earlier elaboration of the relativity theory, Minkowski's exposition of 1908 fired the imagination of physicists, mathematicians, and philosophers. As Infeld (1950) wrote:

> But it was not before 1908 or 1909 that the attention of great numbers of physicists was drawn to Einstein's results. One of the contributing factors in making relativity more widely familiar was the appearance of Minkowski's lecture "Space and Time."

16. The key concept in Minkowski's "world" is the invariant quantity

$$F = c^2 t^2 - x^2 - y^2 - z^2 = c^2 t'^2 - x'^2 - y'^2 - z'^2 \tag{7.28}$$

where (x, y, z, t) and (x', y', z', t') are connected by a Lorentz transformation. The values $F = \pm 1$ define two-sheeted hyperboloids; the value $F = 0$ yields their asymptotes, which are physically the trajectories of a light ray, i.e., $v = c$. Minkowski illustrated his space-time formalism for the case of motion along the parallel x-axes of K and k. Perhaps in an attempt not to becloud the issues in a semipopular presentation, Minkowski used a real fourth coordinate, and so the x- and t-axes of the systems k and K were not both orthogonal. Thus in Fig. A the x- and t-axes of K are orthogonal but the x'- and t'-axes of k are skewed. Minkowski emphasized that this divergence should not be construed as a preference for either k or K, because the two were related by a Lorentz transformation. The line OB is part of the asymptote $x = ct$; the segment Om is the distance $1/c$ to the branch of the hyperbola $c^2 t^2 - x^2 = 1$, for $t > 0$; similarly for the line $Om' = 1/c$. Since the coordinates of K and k are related by a Lorentz transformation, then as $c \to \infty$ the x'- and x-axes coincide and the "axis of t' may have any upward direction whatever." The lines OB and OA are the asymptotes $x = \pm ct$ and in a four-dimensional space-time they lie on light cones. The forward (backward) light cone encloses the region after (before) O, and comprises events occurring after (before) O. Minkowski referred to space-time vectors of the first kind normalized to $+1$, as "time-like vectors," and those normalized to -1 as "space-like vectors." The former vectors are directed into either the forward or backward light cones where $v < c$, and the latter into the lateral light cones where $v > c$. Therefore, acausal phenomena can occur only in the lateral light cones. The Lorentz invariance of the line element $c^2t^2 - x^2$ means that phenomena in the forward or backward light cones cannot be transformed into the lateral light cones, and vice versa; this was Minkowski's geometrical interpretation for the velocity of light as the limiting velocity for ponderable matter. Furthermore, concluded Minkowski, only substances traveling with sublight velocities could be transformed into an instantaneous rest system.

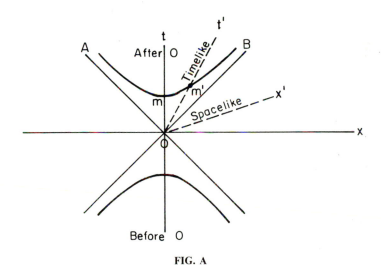

FIG. A

17. This is what Herglotz (1911) did. See Pauli (1958) for discussion of Herglotz's work.

18. Von Laue (1911b) showed conclusively that relativity theory could not reduce a body's number of degrees of freedom. Consider a resting body, and disturb its state at N points within it. After a time t the disturbances spread out like spheres whose maximum radii are ct. For small enough t these spheres do not intersect, and so within the body there are N distinct motions, and N can be arbitrarily large.

19. He had been alerted to Ehrenfest's paradox by the paper of Stead and Donaldson (1910). Instead of a rotating cylinder, Stead and Donaldson investigated the shape of a rotating disk whose axis of rotation was perpendicular to its plane. They arrived at an alternate interpretation of Ehrenfest's paradox: a possible consequence of the

contraction of every line element parallel to the disk's instantaneous linear velocity, while radial line elements remain unchanged, is that stresses deform the spinning disk into a cup. They proposed an optical experiment for deciding between Ehrenfest's contention that the disk remained in a plane while its material is "strained," and the disk being transformed into a cup. They wrote that their conclusions were "in perfect accordance" with Einstein's principle of relativity.

20. Born [(1910d), (1911)] continued to seek approximation schemes that would permit a rigid body its six degrees of freedom within the context of the "Lorentz-Einstein principle of relativity": Born (1910d) investigated restrictions on the size of a rigid body; Born (1911) investigated the possibility of virtual rotations of infinitesimally small rigid bodies.

21. For example, in an earlier paper Lewis (1908) had written that just as non-Euclidean geometry emerged from Euclidean geometry by changing certain axioms, a new non-Newtonian mechanics emerged from giving up the axiom that "the mass of a body is independent of its velocity."

22. One of the few physicists who mentioned Einstein's Part I of the relativity paper, Kaufmann (1906), also considered that Einstein's theory had only one principle. On the other extreme, certain physicists considered that Einstein had proposed three postulates (see Chapter 4, footnote 7).

23. See Winnie (1972) for further analysis of Einstein's *Gedanken* experiment.

24. Holton (1973d) discusses the Mach-Einstein correspondence and Planck's disagreement with Einstein over the central role played by Mach's principle in Einstein's early research on general relativity. Goldberg (1973) surveys Planck's philosophy of science.

25. Einstein had met Langevin at the Solvay Congress of 1911. They had kindred views on physics and world affairs, and became life-long friends. For further discussion of Langevin see the Epilogue to this book, and Biquard (1969), Einstein (1947) and Weill-Brunschvicg (1973).

26. In 1922 Lorentz added a footnote to this passage citing Einstein's lecture at Leiden "Relativity and the ether" (1920). There Einstein spoke of how the geometrical structure of space-time could be considered as playing the role of an ether. Lorentz had already realized this point because in a letter dated 17 June 1916, Einstein in Berlin wrote to Lorentz

> I agree with you that the general relativity theory admits of an ether hypothesis as does the special relativity theory. But this new ether theory would not violate the principle of relativity. The reason is that the state $g_{\mu\nu}$ = Aether is not that of a rigid body in an independent state of motion, but a state of motion which is a function of position determined through the material phenomena.

This is an early indication that Einstein could quantify his ether by means of the higher level general relativity theory. I am grateful to the Center for the History of Physics at the American Institute of Physics for making available to me a copy of this letter. See Miller (1979b) for further discussion of Einstein's view of an ether. Lorentz made many fundamental contributions to the general theory of relativity.

27. Wilson's magnetic dielectric was composed of steel balls embedded in paraffin. Turning the magnetic field on or off produced a transient current between the plates of the spinning capacitor, whose axis of spin was parallel to the external magnetic field. To order v/c Einstein and Laub predicted a transient current proportional to

$$(\varepsilon\mu - 1)v/c,$$

whereas Lorentz's theory of the electrodynamics of moving macroscopic bodies predicted a transient current of

$$\mu(\varepsilon - 1)v/c.$$

Since Lorentz's theory of the electrodynamics of moving bulk matter does not include the relativity of time it cannot treat accurately magnetized bodies, and especially not the magnetic dielectric in Wilson's experiment [see Pauli (1958)].

28. Today Darwin (1957) is usually credited with this demonstration of asymmetric aging [e.g., see French (1966)].

29. To which Wiechert (1915) replied that it is well known that certain mathematical laws have different meanings in different branches of physics.

30. Suppose observer A goes away and then returns along a linear trajectory PQR, while observer B remains at rest in K and so B's trajectory in space-time is the line PR. There are three events occurring at the space-time points P, Q, and R. The distance traveled in space-time by either clock is the Lorentz-invariant quantity

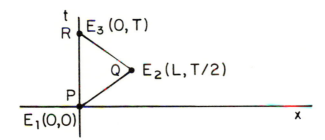

$$(ds)^2 = c^2(dt)^2 - (dx)^2 = c^2(d\tau)^2 - (d\xi)^2. \tag{A}$$

Since k is A's rest system, then this expression becomes

$$d\tau = dt\sqrt{1 - v^2/c^2}. \tag{B}$$

Minkowski defined τ as the "proper time" of A, i.e., the time measured by a clock in A's rest system. In the commentary to Minkowski's lecture "Space and Time," Sommerfeld (PRC) wrote, "As Minkowski once remarked to me, the element of proper time $d\tau$ is not a complete differential" — that is, the time elapsed is a path-dependent quantity, and so its value differs, depending whether $d\tau$ is integrated along the paths PR or PQR. Along PR, $dx = 0$, and so we have

$$\int_{0}^{T} d\tau = \int_{0}^{T} dt = T \tag{C}$$
$$_{PR}$$

which is the result for the case that the astronaut and earth observer remained at relative rest on the earth. Along PQR

$$\underset{PQR}{\int_0^T} d\tau = \underset{PQ}{\int_0^{T/2}} d\tau + \underset{QR}{\int_{T/2}^T} d\tau. \tag{D}$$

But since along PQ, $dx = v\,dt$ and along QR $dx = -v\,dt$, then

$$\underset{PQR}{\int_0^T} d\tau = \sqrt{1 - v^2/c^2}\left[\int_0^{T/2} dt + \int_{T/2}^T dt\right] = T\sqrt{1 - v^2/c^2}. \tag{E}$$

Eq. (E) relates the proper times of A (T') and B (T) as follows

$$T' = T\sqrt{1 - v^2/c^2} \tag{F}$$

which can be rewritten in a form that shows clearly that A's clocks run behind B's

$$T - T' = (1 - \sqrt{1 - v^2/c^2})T \tag{G}$$

and this is Einstein's peculiar result.

31. The Machist philosopher Petzoldt (1914) welcomed von Laue's resolution of the clock paradox because special relativity says "nothing at all" about instantaneous changes in velocity. See Blackmore (1972) and Holton (1973d) for discussions of Petzoldt's views on relativity theory.

32. Einstein (1911c), quoted from Whitrow (1961).

33. A theme discussed by the two interlocutors – one a relativist and the other a learned critic of relativity – was the interpretation of the clock paradox. The particular case of the clock paradox that "they" analyzed was the linear outward and return trajectories of one of the clocks C_2, while the other clock C_1 remained at rest. The problem was determining how the symmetry was broken between the two inertial systems. It is clear from this essay that Einstein was not concerned over whether there were any inconsistencies in the special relativity theory; rather, he wished to discuss how the symmetry was broken. The relativist (speaking for Einstein) demonstrated, by proper application of the relativistic space and time transformations, that there was no paradox. The "critic," however, was not completely satisfied with the account offered by the special theory of relativity. The relativist next proceeded to give a detailed resolution of the clock paradox which turned about the principle of equivalence from the general theory of relativity: taking C_2 as at rest during the entire process requires that its acceleration as observed by C_1 be replaced by an equivalent gravitational field acting upon C_2 thus serving to break the symmetry between their reference systems. The critic, however, replied that the difficulty had simply been removed to the problem of whether the equivalent gravitational field was real, and there ensued a discussion of what real and unreal meant in physical theory. (One of Einstein's goals was to give a lesson in epistemology to such opponents of the relativity theory as Philipp Lenard.) Einstein, consistent with his view of the stratification of scientific theories [which he explicitly set down in his (1936)], explained that as one progressed to higher-level theories, what constituted physical reality became ever-more difficult to relate to the world of sensations. Since in Einstein's viewpoint, the general relativity theory

was closer to the true physical reality than the lower-level special relativity theory, then one can conjecture with confidence that Einstein preferred a dynamical cause for the effect of time dilation.

Einstein concluded by asserting that "the special relativity theory had disavowed the former meaning of the ether [i.e., Lorentz's]." Then, consistent with his letter of 1916 to Lorentz Einstein wrote that it is the gravitational potentials of the general relativity theory that characterize the "physical qualities of empty space [i.e., space devoid of matter]."

Builder (1957) demonstrated that general relativistic treatments of the clock paradox succeeded in only complicating the problem, and often lacked physical significance. He stressed the importance of resolving the clock paradox using only special relativity theory because if the special theory contains a logical inconsistency, then so does the general theory. An interesting sidenote is that subsequently Builder (1958) attempted to formulate an ether-based version of special relativity theory that relied upon a variant of Wiechert's *Schreitung*. Builder sought to *explain* the occurrence of relativistic phenomena [see Prokhovnik (1967) for further discussion]. It turned out that Herbert Dingle was influential in changing Builder's view on relativity theory.

I conclude this footnote by emphasizing that Einstein received the Nobel Prize in 1921 "for his services to Theoretical Physics, and especially for his discovery of the law of the photoelectric effect." This general statement on Einstein's scientific work permitted the Swedish Academy to avoid taking a "stand on the relativity theory" [Frank (1947)]. Einstein did not attend the award ceremony since he was traveling in the Orient. As his Nobel Lecture Einstein later submitted his (1923) which was an epistemological analysis of the special and general theories of relativity.

34. The literature on the clock paradox is voluminous, and increases daily. See Marder (1971) for an extensive bibliography, and for a summary of the widely-read exchanges between Dingle and McCrea; see Holton (1962) for other references and for reprints of certain relevant papers.

Briefly, Bergson (1922) contends that psychological time is the real time and the time from the Lorentz transformations is the mathematical time, i.e., an illusion. The real passage of time, or real duration, is experienced by the observer. Bergson's prescription for learning the real time of inertial observers is to move between their systems, thereby placing them at absolute rest and experiencing their duration. His proof that there is no asymmetric aging is erroneous since it is based on incorrect applications of the Lorentz transformations of the sort that had been pointed out by Lorentz (1910–1912) (see Section 7.3), besides being predicated upon the somewhat uncomfortable circumstance of being concurrently in two inertial systems. Bergson viewed himself as a proponent of "radical relativity" and his resolution of the clock paradox as indicating the "perfect reciprocity of rectilinear motion." (Chapter 9 discusses cases of electromagnetic induction that lead to conundrums owing to other misunderstandings of the meaning of perfect reciprocity.)

Bergson's lucid style of writing gave wide appeal to his philosophical view. On 6 April 1922, in Paris, Einstein responded personally to Bergson [quoted from Gunter (1969)]:

> The question is therefore posed as follows: is the time of the philosopher the same as that of the physicist? The time of the philosopher is both physical and psychological at once; now, physical time can be derived from the time of consciousness. Originally individuals have the

notion of the simultaneity of perception; they can hence understand each other and agree about certain things they perceive; this is a first step toward objective reality. But there are objective events independent of individuals, and, from the simultaneity of perceptions one passes to that of events themselves. In fact, that simultaneity led for a long time to no contradiction due to the high propagational velocity of light. The concept of simultaneity therefore passed from perceptions to objects. To deduce a temporal order in events from this is but a short step, and instinct accomplished it. But nothing in our minds permits us to conclude to the simultaneity of events, for the latter are only mental constructions, logical beings. Hence there is no philosopher's time; there is only a psychological time different from the time of the physicist.

See Gunter (1969) and Tonnelat (1971) for further discussions of Bergson's view of relativity. Not surprisingly Dingle wrote the Introduction to the English translation of Bergson's (1922).

CHAPTER 8

THE THEOREM OF ADDITION OF VELOCITIES

> In short, let us assume that the simple law of the constancy of the velocity of light c (in vacuum) is justifiably believed by the child at school. Who would imagine that this simple law has plunged the conscientiously thoughtful physicist into the greatest intellectual difficulties?
>
> A. Einstein (1917a)

From the axioms (postulates) of the relativity theory, in §5 Einstein deduced an equation for the addition of velocities in which the velocity of light in vacuum is always c. Lorentz's constructive approach to covariance in electrodynamics caused him to maintain a number of interlocking hypotheses in order to support this result in inertial reference systems.

8.1. EINSTEIN'S ANALYSIS

The equations of motion for a particle moving in the ξ-η-plane of k with a uniform velocity w are [§5, 1–7]:

$$\xi = w_\xi \tau, \tag{8.1}$$

$$\eta = w_\eta \tau, \tag{8.2}$$

$$\zeta = 0, \tag{8.3}$$

where w_ξ, w_η are the ξ and η components of the velocity w with respect to k. Using the space and time transformations, inverse to Eqs. (6.52)–(6.55), Einstein deduced the particle's equations of motion in K as:

$$x = \left(\frac{w_\xi + v}{1 + v w_\xi / c^2} \right) t, \tag{8.4}$$

$$y = \left(\frac{\sqrt{1 - v^2/c^2}}{1 + v w_\xi / c^2} \right) w_\eta t, \tag{8.5}$$

$$z = 0. \tag{8.6}$$

Perhaps because it was not germane to the ensuing discussion in §5, Einstein did not deduce from Eqs. (8.4)–(8.6) the explicit "addition of velocities" for the velocity components in k and K. Differentiation of Eqs. (8.4)–(8.6) with respect to the time in K yields

$$w_x = \frac{w_\xi + v}{1 + w_\xi v/c^2}, \tag{8.7}$$

$$w_y = \left(\frac{\sqrt{1 - v^2/c^2}}{1 + w_\xi v/c^2}\right) w_\eta, \tag{8.8}$$

$$w_z = 0. \tag{8.9}$$

Neglecting terms of order v^2/c^2 gives:

$$x \cong (w_\xi + v)t, \tag{8.10}$$

$$y \cong w_\eta t, \tag{8.11}$$

$$z = 0. \tag{8.12}$$

"Thus, the law of the [Newtonian] parallelogram [addition] of velocities is valid according to our theory only to a first approximation [i.e., to order v/c]" [§5, 8–11]. Consequently, at velocities approaching c we can expect the magnitude of the particle's velocity, relative to K, to differ from the Newtonian quantity $V = \sqrt{(w_\xi + v)^2 + w_\eta^2}$. From Eqs. (8.7)–(8.9) Einstein obtained:

$$V = \frac{\sqrt{(v^2 + w^2 + 2vw\cos\alpha) - (vw\sin\alpha)^2/c^2}}{1 + (vw/c^2)\cos\alpha} \tag{8.13}$$

where $V = \sqrt{(dx/dt)^2 + (dy/dt)^2}$, $w^2 = w_\xi^2 + w_\eta^2$, and $\tan\alpha = w_\eta/w_\xi$. Always on the alert for symmetries, Einstein wrote [§5, 12–13]:

> It is worthy of remark that v and w enter into the expression for the resultant velocity in a symmetrical manner.

However, if the relative velocity v is not along the common direction of the x-axes of k and K, then Einstein's symmetry is lost [Møller (1972), Pauli (1958), Silberstein (1914)]. The equations relating individual velocity components between k and K, i.e., Eqs. (8.7)–(8.9), exhibit the symmetry of reciprocity only, and so Einstein did not display them; on the other hand, they appeared in Poincaré's paper (1906) because he sought to deduce the correct transformation law for the charge density and convection current. Poincaré's derivation from the postulated Lorentz transformations of the addition law for velocity was independent of l and also of the relativity of simultaneity. However, in ether-based theories of the electron this law lacked an unambiguous physical interpretation owing to the lack of a well-defined notion of at rest relative to an inertial reference system.

Emphasizing the case in which the particle moves along the common x-axes of k and K permitted Einstein [§5, 13–31] to extract the velocity addition law's most important properties in the limit that $w = w_\xi$, $w_\eta = w_\zeta = 0$, Equation (8.10) becomes

$$V = \frac{v + w}{1 + vw/c^2}. \tag{8.14}$$

(1) Setting in Eq. (8.14) $v = c - \psi$ and $w = c - \lambda$ where λ and ψ are positive and less than c, Einstein demonstrated that two subluminal velocities could not be combined to give a velocity greater than that of light. As Einstein had written in §4: "We shall, however, find in what follows that the velocity of light in our theory plays the part, physically, of an infinitely great velocity."

(2) Einstein continued with a result expected from Eq. (8.14), since it incorporates the second postulate: "It follows, further, that the velocity of light c cannot be altered by composition with a velocity less than that of light. For this case we obtain:

$$V = \frac{c + w}{1 + w/c} = c. \text{''} \tag{8.15}$$

(Clearly Eq. (8.15) contains a trivial oversight because Einstein meant that $w = c$, not $v = c$ since v is the velocity of an inertial reference system composed of rigid rods.) Eq. (8.15) is the mathematical expression for Einstein's Gordian solution to the paradox from Aarau.

(3) Assuming that the moving particle is at the origin of a third inertial reference system k', Einstein used two successive relativistic transformations, from k' to k and from k to K, in order to relate k' to K directly; the result is:

$$x = \gamma(x' + Vt'), \tag{8.16}$$

$$y = y', \tag{8.17}$$

$$z = z', \tag{8.18}$$

$$t = \gamma(t' + V/c^2x), \tag{8.19}$$

where primed quantities refer to k', and the "place of 'v' is taken by the quantity" V in Eq. (8.14).

Thus, not only has Einstein succeeded in deriving the relativistic velocity addition law by another method, but he has also shown that two successive relativistic transformations in a common direction can be replaced by a single relativistic transformation. From §3, the set of relativistic transformations, Eqs. (6.52)–(6.55), have an inverse (set v equal to $-v$ and exchange Greek and Roman letters) and there is an identity transformation (the system K' turned out to be the system K). From these three properties the relativistic transformations Eqs. (6.52)–(6.55) "form a group." So far as I can tell, in 1905 group-theoretical considerations were hardly used by physicists, and, as one may suspect, someone like Poincaré was the only other scientist to apply them to a fundamental problem in the kinematics or dynamics of electrons. Einstein's knowledge of mathematics may well have been of an instinctive sort, because his recollections (1946) of his education at the ETH (i.e., "I really could have gotten a sound mathematical education") can be interpreted as lacking in technical proficiency. He scored high on mathematics in the entrance examinations to the ETH, and in mathematics he received his highest score on the final examination (5.5 out of 6); despite himself, he could well have absorbed a great deal from the mathematical curriculum at the ETH, particularly from Minkowski's courses.[1]

In a matter-of-fact manner Einstein concluded the first part of the relativity paper by asserting that Part "II. Electrodynamical Part," is an "application to electrodynamics [of] the essential theorems of the kinematics corresponding to our two principles." Einstein had turned the programmatic intent of the physics of 1905 sideways.

8.2. POINCARÉ AND THE ADDITION LAW FOR VELOCITY

> No velocity can exceed that of light.
>
> H. Poincaré (1912a)

Poincaré [(1905a), (1906)] derived the addition law for velocity from the general Lorentz transformation Eqs. (1.198)–(1.201). Consequently this law is valid for any theory of a deformable electron, i.e., it is independent of l. It is no surprise then that a version of the addition law for velocity could have been deduced from Lorentz's modified transformations of 1895 (see Section 4.2). Simultaneously and independently, as well as within different conceptual frameworks, Poincaré and Einstein demonstrated this law's validity to any order in v/c; only in Einstein's view could it achieve its full potential.

For example, Poincaré's post-1906 writings on the dynamics of the electron emphasized that the velocity addition law's most important property was proving that c was the ultimate velocity [Poincaré (1912a)]. But he never used this law to prove the independence of the velocity of light from its source because:

(1) Lorentz's theory contained special hypotheses for this purpose;

(2) the primed system was a fictitious system in which occurred "*phénomène idéal.*"

8.3. VON LAUE'S DERIVATION OF FRESNEL'S DRAGGING COEFFICIENT

> Now in practice we can move clocks and measuring-rods only with velocities that are small compared with the velocity of light; hence we shall hardly be able to compare the results of the previous section directly with the reality. But, on the other hand, these results must strike you as being very singular, and for that reason I shall now draw another conclusion from the theory, one which can easily be derived from the foregoing considerations, and which has been most elegantly confirmed by experiment.
>
> A. Einstein (1917a)

[We consider the derivation of Fresnel's dragging coefficient] as an example of Einstein's law of the addition of velocities [consequently] we have eliminated the necessity for introducing an 'ether' into optics...

M. von Laue (1907)

Although the length contraction and time dilation effects of §4 manifest themselves only at velocities near c, a quantity "easily derived" from results of §5 can be spectacularly displayed at low velocities – that is, Fresnel's dragging coefficient.

Einstein had emphasized the importance to his thinking of Fizeau's experimental verification of the Fresnel dragging coefficient; yet in 1905 he made no mention of it. Von Laue (1907) wrote that a search of the literature had convinced him that no one had deduced the Fresnel dragging coefficient from Einstein's theory;[2] he considered its derivation as a further illustration of the Einstein addition theorem for velocities. According to von Laue, light propagating through a moving body, whose rest system was k, would have a phase velocity of magnitude

$$w = c/N' \qquad (8.20)$$

where all quantities were measured in k. Thus, he wrote Eq. (8.13) as

$$V = \frac{\sqrt{\left(v^2 + \dfrac{c^2}{N'^2} + 2v\dfrac{c}{N'}\cos\alpha\right) - \dfrac{v^2}{N'^2}\sin^2\alpha}}{1 + \dfrac{v}{cN'}\cos\alpha}. \qquad (8.21)$$

For $\alpha = 0$, π Equation (8.21) becomes

$$V = \frac{(c/N') \pm v}{1 \pm (v/cN')}. \qquad (8.22)$$

To second order in v/c, Eq. (8.22) is

$$V = \frac{c}{N'} \pm \left(1 - \frac{1}{N'^2}\right)v - \frac{v^2}{cN'}. \qquad (8.23)$$

To first order in v/c, Eq. (8.23) is identical (mathematically) with Lorentz's results of (1892a) and (1895) for the case in which light propagates either parallel or antiparallel to v.[3]

Von Laue considered the derivation of Fizeau's coefficient from a kinematic quantity as another reason for dismissing the notion of an ether. Although Lorentz's theory interpreted the coefficient as arising from the interaction of electrons in motion with the ether, in Einstein's theory it resulted from the relativity of time.

So far as I can learn, von Laue (1911a) was the first to note that the somewhat peculiar form of Einstein's velocity addition law resulted from measurement of the velocities w and v in two different inertial systems. If they had been measured in one inertial system, the usual (i.e., Newtonian) parallelogram rule could have been used.[4]

8.4. WALTHER RITZ'S EMISSION THEORY OF LIGHT

> Do [the Maxwell-Lorentz] equations merit such an excessive trust? The answer is emphatically no.
>
> W. Ritz (1908b)

Walther Ritz was displeased with the efforts of Einstein, Lorentz and Poincaré, on the one hand, to reject the null results of ether-drift experiments, and, on the other, to try to save Maxwell's equations. In Ritz's opinion [(1908a, b, c), (1909)], the equations were not worth preserving, since they possessed unphysical solutions – the advanced potentials. In an effort to preserve the classical notions of mass, length, time and a principle of relativity, Ritz proposed an action-at-a-distance theory of light propagation that evoked the "image" of "fictitious" corpuscles that served as intermediaries for electromagnetic disturbances. Ritz assumed that the velocity of a light corpuscle relative to its emitter was c, but that relative to an inertial observer it was $c \pm v_r$, where v_r was the radial velocity between source and observer. This sort of theory, referred to as an emission theory, automatically accounted for the Michelson-Morley interferometer experiment, but was at variance with Fizeau's 1851 experiment.

In a joint communication of 1909, Einstein and Ritz [Ritz (1909)] briefly stated their principal difference. For Ritz only the retarded potential had physical meaning because of its consistency with the irreversibility of physical processes, which was the root of the second law of thermodynamics. For Einstein "the irreversibility rests exclusively upon grounds of probability;" according to the second law of thermodynamics it is overwhelmingly probable that physical states in nature proceed in the direction of increasing time, and, hence, nature chooses the retarded potential rather than one that would allow an electromagnetic wave to collapse back on its source.[5]

In September 1909 at Salzburg, Einstein (1909b) elaborated on this point. In the kinetic-molecular theory, he said, every reaction had an inverse; however, the wave theory of light considered only the emission of radiation from an oscillating charge as an elementary process. Circa 1909 the collapse of the spherical wave corresponding to the charge absorbing its own radiation, i.e., the inverse process, was not considered an elementary process, although it was a possible mathematical solution to Maxwell's equations. Einstein's reason was that the inverse process depended on collecting the enormous amount of radiation that had propagated to a great distance and then concentrating it again at a point. Both processes were elementary from the point of view of a corpuscular theory of light. In order to remove this asymmetry from electromagnetic theory, Einstein went on to suggest the pursuit of a corpuscular theory of light. But Einstein's corpuscular theory was not an emission theory because the light quantum's velocity was a constant and independent of the emitter's motion. The velocity of Einstein's light corpuscle could be calculated only with the relativistic addition law of velocities, and it always turned out to be c.[6, 7]

FOOTNOTES

1. As a student at the ETH's school for teachers of mathematics and science (mathematical section) during the period wintersemester of 1896/1897 — summersemester of 1900, Einstein was obliged to register for the following courses taught by Minkowski [Seelig (1954)]: Geometry of Numbers, Function Theory, Potential Theory, Elliptical Functions, Analytical Mechanics, Variational Methods, Algebra, Partial Differential Equations, Applications of Analytical Mechanics. Seelig wrote that Einstein was enthusiastic over Minkowski's course "Applications of Analytical Mechanics." Evidently Einstein's enthusiasm did not carry over to Minkowski's other courses. Constance Reid writes (1970) of Minkowski in 1902: "In front of a group, Minkowski suffered from 'Lampenfieber' — in English, *stagefright*. He was still embarassed by attention, even of much younger people; and in Zürich his shy, stammering delivery had completely put off a student named Albert Einstein." Minkowski was unhappy during his sojourn at the ETH (1896–1902) and longed to be in Göttingen with David Hilbert. He described the ETH to Hilbert as a place "where the students, even the most capable among them, ... are accustomed to get everything spoon-fed" [Reid (1970)].

2. It turned out that in the same *Annalen* volume, Laub (1907) published an attempt at deducing Fresnel's dragging coefficient. Laub's result was accurate to order v/c, and he considered only a nondispersive medium. His method of derivation was based upon the independence of Einstein's definition of clock synchronization from the medium in which the clocks were immersed. In a sequel publication Laub (1908) was able to extend his method to dispersive media and to all orders in v/c. Nevertheless, von Laue's results were more general since they could discuss the relative velocity of light rays traveling at any angle relative to K.

3. Von Laue (1907), mentioned that Eq. (8.23) required certain additional terms for a dispersive medium. Laub (1908) calculated the corrections, which to order v/c is the term $-(v/N)T(dN/dT)$ where N is the moving medium's index of refraction as measured in K; T is the period of the light wave as measured in K. This additional term is mathematically equivalent to the one that Lorentz calculated in the *Versuch* (see Section 1.6.3). Zeemann's (1914–1915) high-precision repetition of the Michelson-Morley (1886) experiment verified the additional term for dispersive media.

4. Sommerfeld (1909) demonstrated how Minkowski's four-dimensional formulation of the "Lorentz-Einstein transformations," expressed as a rotation by an

imaginary angle $\tan \phi = i\beta$, led to the result that according to the "relativity theory" velocities add according to a spherical geometry. This was one of the first demonstrations that non-Euclidean geometry could be used to present concisely certain results of relativity theory. See also Herglotz (1910) and Varičak (1910a, b). Einstein (1921a) used Sommerfeld's method for deriving the relativistic addition law of velocity.

5. For discussion of Ritz's emission theory, see Fox (1965) and Panofsky and Phillips (1961). Ritz is perhaps best remembered for his formulation of the principle of combination (1908d).

6. During an interview of February 1950, Shankland (1963) reported Einstein having said that

> he had thought of, and abandoned the (Ritz) emission theory before 1905. He gave up this approach because he could think of no form of differential equation which could have solutions representing waves whose velocity depended on the motion of the source. In this case, the emission theory would lead to phase relations such that the propagated light would be all badly "mixed up" and might even "back up on itself." He asked me, "Do you understand that?" I said no and he carefully repeated it all. When he came again to the "mixed up" part he waved his hands before his face and laughed, an open hearty laugh at the idea!

7. During 1910–1913, emission theories were proposed by Comstock (1910) and Tolman (1910); at the time, both of them were unaware of Ritz's work. Tolman (1912b) asserted the importance of an "experimental decision between the relativity theories of Ritz and Einstein." Ehrenfest (1912), as in the rigid-body controversy, brought his critical manner of thought to bear on this problem. His conclusion as a result of studying the possibility of an experimentum crucis to decide between the theories of Einstein or Ritz was as Klein (1970) writes:

> "Let us suppose," he wrote, "that someone soon manages to plan a crucial experiment which can actually be carried out and which will distinguish between assumptions B and D." (Assumption B was essentially Ritz's assumption that the velocity of light depends upon the velocity of its source, while assumption D was Einstein's assumption that it does not.) "What sort of situation would this lead to? The supporters of the aether hypothesis would have to wish that assumption D turns out to be correct. The supporters of the real emission theory, [Ritz's theory], would have to wish that assumption B is confirmed. The supporters of Einstein's theory of relativity would have to wish that this time the supporters of the aether hypothesis win out over the supporters of the real emission theory."

Klein continues (1970), Ehrenfest touched upon an aspect of the theory of relativity which is too often ignored.

By 1917, Tolman's enthusiasm for an emission theory had been dampened by the result of de Sitter's 1913 observations of the light emitted from eclipsing binary stars [Tolman (1917)]. Indeed, Tolman (1910) himself had suggested this type of experimental test.

In a critical survey of emission theories, Fox (1965) pointed out that the extinction theorem of Ewald and Oseen cast doubt upon de Sitter's lack of detection of any peculiarities in the light received from binary star systems; de Sitter's result had been considered as a principal piece of evidence against emission theories. According to the extinction theorem light emerging from a dispersive medium such as the gas permeating interstellar space assumes the velocity characteristic of the medium after one "extinction length." Using x-ray sources in binary star systems to minimize effects predicted by the extinction theorem, Brecher (1977) found that Einstein's second postulate was consistent with these data to one part in 10^9. Brecher's result indicates the high degree of accuracy of special relativity theory.

CHAPTER 9

THE RELATIVITY OF THE ELECTRIC AND MAGNETIC FIELDS

> Likewise, questions as to the "seat" of electrodynamic electromotive forces (unipolar machines) become meaningless.

A. Einstein (1905d)

> This phenomenon (unipolar induction) seems therefore to lend definite support to the existence of an electromagnetic ether ... It seems, for this reason, to have some importance as a stumbling block in the way of those ultra-relativists who would abandon the conception of an ether altogether.

E. H. Kennard (1917)

Having set the kinematical foundations of his point of view Einstein, in Part II, applied it to electrodynamical problems, and felt that he could audaciously dismiss the long-standing problem of unipolar induction as "meaningless." Earle H. Kennard's (1917) comment on unipolar induction typifies the misunderstanding of Einstein's essentially kinematical approach to a problem that physicists since 1832, and many even today, believe demands a constructive approach. But in addition, Kennard's comment, quoted above, mistakenly assumes that the relativity theory requires every measureable effect to result from the relative motion of ponderable bodies; the latter misunderstanding is not far removed from the one that sometimes surrounds time dilation, except unipolar induction and electromagnetic induction in general, are low-velocity phenomena that necessitate a relativistic interpretation.

9.1. EINSTEIN'S ANALYSIS

Einstein began the "Part II. Electrodynamical Part" with a section whose lengthy title describes its contents excellently [§6, 1–4]:

II. ELECTRODYNAMICAL PART

§6. Transformation of the Maxwell-Hertz Equations for Empty Space. On the Nature of the Electromotive Forces Occurring in a Magnetic Field During Motion.

For §6 Einstein applied the two principles of relativity theory to the equations of electrodynamics for the case of radiation in "free space," or as Lorentz

(1895) had written in the "pure ether," i.e., in the absence of charges. In addition, as Einstein signalled in the section title, he returned to the problem of the "nature of electromagnetic forces," in order to resolve the asymmetry in the case of magnet and conductor in relative inertial motion. Then, in §§7 and 8 he used some of the results from §6 to analyze the characteristics of radiation in the optics of moving bodies – for example, optical Doppler effect, radiation pressure, and the reflection of light from a moving mirror.

But why, since Einstein discussed throughout his paper the modifications necessary in Lorentz's electrodynamics, did he refer to the "Maxwell-Hertz Equations"? After all, he was aware that Abraham and Lorentz had demonstrated insufficiencies in Hertz's theory. From the previous analysis I can conjecture that his principal reasons were: (1) In order to avoid discussing the constitution of matter, Einstein preferred Hertz's neutral nomenclature for electromagnetic quantities over Lorentz's – for example, Hertz's electric force E became in Lorentz's theory the dielectric polarization D. (2) Einstein was impressed with Hertz's insistence that every electromagnetic quantity have an in-principle operational definition; the purification of electromagnetic theory actually began with Hertz. Einstein had license to take advantage of these two essentially epistemological points because, as Abraham [(1902), (1903)] had emphasized, the "Maxwell-Hertz" and "Maxwell-Lorentz" equations were equivalent for dealing with phenomena concerning radiation in the absence of charges. By May 1906 Einstein was referring to the Maxwell-Hertz equations as the Maxwell-Lorentz equations, and the most probable reason was that by that time many physicists were discussing a Lorentz-Einstein theory, and so for consistency Einstein refined his notation (1906c). Yet he persisted in avoiding Lorentz's terminology for the electromagnetic field quantities, referring to them instead as "electric fields." In addition to Hertz's nomenclature, Einstein used his notation as well – $E(X, Y, Z)$ and $B(L, M, N)$. Toward avoiding unnecessary complications I shall refer to Einstein's Maxwell-Hertz equations in §6 as the Maxwell-Lorentz equations.

Owing to the relativity of simultaneity the three equations in the right-hand column of Einstein's Eqs. [§6.1] are mathematically equivalent to Hertz's Eq. (1.1) or Lorentz's Eq. (1.25) [§6, 5–8]. Using vector notation Einstein's version of Faraday's law is

$$\mathbf{V} \times \mathbf{E} = -\frac{1}{c}\frac{\partial \mathbf{B}}{\partial t}. \tag{9.1}$$

Similarly the three equations in the left hand column are mathematically equivalent to Hertz's Eq. (1.2) with $\mathbf{H} = \mathbf{B}, \mathbf{D} = \mathbf{E}$, and $\sigma = 0$, or Lorentz's Eq. (1.26) with $\rho = 0$. I write Einstein's equations as

$$\mathbf{V} \times \mathbf{B} = \frac{1}{c}\frac{\partial \mathbf{E}}{\partial t}. \tag{9.2}$$

Einstein did not exhibit the other two Maxwell-Lorentz equations

$$\mathbf{V} \cdot \mathbf{E} = 0, \qquad (9.3)$$

$$\mathbf{V} \cdot \mathbf{B} = 0, \qquad (9.4)$$

although he subsequently used them. In the *Annalen* version, Einstein wrote V for c, but his units were those of Abraham, i.e., Gaussian (cgs) units. Whereas by 1903 such major electrodynamicists as Abraham and Lorentz freely employed the vector notation, Einstein did not in 1905; his reason may well have been that it would have complicated unnecessarily the relativistic transformations to follow. Einstein assumed that the Eqs. (9.1)–(9.4) "hold for the resting system K." Thus, as in prerelativistic electromagnetic theory, he accepted the exact validity of the field equations, and of the constancy of light velocity, in one reference system. The basic difference for Einstein was that the resting system K could be any inertial reference system.

Using Eqs. (6.52)–(6.55), Einstein [§6, 9–25] transformed the space and time coordinates in Eqs. [§6.1] from K to k, i.e.,

$$\frac{\partial}{\partial x} = \gamma \left(\frac{\partial}{\partial \xi} - \frac{v}{c^2} \frac{\partial}{\partial \tau} \right), \qquad (9.5)$$

$$\frac{\partial}{\partial y} = \frac{\partial}{\partial \eta}, \qquad (9.6)$$

$$\frac{\partial}{\partial z} = \frac{\partial}{\partial \zeta}, \qquad (9.7)$$

$$\frac{\partial}{\partial t} = \gamma \left(\frac{\partial}{\partial \tau} - v \frac{\partial}{\partial \xi} \right), \qquad (9.8)$$

and obtained straightforwardly his Eqs. [§6.2]–[§6.7]. Invoking the principle of relativity as a heuristic aid required Einstein to undertake additional transformations besides those concerning the space and time coordinates, in order to maintain the covariance of the equations of electromagnetism. For if the Maxwell-Lorentz Eqs. (9.1) and (9.2) in k were to be written thus

$$\mathbf{V'} \times \mathbf{E'} = -\frac{1}{c} \frac{\partial \mathbf{B'}}{\partial \tau}, \qquad (9.9)$$

$$\mathbf{V'} \times \mathbf{B'} = \frac{1}{c} \frac{\partial \mathbf{E'}}{\partial \tau}, \qquad (9.10)$$

where primed quantities refer to k, i.e., $\mathbf{V'} = (\partial/\partial\xi, \partial/\partial\eta, \partial/\partial\zeta)$ and $\mathbf{E'} = (E_\xi, E_\eta, E_\zeta)$, then the electric and magnetic quantities in these two inertial systems had to be related:

$$E_\xi = \psi(v)E_x, \qquad (9.11)$$

$$E_\eta = \psi(v)\gamma(E_y - (v/c)B_z), \tag{9.12}$$

$$E_\zeta = \psi(v)\gamma(E_z + (v/c)B_y), \tag{9.13}$$

$$B_\xi = \psi(v)B_x, \tag{9.14}$$

$$B_\eta = \psi(v)\gamma(B_y + (v/c)E_z), \tag{9.15}$$

$$B_\zeta = \psi(v)\gamma(B_z - (v/c)E_y), \tag{9.16}$$

where $\psi(v)$ is analogous with the undetermined function $a(v)$ in §3. The reason why $\psi(v)$ can depend only on the magnitude of v and not on the space-time coordinates or the direction of v is the same as that given for a's depending only on v.

Minkowski [(1907), (1908a)] referred to the identity in form of the equations of electromagnetism in every inertial reference system as "covariance." As Einstein wrote (1917a): this condition is the "mathematical condition that the theory of relativity demands of a natural law...."

On the other hand, Lorentz and Poincaré did not interpret covariance in a heuristic sense; rather, they held that the transformation only of the space and time coordinates and of **E** and **B**, but not of c, was predicated upon a constructive theory of the electron. Before Poincaré's (1905a) Lorentz's interpretation of covariance was to transform the equations of electromagnetism from S_r to another reference system having the properties of one fixed in the ether; he was only partially successful. Even though Poincaré's elimination of S_r facilitated the mathematical procedure for coordinate transformations, it served to obscure matters further because he achieved exact mathematical covariance by relating directly an unobservable ether-fixed system S with the "imaginary system." The tacit assumption of ether-theorists in electrodynamical calculations was that the ratio of velocities relative to the ether to that of light was much less than the ratio of velocities measurable in the laboratory to the velocity of light. Thus the laboratory system S_r could be taken as S, and if future ether-drift experiments failed, Poincaré's principle of relativity could become a convention, thereby legalizing the above mentioned tacit assumption. In 1905 Einstein's approach to covariance was a combination of Hertz's and Lorentz's: Like Hertz, Einstein asserted axiomatically that the laws of electromagnetism maintained their form in an inertial reference system (Hertz's covariance extended to noninertial systems as well); like Lorentz, Einstein sought equations for transforming the space and time coordinates and electromagnetic quantities. Thus Lorentz (1909) had cause to have written that "Einstein simply postulates what we have deduced."

In order to avoid the ambiguities introduced by unknown velocities in the ether-based theories, Einstein in 1905 proposed in-principle operational definitions of the electromagnetic field quantities **E** and **B** in k through their "pondermotive effects on electric or magnetic substances"; most likely the analogy with Hertz's 1890 in-principle operational definition was not accidental.

Toward using covariance for determining a new law of physics, Einstein [§6, 26–32] next determined $\psi(v)$ by calculating the "reciprocal" of Eqs. (9.11)–(9.16) in two different ways, and then comparing the results. (1) He inverted the Eqs. (9.11)–(9.16) in order to solve for the field quantities in K in terms of those in k; the result for E_y is

$$E_y = \frac{1}{\psi(v)}\, \gamma \left[E_\eta + \frac{v}{c} B_\zeta \right]. \tag{9.17}$$

(2) In Eqs. (9.11)–(9.16) he changed v to $-v$, and interchanged Greek and Roman subscripts to obtain

$$E_y = \psi(-v)\gamma \left[E_\eta + \frac{v}{c} B_\zeta \right]. \tag{9.18}$$

Comparison of Eqs. (9.17) and (9.18) yielded

$$\psi(v)\psi(-v) = 1. \tag{9.19}$$

Then Einstein proposed that "from reasons of symmetry"

$$\psi(v) = \psi(-v); \tag{9.20}$$

in a footnote he explained that assertion:

> If, for example, $X = Y = Z = L = M = 0$, and $N \neq 0$, then from reasons of symmetry it is clear that when v changes sign without changing its numerical value, $[E_\eta]$ must also change sign without changing its numerical value.

There is no electric field in K (i.e., $E = 0$), and the magnetic field in K has only a z-component B_z. Then, from Eqs. (9.12) and (9.16), in k there is an electric field component

$$E_\eta = -\psi(v)\gamma(v/c)B_z \tag{9.21}$$

and a magnetic field component

$$B_\zeta = \psi(v)\gamma B_z. \tag{9.22}$$

Einstein's assertion that "from reasons of symmetry" $\psi(v) = \psi(-v)$ can be interpreted as follows. If E_η is the force on a unit charge situated at k's origin, then changing v to $-v$ reverses the force's direction, leaving unchanged its "numerical value" for observers in k and K; therefore, the magnitude of Eq. (9.21) for the cases with v and $-v$ must be the same, and Eq. (9.20) follows. The procedure of deducing general results from symmetry arguments based on the behavior of electric and magnetic field quantities under the inversion of spatial quantities had appeared in Abraham's *Annalen* paper (1903). Einstein's intuitive notion of symmetry may well have received quantitative support from Abraham's work.

In §3 Einstein proved that $\phi(v) = \phi(-v)$ by using a third reference system K', which turned out to be K. In §6 Einstein chose a two-step procedure to prove that $\psi(v)\psi(-v) = 1$, although his reasons for $\phi(v) = \phi(-v)$ and $\psi(v) = \psi(-v)$ were the same — "reasons of symmetry." Thus, combining Eq. (9.20) with Eq. (9.19), Einstein obtained

$$\psi(v) = 1;\tag{9.23}$$

Einstein chose the positive root for reasons to be discussed in a moment. Eqs. (9.11)–(9.16) thus become:

$$E_\xi = E_x,\tag{9.24}$$

$$E_\eta = \gamma(E_y - (v/c)B_z),\tag{9.25}$$

$$E_\zeta = \gamma(E_z + (v/c)B_y),\tag{9.26}$$

$$B_\xi = B_x,\tag{9.27}$$

$$B_\eta = \gamma(B_y + (v/c)E_z),\tag{9.28}$$

$$B_\xi = \gamma(B_z - (v/c)E_y).\tag{9.29}$$

Eqs. (9.24)–(9.29) assert that electric and magnetic fields, as well as length and time, are relative quantities, and this is a new law of nature. Since k and K are equivalent reference systems, then the inverse transformations are obtained by changing v to $-v$ and interchanging subscripts in Eqs. (9.24)–(9.29).

Unknown to Einstein in 1905 was that in 1904 Lorentz had arrived at equations mathematically identical to Einstein's Eqs. (9.24)–(9.29), for k as the electron's fictitious instantaneous rest system S' [see Eqs. (1.166)–(1.171)]. In 1905 Poincaré extended these equations to the fictitious reference system Σ' which is mathematically equivalent to Einstein's physically real system k. For Lorentz and Poincaré the quantities in Eqs. (9.24)–(9.29) lacked unambiguous physical meaning because they related quantities that could not even be defined by in-principle operational procedures; this ambiguity was directly related to the lack of clarity in ether-based theories for the notion of at rest relative to an inertial reference system.

Einstein continued [§6, 33–54] by comparing Eqs. (9.24)–(9.29) with equations available from Lorentz's electromagnetic theory. A point charge q_1 of unit magnitude is at rest relative to K, and at a distance of 1 cm from q_1 there is a second point charge q_2. From Coulomb's law, applied in K, the charges exert forces on each other. If q_2 were in k, then by the "principle of relativity" observers in K and k would measure the same charge on q_2. In §9 Einstein considered briefly, and in a typically succinct Einsteinian manner, the proof of the invariance of electric charge.

In order to avoid any shortcomings of an ether-based theory of electromagnetism Einstein next gave an in-principle operational definition of E as the force that a unit source charge q_1 exerted on a unit charge q_2 which was

1 cm from q_1. Needless to say E is the field attributable to the source charge q_1 and does not have the units of a force. The force of q_1 on q_2 is 1 dyne and is represented in the standard manner as

$$F = q_2 E, \tag{9.30}$$

which is Coulomb's law. Einstein next took q_2 as moving through the Coulomb field of q_1, and he assumed a succession of inertial reference systems with different velocities, with the result that, "at the relevant instant," q_2 could be taken as instantaneously at rest in one of them, e.g., k.[1] Then Einstein defined E' as the force on q_2 due to q_1 as measured in k. These in-principle operational definitions enabled him to compare the interpretations of the force acting on a point charge in an electromagnetic field according to the "old manner of expression" – which turned out to be Lorentz's – and the "new manner of expression."

Neglecting terms of order $(v/c)^2$, then the force on a unit charge q_2 according to the old manner of expression is given by Eqs. (9.24)–(9.26) as [see Eq. (1.84)]

$$E' = E + (v/c) \times B. \tag{9.31}$$

Thus, in addition to the "electric force" E, there is an "electromotive force" $v/c \times B$, because in Lorentz's theory the ether-based system was cavalierly taken as the laboratory system, and the S' system had no physical meaning. On the other hand, in the "new manner of expression" Einstein obtained the force acting upon q_2 by considering the situation from the point of view of the charge's instantaneous rest system. Therefore, all electrodynamical problems could be reduced to those of electrostatics. This result had been achieved by Lorentz in the *Versuch* (1895) using the space and time transformation Eqs. (1.63)–(1.66), and then more exactly in the paper of 1904. However, Lorentz interpreted the transformations as if the electron was placed at rest in a fictitious reference system having all the properties of one fixed in the ether and ensured the "imaginary" electron's sphericity by postulating a Lorentz contraction for the real moving electron in S_r. Then, after completing the electrostatic calculation, Lorentz transformed back to the "true inertial coordinates" (x_r, y_r, z_r, t). Einstein next discussed the remarkable consequences of the "new manner of expression" [§6, 55–59]:

> The analogy holds with "magnetomotive forces." We see that in the developed theory the electromotive force plays only the part of an auxiliary concept, which owes its introduction to the circumstance that electric and magnetic forces do not exist independently of the state of motion of the coordinate system.

Around 1905 the quantity "magnetomotive force" $v/c \times E$ was taken as the analogue of the electromotive force, and Einstein's following discussion applied as well to it. In the "new manner of expression," continued Einstein, there were no preferred observers, and consequently the descriptions of a system's state given by any inertial observer were possible descriptions. Thus, the observer in k described q_2 as acted on only by an electric field, even though an observer in K described q_2 as under the influence of electric and magnetic fields. Needless to say, both observers agreed on q_2's motion in succeeding instants. Thus, concluded Einstein, the "electric and magnetic forces do not exist independently of the state of motion of the coordinate system" – that is, the electric and magnetic fields are relative quantities.[2]

The reader of 1905 could have clarified Einstein's "new manner of expression" further by using the illustrative example in Einstein's only footnote to §6: In K there is a magnetic field B along the z-axis of K, and a charge q is instantaneously at the origin of the inertial system k. Since in k the charge q is instantaneously at rest, the observer in k attributes its subsequent motion along the negative η-axis to an electric field $E_\eta = -\gamma(v/c)B$ giving rise to the force[3]

$$F_\eta = qE_\eta = -q\gamma(v/c)B. \tag{9.32}$$

The observer in K, however, interprets q's motion in the negative y-direction as the electromotive force $\gamma(v/c) \times B$, giving rise to the mechanical force

$$F = -q\gamma(v/c)B.$$

Consequently though both observers agree on q's motion, in the new interpretation, only the observer in K measures an electromotive force, since K is not the instantaneous rest system for the charge q.

Almost certainly it was no coincidence that Einstein's illustrative example concerned motion relative to a constant magnetic field, for in the relativity paper he next stated [§6, 60–64]:

> Furthermore it is clear that the asymmetry mentioned in the introduction as arising when we consider the currents produced by the relative motion of a magnet and a conductor, now disappears.

The relativity of the electromagnetic field quantities permitted him to describe the phenomenon of electromagnetic induction purely in terms of electric and magnetic forces (or if you wish, fields). As Einstein wrote (1907e): "the 'electromotive force' that acts upon an electrical body moving in a magnetic field is nothing else than the 'electric' force to which we are led through considering the electrical body from the viewpoint of its resting system." This observation meshes well with those he made in 1919 ("the existence of an electric field was therefore a relative one . . ."), and in 1952 to the effect that an important ingredient in his thinking toward the special relativity theory was that the electromotive force acting on a body moving in a magnetic field "was nothing else but an electric field."

Having resolved the problem posed at the start of his paper, Einstein concluded §6 by audaciously dismissing the problem of "the 'seat' of electrodynamic electromotive forces" or unipolar induction – as "meaningless."[4] For Einstein the relativity of the electromagnetic field quantities was a law of physics rooted in the relativity of time, and this law rendered meaningless the problem of whether magnetic lines of force rotate.

9.2. FURTHER COMMENTS ON UNIPOLAR INDUCTION

> [Whether lines of force rotate or not is a] fundamental *Anschauung* of immediate importance and technical meaning, and not an academic moot point.
>
> C. L. Weber (1895)

Many others did not consider the problem of unipolar induction as meaningless, and it is useful to review the ensuing controversy because of the lesson drawn in interpreting too literally the canon that all measurable effects are relative.

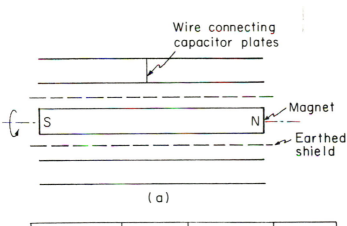

Wire connecting
capacitor plates

S   N   Magnet

Earthed
shield

(a)

Experiment	Magnet	Capacitor	Charge
1	X	R	Yes
2	R	X	No

(b)

FIG. 9.1. The apparatus (a) and data (b) for the system of a permanent conducting magnet and a capacitor in relative rotary motion about a common axis, where × = No Rotation and R = Rotation.

From our discussion of unipolar induction in Chapter 3, it is clear that arrangements of magnet and circuit equivalent to the arrangement in Fig. 3.3 cannot distinguish whether lines of force rotate or not; for an obvious reason, these arrangements were called closed circuits. In the period 1912–1922 Samuel J. Barnett [(1912), (1918)], Earle H. Kennard [(1912), (1913), (1917)], and George B. Pegram (1917) in the United States performed the first truly open-circuit experiments. The set of experiments utilized cylindrical capacitors coaxial with the sources of a magnetic field, which were either a cylindrical conducting magnet (Fig. 9.1) or a solenoid (Fig. 9.2). The systems were arranged so that their parts could rotate separately. During a run the capacitor plates were connected, the rotation was stopped, and the insulated plates were disconnected and then tested for a charge. If the lines of force rotated, then it was interpreted that a charge could be induced on the capacitor plates; if the lines did not rotate, then an electrostatic charge was expected on the magnet's surface. In order to prevent this charge from affecting the capacitor, and also to eliminate any spurious effects, an earthed shield was placed between the magnetic system and the capacitor. The results of the experimental runs are in the tables in Figs. 9.1 and 9.2.

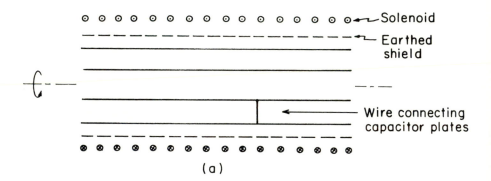

(a)

Experiment	Solenoid	Capacitor	Charge
1	X	R	Yes
2	R	X	No
3	R	'R	Yes

(b)

FIG. 9.2.   The apparatus (a) and data (b) for the system of a current-carrying solenoid and a capacitor in relative rotatory motion about a common axis.

Everyone expected identical results for the first two experiments in Figs. 9.1 and 9.2. Kennard (1917) wrote that there was no reason to expect a "radical difference between induced and permanent magnetism." But in Kennard's opinion the lack of a charge on the capacitors in experiments 2 of Figs. 9.1 and 9.2 contradicted the moving-line theory and supported Lorentz's theory with its all-pervasive resting ether. Although Barnett recognized correctly the importance of electromagnetic induction to Einstein's thinking toward the relativity theory of 1905, he interpreted incorrectly Einstein's analysis of electromagnetic induction in the beginning of Einstein's 1905 special relativity paper. Barnett had concluded that, according to the theory of relativity, all phenomena should be measurably relative. But experiments 1 and 2 in Figs. 9.1 and 9.2 contradict what Barnett considered as Einstein's principle of relativity. So in order to maintain the rotating-line viewpoint, which was consistent with some kind of principle of relative motion, Barnett invoked an ether with the appropriate dielectric properties for neutralizing the induced charge arising from the rotating lines of force in experiments 1 and 2 of the tables of Figs. 9.1 and 9.2. Thereby Barnett considered that he had also explained the lack of symmetry between these two experiments. But Barnett went on to claim that his interpretation precluded even open-circuit experiments from disproving the moving-line hypothesis. For Kennard, on the other hand, the decisive experiment was the third one in Fig. 9.2, which he performed in 1917. Kennard (1917) considered experiment 3 of Fig. 9.2 to be "a stumbling block in the way of those ultra-relativists who would abandon the conception of an ether altogether," i.e., Lorentz's ether, which has no dielectric properties. Kennard's reason was that electromagnetic induction held for the case in which there was no relative motion between the ponderable parts of the system; therefore, electromagnetic induction must depend at least in part on rotation through the ether.

Pegram (1917) reminded Barnett and Kennard, as well as "many well-trained physicists and engineers," that previously Larmor (1895) had shown that the magnetic field of a current-carrying solenoid was unaltered when the solenoid rotated about its axis; therefore a rotating or resting solenoid could affect only a moving charge.[5] Furthermore, Pegram continued, this result was valid in any electron theory. Pegram (1917) concluded that the second experiment in the table of Fig. 9.2 was consistent with the "Lorentz-Einstein relativity theory."[6]

W. F. G. Swann (1920) reiterated that in any electron theory the external field of a rotating conducting magnet was electrostatic in origin and that, therefore, no result could have been expected in experiment 2 of Fig. 9.1 because of the earthed shield around the magnet. Swann continued by emphasizing that the equations of electromagnetism made no assertions on whether lines of force rotate. Representing the permanent magnet as a collection of either Ampèrian current whirls, or magnetic poles, he deduced the electromagnetic quantities arising from a magnet in inertial motion. Although these results, wrote Swann, could also be obtained from the relativity transformations and were therefore independent of the magnet's constitution, he preferred to deduce all results in "as clear a manner as is consistent with a rigorous treatment which may perhaps be more convincing than direct appeal to the relativity transformations." The need for a constructive rather than an axiomatic solution to physical problems was echoed in J. F. Tate's review paper (1922) on unipolar induction — namely, that using relativity theory for analyzing the electromagnetic quantities of moving magnets not only begged the question, but "seems a confession of weakness on the part of the electromagnetic theory."[7]

In concluding this sketch of the work of Barnett, Kennard, Pegram, and Swann, I present for comparison the results of the open-circuit experiments for the magnet and solenoid without the earthed shield (Fig. 9.3). Clearly one must be cautious in asserting, without further explanation, that electromagnetic induction is a matter simply of relative motion. But the asymmetry between the results for the magnet and the solenoid does not violate relativity theory. In the years to come, physicists and engineers would have to be reminded repeatedly of the fundamental difference between the unipolar induction effects arising from a rotating magnet and a rotating solenoid. This is a good illustration of the faulty expectations that result from relying on lines of force only, and not taking into account their source. In fact, the case of unipolar induction serves as a caveat that in scientific research a visual representation, such as lines of force, and more generally an *Anschauung*, need not always be a fruitful method of approach. As for the production of large-scale unipolar generators, a

Experiment	Magnet	Capacitor	Charge
1	X	R	Yes
2	R	X	Yes

(No earthed shield)

Experiment	Solenoid	Capacitor	Charge
1	X	R	Yes
2	R	X	No

(Earthed shield is irrelevant here)

FIG. 9.3.   The results of the electromagnetic induction experiments using the apparatus in Figs. 9.1 and 9.2 *without* the earthed shields.

preference for these modes of thinking led, in German-speaking countries, to an unproductive interaction between scientists and engineers regarding the design of a large-scale unipolar dynamo [Miller (1977b)].

By 1936 the controversy had degenerated into the proponents of either the moving-line or fixed-line hypotheses enumerating the names of well-known physicists who, in their opinion, did not find meaningless the question of whether lines of force have physical characteristics.[8]

Today we know that the electromagnetic theory places stringent restrictions on assumptions concerning the state of motion of lines of force.[9] Minkowski's electrodynamics can be used in the laboratory system to calculate the electromagnetic quantities of moving magnets. In this theory one may use lines of force to discuss effects of the magnet's external electric field on conductors, but the internal state of a rotating magnet can be described by rotating lines of force only in certain limiting cases, e.g., a very long cylindrical magnet.[10] Yet using Minkowski's equations in a rotating reference system leads to an internal electric field, in violation of Gauss' law, because Minkowski's equations are not applicable in noninertial reference systems; rather, general relativity theory with its cosmological implications is required in noninertial reference systems.[11] In addition, a quantum-mechanical description of the rotating magnet's internal state presents further difficulties. Consequently in 1979 the problem of unipolar induction still awaits a complete solution. In the course of 145 years it has moved across many different disciplines, including physics and engineering, and today it enters into cosmology as well.

FOOTNOTES

1. Tolman (1913) is the first systematic relativistic treatment of the force between two moving charges. See also French (1966).

2. The positivistic ring to Einstein's statement [§6, 71–72] is characteristic of this period in his philosophical pilgrimage, in which he mixed versions of Mach's sensationism with a budding realism. Einstein's 1905 conclusion of the relativity of electric and magnetic fields is open to the interpretation that no reality can be granted to a charged particle's electromagnetic fields unless an inertial observer performs a measurement on them. Holton (1973d) has shown that by 1919 Einstein's philosophical pilgrimage was nearly completed. His conclusion is borne out by Einstein's 1919 statement that "a kind of objective reality could be granted only to the *electric and magnetic fields together*, quite apart from the state of relative motion of the observer and coordinate system" (italics in original).

3. Einstein's (1910) survey of special relativity theory developed the footnote from his (1905d) as I do here.

4. Whereas Hertz (1890b) and Poincaré (1900d) arrived at this judgment through much deliberation on electromagnetic theory. Hertz's analysis of the foundation of electromagnetic theory was discussed in Chapter 3, and Poincaré's detailed calculations

turned out to have been incorrect since he applied Hertz's electromagnetic theory to moving magnets and dielectrics.

5. A steady current cannot affect a resting charge, and the uniform rotation of a current carrying solenoid about its axis only superposes another uniform motion on its current.

6. This is indicative of how many American physicists understood special relativity theory circa 1917. See also footnote 7 below.

7. Another of Swann's goals was to demonstrate that, contrary to the belief of many physicists, Faraday's law in conjunction with the other equations of electromagnetic theory could account exactly for the current induced in an open circuit moving through a uniform magnetic field, e.g., unipolar induction. Swann accomplished this by using Lorentz's electron theory supplemented with a principle of relativity and assumptions concerning the constitution of a permanent magnet — for example, that the magnet is composed of either Ampèrian current whirls or magnetic poles. He preferred to deduce the electromagnetic field quantities of a moving magnet from the Lorentz electron theory rather than the more straightforward methods employing the transformation formulae of Einstein's special relativity theory. Swann's reason was that although derivations dependent upon assumptions concerning the constitution of matter are more laborious, nevertheless they were more rigorous and hence more convincing. This attitude toward Einstein's special relativity theory was shared by other American physicists — for example, Pegram (1917) and Tate (1922).

8. For example, Cramp and Norgrove (1936) believed in the fixed-line hypothesis. In support of their viewpoint they offer the following experiment: A bar magnet, free to rotate, is hung by a string above another bar magnet whose rotation can be measured. The axes of both magnets are collinear. Rotating the upper magnet produces no effects on the lower magnet, other than the expected longitudinal force between the opposing poles. Thus, Cramp and Norgrove conclude that lines of force remain fixed because if not, then the lines belonging to the upper magnet would have grabbed hold of those generated by the lower one, thereby causing the latter magnet to turn.

9. For example, even in a closed-circuit arrangement, although the EMF can be described by the rotating-line viewpoint, the conducting magnet's volume and surface-charge distributions cannot. Furthermore, any relativistic theory of the electrodynamics of moving bodies contains the apparent polarization that cannot be interpreted from moving lines.

10. Briefly, in a very long cylindrical magnet that is rotating about its axis there is no displacement of charge, i.e., the Lorentz force on each of the magnet's electrons vanishes. Consequently, one can say that the long rotating magnet does not rotate through its lines of force. For details see Cullwick (1959).

11. For example, the external electric field E' in a reference system Γ' rotating with the magnetized sphere is

$$E' = (v/c) \times B + E.$$

Since in Γ' there is no electric dipole moment, the electric displacement $D' = E'$. Since the conducting sphere is isolated, an observer in Γ' should measure no net charge, i.e.,

$\mathbf{V}' \cdot \mathbf{D}' = 0$, where the divergence is calculated in Γ'. But instead we obtain $\mathbf{V}' \cdot \mathbf{D}' = \mathbf{V}' \cdot \mathbf{E}' = 2(\omega/c) \cdot \mathbf{B}$, and charge conservation is violated. For further discussions of electrodynamics in rotating reference systems, see Cullwick (1959) and Rosser (1964), Sommerfeld (1964), T. Schlomka and G. Schenkel (1949), M. G. Trocheris (1949). For examples of electrodynamical calculations using general relativity see Schiff (1939), A. Yildiz and C. H. Tang (1966).

CHAPTER 10

DOPPLER'S PRINCIPLE AND STELLAR ABERRATION

> ...the optics of moving bodies may be treated in quite an elementary way and we shall describe them here as one of the most beautiful applications of Einstein's theory.
>
> Max Born (1920)

In §7 Einstein applied the transformation equations for the space and time coordinates and the electromagnetic fields to two long-standing problems in the optics of moving bodies: optical Doppler effect and stellar aberration. In Lorentz's electromagnetic theory, as in any ether-based theory, these problems resisted a solution accurate to every order in v/c: (1) the case of source and observer in motion relative to the ether could not be solved in terms only of their velocity relative to each other; (2) although Lorentz's theorem of corresponding states asserted that to order v/c stellar aberration was independent of the earth's motion relative to the ether, the constructive character of this theorem demanded a two-step solution for a water-filled telescope — furthermore, although to first order in v/c stellar aberration depended on only the relative velocity between the earth and the star, this phenomenon was interpreted differently in the ether-fixed and geocentric reference systems; (3) the constructive character of the theorem of corresponding states required that the calculation of the transformation properties of a plane wave required two steps even though to order v/c the wave normal and frequency depended on only the relative velocity between two reference systems. The similarity of the redundancy in the second problem to the one in Einstein's description of Maxwell's theory of electromagnetic induction is unlikely to have escaped Einstein's keen sense of aesthetics because it was another instance in which Maxwell's electrodynamics led to an asymmetry that did not "appear to be inherent in the phenomena." Consequently unlike most physicists of 1905, Einstein realized that the difficulties that Lorentz's theory was having with problems in the optics of moving bodies transcended the merely calculational and concerned the foundations of physical theory. In fact, Einstein's approach to the optics of moving bodies was similar to his approach to electromagnetic induction: the principle of relativity dictated the equivalence of the two viewpoints. Since for Einstein there was no Lorentzian ether to interact with charges constituting matter, then stellar aberration could be viewed as a kinematical phenomenon; thus it is not surprising that Einstein's (1907e) discussed stellar aberration in the "*Kinematischer Teil*," where it appeared in the section, "Application of the Transformation Equations to a Problem in Optics."

In the reference system K [§7, 1–10] there is a light source at rest and far enough away from the origin so that, when its wave fronts arrive there they can be "represented to a sufficient degree of approximation" by plane waves – that is, by solutions to the homogeneous wave equations obtained from the Maxwell-Lorentz equations of §6:

$$E = E_0 \sin \Phi, \tag{10.1}$$

$$B = B_0 \sin \Phi, \tag{10.2}$$

where E_0 and B_0 are the amplitudes of the electromagnetic disturbances, and they are equal in magnitude. The phase Φ is

$$\Phi = \omega\{t - (1/c)(lx + my + nz)\} \tag{10.3}$$

where $\omega = 2\pi/T$ is the angular frequency, and l, m, n are the direction cosines of the wave normal. Applying the electromagnetic field transformation Eqs. (9.24)–(9.29) to Eqs. (10.1) and (10.2) Einstein [§7, 11–14] investigated how the quantities characterizing the wave in K were measured by an observer moving with velocity v relative to the source.

$$E_\xi = E_{ox} \sin \Phi', \tag{10.4}$$

$$E_\eta = \gamma(E_{oy} - (v/c)B_{oz}) \sin \Phi', \tag{10.5}$$

$$E_\zeta = \gamma(E_{oz} + (v/c)B_{oy}) \sin \Phi', \tag{10.6}$$

$$B_\xi = B_{ox} \sin \Phi', \tag{10.7}$$

$$B_\eta = \gamma(B_{oy} + (v/c)E_{oz}) \sin \Phi', \tag{10.8}$$

$$B_\zeta = \gamma(B_{oz} - (v/c)E_{oy}) \sin \Phi', \tag{10.9}$$

where Φ' is the phase in k. The exact equivalence of k and K, asserted by the axiomatic principle of relativity, permitted Einstein to take advantage of the invariance of the plane wave's phase. Consequently,

$$\omega[t - (1/c)(lx + my + nz)] = \omega'[\tau - (1/c)(l'\xi + m'\eta + n'\zeta)] \tag{10.10}$$

where primed quantities refer to k. In order to determine ω', l', m', n', Einstein applied to the left-hand side of Eq. (10.10) the transformation equations inverse to Eqs. (6.52)–(6.55) and obtained

$$\omega' = \gamma(1 - lv/c)\omega, \tag{10.11}$$

$$l' = \frac{l - v/c}{1 - lv/c}, \tag{10.12}$$

$$m' = \frac{m}{\gamma(1 - lv/c)}, \tag{10.13}$$

$$n' = \frac{n}{\gamma(1 - lv/c)}. \tag{10.14}$$

These equations relate the frequencies and the direction of the wave normals between k and K. To order v/c, and for a plane wave propagating along the common x-axes of k and K, Eqs. (10.11)–(10.14) resemble mathematically Lorentz's Eqs. (1.88) and (1.93) when S_r is taken as the observer's rest system. However, Lorentz's results for the direction cosines relative to the y- and z-axes turned out to be incorrect.

Without loss of generality Einstein specialized to a plane wave whose wave normal subtends in the x-y-plane an angle ϕ with the relative velocity v, i.e., $l = \cos \phi$; then, Eq. (10.11) becomes

$$v' = v \frac{1 - \cos \phi \cdot v/c}{\sqrt{1 - v^2/c^2}}, \tag{10.15}$$

which is the relativistic version of the optical Doppler effect for any velocity v less than c, and not, as Einstein carelessly wrote, "for arbitrary velocities." Equation (10.15) was a new result, and to illustrate it further Einstein specialized to the case of $\phi = 0$ [§7, 15–23], then

$$v' = v \sqrt{\frac{1 - v/c}{1 + v/c}}. \tag{10.16}$$

Consequently the observer in k measures a smaller frequency than the one in K because k is moving away from the source; the result for k moving toward the source is obtained by replacing v with $-v$, with the consequence that the observer in k measures an increase in frequency. Thus Einstein restored the symmetry he felt was appropriate to this phenomenon. When $v = -c$ the measured frequency in k becomes infinite; it was unnecessary for Einstein to demonstrate this result using $\phi = 0$, because the unphysical result was contained in Eq. (10.15). However, he might well have wanted to show here that the relativity theory displayed a limitation on the velocity of ponderable matter that was absent from the "customary conception"; with great certainty we can take this "customary conception" as Lorentz's (1895) application of the theorem of corresponding states to the Doppler effect and to stellar aberration. According to Lorentz's Eq. (1.92) the frequency measured by a moving observer for the case $\phi = 0$ is to order v/c

$$v_r = v_0(1 + v/c) \tag{10.17}$$

and thus when $v = -c$, $v_r = 0$ – the moving observer sees an *electromagnetic wave* frozen in space, which could only have *doubly-contradicted* the intuition of Einstein's 1895 *Gedanken* experimenter; I conjecture that it was this point that Einstein emphasized by specializing to $\phi = 0$, when $v = -c$.

Supposing that the source and observer move with velocities v_1 and v_2, respectively, relative to the laboratory, then relativity theory predicts for the frequency measured by the moving observer

$$v' = v_0 \sqrt{\frac{1 - v_1^2/c^2}{1 - v_2^2/c^2}} \left(\frac{1 - \boldsymbol{v}_2 \cdot \hat{n}_2/c}{1 + \boldsymbol{v}_1 \cdot \hat{n}_1/c} \right) \qquad (10.18)$$

(where $\hat{n}_1$ and $\hat{n}_2$ are the wave normals relative to v_1 and v_2) instead of Eq. (1.99); so Einstein offered a systematic and exact solution to the optical Doppler effect. This elimination of the serious shortcoming in Eq. (1.99), emphasized by Drude (1900), among others, could only have pleased Einstein.

If the line of observation in K is normal to the relative velocity which is along K's axis (i.e., $\phi = 90°$) Lorentz's theory predicted no Doppler shift [see Eq. (1.92), where $\hat{n} \cdot \boldsymbol{v} = v \cos \phi$]. But according to Eq. (10.11)

$$v' = v_0 \sqrt{1 - v^2/c^2} \qquad (10.19)$$

and in 1907 Einstein posed the transverse Doppler effect as a test for time dilation. This is another demonstration that the choice between the Lorentz and Einstein theories was not one of epistemologies; furthermore, both key experiments were proposed by Einstein [the other was by Einstein-Laub (1908a), see Chapter 7, footnote 27].

Einstein [§7, 24–28] pursued his investigation of the characteristics of plane waves emitted from a source at rest in K with the case of stellar aberration. The observers in K and k measure two different angles defined by the line drawn from the source to k's origin, and the x-axis of K; thus $l' = \cos \phi'$ and $l = \cos \phi$, and so Eq. (10.12) becomes (Fig. 10.1)

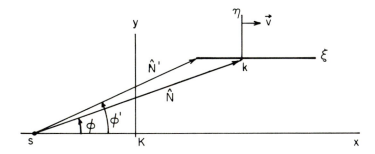

FIG. 10.1. Taking k as the geocentric system, starlight from a source S in K is observed in k at an angle ϕ' that exceeds ϕ [see Eq. (10.20)]. The quantities $\hat{N}'$ and $\hat{N}$ are relativistic wave normals whose components are the direction cosines in k and K [Eqs. (10.11)–(10.14)]. Owing to the relativity of simultaneity, only to accuracy v/c, do the unit vectors $\hat{N}'$ and $\hat{N}$ combine with the quantity $\boldsymbol{v}$ according to the Newtonian addition rule for velocities.

$$\cos \phi' = \frac{\cos \phi - v/c}{1 - \cos \phi \cdot v/c} \tag{10.20}$$

which, wrote Einstein, "is the law of aberration in its most general form."

Einstein's exact treatments of the Doppler effect and of aberration were not immediately appreciated, perhaps because terms higher than first order in v/c were beyond the experimentalists' accuracy; furthermore everyone's attention was focused on the various predictions for the electron's transverse mass.

Upon omitting the denominator in Eq. (10.20), which is due to the relativity of simultaneity, we obtain

$$\cos \phi' = \cos \phi - v/c. \tag{10.21}$$

This is mathematically equivalent with Lorentz's Eqs. (1.13) or (1.89) which were based upon the notion of absolute and relative rays propagating with different velocities. As we discussed in Chapter 3, to all orders in v/c Einstein's second axiom resolved the lack of agreement between experiment and Eq. (1.13) because *ab initio* special relativity theory does not contain the notions of absolute and relative rays. Einstein continued in §7 with a special example of Eq. (10.20): to receive light rays falling normally on k, as observed in K, a telescope in k would have to be oriented in the direction $\cos \phi' = -v/c$.

Einstein (1907e) developed his 1905 result for aberration phenomena to the case where k contained a resting medium of refractive index N'. By equating the phases in k and K for light propagating along the common direction of the x-axes of k and K, i.e.,

$$\omega'(\tau - \xi/u') = \omega(t - x/u), \tag{10.22}$$

where $u' = c/N'$, and then transforming the coordinates in k to those in K, Einstein obtained the result that explains Fizeau's 1851 experiment – that is, the phase velocities in K and k are related through the equation

$$u = \frac{u' + v}{1 + u'v/c^2}. \tag{10.23}$$

Eq. (10.23), continued Einstein, had been deduced by von Laue "directly through application of the addition theorem for velocities."[1] Einstein did not have to discuss the experiments of Airy and Arago because special relativity theory reduced their observations to a foregone conclusion. With his typical pungency Pauli (1958) wrote: "The Airy experiment, as seen from the rest system of the observer (earth), therefore only demonstrates the (relativistically) trivial fact that for a zero angle of incidence (normal incidence) the angle of refraction is zero too."

In an effort to obtain results useful in §8 of the relativity paper, Einstein [§7, 29–33] concluded §7 by deriving the transformation equations for the electromagnetic field amplitudes E_0 and B_0; since Einstein gave no indication of how he deduced these results, I describe his (1907e) method. A plane wave propagates in the x-y-plane of K at an angle ϕ with respect to K's x-axis. Assume that the electric field is along the z-axis, and the magnetic field thus lies in the x-y-plane and subtends an angle ϕ relative to the negative y-axis. Consequently the components of electric and magnetic field amplitudes in K are:

$$E_{ox} = E_{oy} = 0, \tag{10.24}$$

$$E_{oz} = A, \tag{10.25}$$

$$B_{ox} = A \sin \phi, \tag{10.26}$$

$$B_{oy} = -A \cos \phi, \tag{10.27}$$

$$B_{oz} = 0, \tag{10.28}$$

where A is the resultant amplitude in K, i.e.,

$$A^2 = E_{ox}^2 + E_{oy}^2 + E_{oz}^2 = B_{ox}^2 + B_{oy}^2 + B_{oz}^2. \tag{10.29}$$

Substituting Eqs. (10.24)–(10.28) into Eqs. (10.4)–(10.9) yields

$$E_\xi = E_\eta = 0, \tag{10.30}$$

$$E_\zeta = \gamma A[1 - (v/c)\cos\phi]\sin\Phi, \tag{10.31}$$

$$B_\xi = A \sin\phi \sin\Phi, \tag{10.32}$$

$$B_\eta = \gamma A[-\cos\phi + v/c]\sin\Phi, \tag{10.33}$$

$$B_\zeta = 0. \tag{10.34}$$

In k the resultant electric field E' is

$$E' = E_\zeta = A' \sin\Phi; \tag{10.35}$$

the resultant magnetic field B' is

$$B' = \sqrt{B_\xi^2 + B_\eta^2 + B_\zeta^2} = A' \sin\Phi. \tag{10.36}$$

Owing to the relativity of electric and magnetic fields, observers in k and K measure different amplitudes, although the phase is an invariant quantity. Comparison of Eqs. (10.35) and (10.31) gives directly

$$A' = A\gamma[1 - (v/c)\cos\phi]. \tag{10.37}$$

Since the direction of the z-axis is arbitrary, Eq. (10.37) is generally valid and Einstein wrote it as

$$A'^2 = A^2 \frac{(1 - \cos\phi \cdot v/c)^2}{1 - v^2/c^2}, \tag{10.38}$$

perhaps wishing to suggest that Eq. (10.38) was also the transformation equation for a quantity of importance in §8 – the time-average energy density

− or perhaps Einstein simply lapsed into sloppy notation. The former conjecture is probably closer to the truth.

Einstein [§7, 34–36] then offered another example of the unphysical results that occur when $v = c$, i.e., when an observer overtakes a light wave and moves alongside it at relative rest. For in this case A' becomes infinite, as does the light intensity to the observers in k since the light intensity is proportional to the square of the light amplitude.

FOOTNOTE

1. For light propagating through k at the angle ϕ' relative to v, the phase velocities in K and k are related by the equation:

$$u = \frac{u' + v}{\sqrt{1 + 2\dfrac{vu'}{c^2}\cos\phi' + \dfrac{v^2 u'^2}{c^4} - \dfrac{v^2 \sin^2\phi'}{c^2}}}$$

which reduces to Eq. (10.23) for $\phi' = 0$. Consequently, only for $\phi' = 0$ (and also $\phi' = \pi$) does the phase velocity transform like a particle's velocity [see Eqs. (8.21) and (8.22) and Chapter 8, footnote 2]. Møller (1972) proves that the ray velocity transforms like a particle velocity in relativity theory as it did in the ether theories. Thus, Eq. (10.20) can also be deduced from the addition theorem for velocities. In general the ray and phase velocities of light propagating through a medium with $N' \neq 1$ that is at rest in k are both equal to c/N' in k, but an observer in K measures them to be to different in magnitude and direction. Needless to say, in special relativity theory the ray and phase velocities are identical in vacuum, i.e., $u' = u = c$.

CHAPTER 11

LIGHT QUANTA, RADIATION AND RELATIVITY

More than any other section of the relativity paper, §8 demonstrates the relationship between the light quantum and relativity papers. The §8 is also an excellent place to realize the calculational power offered by an axiomatic approach to the electrodynamics and optics of moving bodies – in short, §8 is Einstein's *tour de force*.

In the three pages of §8 Einstein deduced the transformation equation for the energy of a light complex, and then solved exactly the problems of determining the characteristics of plane waves incident on a perfectly reflecting surface that is in inertial motion with velocity v, and of calculating the pressure of light on this surface [§8, 1–2]. Abraham (1904b) spent forty pages to solve these problems.

11.1. EINSTEIN'S ANALYSIS

The electromagnetic field energy densities $A^2/(8\pi)$ and $A'^2/(8\pi)$ relative to two inertial reference systems K and k, respectively, that Einstein discussed, we can develop as follows [§8, 3–5]: In an inertial system K the energy density w in an electromagnetic wave is

$$w = (1/8\pi)(\boldsymbol{E} \cdot \boldsymbol{E} + \boldsymbol{B} \cdot \boldsymbol{B}). \tag{11.1}$$

The time average of w in the reference system K over one period of oscillation for an electromagnetic disturbance of amplitude A is

$$\bar{w} = A^2/8\pi. \tag{11.2}$$

The time average of the energy density relative to an inertial system k moving with velocity v relative to K must "by the principle of relativity" have the same form as that of $\bar{w}$; thus, the time average of the energy density in k is

$$\bar{w}' = A'^2/8\pi. \tag{11.3}$$

Since the laws of electromagnetism must have the same form in K and k, then so must the energy densities in K and k.

Einstein's subsequent discussion [§8, 5–8] does not concern the plane-wave representation of the electromagnetic field from §7 of the relativity paper; rather, he focuses on "a given light complex," which is a plane wave bounded in space. Einstein calculates the total energy of this light complex by multiplying $\bar{w}$ and $\bar{w}'$ by the volumes of the light complex measured by observers in K and k, respectively. However, due to the relativity of time and length, observers in K and k do not measure the same volumes of the light complex. Thus, the ratio of the total energies in K and k is not A'^2/A^2.

Einstein next [§8, 8–14] calculates the ratio of the total energies in K and k. Let l, m, n be the direction cosines relative in K of a vector defining the direction of motion of a spherical light pulse of radius R, that was emitted from a source at rest in K. No energy passes through the boundary, or envelope, of the light pulse because the boundary and pulse move together. In K, Einstein writes the equation of the spherical boundary surface as

$$(x - lct)^2 + (y - mct)^2 + (z - nct)^2 = R^2. \tag{11.4}$$

To better discuss Einstein's calculation of "the amount of energy enclosed by this surface from the viewpoint of the system k," we define the total time averaged energy $\overline{U_T}$ as measured in K of this light complex as

$$\overline{U_T} = (A^2/8\pi)(\tfrac{4}{3}\pi R^3), \tag{11.5}$$

where $\tfrac{4}{3}\pi R^3$ is the volume of the light complex as measured in K. However, calculating the energy of this light complex as measured in k necessitates calculating its volume as measured in k.

Using the relativistic transformations for the space and time coordinates from §3, Einstein [§8, 15–18] transforms Eq. (11.4) to k and obtains at time $\tau = 0$ the result Eq. [§8.1], which can be rewritten as

$$\frac{\xi^2}{[R\sqrt{1 - v^2/c^2}/(1 - lv/c)]^2} + \frac{(\eta - m\gamma\xi v/c)^2}{R^2} + \frac{(\zeta - n\gamma\xi v/c)^2}{R^2} = 1. \tag{11.6}$$

Einstein wrote that according to Eq. [§8.1]: "The spherical surface – viewed in the moving system – is an ellipsoidal surface..." Eq. (11.6) describes an ellipsoidal light pulse with semiminor axis

$$R\sqrt{1 - v^2/c^2}/(1 - lv/c), \tag{11.7}$$

and semimajor axis R. The volume S' of the ellipsoidal light pulse as measured in k is

$$S' = \frac{4}{3}\pi \left(\frac{\sqrt{1 - v^2/c^2}}{1 - lv/c}\right) R^3. \tag{11.8}$$

The volume S of the light pulse with respect to K is

$$S = \tfrac{4}{3}\pi R^3, \tag{11.9}$$

since the light pulse is observed to have a spherical shape relative to K. The ratio of the volumes in k and K is "by a simple calculation" Eq. [§8.2], where Einstein set $l = \cos\phi$.

Einstein continued [§8, 19–21] by calculating the total time-averaged energy in k, $\overline{U'_T}$, is

$$\overline{U_T'} = (A'^2/8\pi)\,S',\qquad (11.10)$$

where $\overline{U_T}$ and $\overline{U_T'}$ indicate the total "light energy enclosed by" the surfaces in K and k. Einstein's Eq. [§8.3] can be obtained from Eqs. (11.5), (11.10) and (10.38). The result is

$$\frac{\overline{U_T'}}{\overline{U_T}} = \frac{A'^2 S'}{A^2 S} = \frac{1 - \cos\phi \cdot v/c}{\sqrt{1 - v^2/c^2}}.\qquad (11.11)$$

Equation (11.11) is the transformation equation for the energy of a light complex, which I rewrite as

$$\overline{U_T'} = \overline{U_T}\,\frac{(1 - \cos\phi \cdot v/c)}{\sqrt{1 - v^2/c^2}}.\qquad (11.12)$$

Next [§8, 22–23] Einstein made what must be considered to be one of the great understatements in the history of science:

> It is remarkable that the energy and the frequency of a light complex vary with the state of motion of the observer in accordance with the same law.

Comparing Eq. (11.12) with the transformation equation for the frequency of light that Einstein had derived in §7, i.e.

$$v' = v\,\frac{1 - \cos\phi \cdot v/c}{\sqrt{1 - v^2/c^2}}\qquad (11.13)$$

yields the result

$$\overline{U_T'}/v' = \overline{U_T}/v.\qquad (11.14)$$

Equation (11.14) predicts that the ratio of energy to frequency for a light pulse is the same in every inertial system. This can only have been what Einstein meant by the "same law." Indeed, as Einstein wrote of this "same law" – "It is remarkable..." – because in his paper on light quanta (1905b) he had shown that "Monochromatic radiation of low density (within the range of validity of Wien's radiation formula) behaves thermodynamically as though it consisted of a number of independent energy quanta of magnitude $R\beta v/N$." The group of constants $R\beta/N$ is Planck's constant h.

Einstein, in the paper on light quanta (1905b) argued from the Wien radiation law to light quanta, showing that their energy is $\varepsilon = hv$. In the relativity paper, he proved that the ratio of the energy to frequency of a "light complex" is invariant; this demonstration involved no assumptions on the constitution of light. The conjunction of these two results for light quanta provides additional evidence for the truly universal character of Planck's constant. Einstein did not take this step, nor did he reference his paper on light quanta. We do not know Einstein's reasons for this omission, perhaps with deliberate understatement he left it to the reader to make the connection. One might note that such understatement is customary in literary works of high distinction.

Alternatively one may conjecture that Einstein's reason for the omission of any mention of light quanta from the relativity paper was that he considered the work on light quanta as "very revolutionary" as he wrote to his friend Conrad Habicht, early Spring 1905 [Seelig (1954)]; for with regard to the relativity paper, Einstein would have had to elaborate on a particle whose motion was independent of its source. In this same letter Einstein described the paper on relativity theory (still in manuscript at this time) in more neutral terms as discussing the "electrodynamics of moving bodies making use of a modification of the doctrine of space and time..." Einstein later remarked to his biographer Seelig: "With respect to the theory of relativity it is not at all a question of a revolutionary act, but of a natural development of a line which can be pursued through centuries." However, Einstein in 1905 may well have realized that a theory containing a modification of the customary notions of space and time would be considered by the physics community as very revolutionary. It would seem "logical," therefore, that to be able to connect the results of the relativity paper with his researches on light quanta, no matter how revolutionary he thought they were, would have been to Einstein's advantage; yet he chose not to explicitly make this connection. Hence, in my opinion, the first reason for Einstein's explicit omission of any mention of light quanta from the relativity paper is closer to the truth, i.e., it was a deliberate understatement.

Einstein continued in the relativity paper [§8, 24–32] by discussing the problem of the reflection of unbounded plane waves, i.e., light rays, from a perfectly reflecting moving mirror (see Fig. 11.1).

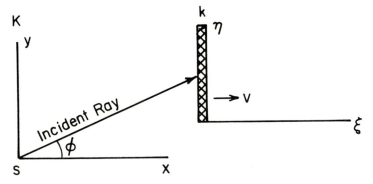

FIG. 11.1.   The plane η-ζ that is normal to the plane η-ξ contains a perfectly reflecting mirror (cross-hatched region in the figure). A source S in K emits a light ray that is incident on the moving mirror at an angle ϕ relative to the x-axis of K.

Einstein's Eqs. [§8.5] to [§8.7] were derived in §7. In Eq. [§8.6], ϕ and ϕ' are the angles subtended by the normals to the plane wave relative to the x- and ξ-axes, respectively. The mathematics of the derivation of Eqs. [§8.11] to [§8.13] is straightforward. The method Einstein used is made possible by his viewpoint based upon the two axioms of the relativity theory: first he transformed into the mirror's rest system k, thereby permitting him to use the simple relationships given by Eqs. [§8.8]–[§8.10] between the incident and reflected light; then he transformed back to K for the solution in terms of quantities measured in the laboratory system, i.e., K.

In order to calculate the pressure as measured in K by the light incident upon the moving mirror, Einstein [§8, 33–40] first calculated the "energy (measured in the resting system) incident per unit time upon the unit of surface of the mirror" and the "energy leaving per unit time upon the unit of surface of the mirror." These quantities are the components of the Poynting vectors normal to the surface of the mirror, where the Poynting vector is

$$\mathscr{S} = (c/4\pi)(\boldsymbol{E} \times \boldsymbol{B}). \tag{11.15}$$

Owing to the brevity of Einstein's statements in this passage, some reconstruction is necessary in attempting to determine what might have been his reasoning in arriving at Eq. [§8.14]. The magnitude of the Poynting vector for a plane wave is

$$\mathscr{S} = (c/4\pi)E^2. \tag{11.16}$$

The time average value of $\mathscr{S}$ over one period of oscillation is

$$\bar{\mathscr{S}} = (c/8\pi)A^2. \tag{11.17}$$

In a matter-of-fact manner, Einstein asserted that the time-averaged incident Poynting vector normal to the surface was "evidently"

$$(A^2/8\pi)(c \cos \phi - v) \tag{11.18}$$

and the time-averaged reflected Poynting vector normal to the surface was

$$(A'''^2/8\pi)(-c \cos \phi''' + v). \tag{11.19}$$

He left the derivation of these equations to the reader. A clue to Einstein's method of deriving Eqs. (11.18) and (11.19) may very well be in the concluding sentence of §8:

> By this means [transforming to the rest system of the moving body] all problems in the optics of moving bodies are reduced to a series of problems in the optics of bodies at rest.

Einstein used this method to solve the problem of determining the characteristics of light reflected from a moving perfectly reflecting surface.

Thus, let us derive Eqs. (11.18) and (11.19) by placing ourselves in the reference system of the moving mirror, i.e., the system k. In k the incident Poynting vector normal to the mirror's surface is

$$\overline{\mathcal{S}'} = (A'^2/8\pi)c \cos \phi', \tag{11.20}$$

where as before primed quantities refer to the incident ray relative to k (see Fig. 11.2). Using Eqs. [§8.5] and [§8.6] we can write $\overline{\mathcal{S}'}$ in terms of quantities measured in K as:

$$\overline{\mathcal{S}'} = \frac{A^2}{8\pi}(c \cos \phi - v)\left[\frac{1 - \cos \phi \cdot v/c}{1 - v^2/c^2}\right]. \tag{11.21}$$

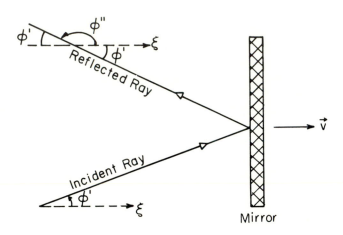

FIG. 11.2. In k the angles of incidence and reflection are equal. All angles are specified with respect to the direction of the relative velocity v between k and K. This direction in k is the ξ-axis. The angle that the incident ray makes with the ξ-axis is ϕ'. The angle that the reflected ray makes with the ξ-axis is ϕ''. From the diagram we see that $\phi' + \phi'' = 180°$.

I have grouped the terms in Eq. (11.21) so as to show that Einstein — if he used this method of derivation — assumed correctly that Eq. (11.21) is the transformation equation that relates the time averages of the normal components of the Poynting vectors in k and K.[1] From Eq. (11.21) the component of the Poynting vector normal to the surface in K is

$$\overline{\mathcal{S}} = (A^2/8\pi)(c \cos \phi - v); \tag{11.22}$$

we may rewrite Eq. (11.21) as

$$\overline{\mathcal{S}'} = \overline{\mathcal{S}}\left[\frac{1 - \cos \phi \cdot v/c}{1 - v^2/c^2}\right]. \tag{11.23}$$

The reflected Poynting vector normal to the surface in k is

$$\overline{\mathscr{S}''} = -(A''^2/8\pi)\,c\cos\phi'' \tag{11.24}$$

where as before double primed quantities refer to the reflected wave as observed in k. To express $\overline{\mathscr{S}''}$ in terms of quantities measured in K we use the transformation equations inverse to Eqs. [§8.11] and [§8.12] to obtain the result:

$$\overline{\mathscr{S}''} = \overline{\mathscr{S}'''}\left[\frac{1 - \cos\phi''' \cdot v/c}{1 - v^2/c^2}\right]; \tag{11.25}$$

hence the time average of the reflected Poynting vector in K normal to the surface is

$$\overline{\mathscr{S}'''} = (A'''^2/8\pi)(-c\cos\phi''' + v), \tag{11.26}$$

which is Einstein's stated Eq. (11.19). Einstein's Eq. [§8.14] follows from taking the difference between Eqs. (11.26) and (11.22). Einstein's Eq. [§8.15] follows from neglecting terms of order v/c in Eq. [§8.14]. Eq. [§8.15] represents the pressure exerted by radiation on a mirror at rest, a result deduced in Maxwell's *Treatise* (1873), and verified experimentally to within one percent by Lebedew (1901) and again by Nichols and Hull (1903). These experiments were well known in 1905 and had been referenced in Abraham's (1904b). Very likely Einstein had them in mind when he wrote: "In agreement with experiment...".

Let us next discuss the latter portion of this sentence: "and with other theories." What other theories? – Einstein did not refer to any. Yet problems concerning the reflection of light from the moving mirror, due to their fundamental relationship to both electromagnetism and thermodynamics, were the theme of many papers. Some prominent pre-1905 works, besides that of Abraham, are: Lorentz's (1886) where this problem is discussed in connection with his analysis of Michelson's interferometer experiment of 1881; Lorentz's paper (1892c), "On the Reflection of Light from Moving Bodies," where he calculated (to order v/c) the amplitude of the reflected light and the radiation pressure on a moving perfect reflector. Lorentz (1892c) noted that his result for the radiation pressure agreed with Maxwell's. Lorentz returned to this problem in the *Versuch*, once again solving it to first order in v/c. Drude (1900) discussed this portion of the *Versuch* as well as reviewing Wien's original derivation in 1893 of his displacement law. Of the aforementioned works Einstein had read Lorentz's *Versuch*, and possibly also Drude's book. Thus, by "other theories" Einstein very probably meant those of Maxwell and Lorentz. Furthermore, Einstein may very well have been emphasizing, as did others, e.g. Abraham (1904b), that Eq. [§8.15] must be reproducible by any theory of radiation.

Einstein did not explicitly state that he had arrived at new results for the class
of problems pertaining to the reflection of light from a moving mirror; but his
viewpoint offered a simple method to treat problems in the optics of moving
bodies. He concluded §8 by drawing attention to this method:

> All problems in the optics of moving bodies can be solved by the method
> here employed. The essential point is that the electric and magnetic force
> of the light, which is influenced by a moving body, be transformed to a
> coordinate system at rest relative to the body. By this means all problems
> in the optics of moving bodies are reduced to a series of problems in the
> optics of bodies at rest.

Thus, according to Einstein: "All problems in the optics of moving bodies"
can be transformed into problems involving the "optics of bodies at rest."
These problems are easier to solve. Using the relevant transformation
equations one simply transforms all electromagnetic field quantities, frequencies and angles into the body's rest system.

11.2. MAX ABRAHAM'S PAPER OF 1904 "ON THE THEORY OF RADIATION AND OF THE PRESSURE OF RADIATION"

Abraham (1904b) deduced equations valid to every order in v/c for the
characteristics of radiation reflected from a perfectly reflecting surface in
inertial motion relative to the ether, and for the light pressure on this surface,
and then commented on Lorentz's work on the optics of moving bodies. He
could obtain exact results for these problems because he used only one
reference system, fixed in the ether, thereby avoiding any calculational
problems whose solution awaited Einstein's relativity of simultaneity. But
Abraham's route was a torturous one, comprised of calculations and
explanations covering forty pages. I emphasize, however, that Abraham's
results on radiation theory were only *mathematically equivalent* to those in
Einstein's §8 of the relativity paper. If he had read Abraham's paper before
mid-1905, Einstein could only have been greatly impressed with Abraham's
highly physical reasoning.
    Let us survey the content of Abraham's introductory remarks to gain some
further insight into the importance of the subject matter of Einstein's §8 to the
physics of 1905. Almost immediately in (1904b), Abraham drew a distinction
between the pressure of light and of radiation: the former having been deduced
by Maxwell and then established directly by Lebedew (1901) and by Nichols
and Hull (1903), while for the latter there existed evidence that was exact,
although indirect, being supported by the *thermodynamic* (i.e., not electromagnetic) law for cavity radiation, a problem which Abraham, among
others, did not consider as solved adequately. He asserted that the theoretical
foundation of these laws, i.e., of Maxwell's light pressure and of cavity

radiation, was an important problem for the "electromagnetic mechanics" — the electromagnetic world-picture. Almost certainly a basic reason for this statement was that the quantity fundamental to theories concerning light and radiation pressure *and* a theory of the electron was the Poynting vector. An assumption basic to both Abraham's and Lorentz's theories of the electron was that the electron's momentum is the electromagnetic momentum of its own electromagnetic field.

Abraham (1904b) continued by reviewing briefly the optics of moving bodies. He wrote of Lorentz having been able to explain the results of optical experiments accurate to first order in v/c, and of Lorentz's difficulty in explaining the null result of "Michelson's interferometer experiment." He then emphasized that if the electron's mass were velocity dependent, then an isotropic body in motion should exhibit double refraction (a second-order effect); such a phenomenon would enable detection of the earth's motion relative to the ether. Bodies composed of Abraham's electrons would indeed become doubly refracting, while this is not the case for bodies composed of Lorentz's electrons. However, the precision experiments of Rayleigh (1902), repeated in 1904 by Brace, had revealed no double refraction in isotropic bodies in motion. Abraham's response to this piece of evidence at variance with his theory was to assert the necessity for more work on the optics of moving bodies.

For the moment Abraham wanted to avoid problems concerning the constitution of matter, problems involving assumptions on the nature of molecular forces: "The researches to follow concern problems in the optics of moving bodies whose solution is possible without being drawn into making such assumptions; namely, the action of radiation upon perfectly black and upon perfectly reflecting surfaces." Abraham's statement of intent may have struck a responsive chord with Einstein because by mid-1904 Einstein already may have realized that avoiding assumptions on the constitution of matter was the proper one, although not within an ether-based framework.

Since problems concerning radiation in the so-called free ether involve investigating the nature of the Poynting vector, a quantity indigenous to any field-theoretical description of the electron, then Abraham's results must necessarily be reproducible by any theory of the electron. Moreover, Abraham stated problems in the optics of moving bodies in such a manner that he could treat simultaneously problems concerning radiation contained in a *Hohlraum* with perfectly reflecting walls. Assuming Kirchhoff's law and the second law of thermodynamics he deduced the Wien displacement law, the Stefan-Boltzmann law, and the general functional form for the frequency distribution of cavity radiation.[2] This is as far as he went on this problem, concluding with the following comment on Planck's researches: "The molecular-theoretical researches [of Planck] fall outside the scope of our considerations"; consequently, Abraham could politely avoid mentioning Planck's results on cavity radiation.

Abraham's derivation of the pressure of radiation on a perfectly reflecting mirror from Lorentz's electromagnetic theory is based upon considerations of the velocity of light relative to the inertial system S_r, i.e., u_r which is the velocity of the relative ray in Eq. (1.13) with $N = 1$. Thus, the quantities $(c \cos \phi - v)$ and $(-c \cos \phi''' + v)$ that multiply the Poynting vectors in Abraham's paper, are interpreted as the components of the incident and reflected light with these relative light velocities normal to the reflecting surface [see Eqs. (11.22) and (11.19)].[3] Abraham's route to Eq. [§8.14] is in outline as follows: via trigonometric relationships based upon the Newtonian addition rule for velocities, Eq. (1.13); defining the work done on unit area per second on the moving surface as $\mathscr{S}_i u_r/c$, where $\mathscr{S}_i$ is the magnitude of the incident Poynting vector associated with c; the use of Doppler's principle to assist him in the elimination of quantities referring to u_r; the principle of the conservation of energy. For completeness, Abraham rederived his version of Eq. [§8.14] for the case of normal incidence by considerations of the boundary conditions on the electromagnetic field quantities at the reflecting surface, and then using the Maxwell stresses.

Abraham discussed all of his results in great detail, particularly the limiting case of $v/c = 1$ for his version of Eq. [§8.14]. He considered this limit as unphysical since an infinite amount of work was necessary to overcome the radiation pressure and accelerate the mirror to the velocity of light. Although he concluded that the mirror could never attain the velocity of light, nevertheless, he considered it feasible for electrons to attain the velocity of light.

It is an indication of how well known were Abraham's results for the problem of the reflection of light from a perfectly reflecting moving surface, and how little known, misunderstood and unappreciated were the contents of Einstein's §8, and indeed the entire relativity paper, that Planck in his *Theory of Radiation* (1906c) referred only to Abraham's results.[4] Planck, one of Einstein's earliest supporters, considered as of primary importance in the relativity paper the results contained in the final section "The Dynamics of the (Slowly Accelerated) Electron." Planck interpreted them as an important generalization of Lorentz's theory of the electron.

Analysis in Section 11.1 of Einstein's derivation of the incident and reflected Poynting vectors normal to the surface allows us to compare further the Abraham-Lorentz concept of a reference system at rest relative to the reflecting surface with that of Einstein. According to the theories of electromagnetism based upon a Lorentzian ether, a reference system S_r on the earth and at rest relative to the reflecting surface could not be considered as the rest system of the surface. This concept assumed clarity only within the context of Einstein's special relativity theory. The reason, as Kaufmann (1906) noted (Section 7.4.1) was that according to Lorentz's theory of electromagnetism the earth had to be considered as in motion relative to an immobile ether, and effects of this motion should be present: ether-based theories were dynamical theories of matter. One avenue of Lorentz's elaborations upon his electromagnetic theory of 1892 was to formulate a set of mathematical transformations upon the space and time coordinates of S_r (i.e., x_r, y_r, z_r, t_r) which (along with certain transformations of the electromagnetic-field quantities) would serve to transform the equations of electromagnetism into a mathematical coordinate system having the same properties as the ether-fixed system S. At the time of publication of Abraham's (1904b) Lorentz had succeeded in this research effort only to order v/c.

According to Einstein's view the reference system k can be considered as the surface's rest system, where it is axiomatic that the velocity of light is exactly c. There are no relative light velocity (u_r) and absolute light velocity (c), and their associated Poynting vectors; thus in k the solution to the problem of the reflection of light from a moving surface can be carried out as if the surface were at rest. The result of this simple calculation is then transformed back to the laboratory system K, where the velocity of light is also exactly c.

In 1905 Abraham once again treated in depth problems concerning radiation in his *Theorie der Elektrizität: Zweiter Band, Elektromagnetische Theorie der Strahlung*.[5] There he also discussed Lorentz's new theory of the electron, applying some of its postulates to the problem of the reflection of light from a moving surface. The "Foreword" to this book published in 1905 is dated March 1905, Wiesbaden; hence, it is not likely that Einstein could have read it before 30 June 1905, the date that his relativity paper was received at the *Annalen*. Even if he had seen it, it is unlikely that its contents would have had any direct influence upon his thought. By March 1905, Einstein was already in the process of preparing his series of three papers, from a point of view quite different from that of Abraham's (1904b), the import of which would not immediately be fully appreciated, yet which moved physics into the 20th century.

FOOTNOTES

1. Although Einstein may not have realized that the Poynting vector does not have the transformation properties of a vector, Sommerfeld (1910a, b) and von Laue (1911a) used Minkowski's mathematical formalism to develop the transformation properties of quantities such as the Poynting vector (see Section 12.5.8).

2. From radiation theory, optics, and Kirchhoff's law, Abraham showed that the reflection of light from a moving, perfectly reflecting surface is, in the sense of thermodynamics, a reversible process. Then employing the second law of thermodynamics to obtain an additional relationship between the components of the Poynting vectors normal to the moving surface, frequencies and absolute temperatures, as these quantities refer to the incident and reflected radiation, he obtained the Wien displacement law.

3. Abraham found that his results for radiation, whose source was at rest on the earth, and was incident normally upon a perfectly black surface (also at rest on the

earth's surface) were difficult to reconcile with Lorentz's theorem of corresponding states. This theorem asserted the equality of the energies, associated with the relative ray of velocity u_r, incident normally per second on unit area of a perfectly black surface where u_r was parallel and then antiparallel to the earth's motion through the ether (these Poynting vectors denoted by Abraham as $\mathscr{S}'_+$ and $\mathscr{S}'_-$, respectively, he referred to as the "relative radiation"). Abraham deduced that this was not the case for similarly defined Poynting vectors associated with the absolute ray of velocity c (Abraham denoted these Poynting vectors as $\mathscr{S}_+$ and $\mathscr{S}_-$, respectively, and referred to them as the "absolute radiation"). Rather the ratio $\mathscr{S}_+ : \mathscr{S}_-$ was a function of v/c. Abraham asserted that he did not understand this result and that further theoretical investigation was necessary – Abraham's import was that something was amiss with Lorentz's theorem of corresponding states, and thus with Lorentz's theory of the optics of moving bodies as well. Symptomatic of the lack of clarity in ether-based theories of optics was Abraham's comment that the v/c dependence of the ratio $\mathscr{S}_+ : \mathscr{S}_-$ was not experimentally noticeable; the velocity v is, strictly speaking, the velocity of the earth relative to the ether, a quantity that resisted measurement and thus similarly did the quantities $\mathscr{S}_+$ and $\mathscr{S}_-$. I add that the concepts of relative and absolute radiation can be generalized to oblique incidence of the relative and absolute rays.

4. Planck rewrote large portions of the first edition, yet he did not mention the results of Einstein's §8 of the relativity paper. Planck's edition of 1906 referred once to Einstein's light quantum paper, noting that Einstein had shown that a usually accepted result of the kinetic theory of gases (that the average energy of the x, y and z components of the motion of an oscillator interacting with radiation is kT) is valid only for large values of λT, where λ is the wavelength of the radiation. In the 1913 edition this footnote was removed, the relevant section having been rewritten; thus, no mention of the light quantum paper was made in the second edition. In the second edition Planck did make brief mention of Einstein's work of 1907 on specific heats. In this footnote Planck avoided further mention of Einstein's light quantum hypothesis: "...a closer discussion [of Einstein's work on specific heats] would be beyond the scope of the investigations to be made in this book." In fact, what can be construed as a subtle rebuke of Einstein's light quantum hypothesis is Planck's calculation (in both editions) on the basis of "Newton's (emission) theory of light," of the pressure of radiation on a perfectly reflecting surface. Planck deduced that "light particles" of mass m traveling with velocity c and incident obliquely on such a surface would produce a pressure that is twice as great as the Maxwell radiation pressure. He carried out this trivial calculation in painstaking detail. Thus, Planck emphasized, the magnitude of the radiation pressure was deducible only from a wave theory of electromagnetism. While this demonstration disposed of a Newtonian corpuscular theory of light (as Planck demonstrated), such was not the case for Einstein's corpuscles of light.

With Planck's comment in mind, let us speculate on why in Einstein's Eq. [§8.15] the factor of 2 is a separate term. Perhaps he wanted to emphasize that the pressure of light on a moving mirror can be conceptualized in a manner similar to that familiar from mechanics and kinetic theory, i.e., that of the pressure on a moving, perfectly reflecting wall that is being hit successively by particles fired from a cannon at rest in K. The factor of 2 designates the fact that the particles rebound perfectly elastically off the moving wall and transfer momentum to it.

With this result let us return to the question of the pressure of light on a moving mirror. If, for example, the mirror were at rest relative to the source of light then [§8.15] becomes

$$P = 2(A^2/8\pi) \cdot \cos^2\phi. \qquad \text{(A)}$$

If $\phi = 0$, i.e., normal incidence, this equation becomes

$$P = 2 \cdot A^2/8\pi. \qquad \text{(B)}$$

The time averaged momentum density $\bar{g}$ of a plane wave is

$$\bar{g} = A^2/8\pi c. \qquad \text{(C)}$$

Using Eq. (C) we can rewrite Eq. (B) as

$$P = 2\bar{g}c \qquad \text{(D)}$$

which is analogous to the pressure exerted on a (motionless in K) perfectly reflecting wall by a projectile traveling with velocity c. Einstein had discussed projectiles of this type in his paper on light quanta. Thus, Einstein may have written the factor of 2 separately in [§8.15] (instead of writing $2 \cdot A^2/8\pi$ as $A^2/4\pi$) to draw attention to the possibility that light has a granular structure. Once more, for reasons that I can only speculate upon, Einstein made no reference even to his own work at that point.

5. Abraham (1905) referred to Lorentz's Σ' system as an "auxiliary system (*Hilfssystem*)," — Lorentz referred to it as an "imaginary" system. Abraham used Σ' to rederive more economically certain results from his (1904b). As part of his criticism of Lorentz's new theory of the electron, Abraham pointed out that there still remained the problem of the velocity dependence of the ratio $\mathscr{S}_+ : \mathscr{S}_-$. Abraham went on to prove that the pressure of radiation upon a moving, perfectly reflecting surface is the same in Σ' and S. However, this result had little physical content in an ether-based theory of radiation since Σ' was not a physically real system and observers in S could not be communicated with. According to Einstein's viewpoint there exist only the reference systems k and K that bear a mathematical resemblance to Σ' and S, respectively. Abraham's result on the equality of radiation pressures in Σ' and S becomes in Einstein's viewpoint the invariance of the radiation pressure. Evidently Einstein in 1905 did not take note of the invariance of the radiation pressure; it was first pointed out by Planck (1907a). That the invariance of radiation pressure is contained implicitly in Einstein's §8 of the relativity paper can be demonstrated as follows: In k the radiation pressure P' is

$$P' = 2(A'^2/8\pi)\cos^2\phi'. \qquad \text{(A)}$$

Using Eqs. [§8.5] and [§8.6] to transform A' and $\cos\phi'$ to K yields

$$P' = 2\frac{A^2}{8\pi}\frac{(\cos\phi - v/c)^2}{1 - v^2/c^2}. \qquad \text{(B)}$$

Comparison of Eq. (B) with Eq. [§8.14] completes the proof, i.e., $P' = P$. Contrary to certain authors, e.g., Pauli (1958), relativity theory did add something new to the problem of the reflection of light from a moving mirror — namely, an unambiguous solution. Although Abraham's solution was mathematically identical to Einstein's, it could be used in the laboratory reference system only under the tacit assumption that the ratio of the laboratory velocity relative to the ether to the velocity of light was much smaller than the ratio of the mirror's velocity to the velocity of light.

CHAPTER 12

ON THE ELECTRODYNAMICS OF MOVING BODIES

> [Concerning his methods in §10 Einstein wrote that it was] logically unjustifiable to base all physical consideration on the rigid or solid body and then finally reconstruct that body atomically by means of elementary physical laws which in turn have been determined by means of the rigid measuring body.
>
> A. Einstein (1923)

Although all published accounts of Einstein's theory circa 1905 focused on §§9 and 10, it was §10 in particular that satisfied him least from the point of view of logical justifiability. Yet, ironically, Einstein realized shortly after publication that §10 contained the result that transcended the special relativity theory — the mass-energy equivalence.

Whereas in §9 Einstein discussed a macroscopic charged body, in §10 he specialized to the motion of a *point* electron moving according to the kinematics derived in Part I of the relativity paper. As Einstein wrote (1923), §10 of the relativity paper contained a deficiency in method, because it was "logically unjustifiable" to have heretofore based all derivations on the "rigid or solid body" as an irreducible entity and then to "reconstruct that body atomically." This logically impure application of the stipulation of meaning displeased Einstein, but the "logically purer method" could not have been applied in 1905 "without going out of our depth."

12.1. EINSTEIN'S §9: MACROSCOPIC CHARGED BODIES

Unlike the approach in §6, Einstein here [§9, 1–9] discussed the "Maxwell-Hertz equations" taking acount of the sources of radiation and of the effects of an external electromagnetic field on the simplest possible macroscopic charged body of charge density ρ — that is, one that was nonmagnetic, nondielectric, and nonconducting. In this case there was sufficient similarity of form between the theories of Hertz and Lorentz for Einstein to feel himself on safe enough grounds to continue using Hertz's neutral nomenclature. Einstein's Maxwell-Hertz equations, when convection currents are taken into account, are:

$$\mathbf{V} \cdot \mathbf{E} = 4\pi\rho, \tag{12.1}$$

$$\mathbf{V} \times \mathbf{E} = -\frac{1}{c}\frac{\partial \mathbf{B}}{\partial t}, \tag{12.2}$$

$$\nabla \times \mathbf{B} = \frac{1}{c}\frac{\partial \mathbf{E}}{\partial t} + \frac{4\pi}{c}\rho\mathbf{u}, \tag{12.3}$$

$$\nabla \cdot \mathbf{B} = 0; \tag{12.4}$$

with the exception that Einstein wrote ρ for $4\pi\rho$, and used V for c, his units were those that Abraham had used (1903) where he had stressed their displaying the critical nature of the velocity $v = c$. If one chooses to imagine that the body's charge has an atomistic basis, continued Einstein, "then these equations are the electromagnetic basis of Lorentz's electrodynamics and optics of moving bodies."

Thus it would have been apropos here for Einstein to replace the name of the Maxwell-Hertz equations with the name Maxwell-Lorentz equations; in (1906c) he did just that.

An indication that the relativity theory was still developing was Einstein's reference [§9, 10–20] to "electrons" as "rigid bodies," a notion that he later criticized (1907d). In order to transform Eqs. (12.1)–(12.4) from K to k, Einstein used the principle of relativity as a heuristic principle as he had in §6. But he found that Eqs. (6.52)–(6.55), and the transformation equations for the electromagnetic field quantities were insufficient to maintain the form of Eqs. (12.1)–(12.4) in k; covariance required separate transformation equations for the charge density, i.e.,

$$\rho' = \gamma(1 - u_x v/c^2)\rho, \tag{12.5}$$

and in order for the convection current to transform properly, the charge's velocity had to transform as:

$$u_\xi = \frac{u_x - v}{1 - u_x v/c^2}, \tag{12.6}$$

$$u_\eta = \frac{u_y}{\gamma(1 - u_x v/c^2)}, \tag{12.7}$$

$$u_\zeta = \frac{u_z}{\gamma(1 - u_x v/c^2)}, \tag{12.8}$$

which can be related to Eqs. (8.7) and (8.8), and Eq. (12.8) replaces Eq. (8.9). Since according to Eqs. (12.6)–(12.8) the charged body transformed properly under the relativistic equations, Einstein wrote that according to "our kinematical principles, the electrodynamic foundation of Lorentz's theory of the electrodynamics of moving bodies" agreed "with the principle of relativity."

Einstein [§9, 21–25] concluded §9 by describing briefly a proof for the invariance of electric charge which he developed in (1907e).[1] Lorentz and Poincaré had arrived independently at results identical symbolically with those contained in Einstein's §9. However, Lorentz's *derivation* of the covariance of the Maxwell-Lorentz equations with source terms, and his transformation equation for the charge density were valid only in the electron's (fictitious) instantaneous rest system; his transformation equation for the velocity was incorrect. Although Poincaré [(1905a), (1906)] corrected Lorentz's mathematical errors the ambiguities lurking in the physical interpretation of the formalism awaited removal by the special relativity theory. Thus rightly, as Guillaume (1924) and Whittaker (1973, vol. 2) wrote, Poincaré deserved credit for having also obtained in 1905 the correct velocity addition law and the correct transformation equation for the charge density; but the similarity was only mathematical because the conceptual frameworks of Einstein and Poincaré were entirely dissimilar.[2]

12.2. EINSTEIN'S §10: MOVING MICROSCOPIC BODIES

In order to neglect energy losses due to the emission of radiation, Einstein [§10, 1–4] restricted discussion in §10 to the "Dynamics of the (Slowly Accelerated) Electron"; thus, he could attribute changes in the charged particle's kinetic energy solely to the work done on the particle by an external electromagnetic field. Einstein did not intend this restriction to signal use of the quasi-stationary approximation, because he went on in §10 to assume that Newton's second law retained its mathematical form from classical mechanics in the electron's instantaneous rest system; owing to the relativity of simultaneity the symbols in Newton's law had to be reinterpreted.

12.2.1. The Moving Electron's Mass

Einstein next [§10, 4–8] considered an electron in the region of uniform electric and magnetic fields. If at an "instant of time" t_0 in K the electron was at rest, it could be acted upon only by the external electric field E which would set it into motion. At the "next instant of time" t_1, differing from t_0 by an infinitesimal amount, Einstein postulated that the electron would move according to Newton's law $F = ma$, where the net force F was the force on a particle of charge e due to an external electric field E. Then in K the equations of motion in the infinitesimal time interval $t_1 - t_0$ are:

$$m_0 \, d^2x/dt^2 = eE_x, \tag{12.9}$$

$$m_0 \, d^2y/dt^2 = eE_y, \tag{12.10}$$

$$m_0 \, d^2z/dt^2 = eE_z, \tag{12.11}$$

where m_0, "the mass of electron as long as its motion is slow," is the constant inertial mass m that is customarily used in Newton's second low, and e is the magnitude of the electron's charge. (Einstein's (1905d) notation ε for the elementary unit of charge should not be confused with the symbol ε for the charge in emu. Einstein (1905d) used μ for m_0, even though in the German-language scientific literature this symbol was usually reserved for the electromagnetic mass.) Einstein's cautionary remark that m_0 is "the mass of the electron, as long as its motion is slow" implies that, as a result of the relativity of simultaneity, we can expect certain changes to occur in the mathematical formulation of this law.

Thus, [§10, 9–25] whereas advocates of an electromagnetic world-picture attempted to deduce Newton's second law from Lorentz's force equation (and with only approximate success), Einstein boldly postulated this law's exact validity for every inertial observer. Suppose, however, that at time $t = 0$ the electron moved with velocity v relative to K, and at this instant k was its instantaneous rest system. According to the principle of relativity, "in the immediately following time" interval in k the electron's equations of motion have the same form as Eqs. (12.9)–(12.11),

$$m_0\, d^2\xi/d\tau^2 = eE_\xi, \tag{12.12}$$

$$m_0\, d^2\eta/d\tau^2 = eE_\eta, \tag{12.13}$$

$$m_0\, d^2\zeta/d\tau^2 = eE_\zeta. \tag{12.14}$$

As Einstein had asserted in §9, electric charge was conserved under the relativistic transformation. The mass m_0 was a constant in k as it was for the previous case where the electron was instantaneously at rest relative to K; at other instants of time other inertial systems could become the electron's instantaneous rest system. Straightforward application of the relativistic transformations for the space and time coordinates, and the electric field permitted Einstein to deduce the electron's equations of motion in K :[3]

$$\frac{d^2x}{dt^2} = \frac{e}{m_0\gamma^3} E_x, \tag{12.15}$$

$$\frac{d^2y}{dt^2} = \frac{e}{m_0\gamma}\left(E_y - \frac{v}{c} B_z\right), \tag{12.16}$$

$$\frac{d^2z}{dt^2} = \frac{e}{m_0\gamma}\left(E_z + \frac{v}{c} B_y\right). \tag{12.17}$$

Consequently Einstein solved this problem in the electrodynamics of moving bodies by reducing it to a problem about the forces acting on an electron at rest. He could now extend to electrodynamics the procedure he had applied to problems in the optics of moving bodies: "By this means [transforming the problem to the body's instantaneous rest system] all problems in the optics of moving bodies are reduced to a series of problems in the optics of bodies at rest [§8]." Abraham, Lorentz and Poincaré, among others (e.g., Thomson and Searle), also made use of this approach; however, they ascribed to it no physical interpretation.

Einstein then [§10, 26–39] turned to the "customary point of view [to inquire into] the 'longitudinal' and the 'transverse' mass of the moving electron." He rewrote Eqs. (12.15)–(12.17) in the form:

$$m_0\gamma^3\frac{d^2x}{dt^2} = eE_x = eE_\xi, \tag{12.18}$$

$$m_0\gamma^2\frac{d^2y}{dt^2} = e\gamma\left(E_y - \frac{v}{c}B_z\right) = eE_\eta, \tag{12.19}$$

$$m_0\gamma^2\frac{d^2z}{dt^2} = e\gamma\left(E_z + \frac{v}{c}B_y\right) = eE_\zeta. \tag{12.20}$$

It is reasonable to conjecture that the "customary view" was Abraham's in which the electron's mass was a two-component quantity that depended upon the orientation of the external forces and was totally electromagnetic in origin. Einstein modified the "customary point of view," however, by taking the electron as a primitive quantity, so that Newton's second law was valid axiomatically in the electron's instantaneous rest system. Then the electron's "'longitudinal' and 'transverse' mass" (terms invented by Abraham) became the coefficients of its acceleration, relative to K, arising from an external electromagnetic field; Einstein did not speculate on the origin of the electron's inertial mass in its rest system.

According to Einstein's interpretation of Eqs. (12.18)–(12.20), in k the electron was acted upon by a force F' with components

$$F' = (eE_\xi, eE_\eta, eE_\zeta), \tag{12.21}$$

possessing an in-principle operational definition. These components, wrote Einstein, could be "measured, for example, by a spring balance at rest in" k, and he defined the quantities in Eq. (12.21) as the "force acting upon the electron." Consequently, from the relativity of the electromagnetic field quantities, the force F acting upon the electron relative to K had components (see Eqs. (12.18)–(12.20)):

$$F = \left[eE_x, e\gamma\left(E_y - \frac{v}{c}B_z\right), e\gamma\left(E_z + \frac{v}{c}B_y\right)\right]. \tag{12.22}$$

If Einstein had equated separately the force components in k and K, then he would have deduced kinematically the relationship between the forces in Abraham's and Lorentz's reference systems S and S'', which they had arrived at from electrodynamical considerations based upon transforming an inhomogeneous wave equation to a Poisson equation. But in the special relativity theory, Lorentz's force equation was not an independent axiom of electromagnetic theory; rather, it followed from the relativity of the electromagnetic fields; this was another new result of Einstein's theory.

Those who by 1911 realized the difference between Einstein's and Lorentz's viewpoints appreciated this result: for example, Tolman (1911a) considered it as "one of the chief pieces of evidence which support the theory of relativity." Einstein (1923) listed his deduction of Lorentz's force law as one of the "appreciable advances" of special relativity theory which "reconciled mechanics and electrodynamics [and] reduced the number of logically dependent hypotheses regarding the latter." By eliminating Lorentz's ether Einstein had removed another redundancy in electromagnetic theory.

Some measure of Einstein's elation at deriving the Lorentz force law can perhaps be gleaned from Sauter's remembrances of things past at the Patent Office. Ever attempting to stir Einstein's interest in Hertz's mechanics, Sauter described to Einstein Hertz's derivation of "Newton's famous lex tertia." Sauter was struck by his colleague's reaction: "Einstein could not conceal his admiration and declared, 'To diminish by one the number of premises of mechanics, c'est un tour de force.'"

From Eqs. (12.18)–(12.20), Einstein obtained the longitudinal and transverse masses, respectively, from the coefficients of the accelerations as

$$m_L = \frac{m_0}{(1 - v^2/c^2)^{3/2}}, \tag{12.23}$$

$$m_T = \frac{m_0}{(1 - v^2/c^2)}. \tag{12.24}$$

Since most likely Einstein had read Abraham's classic paper of 1903, "*Prinzipien der Dynamik des Elektrons*," and had used Abraham's nomenclature for the components of the electron's mass, why did he not refer to Abraham's work? Perhaps Einstein's disagreement with the electromagnetic world-picture was so basic that he chose to criticize it, albeit implicitly, in the two other papers in vol. 17, while in the relativity paper he played the part of the maverick physicist. By 1907 Einstein had mellowed somewhat, criticizing this research effort in "The Consequences of the Principle of Relativity for the Inertia of Energy," and then surveying it in his (1907e) paper complete with references to Abraham, Lorentz and Bucherer.

But there was a subtlety in "taking the customary point of view" because maintaining the "equation − mass × acceleration = force" − stressed Einstein, left open the functional form for the force. For example, Einstein could have defined the components of the "force acting upon the electron" as

$$\left[\frac{eE_x}{\gamma^3}, \frac{e}{\gamma}\left(E_y - \frac{v}{c}B_z\right), \frac{e}{\gamma}\left(E_z + \frac{v}{c}B_y\right) \right], \tag{12.25}$$

thereby rendering the longitudinal and transverse masses equal and constant; but this choice violated Kaufmann's data. Clearly many choices existed for which the "force acting upon an electron" depended only upon the powers of γ

that multiply both sides of Eqs. (12.18)–(12.20). Perhaps Einstein wrote the same equations in two different ways, i.e., Eqs. (12.15)–(12.17) and (12.18)–(12.20), in order to draw the reader's attention to this subtlety. Einstein, however, made a poor choice for the force acting upon the electron in K, and in 1906 Planck showed that a proper mechanics could best be formulated from the equations

$$m_0\gamma^3\frac{d^2x}{dt^2} = eE_x, \tag{12.26}$$

$$m_0\gamma\frac{d^2y}{dt^2} = e\left(E_y - \frac{v}{c}B_z\right), \tag{12.27}$$

$$m_0\gamma\frac{d^2z}{dt^2} = e\left(E_z + \frac{v}{c}B_y\right). \tag{12.28}$$

Equation (12.26) is identical to Eq. (12.18), but Eqs. (12.27) and (12.28) differ from Eqs. (12.19) and (12.20) by one factor of γ, thereby making mathematically equivalent the predictions of Einstein and Lorentz for m_T. To the best of my knowledge, Kaufmann's (1906) was the first to mention in print that Lorentz's and Einstein's predictions were equivalent because, he wrote, the difference between their predictions was only "apparent." Denoting the right-hand side of Eqs. (12.26)–(12.28) as components of the force F in K, Planck (1906a) found that he could write these three equations as

$$F = \frac{d}{dt}\left[\frac{m_0v}{\sqrt{1 - v^2/c^2}}\right], \tag{12.29}$$

thereby demonstrating the mathematical equivalence of Einstein's and Lorentz's predictions for the longitudinal and transverse masses. Thus, although Einstein was correct in pointing out the subtlety involved in defining force through the procedure of coordinate and field transformations, he had erred on the one hand in demonstrating the velocity dependence of mass, and on the other in defining force as "mass × acceleration," instead of as the rate of change of momentum, as he should have done in a system whose mass was variable.

Planck (1906a) went on to show that this definition of force, analogous to the one in Newtonian physics, enabled formulation of a Hamilton-Lagrange version of the relativistic mechanics of "mass points." Identifying the Lagrangian L (which Planck denoted as the "kinetic potential" H) as

$$L = -m_0c^2\sqrt{1 - v^2/c^2} + \text{constant} \tag{12.30}$$

led Planck to define the quantity

$$p = \frac{m_0v}{\sqrt{1 - v^2/c^2}} \tag{12.31}$$

as the "impulse coordinate," because in the Lagrange formulation of mechanics

$$p = \partial L / \partial v. \tag{12.32}$$

Planck's detailed sequel (1907a) contained a more complete mechanics of mass points as well as a virtually completed relativistic thermodynamics, where Planck referred to p as the "momentum."

Referring to the many possible definitions of force, Einstein in 1905 wrote that "in comparing different theories of the motion of the electron we must proceed very cautiously." What other "theories of the electron" could Einstein have had in mind? Most likely he was not referring directly to Abraham's theory because Abraham's prediction for the transverse mass could not be expressed as m_0^e multiplied by a power of γ; thus, it did not fit into the class of electron theories to which Einstein addressed himself. We can conjecture that by different electron theories Einstein meant every electron theory obtainable from the various definitions of force in Eqs. (12.18)–(12.20) – for example, the one of Langevin-Bucherer. Einstein's decision not to pursue this point could be construed as further evidence for his considering the predictions for m_L and m_T as merely one more result of this theory; he may also have wished to point out the arbitrariness of the theory of Langevin and Bucherer, for he could have come across Bucherer's 1904 book or his (1905) paper. But Einstein's ignoring the difference between his prediction for m_T and Lorentz's is a mystery, because it was not necessary for him to study Lorentz's 1904 paper in order to know Lorentz's prediction for this quantity. Abraham's (1904a) mentioned it; Lorentz's (1904b) had discussed it as well as his (1899a) (although Einstein may not have read the earlier paper before (1905); Wien's (1904b) in the *Annalen* had discussed it; Cohn (1904a, b) had explored Lorentz's (1904c) theory in some detail.

Perhaps Einstein simply wished to avoid mentioning Lorentz's m_T because he eschewed conjectures on the constitution of matter. In the next passage [§10, 40–43] Einstein stated this preference clearly:

> We remark that these results as to the mass are also valid for ponderable material points, because a ponderable material point can be made into an electron (in our sense of the word) by the addition of an *arbitrarily small* electric charge.

By "an electron (in our sense of the word)" Einstein could only have meant accepting the electron as a primitive body whose trajectory was describable as the motion of its center of mass; therefore, for the purpose of the relativity theory, electrons could be taken, as Einstein wrote, as "ponderable material points." Then, when a macroscopic ponderable body was charged with "an *arbitrarily small* electric charge," it became inbued with the properties of electrons, each of charge density ρ. Hence this passage proposed a bold, unifying generalization – for all matter, charged or uncharged, microscopic or macroscopic – of Einstein's prediction that the mass of the point electron was not a constant, as Newtonian mechanics predicted, but rather a velocity-dependent quantity. With this simple sentence, Einstein built another bridge between what many physicists at the time saw as two very different fields – ordinary mechanics and electrodynamics. It was a typically Einsteinian way of seeing unities.[4]

12.2.2. The Moving Electron's Kinetic Energy

Einstein continued [§10, 44–56] with a pregnant passage that begins with a rather low-key, commonplace statement: "Next we determine the electron's kinetic energy," and goes on to what Lorentz has called (1910–1912) "one of the most remarkable conclusions of the theory of relativity" – the equivalence of mass and energy.

Assuming that the electron was accelerated slowly from rest to a velocity v by an electrostatic field, Einstein neglected energy losses due to radiation and attributed the change in the electron's kinetic energy to the work done by the electrostatic field; his result was

$$K = m_0 c^2 \left[\frac{1}{\sqrt{1 - v^2/c^2}} - 1 \right]. \tag{12.33}$$

Einstein wrote that when "$v = c$, [K] becomes infinite"; I add that when $v > c$, γ became imaginary; both these results were unphysical. Consequently, continued Einstein, "velocities greater than that of light have – as in our previous results – no possibility of existence," that is, of physical existence.

One of Lorentz's (1904c) assumptions was that the velocity of material bodies could not exceed c, and this precluded his electron from becoming a hyperboloid (for $v = c$) or its dimensions from becoming mathematically imaginary (for $v > c$). On the other hand, Einstein's point of view excluded the domain of velocities $v \geq c$ on strictly kinematical grounds that were *deducible* from the two axioms of relativity theory. As Einstein wrote (1923), the special relativity theory leads "in particular to a modification of the Newtonian point motion law in which the velocity of light in vacuum is considered the limiting velocity."

Einstein, by proposing another bold conjecture, provided a basis for the previous one (that the mass of all bodies behaves like the electron's mass):

> This expression for the kinetic energy must also, by virtue of the argument stated above, apply to ponderable masses as well.

Thus the kinetic energy of all bodies behaved in the manner specified in Eq. (12.33). A variant of this equation became known in the more familiar form

$$E_T = m_0 c^2 / \sqrt{1 - v^2/c^2} \tag{12.34}$$

where E_T was the body's total energy measured in K; its meaning was that mass and energy were equivalent. The calculation of the electron's kinetic energy was simple. The result was earthshaking.

Einstein did not make the observation, that Eq. (12.33) was consistent with the Newtonian notion of kinetic energy as energy due to motion because then $v = 0$, $K = 0$, and for $v \ll c$, $K = \frac{1}{2}mv^2$. In unpublished calculations before 1905, Kaufmann had also defined the electron's kinetic energy as the difference between the electron's total and rest energies – see Chapter 1, footnote 45. It seems, therefore, that this definition of kinetic energy for systems moving with velocities approaching c was in use before the appearance of Einstein's Eq. (12.33). The only published statement of Kaufmann's definition of kinetic energy was in Abraham's (1905), which Einstein probably had not read before submitting the relativity paper for publication.

12.3. THE CONCLUSION OF THE RELATIVITY PAPER

Now, almost as an afterthought at the conclusion of his 31-page paper, Einstein presented three predictions that followed from his theory for the motion of an electron in an electromagnetic field, Eqs. (12.15)–(12.17) [§10, 57–81].

(1) From Eq. (12.16), choosing E_y and B_z so that their ratio was v/c yielded a result that followed from theories based on Lorentz's electromagnetic theory (except for Bucherer's which was not published until 1906), and so it is unclear what Einstein meant by the "magnetic deflection A_m" and the "electric deflection A_e," whose ratio is

$$A_m / A_e = v/c. \tag{12.35}$$

So far as I know, the quantities A_m and A_e were invented by Einstein, and further discussion of them is deferred until after presentation of his other two experimental tests.

(2) "From the derivation for the electron's kinetic energy, it follows that between the potential difference [P] traversed and the acquired velocity v of the electron there must be the relationship

$$P = \int X \, dx = \frac{m}{\varepsilon} c^2 \left\{ \frac{1}{\sqrt{1 - v^2/c^2}} - 1 \right\}.\text{"}$$

(3) "We calculate the radius of curvature R of the path of the electron when a magnetic force N is present (the only deflective force), acting perpendicularly to the electron's velocity. From the second of the equations (A) we obtain

$$-\frac{d^2 y}{dt^2} = \frac{v^2}{R} = \frac{\varepsilon}{m} \frac{v}{c} N \sqrt{1 - v^2/c^2}$$

or

$$R = \frac{mc^2}{\varepsilon} \cdot \frac{v/c}{\sqrt{1 - v^2/c^2}} \cdot \frac{1}{N}. \text{"}$$

As Einstein specified, the details for the second test were the same as the ones that led to Eq. (12.33); those for the third test were straightforward, predicting that an electron injected normally into a constant magnetic field executed circular motion with a radius that was a function of its velocity.

For Einstein, these three relationships ($A_m/A_e = v/c$, together with those for potential difference and radius) followed from the relativity of simultaneity ("by the theory here advanced") and constituted "a complete expression for the laws according to which ... the electron must move." To most of the readers in 1905 who were steeped in the absolute primacy of experiment over theory, this statement may well have sounded like a haughty command. What was expected at the time was either a frank bow (such as: if the electron was found to move thus, then the theory was vindicated), or reference to existing experimental data that seemed to bear out the theory (in the manner of the end of Lorentz's (1904c) paper, or Abraham's papers [(1902), (1903)]. In fact, Einstein's choice of experimental tests is, to say the least, baffling, because he was probably aware of the methods and results of Kaufmann's (1901–1903) experiments on injecting electrons normally into parallel electric and magnetic fields. Consequently Einstein's predictions in the second and third tests could not be carried out in Kaufmann's laboratory, and the ratio A_m/A_e seemed to be indigenous to any theory based on Lorentz's force equation. I believe that the clue to Einstein's proposal of these three tests is that he was searching out "laws according to which ... the electron must move," and not merely specialized predictions like those for the electron's mass. For Einstein had convinced himself that the Maxwell-Lorentz electrodynamics was not sufficient for investigating the electron's structure and that theories so based harbored weaknesses such as the ambiguous definition of force. Thus, in seeking to assess the laws of nature, Einstein avoided proposing a specific test for his predictions of m_L and m_T.

As to the question—why did Einstein not compare his theory with Kaufmann's available data?—I submit that the omission was intentional because Einstein was aware that his 1905 prediction for the electron's transverse mass disagreed with those data. If we assume that Einstein had read Kaufmann's (1902b) or (1903) before he wrote the relativity paper, then Einstein could have reached this conclusion as follows. Einstein could have substituted into Eq. (1.156) the value of $\psi(\beta)$ from his prediction for m_T, i.e., $\psi(\beta) = \frac{4}{3}(1 - \beta^2)^{-1}$ and then used Eq. (1.155) to eliminate β to obtain

$$z'^2 = y'^2 \left[k_1^2 + \frac{4}{3k_2}y' \right]^{-1}.$$

I cannot imagine Einstein having performed a least squares fit à la Runge to k_1, k_2 and z', taking oberved values for y'. Einstein very probably took the direct approach of using Kaufmann's values for k_1 and k_2 either from Kaufmann's numerical analysis or calculated from the apparatus' dimensions and the external electric and magnetic fields. Then, taking observed values for y' from Plate 19, Einstein could have used his prediction for the curve of z' as a function of y' to calculate the corresponding values of z'. Einstein would have obtained values for z' which were of the order of 13% less than those obtained empirically by Kaufmann. This is not good agreement between theory and experiment.

12.4. AN EXAMPLE OF THE DELICATE INTERPLAY BETWEEN THEORY AND EXPERIMENT

12.4.1. The Responses of Lorentz and Poincaré to Kaufmann

> Unfortunately my hypothesis of the flattening of electrons is in contradiction with Kaufmann's results, and I must abandon it. I am, therefore, at the end of my Latin.
>
> H. A. Lorentz to H. Poincaré, 8 March 1906

By mid-1906 a revealing episode in the history of relativity was in progress. Just how devastated Lorentz was by Kaufmann's results is clear from his unpublished letter to Poincaré, dated 8 March 1906 (Fig. 12.1). After congratulating Poincaré for the essay "On the Dynamics of the Electron," Lorentz wrote that, nevertheless all their work may have been for nothing:

> Unfortunately my hypothesis of the flattening of electrons is in con-tradiction with Kaufmann's new results, and I must abandon it. I am, therefore, at the end of my Latin. It seems to me impossible to establish a theory that demands the complete absence of an influence of translation on the phenomena of electricity and optics. I would be very happy if you would succeed in clarifying the difficulties which arise again.

What a remarkable confession, and what a clear-cut example of falsification. After all those years of work Lorentz was willing to abandon his theory because of the report of a single experiment. Lorentz repeated this message in his lectures at Columbia University, March-April, 1906 that appeared in 1909 as *The Theory of Electrons*: "But, so far as we can judge at present, the facts are against our hypothesis." For his part, Poincaré, too, was strongly affected by Kaufmann's new results; but his reaction was not as radical as Lorentz's. In the introductory paragraphs to the 1905 version of "On the Dynamics of the Electron," Poincaré had written that the theories of Abraham, Langevin and Lorentz "agreed with Kaufmann's experiments." Of course he meant those data prior to 1905. But in the last paragraph of the Introduction to the 1906 version, Poincaré wrote, concerning Lorentz's theory, that at "this moment the entire theory may well be threatened" by Kaufmann's new 1905 data.[5] Poincaré was still willing for the sake of argument to consider the principle of relativity for the moment as valid in order "to see what consequences follow from it." He called for someone to repeat Kaufmann's experiment (1908a, b). Whereas for Lorentz, Kaufmann's results threatened a theory, for Poincaré, a philosophic view was also at stake, one which emphasized a principle of relative motion.

12.4.2. Bestelmeyer's 1906 Data

Circa 1906 a widely quoted measurement of the charge-to-mass ratio of high-velocity cathode rays was that of August Becker (1905). At Lenard's laboratory in Kiel, Becker employed apparatus similar to that used in 1898–1899 by Kaufmann and Simon. Becker accelerated cathode rays to velocities of the order of $c/3$. Using Abraham's expression for m_T for separating out the velocity dependence of ε/m_T, Becker claimed to have established "definitively" that to the first decimal place $\varepsilon/\mu_0^e = 1.8 \times 10^7$.

In 1906 the Göttingen experimentalist Adolf Bestelmeyer decided to show conclusively that the velocity of the cathode rays produced when Röntgen rays struck a heavy metal target was independent of the intensity of the Röntgen rays. An important by-product of this experiment was a determination of the charge-to-mass ratio of fast cathode rays, under what Bestelmeyer considered to have been "very pure conditions," and his "experiment yielded a noticably smaller value" for ε/μ_0^e than Becker's. Bestelmeyer (see Fig. 12.2) allowed Röntgen rays to enter his apparatus through the aluminium window A, and strike a Platinum plate labelled Pt. An inhomogeneous beam of cathode rays was generated with velocities of the same order as the particles that gave rise to the Röntgen rays.

In order to select electrons with a definite velocity, Bestelmeyer returned to Thomson's 1897 method of crossed electric and magnetic fields for determining the charge-to-mass ratio of cathode rays. Bestelmeyer's magnetic field was perpendicular to an electric field whose source was the parallel capacitor plates C of length 6.68 cm and separation 0.058 cm. The magnetic field's source was a solenoid, which ensured a higher degree of homogeneity than Kaufmann's electromagnets (see Fig. 1.7). These crossed fields served as a velocity filter that limited the electrons emerging from the diaphragm B_2 to those that had traveled a straight-line trajectory through the capacitor. Thus, the electrons that emerged from B_2 had been acted on by no net force under the influence of the combined electric and magnetic fields, and so their common velocity was

$$v/c = E/B. \tag{12.36}$$

In the region between the diaphragm B_2 and the photographic plate S (6.65 cm), the electrons of velocity v moved in a circular arc in the x-y-plane of radius

Leiden, le 8 mars 1906.

Monsieur et très honoré Collègue,

C'est déjà trop longtemps que j'ai négligé de vous remer- cier de l'important mémoire sur la dynamique de l'électron que vous avez bien voulu m'en- voyer. Inutile de vous dire que je l'ai étudié avec le plus grand intérêt et que j'ai été très heureux de voir mes conclusions confirmées par vos considérations. Malheureuse-

[Handwritten letter in French:]

ment mon hypothèse de l'aplatisse-
ment des électrons est en contradiction avec les résultats des nou-
velles expériences de M. Kaufmann et je crois être obligé de
l'abandonner; je suis donc au
bout de mon latin et il ne
semble impossible dès d'établir
une théorie qui exige l'absence
complète d'une influence de la
translation sur les phénomènes
électromagnétiques et optiques.
Je serais très heureux si vous
arriviez à éclaircir les difficul-
tés qui surgissent de nouveau.
Veuillez agréer, cher collègue,
l'expression de mes sentiments
sincèrement dévoués.

                    H. A. Lorentz

FIG. 12.1. H. A. Lorentz's letter of 8 March 1906 to H. Poincaré. Reproduced with permission from the Estate of Henri Poincaré.

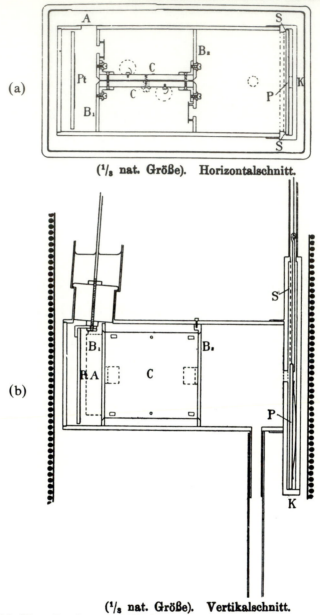

(a)

(¹/₈ nat. Größe).   Horizontalschnitt.

(b)

(¹/₈ nat. Größe).   Vertikalschnitt.

FIG. 12.2. [From Bestelmeyer's (1907).] In (a) is a horizontal cross-section of Bestelmeyer's apparatus drawn to $\frac{1}{3}$ the actual size. The capacitor plates C are 6.68 cm in length. The distance from the diaphragm B_2 to the photographic plate P is 6.65 cm. The coordinate axes are to be taken

as $\overset{y}{\underset{z}{\rule{0.4pt}{18pt}}\!\!-\!\!x}$. The crossed electric and magnetic fields are $\boldsymbol{E} = E\hat{j}$ and $\boldsymbol{B} = B\hat{k}$. In (b) is a vertical cross-section that shows the windings of the solenoid which are designated as solid circles. (From *Ann. Phys.*)

$$\rho = \frac{l^2}{2y}\left(1 + \frac{y^2}{l^2}\right)^{-1} \qquad (12.37)$$

where $l = 6.65$ cm is the distance over which the electron is acted on by only the magnetic field before striking the photographic plate at the coordinate y. From Eqs. (12.36), (12.37) and the Lorentz force law, Bestelmeyer's expression for ε/m_T was

$$\frac{\varepsilon}{m_T} = \frac{2yEc}{l^2 B^2}\left(1 + \frac{y^2}{l^2}\right)^{-1}, \qquad (12.38)$$

and he did not assume small deflections, i.e., the largest deflection was $y = 2.274$ cm compared with $l = 6.68$ cm. Had he assumed small deflections he would have dropped the term y^2/l^2 in the bracket in Eq. (12.38).

The smaller gap between the capacitor plates in Bestelmeyer's experiment (0.058 cm compared with Kaufmann's $\delta = 0.1243$ cm) led to the following advantages over Kaufmann's experiment: (1) an increased accuracy in the uniformity of the electric field; (2) a smaller plate voltage was required (an average of 200V compared to Kaufmann's 2500V). These two advantages, in conjunction with Bestelmeyer's improved vacuum system, decreased the probability of sparking. Another advantage of the crossed-fields arrangement, emphasized Bestelmeyer, was that the apparatus' dimensions (i.e., apparatus constants) entered into only the empirical determinations of v and ε/μ_0^e, but not $\psi(\beta)$. I have already mentioned the advantage of using a solenoidal source for the magnetic field.

A possible major source of error in Bestelmeyer's experiment was that electrons with $v/c \lesssim E/B$ could have the proper trajectories to pass through the capacitor plates and reach the photographic plate. He proved that these electrons arrived at the photographic plate with a velocity within 2% of Ec/B if the length of the capacitor plates was nearly equal to the distance B_2, which was the case in his experiment.

Nr.	β	ε/μ beob.	ε/μ berechnet nach			Differenzen beob.-ber.		
			A.	L.	B.	A.	L.	B.
	0		1,720	1,733	1,713			
3	0,195	1,697	1,694	1,700	1,690	+ 3	− 3	+ 7
1	0,247	1,678	1,678	1,679	1,677	0	− 1	+ 1
4	0,322	1,643	1,647	1,640	1,651	− 4	+ 3	− 8

FIG. 12.3. Bestelmeyer's data from his three best runs, where Nr. = number of run, *beob.* = observed, *berechnet nach* = calculated from and *Differenzen beob.-ber.* = differences between observed and calculated, A. = Abraham, L. = Lorentz, and B. = Bucherer. (From *Ann. Phys.*)

An exposure time of 90 minutes for each value of v resulted in a straight line on the photographic plate whose displacement below the null point could be measured with a micrometer. Thus Bestelmeyer avoided Kaufmann's complex data analysis that depended on a fuzzy photographic curve of complex shape which required 48 hours of exposure. The results from Bestelmeyer's three best runs are in Fig. 12.3. Bestelmeyer determined ε/m_T by substituting empirical values for y into Eq. (12.38). Then he wrote the quantity ε/m_T as

$$\varepsilon/m_T = \varepsilon/\mu_0^e \psi(\beta) \tag{12.39}$$

where the quantities $\psi(\beta)$ for the theories of Abraham, Bucherer and Lorentz were in Eqs. (7.21)–(7.23). Using empirical values for β and ε/m_T, Bestelmeyer calculated ε/μ_0^e from Eq. (12.39). His results are in Fig. 12.3.

The quantities in the row labelled $\beta = 0$ were the averages of Bestelmeyer's calculations of ε/μ_0^e for each theory using empirical quantities from his three best runs. From the runs 3, 1, and 4 we see that Abraham's and Lorentz's theories agreed best with these data. In Bestelmeyer's opinion the average of ε/μ_0^e for the three electron theories, i.e.,

$$\varepsilon/\mu_0^e = 1.72 \times 10^7 \tag{12.40}$$

was a good approximation to the charge-to-mass ratio for the electron. As we see from Fig. 12.3, owing to the low velocity of the cathode rays, Bestelmeyer could not offer his experiment as deciding between the theories of Abraham, Bucherer and Lorentz. Nevertheless, Bestelmeyer emphasized that the sharp lack of agreement with Kaufmann in the range of low velocities demonstrated the necessity for further experimental investigations on the mass of moving electrons.

Bestelmeyer's low values obtained for ε/μ_0^e compared to the one used by Kaufmann favored the Lorentz-Einstein theory, because it was 7.8% less than Simon's and 8.5% less than Kaufmann's corrected value of 1.878×10^7. Bestelmeyer's comment that his experiment decreased the probability for sparking, led Planck (1907b) to return to his earlier criticism of 1906, namely, that since one point of Kaufmann's reduced curve led to $\beta > 1$, then the theoretical interpretation of the measured y_0 and z_0 was unclear. Planck used the special form of his (1906b) equations for (y_0, z_0), and Bestelmeyer's ε/μ_0^e, to calculate the electric field from Kaufmann's observed coordinates. He found that, on the average, the sphere theory and the *Relativtheorie* deviated from Kaufmann's assumed constant value for the electric field by 8% and 11%, respectively; and that the differences were velocity dependent, decreasing with decreasing electron velocity. Furthermore, Planck's comparison of the differences between the largest and smallest values for the calculated electric fields from their assumed constant value, revealed deviations of 5% and 2% for the sphere theory and the *Relativtheorie*, respectively. "This circumstance," wrote Planck, "weighs strongly for the *Relativtheorie*, naturally without being decisive." Planck concluded his (1907b) by suggesting ionization of the gas residue in Kaufmann's evacuated apparatus, i.e., sparking, to be the cause of the electric field's variation. He called for new measurements on β-rays. In a postcard dated November, 1907 Planck notified Einstein that he had sent to Einstein under separate cover copies of his papers (1906b) and (1907b). Planck added that, in a soon to be published paper, Kaufmann had shown that his "electric field comes extraordinarily close to homogeneity." In fact, in the strongest of terms Kaufmann (1907) contended that an electric field variation of 8% in his experiment was impossible, and that any small deviations could not alter the fact that his data weighed decisively against the "Lorentz-Einstein relativity theory."

12.4.3. Einstein Replies to Kaufmann

> With admirable care Mr. Kaufmann has ascertained the relation between A_m and A_e for β-rays emitted from a radium-bromide source.... The theories of the electron's motion of Abraham and Bucherer [agree better with Kaufmann's data] than the relativity theory. In my opinion both theories have a rather small probability....
>
> A. Einstein (1907e)

Einstein's first response to Kaufmann was to ignore him. In a short 1906 *Annalen* note (received 6 August 1906), Einstein (1906d) mentioned Kaufmann's recent investigations with no reference and with no comment on Kaufmann's disconfirming data. Instead Einstein proposed an experiment using cathode rays for deciding between the theories of Abraham, Bucherer and "Lorentz and Einstein." Einstein noted that there were three quantities in cathode-ray experiments that could be measured precisely – the accelerating potential and the electric and magnetic deflections, and it was possible to relate any two of these quantities. For example, continued Einstein, "one of these connections was studied for β-rays by Mr. Kaufmann, namely, the relation between the magnetic and electrostatic deflections." Thus, by August of 1906 Einstein was aware of Kaufmann's complex data analysis that required determining v and ε/m_T from curve coordinates. Einstein suggested performing nothing less than a new class of experiment which altogether eliminated measuring the electron's velocity. Einstein proposed to relate the accelerating potential and the electrostatic deflection "or, equivalently, the ratio of the transverse to longitudinal masses of the electron as a function of the accelerating potential." Einstein's description of his ingenious experiment is in Fig. 12.4, and is, as he wrote, just a "*schematische Skizze*" for he did not even indicate that the apparatus should be in an evacuated enclosure.

If P was the potential difference that caused the shadow of the wire D to shift, then so was P the potential difference that caused a group of electrons to have a kinetic energy $eP = \frac{1}{2} m_L v^2$, where P was a linear function of P_1 and P_2 of the form $P = P_2 - \alpha(P_2 - P_1)$ and α was a constant that depended on the

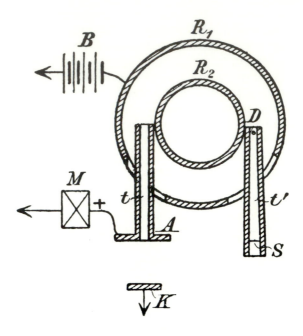

FIG. 12.4. Einstein's schematic from his (1906d). The cathode rays were accelerated from the cathode K to the anode A by an accelerating potential P_2. M was the current source whose positive wire was attached by a clamp to A and whose negative wire was earthed. A diaphragm in the anode A admitted electrons through the tube t into the region between the cylindrical capacitor with plates R_1 and R_2. R_2 was charged positively by M to the potential P_2. R_1 was charged negatively to a potential P_1 by means of the battery B whose other wire was earthed. Electrons moving sufficiently slowly traversed a circular path from the tube t to the tube t', and caused the shadow of the wire D to be cast on the phosphorescent screen S. Increasing P_2 caused the shadow to shift, whereupon the battery was adjusted so that the shadow moved back to its original position. (From *Ann. Phys.* by permission of the Estate of Albert Einstein.)

apparatus' dimensions and was small compared to one. In the circular orbit of radius ρ these electrons were acted on by a force $eE = m_T v^2/\rho$, where E was the radial electric field directed from R_2 to R_1. Consequently, with this experimental arrangement the ratio of m_T to m_L is independent of the electron's velocity and is $m_T/m_L = \rho E/2P$. Since P was directly proportional to $P_2 - P_1$, then neglecting terms of higher than the first power in α and P_1/P_2 Einstein obtained $m_T/m_L = \text{constant } (1 - (1 + \alpha)P_1/P_2)$. Einstein claimed that the apparatus could reveal a deviation of m_T/m_L from unity to 0.3% accuracy. Perhaps having in mind Kaufmann's elaborate procedures for determining the electric field's degree of homogeneity, Einstein emphasized that the "inevitable fluctuations" of P_2 could have only a negligible effect on the experiment's accuracy. The theoretical predictions to be compared with the empirical data Einstein listed as follows (to order β^2): $m_T/m_L = 1 - 0.0070 \ P/10,000$ ("theory of Bucherer"); $m_T/m_L = 1 - 0.0084 \ P/10,000$ ("theory of Abraham"); $m_T/m_L = 1 - 0.0104 \ P/10,000$ ("theory of Lorentz and Einstein"). (In Einstein's notation the potential P is in volts and, for example, to order β^2, for the theory of Lorentz and Einstein $m_T/m_L = 1 - \beta^2/4$, where $\beta^2 = 2eP/m_Lc^2 \simeq 2eP/\mu_0^e c^2$; Einstein used Kaufmann's value for $\varepsilon/\mu_0^e = 1.878 \times 10^7$ which he converted to mks units.)

Einstein concluded the 1906 paper by writing that, since he was not in a position to carry out the experiment, it would "please" him if a "physicist" became interested in it. As far as I know, circa 1906, no physicist did. (In an article tainted with the politics of the time, Starke (1934) declared Einstein's 1906 experimental suggestion to be "unfeasible.")

Einstein concluded the introductory portion to his (1907d) by offering a "word of thanks" to Planck and Kaufmann who, as a "circumstantial examination of the literature" revealed to him, had been the first to mention "my first work on the principle of relativity." The reference to Kaufmann could only have been to the lengthy 1906 *Annalen* paper which, although professing empirical disproof of the principle of relativity, contained a good survey of Einstein's views on space and time, in addition to referring to a Lorentz-Einstein theory. Once again Einstein ignored Kaufmann's conclusions that had devastated Lorentz and caused Poincaré to frame his support of the principle of relativity in somewhat cautious terms. Einstein's appraisal of Planck's and Kaufmann's papers was nicely put. For although they both cited Einstein's first work on a principle of relativity, they interpreted it only as a useful contribution to electrodynamics.

The Stark-Einstein correspondence of 1907 (see Section 1.15.1) is clear-cut evidence that by the fall of 1907 Einstein was aware of both Planck's (1907b) criticism of Kaufmann's and Bestelmeyer's results. Einstein's decision to have ignored Kaufmann's disconfirming data in his (1906d) and (1907d) publications could be attributed to Einstein's intuition that Kaufmann was wrong. But Patent Clerks have not the license to publish speculations of that sort. When Einstein wrote the (1907e), however, the scientific climate had changed with the appearance of Bestelmeyer's data, and then came Planck's (1907b). In short, the time was ripe for Einstein to indicate in print his own opinion of Kaufmann's data, preceded, of course, with a description of the empirical situation.

Einstein (1907e) again proposed the second test from the 1905 relativity paper, and then went on to define explicitly quantities that he referred to as A_m and A_e. If an electron with a velocity $v = v\hat{i}$ were injected into crossed electric and magnetic fields $E = E\hat{k}$ and $B = B\hat{j}$, respectively, then from Eq. (12.28)

$$-\frac{d^2z}{dt^2} = \frac{e}{m_0\gamma}\left(E + \frac{v}{c}B\right). \tag{12.41}$$

The resultant electron orbit in the x-z-plane, continued Einstein, had a curvature R where $d^2z/dt^2 = v^2/R$. Then Eq. (12.41) became

$$-\frac{1}{R} = \frac{e}{m_0\gamma v^2}E + \frac{e}{m_0\gamma vc}B; \tag{12.42}$$

Einstein defined A_e and A_m as the coefficients of E and B, respectively:

$$A_e = e/m_0\gamma v^2, \tag{12.43}$$

$$A_m = e/m_0\gamma vc. \tag{12.44}$$

These definitions of A_e and A_m lead me to conjecture that Einstein had them in mind when he wrote the 1905 relativity paper because his first experimental test in (1905d) concerned the deflection of an electron in crossed electric and magnetic fields. However, bearing in mind the ambiguity in the definition of force, and perhaps also the lack of agreement with empirical data of his prediction for m_T, in (1905d) Einstein offered only the ratio of A_e and A_m as an experimental test.

The important point in Einstein's (1907e) was not that the ratio $A_m/A_e = v/c$, but that these factors were dependent on the different theories of the electron's mass; for example, the quantity $m_0\gamma$ in Eqs. (12.43) and (12.44) for the Einstein-Lorentz theory could be replaced with Abraham's prediction for m_T.

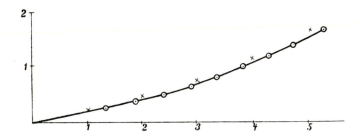

FIG. 12.5. Einstein's (1907e) reproduction of Kaufmann's reduced curve (see Fig. 7.1). Instead of Kaufmann's reduced coordinates y' and z', Einstein calibrated the ordinate and abscissa with A_e and A_m, respectively, which were directly proportional to y' and z'. The marking ⊙ designates Kaufmann's points, and × the predictions for the "relativity theory." (From *Jahr. Radioakt.* by permission of the Estate of Albert Einstein.)

Einstein then turned to Kaufmann's new data from late 1905, those same data that, in 1906, had driven Lorentz to the "end of [his] Latin." Since Einstein had assumed infinitely small deflections, he could apply his analysis to Kaufmann's reduced coordinates even though Kaufmann's electric and magnetic fields were parallel to each other. Einstein's (1907e) has a version of Kaufmann's empirical reduced curve with only the "relativity theory" predictions (here Einstein refrained from using the name "Lorentz-Einstein theory")—see Fig. 12.5. Einstein emphasized that the z' and y'-axes (see Kaufmann's Fig. 7.1) "demonstrated the relation between A_m and A_e." The reason is that Einstein's A_m and A_e are the coefficients of $\mathbb{B}$ and $\mathbb{E}$ in Kaufmann's Eqs. (7.15) and (7.14). In the face of Kaufmann's assessment of the Lorentz-Einstein theory, Einstein's comment on Fig. 12.5 was: "Considering the difficulty of the experiment, we may be inclined to regard the agreement as sufficient." Perhaps to support his opinion Einstein omitted from his figure the more than sufficient agreement with the data of the theories of Abraham and Bucherer. Einstein next referred to Planck's (1906b) and (1907b) analyses of Kaufmann's calculations and mentioned only that Planck's calculations "agree completely with those of Mr. Kaufmann." In a footnote, however, Einstein pointed out that Kaufmann's reduced curve was "not the same as the observed one." This could be interpreted as a caveat that was prompted by Planck's (1906b) calculation of $\beta > 1$ which at the time did not strike Planck as important. But Bestelmeyer's data led Planck to return to this difficulty with Kaufmann's reduced curve.

Einstein then acutely emphasized a facet of Kaufmann's reduced curve that had escaped almost everyone else, namely, that the "systematic deviation" between the relativity theory and Kaufmann's data could indicate a hitherto "unnoticed source of error." Then, after agreeing with Planck's 1906–1907 conclusion that further data were necessary to decide between the various electron theories, Einstein gave vent to his intuition:

> In my opinion both theories have a rather small probability, because their fundamental assumptions concerning the mass of moving electrons are not explainable in terms of theoretical systems which embrace a greater complex of phenomena.

Einstein's intuition served him well here because it turned out that Kaufmann's data were incorrect.

12.4.4. Bucherer's 1908 Data

> ...by means of careful experiments I have elevated the principle of relativity beyond any doubt.
>
> A. H. Bucherer to A. Einstein, 7 September 1908

Stark (1908) noted the possibility of Kaufmann's accuracy having been further reduced owing to contamination of the empirical curve from electrons that were deflected from the capacitor plates, i.e., secondary electrons. Kaufmann (1908) offered a proof that errors of this sort were not large enough to bring the relativity theory into agreement with his data. He did, however, admit the possibility of "unknown sources of error." Kaufmann concluded by calling for "new experiments on β-rays."

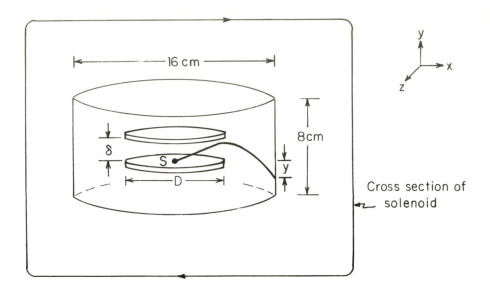

FIG. 12.6. Bucherer's experimental apparatus for deciding between the electron theories of Abraham and Lorentz. The parallel-plate capacitor is made of two discs of diameter $D = 8$ cm, and separated by a distance $\delta = 0.25$ mm. Bucherer placed the capacitor in a cylindrical evacuated container, of height 8 cm and diameter 16 cm, which is situated in the center of a 103 cm long solenoid. The electric and magnetic fields are normal to each other, i.e., $\boldsymbol{E} = E\hat{\jmath}$ and $\boldsymbol{B} = -B\hat{k}$. Photographic film is pressed against the inner walls of the cylinder. The source S for β-rays is a sphere of radium fluoride with a diameter of 0.5 mm, that is placed in the center of one of the capacitor's plates. Without the electric and magnetic fields the electrons that are emitted from the radium fluoride travel in a straight line and strike the photographic film to form a circle. When the film is unrolled the circle becomes a straight line (see Fig. 12.9). Bucherer adjusted the electric and magnetic fields so that certain electrons maintained their linear trajectory between the plates. These electrons have a velocity $v = cE/B \sin\theta$. After emerging from between the capacitor plates, the electrons execute approximately circular motion of radius ρ under the influence of the magnetic field $\boldsymbol{B}$, and strike the photographic film at the distance y below the lower plate of the capacitor.

Kaufmann did not have long to wait because a colleague in Bonn, Alfred Bucherer, took up the challenge with apparatus that was based on the "Thomson-Bestelmeyer principle . . . of the crossed fields" (see Figs. 12.6, 12.7, 12.8 and 12.9). Bucherer wanted the electrons to enter the combined field region at all angles θ, measured relative to the magnetic field direction, in order to test also his own version of the force exerted on an electron in a magnetic field. When Bucherer unwound the photographic film the angle θ became the abscissa. According to the Maxwell-Lorentz theory only those electrons emerge from the crossed fields that have the velocity

$$v = cE/B \sin\theta. \tag{12.45}$$

Bucherer's principle of relativity predicted

FIG. 12.7. In a is Bucherer's parallel-plate capacitor. In b is the cylinder into which he placed the capacitor. In c is the guard ring that Bucherer employed to reduce any extraneous deviations of the electrons owing to the fringing of the capacitor's electric field. (From *Ann. Phys.*)

$$v = \frac{cE}{B}\frac{(1 - \beta^2 \cos^2\theta)}{\sin\theta} \tag{12.46}$$

(see Chapter 7, footnote 10). If in Eq. (12.45), $E/B = \frac{1}{2}$ and $\sin\theta = \frac{1}{2}$, then electrons moving at angles relative to the magnetic field of $\theta = 30°$ and $150°$ were undeflected. Similarly, if in Eq. (12.46), $E/B = \frac{1}{2}$ then electrons moving at $\theta = 0°$ suffered no deflection. Thus, the Maxwell-Lorentz and Bucherer theories predicted that electrons moving at certain angles relative to magnetic field were undeflected, and on the photographic film their trace should overlap the one that occurred when no fields were in operation. Bucherer's data verified Eq. (12.45).

From Lorentz's force law, Eqs. (12.37) and (12.45), Bucherer obtained

$$\frac{\varepsilon}{m_T} = \frac{2yEc}{l^2 B^2 \sin^2\theta}\left(1 + \frac{y^2}{l^2}\right)^{-1} \tag{12.47}$$

where $l = 4\,\text{cm}$, and to simplify the data analysis Bucherer took $\theta = 90°$ so that y is the maximal deflection. Bucherer's exposures for the range $0.7 \leqslant \beta < 1$ took four hours, and for $0.32 \leqslant \beta < 1$ took sixty hours. Like Bestelmeyer's empirical curve, Bucherer's was a straight line; consequently, Bucherer could determine the deflections with great accuracy. (Depending on the magnetic field—average value 115 Gauss—the deflections were in the range 6 mm $< y <$ 16.4 mm.) Instead of using his data to calculate a value for ε/μ_0^e Bucherer sought the electron theory that yielded a constant ε/μ_0^e over the range of velocities in his runs. Bucherer had *a priori* eliminated his own theory of the electron because of its lack of agreement with the Rayleigh- and Brace experiments. Bucherer's data are shown in Fig. 12.10. These data show best agreement for Lorentz's theory, and a value for ε/μ_0^e in the vicinity of Bestelmeyer's.

FIG. 12.8. An example of how experimental elementary-particle physics was done in 1908. On the table is Bucherer's 103 cm length solenoid which contained the cylinder shown in Figs. 12.6 and 12.7b. (From *Ann. Phys.*)

In a letter of 7 September 1908, A. H. Bucherer in Bonn wrote Einstein in Bern of his recent experimental results concerning the mass of high-velocity electrons — "... by means of careful experiments, I have elevated the validity of the principle of relativity beyond any doubt." The principle of relativity for the electromagnetic world-picture, that is, for when on 22 September 1908 Bucherer presented his results to the *Naturforscherversammlung* at Cologne, he entitled his communication, "Measurements on Becquerel-rays. The Experimental Confirmation of the Lorentz-Einstein Theory."[6]

FIG. 12.9. One of Bucherer's photographic plates. Bucherer obtained the symmetric curves by reversing the directions of the electric and magnetic fields and the straight line results from undeflected gamma radiation. The Lorentz force law gives the maximum deflection at $\theta = 90°$. (From *Ann. Phys.*)

β	ε/μ_0^e Lorentz	ε/μ_0^e Abraham
0.3173	1.752	1.726
0.3787	1.761	1.733
0.4281	1.760	1.723
0.5154	1.763	1.706
0.6870	1.767	1.642

FIG. 12.10. Bucherer's published results.

Bucherer prefaced his response to Wien, who had requested him to compare his results to Kaufmann's, by professing admiration for Kaufmann's "pioneering work" and emphasizing the "difficulty of the measurements on such small curves." Then Bucherer launched into criticisms of Kaufmann in addition to those in his paper and in Bestelmeyer's. In particular, Bucherer had found an asymmetry of 5% on Kaufmann's "very small" curves with which Kaufmann had concurred upon remeasurement. Bucherer left unsaid that this asymmetry was further evidence for the inhomogeneity of Kaufmann's electric field. Neither Kaufmann nor Abraham were present to reply to this and other criticisms. Exit Kaufmann whose innate attraction for the complex turned out to be fatal.

12.4.5. Aftermath

The rigid electron is in my view a monster in relation to Maxwell's equations, whose innermost harmony is the principle of relativity ... the rigid electron is no working hypothesis, but a working hindrance.

Approaching Maxwell's equations with the concept of the rigid electron seems to me the same thing as going to a concert with your ears stopped up with cotton wool. We must admire the courage and the power of the school of the rigid electron which leaps across the widest mathematical hurdles with fabulous hypotheses, with the hope to land safely over there on experimental-physical ground.

H. Minkowski, from the discussion session that followed Bucherer's (1908a).

At the moment of going to press we learn that M. Bucherer had repeated the experiment, surrounding it with new precautions, and that, unlike Kaufmann, he has obtained results confirming Lorentz's views.

H. Poincaré (1908a)

Recent experiments by Bucherer on the electric and magnetic deflexion of β-rays, made after a method that permits greater accuracy than could be reached by Kaufmann, have confirmed the formula [m_T], so that, in all probability, the only objection that could be raised against the hypothesis of the deformable electron and the principle of relativity now has been removed.

H. A. Lorentz (1909)

On the basis of the foregoing discussion it seems fair to say that the Bucherer-Neumann experiments actually proved very little, if anything more than the Kaufmann experiments, which indicated a large qualitative increase of mass with velocity. Indeed, it seems remarkable that [Bucherer and Neumann] were able to obtain lines at all for higher velocities in consideration of the exceedingly poor performance of their velocity filters.

A. T. Zahn and C. H. Spees (1938c)

The first point of view is obvious [to criticize a physical theory]: the theory must not contradict empirical facts. However evident this demand may in the first place appear, its application turns out to be quite delicate.

A. Einstein (1946)

In the discussion session that followed Bucherer's (1908a) presentation, the normally taciturn Minkowski waxed enthusiastic with "joy" over the vindication of "Lorentz's theory." Then, as we see from his statement above, he lambasted Abraham's theory.

"At the moment of going to press," Poincaré added a note to the passage in his (1908a) where he called for others to repeat Kaufmann's measurements, that congratulated Bucherer on "confirming Lorentz's views."[7] Bucherer's results supported Poincaré's dazzling pronouncement from earlier in his (1908a, b), namely that a consequence of the principle of relativity was that there "is no more matter, since the positive electrons have no longer any real mass."[8]

Lorentz, too, added a note to a text about to go to press. To a passage in the 1909 edition of *The Theory of Electrons*, in which he discussed how well

Abraham's theory agreed with Kaufmann's 1905 data, Lorentz appended the Note 87 which stressed that Bucherer had removed the "only objection that could be raised against the hypothesis of the deformable electron . . . ."

Lorentz and Poincaré welcomed Bucherer's results, and did not call for someone to repeat his measurements. Others, such as Planck[9] and Bestelmeyer, called for further tests. Bestelmeyer (1909) criticized Bucherer particularly on how he had accounted for electrons of incorrect velocity that had managed to pass through the velocity filter, i.e., noncompensated electrons. Neumann (1914) was credited with having shown that noncompensated electrons had a negligible effect on Bucherer's data.

Using apparatus similar to Bucherer's, Neumann (1914) was considered by most physicists to have removed any reasonable doubts concerning the validity of the Lorentz-Einstein prediction for m_T. [Lorentz (1910–1912) wrote that a "good supplement" to Neumann's results were those of Guye and Lavanchy (1916) who used cathode rays of $0.2c < v < 0.5c$, where Neumann's β-rays were of $0.5c < v < 0.8c$.] Thus, in the 1915 tome, *Kultur der Gegenwart: Physik* (ed., E. Warburg), that contained summary papers on the state of physics in 1914 by Einstein, Kaufmann, Lorentz, Wiechert and Planck, among others, did not discuss the great β-ray deflection experiments that were deemed to have been decisive for the physics of 1905.[10]

By a curious twist, in 1938, two experimentalists at the University of Michigan had the occasion to reanalyze the class of electron-deflection experiments using velocity filters. Whereas experimenters at the time of Kaufmann, Bucherer and Neumann had no cause to distinguish between the electrons from β-decay, whose momentum spectrum was continuous, and the secondary electrons from internal conversion with discrete momenta, by 1938 the continuous spectrum had opened new vistas in the physics of the nucleus. Zahn and Spees (1938a, b, c) attempted to "distinguish between ordinary Lorentz electrons and the widely differing type of heavy electrons required by the speculation that the well-known beta-ray paradox might be explained by variations, with velocity, of the rest mass of the electrons created in the nucleus, rather than by the neutrino hypothesis" (1938b).

For a fixed position of the Geiger counter (Fig. 12.11) and for a magnetic field of 120.85 Gauss, Zahn and Spees increased the voltage on the capacitor. They found that to within 1.5% accuracy the Geiger counter readings peaked at the voltage expected for electrons with Lorentz's m_T, i.e., $m_0 \gamma$. They concluded "that the assumption of the previously mentioned type of heavy electron is untenable" (1938b). Zahn and Spees next compared the result of their analysis of a velocity filter's resolution with the original papers of Bucherer and Neumann, and concluded that "the general behavior of a condenser in crossed fields seems not to have been understood at the time." For example, the relationship between the apparatus' geometry and its velocity resolution had not been fully explored previous to Zahn and Spees. It turned out that the resolution of the velocity filters used in the early experiments were inadequate

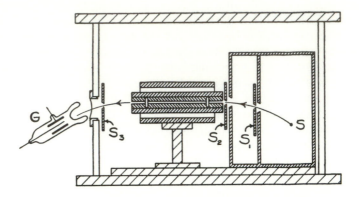

FIG. 12.11. The schematic diagram used by Zahn and Spees (1938b) to describe their apparatus. Beta-rays from a Radium E source S were collimated twice by slits S_1 and S_2 before they passed through crossed electric and magnetic fields. The magnetic field of strength 120.85 Gauss, whose source was a solenoid, was normal to the plane of the paper. They varied the voltage of the parallel plate capacitor (shaded region of length 12 cm and plate separation $\frac{1}{2}$ mm) from zero to 2,000 V. When a β-ray emerged from the velocity selector, its trajectory became circular under the action of only the magnetic field. The Geiger counter G, preceded by a third slit, replaced the photographic plates of early investigators such as Bestelmeyer, Bucherer and Neumann. The use of a Geiger counter greatly increased the accuracy of Zahn and Spees' experiment over the earlier ones. For example, the counter eliminated observational errors from microphotometric measurements, and it permitted them to work with lower intensities of electrons, thereby drastically decreasing errors arising from data due to electrons that had been scattered from the capacitor plates. (From *Phys. Rev.*)

for values of β approaching 0.7, which is the region where the velocity dependence of the electron's mass became important. They concluded that "it seems fair to say that the Bucherer-Neumann experiments proved very little, if anything more than the Kaufmann experiments, which indicated a large qualitative increase of mass with velocity" (1938c).

As Einstein wrote in 1946, the comparison between theory and experiment is indeed "delicate."

12.5. FURTHER DEVELOPMENTS OF EINSTEIN'S VIEW

12.5.1. Einstein's Fourth Annalen Paper of 1905

> However, a consequence of the work on electrodynamics has suddenly occurred to me.
>
> A. Einstein to C. Habicht, late 1905 [Seelig (1954)]

Shortly after completing the relativity paper, Einstein became aware of the far-reaching implications of his result for the kinetic energy of a moving electron; later in 1905 he wrote to Habicht [Seelig (1954)]:

> However, a consequence of the work on electrodynamics has suddenly
> occurred to me, namely, that the principle of relativity in conjunction with
> Maxwell's fundamental equations requires that the mass of a body is a
> direct measure of its energy content – that light transfers mass. An
> appreciable decrease in mass must occur in radium. This thought is both
> amusing and attractive; but whether or not the good Lord laughs at me
> concerning this notion and has led me around by the nose – that I cannot
> know.

Einstein elaborated upon this "amusing and attractive thought" in the
fourth paper that he published in the *Annalen* of 1905 (received 27 September
1905), "*Ist die Trägheit eines Körpers von seinem Energieinhalt abhängig* [Does
the inertia of a body depend upon its energy content]?" The answer to the
question posed in the title was a resounding yes, and for the first time Einstein
stated the equivalence of mass with any sort of energy (e.g., light or mechanical
energy) whose mathematical statement is Eq. (12.34). In the vol. 18 paper,
Einstein derived the mass-energy equivalence in a manner that displayed its
far-reaching consequences and focused on the problem that he thought to be of
basic importance – the nature of radiation. He considered the case of a moving
body that emits a pulse of radiation at an angle ϕ with respect to the x-axis of
K. Using the transformation equation for the energy of a light pulse Eq.
(11.12), Einstein compared the energy of the emitter before and after emission
to obtain the following result, valid to order $(v/c)^2$:[11]

> If a body gives off the energy L in the form of radiation, its mass
> diminishes by L/c^2.

During 1906–1907, Einstein published three derivations of the mass-energy
equivalence from which we can glean that he was very much immersing himself
in the mainstream of the physics of 1905.

12.5.2. Einstein's 1906 Derivation of the Mass-Energy Equivalence

> [Lavoisier's principle] is not to be touched without unsettling all mechanics.
>
> H. Poincaré (1904)

Einstein (1906c) demonstrated that, to first order in v/c, the validity of the
principle of conservation of center of mass motion depended on the theorem of
the "inertia of energy," i.e., the mass-energy equivalence. As in the 1905 vol. 18
paper, Einstein's (1906c) two examples of the mass-energy equivalence were
based on electromagnetic theory in conjunction with the principle of relativity.

Since the law of conservation of center of mass motion was connected with
the law of conservation of momentum, Einstein wrote that his results were "in
principle" contained in Poincaré's (1900b) in the *Lorentz Festschrift*. Einstein

asserted the independence of his results from those of Poincaré who had neither ascribed an actual decrease in mass to an emitter of radiation, nor considered the complete equivalence of mass and energy.

Einstein's (1906c) first example became an often-used *Gedanken* demonstration of the "inertia of energy." [See von Laue (1949), French (1966).] Floating freely in space was a cylinder of mass M and length l, and the cylinder's end faces A and B were of mass $m \ll M$. Einstein analyzed the motion of the system that resulted when radiation was emitted from A and absorbed by B. He wrote that "at least to a first approximation" the motion of the center of mass of a system interacting with radiation was conserved only upon attributing a mass to the radiation emitted or absorbed. In this demonstration Einstein set the momentum of the cylinder's center of mass as $m_0 v$, hence his qualification, "to a first approximation."

Einstein next considered a collection of charged material particles, each of mass m_v and x-coordinate x_v, interacting among themselves through electromagnetic forces (the particles' self-fields were ignored) as well as conservative forces.

Then, to first order in v/c, Einstein used the Maxwell-Lorentz equations[12] to define the system's center of mass coordinate $\bar{X}$ as

$$\bar{X} = \frac{\sum_v m_v x_v + (1/4\pi) \int \rho_e x \, dV}{\sum_v m_v + (1/4\pi) \int \rho_e \, dV}, \qquad (12.48)$$

where ρ_e is the electromagnetic field's mass density (i.e. $\rho_e c^2 = w$), and x is the center of mass for the ensemble of particles. The time independence of the denominator in Eq. (12.48) which is the system's total mass, led Einstein to realize that the law of constancy of mass from chemical reactions (Lavoisier's law) was a "special case of the energy principle."[13]

12.5.3. Einstein's First 1907 Derivation of the Mass-Energy Equivalence

> We do not yet possess a *Weltbild* according to the principle of relativity.
>
> A. Einstein (1907d)

Einstein (1907d) continued to discuss the mass-energy equivalence, because he believed that "an assumption of such unusual generality inevitably invites challenge for proving its necessity or correctness in the most general manner." He posed the question: "Are there not special cases in which this assumption leads to incompatible consequences?" In Einstein's opinion, it was necessary to pursue these special cases further because "the general reply to the question posed above is presently not possible since we do not yet possess a world-picture [*Weltbild*] according to the principle of relativity." He had aleady investigated one special case in 1906 and had demonstrated to order v/c the compatibility of the mass-energy equivalence with the constancy of motion of the center of mass. He now considered two cases amenable to treatment "from the point of view of the present *Relativitätselektrodynamik"* — (1) a rigid macroscopic charged body, (2) a collection of mass points. In order to emphasize his disillusionment with constructive theories, Einstein next launched into a critique of applying electromagnetism, mechanics and thermodynamics to spatial volumes in which fluctuations could not be ignored (see Section 2.5). But, he continued, if we avoid this regime these branches of science can be used "with confidence"; consequently his investigations of the two special cases was justified.

Einstein's first special case extended the 1905 result in Eq. (12.33). He began with what can be construed as a two-fold snub of the nose at contemporaneous research toward an electromagnetic world-picture. He rewrote Eq. (12.33) with the symbol μ for the inertial mass instead of m_0, knowing full well that the symbol μ was reserved for the mass derived from the electron's electromagnetic quantities. Then he referred to μ as the mass "in the customary sense." Einstein wrote that "we shall now show that according to the theory of relativity this expression [Eq. (12.33)] is no longer valid for the case in which a body maintains its equilibrium when acted on by an external force." For the purpose of demonstration Einstein considered a rigid electrified body ("*Körper starr elektrisiert*"), with a continuous volume distribution of charge, that was acted on by an externally imposed electromagnetic field, and yet persisted in a uniform translational motion. Einstein assumed that the body's charge density was small enough so that the mutual interaction among the body's constituent charges could be neglected compared with their interaction with the external electromagnetic field. (This assumption, continued Einstein in a footnote, enabled him to avoid any restrictions on how the external electromagnetic field was produced.) The external electromagnetic field acted on the charged body during the interval $t_0 \leqslant t \leqslant t_1$, thereby increasing the body's energy relative to K by the amount

$$\Delta E = \int_{t_0}^{t_1} dt \int \rho v E_x \, dx \, dy \, dz. \tag{12.49}$$

In k, the charged body's rest system, Eq. (12.49) became

$$\Delta E = \gamma v \int d\tau \int \rho' E_\xi \, d\xi \, d\eta \, d\zeta. \tag{12.50}$$

In order to continue discussing the charged body from the viewpoint of an inertial reference system at relative rest (k), Einstein further assumed that the external electromagnetic field to be such that for every time τ in k every element of the charged body remained in equilibrium, i.e.,

$$\int \rho' E_\xi \, d\xi \, d\eta \, d\zeta = 0 \tag{12.51}$$

and, wrote Einstein, this equilibrium condition was necessary but not sufficient owing to the omission of the condition for rotational equilibrium. Perhaps having an inkling of things to come, Einstein wisely avoided discussing the possibility of a rigid body undergoing a finite rotation (see Sections 7.4.5 and 7.4.8–7.4.11). If the limits on τ were independent of ξ then, continued Einstein, $\Delta E = 0$ in K too. He emphasized, "this is, however, not the case" because owing to the relativity of simultaneity, Eq. (12.50) did not vanish because the integration limits on τ were functions of ξ; the upper and lower limits on τ were $t_0/\gamma - (v/c^2)\xi$ and $t_1/\gamma - (v/c^2)\xi$, respectively. In order to evaluate the integral in Eq. (12.50) Einstein assumed that at $t = t_0$ no external force acted, and during the interval $t_0 < t \leqslant t_1$ the external force could be assumed to be either constant or so slowly varying that it could be taken outside of the integral. Einstein evaluated Eq. (12.50) as

$$\Delta E = -(v/c)^2\gamma \left[\sum \xi F_\xi \right]_{t_0}^{t_1} \tag{12.52}$$

where

$$F_\xi = \rho' E_\xi \, d\xi \, d\eta \, d\zeta \tag{12.53}$$

and the summation took into account that the force F_ξ acted on every element of the body. The additional energy $\Delta E > 0$ because ξ and the external force F_ξ are directed oppositely. Einstein described the Eq. (12.52) thus: "We have, therefore, obtained the following unexpected result. If we subject a rigid body, on which originally no forces acted, to the influence of forces that impart no acceleration to the body, these forces perform — from the viewpoint of a coordinate system moving relatively to the body — a work ΔE upon the body which just depends on the force's distribution over the body and the velocity v. According to the energy principle the kinetic energy of a rigid body acted on by an external force is ΔE greater than the same kinetic energy of a fast-moving body that is acted on by no force." (This purely relativistic phenomenon was further clarified, and stated in the correct mathematical form, in 1911 by Max von Laue. Von Laue described the additional energy as originating in a change of the moving body's internal state — see Section 12.5.8.)

In order to investigate exclusively "the inertia of an electrically charged rigid body," Einstein next assumed (1907d) that the charged body persisted in inertial motion in the absence of an external electromagnetic field. Then, relative to K, this body possessed an amount of energy E^e originating in the charged body's self-electromagnetic fields E and B of the amount

$$E^e = \frac{1}{8\pi} \int (E^2 + B^2) \, dV \tag{12.54}$$

Transforming Eq. (12.54) to the charged body's rest system k, Einstein obtained

$$E^e = \frac{1}{8\pi\gamma} \int \left[E_\xi^2 + \left(\frac{1+\beta^2}{1-\beta^2}\right)(E_\eta^2 + E_\zeta^2) \right] d\xi \, d\eta \, d\zeta. \qquad (12.55)$$

Consequently, the charged body's total kinetic energy was $K_0 + K^e$, where K_0 was the kinetic energy for the "case when no electrical charges are present," i.e., K_0 was due to the body's inertial mass m_0, and $K^e = E^e - E_0^e$ was the "excess of the moving body's electromagnetic energy over the case for the body at rest" (e.g., $E_0^e = e^2/2R$ for a spherical body with a surface charge distribution). Thus, K^e depends on the charged body's orientation and Einstein proceeded to demonstrate that this was a contradiction: Consider that relative to k the charged body executes an infinitely slow rotation without any external influences. According to the principle of relativity this "must be a possible force-free motion" because the body's comoving system k remains an inertial system. Relative to K the infinitely slow rotation contributes nothing to the body's kinetic energy which has only the translational contribution. Since the body can move in any direction relative to K, then "a dependence on the orientation of the kinetic energy of an electrical body in translational motion is impossible." Of the result in Eq. (12.55) Einstein wrote that "this contradiction is removed by means of the result from the previous paragraph," i.e., Eq. (12.52). The reason, continued Einstein, was that "the kinetic energy of the body under consideration cannot be calculated as if the rigid body were acted on by no force." In this case the force "has its source in the mutual interaction between the electrical substances" that constitute the charged body. Therefore, continued Einstein the body's total kinetic energy K_T in K should have been written as

$$K_T = K_0 + \Delta E + (E^e - E_0^e), \qquad (12.56)$$

where Einstein designated the last term as the excess of electrical energy over the electrostatic energy possessed by the body when it was in a state of rest. From Eq. (12.53) and Gauss' law for replacing ρ, he calculated the quantity ΔE as

$$\Delta E = \left(\frac{v}{c}\right)^2 \frac{\gamma}{8\pi} \int (E_\xi^2 - E_\eta^2 - E_\zeta^2) \, d\xi \, d\eta \, d\zeta, \qquad (12.57)$$

and Eq. (12.56) became

$$K_T = \left(m_0 + \frac{E_0^e}{c^2}\right) c^2 \left[\frac{1}{\sqrt{1 - v^2/c^2}} - 1 \right] \qquad (12.58)$$

which was the desired extension of the 1905 result [Eq. (12.33)] which held for the case of an uncharged body in inertial motion: "the electrostatically charged body possesses an inertial mass which exceeds that of the uncharged body by the electrostatic energy divided by the square of the velocity of light."

Toward understanding a general result physicists often apply it to a particular case. Let us conjecture that Einstein applied Eqs. (12.54)–(12.57) to the case of a deformable electric body with a uniform surface charge distribution that was always spherical to an observer in its rest system, and to other inertial observers the body transformed properly under the relativistic transformations, i.e., Lorentz's electron. In this case $E_\xi^2 = E_\eta^2 = E_\zeta^2 = \frac{1}{3}E'^2$ and Einstein would have obtained

$$E^e = \frac{\gamma}{8\pi}\left[\int E'^2\, d\xi\, d\eta\, d\zeta\right]\left[1 + \frac{1}{3}\left(\frac{v}{c}\right)^2\right] \tag{12.59}$$

and

$$\Delta E = -\frac{1}{3}\left(\frac{v}{c}\right)^2 \frac{\gamma}{8\pi}\left[\int E'^2\, d\xi\, d\eta\, d\zeta\right]. \tag{12.60}$$

Consequently the total energy of Lorentz's electron is

$$E_T^e = E^e + \Delta E = \frac{m_0^e c^2}{\sqrt{1 - v^2/c^2}}. \tag{12.61}$$

But where is the $\frac{4}{3}$-factor (see Eq. (1.212))? It is reasonable to conjecture that by May 1907, when Einstein submitted (1907d) for publication, he knew full well that the electron's mass occurred in kinematical quantities deduced from its self-fields as $\frac{4}{3}$ times its electrostatic mass — for example, by 1905 he was aware of at least Abraham's (1903) and by May 1907 of Kaufmann's (1906) which mentioned the role of Poincaré's stress and very probably of Abraham's (1905) which contained a detailed discussion of the necessity for an extra energy to correct the Lorentz-electron's total energy. In fact, Einstein may well have avoided the particular example of Lorentz's electron because of his having been unable to deduce the $\frac{4}{3}$-factor from the relativistic kinematics. We can trace this failure to Eq. (12.52) which yields an $E > 0$ only for external forces. In other words, the formalism that Einstein presented for the "unexpected result" in (1907d) was not always Lorentz covariant. (We shall see in Section 12.5.8 that Max von Laue's explication of the relativistic transformation properties of the energy and momentum densities revealed that in (1907d) Einstein had considered incompletely how deformable bodies were stressed in their rest system.) For Einstein it may well have been that the main point of his considerations thus far in (1907d) transcended a mere numerical factor, and the electromagnetic world-picture as well: a quasi-aesthetic argument, in conjunction with the relativity of simultaneity, i.e., kinematics, had yielded both the "unexpected result" concerning moving charged bodies acted on by an external force whose resultant vanished in the body's rest system, as well as a part of the energy from the Poincaré stress. Poincaré's elegant (1906) derivation of this quantity had depended on a Lorentz-covariant Lagrangian formulation of the dynamics of the electron.

Einstein's second special case was to deduce the total energy of a force-free mass point. The result was Eq. (12.34), and he emphasized that the null point for the energy in classical mechanics disappeared here, because the relativistic total energy contained also the energy for the resting particle.

Einstein went on to emphasize that the results of the first special case should not be construed as a major step toward a dynamics of parallel translation, since (1) a particularly simple assumption on the external force was necessary for calculating the additional energy, and (2) the derivation turned on the notion of a rigid body which Einstein criticized with the example of a rigid rod in inertial motion that was acted on by equal and opposite forces at its ends (see Section 7.4.5). In this example, the additional kinetic energy necessary for the energy and momentum of the rod to have the properties associated with an object with mass could not be ascribed to electromagnetic forces. Einstein's physically permissible comment was that something flowed in the rod in order to maintain the rod's equilibrium relative to K and k. After all, according to the principle of relativity if a body was in equilibrium in one inertial system, then it was in equilibrium for every inertial system because equilibrium could not be Lorentz-transformed away. But exactly what conditions had to be satisfied for maintaining the rod's equilibrium were not yet clear.

12.5.4. Some Others who Discussed an Association Between Energy and Mass

> [Light] energy has no mass...
>
> H. Poincaré (1908a)

> Whether the principle of relativity [for material bodies as well] is correct remains an open question, which does not affect the results of this work.
>
> K. von Mosengeil (1907)

As I discussed in Chapter 1, the association of mass with energy had been explicitly introduced twice into electromagnetic theory: in 1881 by Thomson, for the purpose of interpreting the increased resistance to motion of a moving charged body due to its self-induction, and in 1900 by Poincaré who compared the electromagnetic field to a "*fluid fictif*" in order to rescue the principle of action and reaction in Lorentz's electromagnetic theory.[14] Proponents of the electromagnetic world-picture assumed that the electron's mass was deduced from the energy density of its self-fields, but the converse statement had no physical meaning for them.

For the purpose of further investigating the relation between the laws of thermodynamics and the pressure of radiation on a moving mirror, Fritz Hasenöhrl (1904) assumed that within a moving radiation cavity was a mirror, and for the purpose of calculating the properties of black-body radiation emitted from this cavity he added the well-known Lambert cosine law to Abraham's (1904b) methods for determining the characteristics of light reflected from a moving mirror. Hasenöhrl went on to show that if the cavity was accelerated from rest in a reversible manner (i.e., adiabatically and isothermally), then the cavity radiation could possess an "apparent mass... which is completely analogous to the so-called electromagnetic mass" of Abraham's spherical electron. However, instead of a result proportional to $\frac{4}{3}c^2$ Hasenöhrl obtained $\frac{8}{3}c^2$. Abraham quickly located the error and Hasenöhrl (1905) emphasized that the result obtained from considering cavity radiation, $\frac{4}{3}c^2$, was independent of the assumption of quasi-stationary motion. But there was another more serious discrepancy in Hasenöhrl's (1904) work; namely, he found that if he placed the cavity in contact with a heat reservoir, and first accelerated the cavity to a velocity v, and then decelerated it back to rest in a reversible manner, the result was a net increase in the cavity's temperature, thereby violating the second law of thermodynamics. Hasenöhrl (1905) found that the incorrect term could be cancelled by assuming, among other hypotheses, that the cavity underwent a Lorentz contraction.[15] As a result of a suggestion by Abraham for simplifying his calculations, Hasenöhrl established (1905) that, in addition to Lorentz's contraction, certain other assumptions concerning the dependence of the relative radiation on the cavity's velocity could remove any violations of the second law of thermodynamics.

Kurd von Mosengeil, a student of Planck's, brought a new approach to the problem of calculating the characteristics of black-body radiation in a moving cavity that avoided both the Lorentz contraction and the Lambert cosine law, whose validity he questioned for this class of problems. In a 1907 *Annalen* paper, von Mosengeil combined appropriate results of Abraham (1904b) with a suitable definition of the temperature T in a moving cavity (i.e., $T = T_0\sqrt{1 - v^2/c^2}$, where T_0 was the resting cavity's temperature), to obtain the energy distribution, total energy, pressure and apparent mass for radiation in a cavity undergoing inertial motion as these quantities were measured in the laboratory; his results were compatible with the second law of thermodynamics. Von Mosengeil went on to show that these characteristics of black-body radiation could also be deduced from Einstein's 1905 solution for the problem of radiation reflected from a moving mirror; but he emphasized that "whether the principle of relativity (for material bodies as well) is correct remains an open question, which does not affect the results of this work."[16]

12.5.5. Planck's 1907 Derivation of the Mass-Energy Equivalence

> The principle of relativity stated by H. A. Lorentz and in more general wording by A. Einstein.
>
> M. Planck (1907a)

Von Mosengeil's unexpected death had led to Planck's writing the 1907 *Annalen* paper bearing von Mosengeil's name and based on his doctoral dissertation. Soon after, Planck published a long paper "On the Dynamics of

Moving Systems" (1907a) in which he essentially put the finishing touches on a relativistic thermodynamics, as well as elaborating upon his own 1906 Hamilton-Lagrange version of relativistic mechanics. Planck assumed the "general and absolute validity" of the principle of relativity, although, echoing von Mosengeil's caution that it had not yet been proved empirically; for, continued Planck, in his opinion the supporting data came only from the Michelson-Morley experiment, though no experiment contradicted it. Planck's goal was to formulate a relativistic mechanics of ponderable bodies and then put it to "direct experimental proof." Planck's anchor in the new domain was von Mosengeil's results which, besides being independent of relativity theory, were also model independent because they were based upon cavity radiation. Planck's guide through the new domain was the principle of least action which he came to view as occupying the "highest position among physical laws" [Planck (1915)].[17] Planck's enthusiasm over the principle of least action was rooted in its invariance under the principle of relativity, whose content he interpreted now as he had in 1906 as purely heuristic: as asserting the equivalence of inertial reference systems when they were Lorentz-transformed. The principle of relativity, Planck continued in 1907, permitted extending the principle of least action beyond the domain of applicability set down by von Helmholtz in 1895 — for example, to a dynamics of ponderable bodies capable of discussing decay processes in which energy was produced.

From the invariance of the principle of least action, in conjunction with the principle of energy conservation and the Lorentz transformations, Planck deduced, among other quantities, the transformation equations for the Lorentz force:

$$F_x = F'_{x'} + \frac{v^2/c^2}{1 + v\dot{x}'/c^2}[F'_{y'}\dot{y}' + F'_z\dot{z}'], \tag{12.62}$$

$$F_y = F'_{y'}\frac{\sqrt{1 - v^2/c^2}}{1 + v\dot{x}'/c^2}, \tag{12.63}$$

$$F_z = F'_{z'}\frac{\sqrt{1 - v^2/c^2}}{1 + v\dot{x}'/c^2}, \tag{12.64}$$

where primed quantities refer to a k-system that is not the instantaneous rest system of the moving body (analogous results had been deduced in 1905 by Poincaré),[18] and new results for the momentum and energy

$$G_x = \frac{1}{\sqrt{1 - v^2/c^2}}\left[G'_{x'} + \frac{v}{c^2}(E' + p'V')\right], \tag{12.65}$$

$$G_y = G'_{y'}, \tag{12.66}$$

$$G_z = G'_{z'}, \tag{12.67}$$

$$E = \frac{1}{\sqrt{1 - v^2/c^2}}\left[E' + vG'_{x'} + \frac{v}{c^2}\frac{(\dot{x}' + v)}{1 + v\dot{x}'/c^2}p'V'\right], \qquad (12.68)$$

where p' was a scalar or hydrodynamic pressure, that is p' could be represented by a stress tensor having only diagonal components and also, according to Planck, a Lorentz-invariant quantity [which result was implicit in previous papers of Abraham (1904b), Poincaré (1906) and Einstein (1907e)].[19] The quantity pressure appeared in Planck's results because a primary consideration in 1907 was the thermodynamics of a moving cavity; thus, in certain limiting cases, Planck could check his results against von Mosengeil's. With the k system taken as the body's instantaneous rest system, Eq. (12.65) became

$$G_x = \frac{1}{\sqrt{1 - v^2/c^2}}\frac{v}{c^2}(E' + p_0 V') \qquad (12.69)$$

where $p' = p = p_0$. Planck deduced the mass-energy equivalence as

$$m_0 = \left(\frac{G_x}{v}\right)_{v=0} = \frac{E' + p_0 V'}{c^2} \qquad (12.70)$$

where m_0 is the body's rest mass; Planck referred to the "Gibb's heat function at constant pressure," $E' + p_0 V'$, as the body's "heat function," and his derivation was model independent.[20] Planck went on to deduce a more general form of Eq. (12.62):

$$F_x = F'_{x'} + \frac{v/c^2}{1 + v\dot{x}'/c^2}\left[F'_{y'}\dot{y}' + F'_{z'}\dot{z}' + V\frac{dp_0}{dt'} + T'\frac{dS}{dt'}\right] \qquad (12.71)$$

where S was the entropy, for whose Lorentz-invariance Planck had argued convincingly. Thus, the motion of a charged body had the "characteristic properties" of a process that was isobaric and adiabatic, i.e., $dp_0 = dS = 0$; although Planck did not pursue this point, he had taken a step toward clarifying the basis for quasi-stationary motion. Another case left unclarified by Planck was one in which, even though no net forces acted in k, nevertheless a change of internal state in k led to a measurable force in K in order to maintain k's inertial motion; this was Einstein's (1907d) "remarkable result," and in 1911 von Laue pushed it to a startling conclusion (Section 12.5.8).

12.5.6. Einstein's Second 1907 Derivation of the Mass-Energy Equivalence

> For radioactive decay the quantity of free energy becomes enormous.
>
> A. Einstein (1907e)

Einstein (1907e) reviewed Planck's relativistic mechanics of material points and, perhaps to emphasize his view of theories of the electron, Einstein entitled this section "Mechanics of Material Points (Electrons)." Whereas Planck considered Einstein's results as a generalization of Lorentz's, Einstein stressed his own point of view by noting that Planck's Eqs. (12.62)–(12.64) held also for the case in which the forces were not electromagnetic: "In this case the equations have no physical content, rather they are interpreted as the defining equations for force."[21] What Einstein meant was that if F was not an electromagnetic force, then Eqs. (12.62)–(12.64) would lose their electromagnetic content, thereby becoming instead a definition for how forces of any sort transform.

After his effective use of notions of thermodynamics, it might at first surprise us that when Einstein explored the mass-energy equivalence in the 1907 *Jahrbuch* paper he used electrodynamics rather than Planck's more general derivation based on concepts from cavity radiation. But in the 1907 *Jahrbuch* paper Einstein may well have wished to keep the notion of quanta, with their attendant fluctuations, separate from a theory of principle. Einstein's *Jahrbuch* derivation of the mass-energy equivalence was an improvement over the earlier derivations of 1907 (particularly (1907d)), because he now substituted for the rigid body a deformable envelope containing charged bodies, whose total mass was μ (in Einstein's notation); and the envelope was impermeable to radiation. Assuming that the envelope, with its enclosed physical system, moved with the velocity v relative to K, and was acted upon by external electromagnetic forces while remaining in inertial motion, Einstein deduced its energy relative to K as

$$E = [(\mu + E_0/c^2)c^2 + (v^2/c^2)p_0 V']\gamma \tag{12.72}$$

where E_0 is the system's energy content relative to k, and Einstein replaced the work done by the *external* force $\sum \xi F_\xi$ with $-p_0 V'$ in Eq. (12.52). Since k was the envelope's rest system this was Planck's result, Eq. (12.68), and Einstein also obtained Planck's transformation equations for momentum.[22]

Einstein applied the energy and momentum transformation equations to three cases, one of them was "the case of an electrically charged massless body [that is] acted on by no external forces." From Eqs. (12.69) and (12.72) Einstein obtained $G = \frac{4}{3}\gamma E_0/c^2$ and $E = \gamma E_0$. He then wrote: "In both these quantities a part has to be attributed to the electromagnetic field and the remainder to the massless body on account of the forces due to its charges." Einstein could only have meant that E_0 was composed of two parts: E^e and ΔE from Eqs. (12.59) and (12.60). Recalling my conjecture that in (1907d) Einstein knowingly omitted specializing his "unexpected result" to Lorentz's electron, then the $\frac{4}{3}$-factor in the momentum is a wonderful example of a Freudian slip. In fact, in the "Corrections" to his (1907e), Einstein wrote that the momentum equation in this illustrative example should not have included the $\frac{4}{3}$-factor.[23]

By 1909 Einstein still was unable to complete the relativistic formalism for stresses. We are aware of this point from the discussion session to Arthur Szarvassi's (1909) at the 1909 *Naturforscherversammlung*. Szarvassi asserted that Lorentz's theory for macroscopic moving electric and magnetic bodies required new forms of energy in order to be consistent with the energy principle. Gustav Mie queried whether Szarvassi's criticism of "Lorentz's theory of relativity" was analogous to Abraham's criticism of Lorentz's deformable electron, i.e., it could not be described by purely electromagnetic forces. To which Einstein, who was making his debut at these meetings, replied that the extra energy cited by Abraham arose owing to forces operating in the rest system of an extended body, and the inclusion of this additional energy made possible agreement with the energy principle. "Is it clear what I mean?," Einstein concluded. "Not fully," replied Szarvassi. Einstein explained further his (1907d) result which he emphasized was a consequence of the "*Relativitätstheorie*." This was a rather bold statement because in 1909 the "*Relativitätstheorie*" still could not account for Abraham's extra energy, i.e., $m_0^e c^2/3\gamma$ [see Einstein (1907d, e)]. But, as I conjectured, Einstein was undeterred by mere numerical factors because he believed that he was in possession of methods and results more general than those of the electromagnetic world-picture. Szarvassi next claimed that his formulation of Lorentz's theory allowed for the admission of an extra energy and he looked to Einstein for agreement. Einstein replied: "To this point I have no reply since I have not penetrated sufficiently into these considerations."

Einstein continued in (1907e) by specializing to the case in Eq. (12.49) where forces acting in k on the deformable envelope vanished at $t = t_1$ and at $t = t_2$, and obtained

$$E = \frac{\mu + (E_0/c^2)c^2}{\sqrt{1 - v^2/c^2}} .$$
(12.73)

This system, wrote Einstein, behaved like a "material point" with mass

$$M_0 = \mu + E_0/c^2$$
(12.74)

and, he continued, this "result is of extraordinary importance because a physical system's inertial mass and energy content appear to be the same thing. An inertial mass μ is equivalent with an energy content μc^2"; thereby Einstein extracted the general meaning from his earlier 1907 result which depended on the model of a charged rigid body. Consequently, wrote Einstein, the law of the "constancy of mass" (Lavoisier's law) was valid only for systems whose energy content remained constant. But, stressed Einstein, the mass-energy equivalence was difficult to demonstrate empirically because, for example, a mass loss of 1000 gm-cal corresponded to 4.6×10^{-11} gm; however, "for radioactive decay the quantity of free energy becomes enormous." With this prophetic statement he went on to consider whether the mass-energy equivalence could be tested in radioactive processes, and for this purpose he elaborated upon an example given by Planck (1907a) from recent data of Precht (1906) from the decay of radium. Planck had shown that the rate of decrease in mass of 1 gm-mole per year of radium undergoing decay was

$$\varDelta m = 1.2 \times 10^{-5}\,\text{gm} \tag{12.75}$$

and this small quantity was difficult to detect. Einstein suggested an indirect method for detecting this small loss of mass that depended upon comparing the atomic masses of the decay products with the original radium sample: the energy released when radium of mass M decayed into its two end products of masses m_1 and m_2, i.e., lead and helium, was

$$M - (m_1 + m_2) = E/c^2. \tag{12.76}$$

Taking the mean lifetime of radium to be 2,600 years, Einstein calculated that

$$\frac{M - (m_1 + m_2)}{M} = 1.2 \times 10^{-4}, \tag{12.77}$$

and he observed that verification of Eq. (12.77) required knowledge of atomic masses accurate to five decimal places, but that information was not available in 1907. Einstein did not belabor this point because so sure was he of the validity of the mass-energy equivalence that he immediately suggested a new, more general theory of relativity: "Hitherto we have assumed implicitly that [the mass-energy equivalence] holds good for inertial as well as for gravitational mass — that is, that a system's inertia and weight are exactly proportional. Thus, we would have to assume, for example, that in a cavity the enclosed radiation possesses not only inertia but also weight. The proportionality between inertial and gravitational masses is valid universally for all bodies with an accuracy as precisely as can currently be determined; so that unless proven otherwise, we must accept it as universally valid. We shall find in the final section of this communication a new argument supporting this assumption." Einstein entitled this section, "Principle of Relativity and Gravitation," and the assumption of the exact equality of the gravitational and inertial masses was the basis of his extension of a principle of relativity to noninertial reference systems:[24] *Despite the lack of corroborating data, by 1907 Einstein had assumed that the mass-energy equivalence transcended the logical deficiency of the special relativity theory — the inertial reference system.*[25]

12.5.7. Planck's Law of Inertia of Energy

Although by the end of 1907 Einstein was moving in other directions, some new developments occurred which further revealed the fecundity of the special relativity theory.

At the 80th *Naturforscherversammlung* at Cologne, in September of 1908, Planck showed how the principle of action and reaction could reenter physical theory as the "law of inertia of energy." On 23 September Planck (1908) reviewed how, despite Poincaré's valiant attempts to save this principle in Lorentz's theory, Abraham had demonstrated that it was transcended by the principle of conservation of momentum. Since electromagnetic theory associated a flow of energy density (the Poynting vector) with a momentum density, Planck asserted that the flow of any sort of energy, e.g., heat, chemical, elastic, gravitational, could also be associated with a momentum density; like Poincaré, Planck used an example from fluid mechanics to illustrate this point. Consider, continued Planck, that the work W done in a time dt by a fluid with energy density ε moving with velocity v, and exerting a pressure p_0 on a surface element $d\sigma$ oriented normally to v was

$$W = v(\varepsilon + p_0) \, d\sigma \, dt, \tag{12.78}$$

and from dimensional considerations Planck could define a momentum density

$$g = v(\varepsilon + p_0)/c^2. \tag{12.79}$$

Consequently the quantity $(\varepsilon + p_0)/c^2$ was a mass density

$$\alpha = (\varepsilon + p_0)/c^2 \tag{12.80}$$

and Eq. (12.80) was Planck's generalization of Einstein's mass-energy equivalence which included the flow of any sort of energy, and not just the total mechanical energy; Planck referred to Eq. (12.80) as the "law of inertia of energy," which asserted that the effect of forces acting on a body was transmitted by a momentum density whose source was a flow of energy; for example, in electromagnetic theory the quantity $v(\varepsilon + p_0)$ was the Poynting vector. Planck went on to write that Eq. (12.80) was a "known relation from the relativity theory" because it could also have been obtained from his Eq. (12.69) by setting $E' = \varepsilon'V'$.

Just as the effect of the electromagnetic field on matter could be calculated from a stress tensor representable by a 3×3 matrix, thus, continued Planck, in a vacuum the effect of the momentum current on a surface could also be associated with the customary stress tensor from the theory of elasticity, because according to Eq. (1.101) if a system was in equilibrium ($f = 0$), then energy impinged upon its enclosing surface at the rate $\mathbf{V} \cdot \tilde{\mathbf{T}}$. Consequently whereas in Lorentz's "theory of the resting ether" the Maxwell stresses were "mathematical auxiliaries," the *Relativitätstheorie* rendered them "legitimized for the resting ether."

The transformation properties of the momentum density and Poynting vector were clarified through the results of an epoch-making 1908 paper of someone in the audience, Hermann Minkowski. Minkowski (1908a) demonstrated that, in a space-time description of the Lorentz-Einstein theory, the Lorentz force Eq. (1.101) and energy conservation law for electromagnetic processes

$$-f \cdot v = \mathbf{V} \cdot \mathscr{S} + \frac{\partial w}{\partial t} \tag{12.81}$$

could be joined into the single equation

$$f_\mu = \sum_{v=1}^{4} \frac{\partial T_{\mu v}}{\partial x_v} \qquad (12.82)$$

where f_μ is a space-time vector of the first kind with components $(\boldsymbol{f}, i\boldsymbol{f} \cdot \boldsymbol{v}/c)$, and $T_{\mu v}$ is a traceless, symmetric tensor representable as a 4×4 matrix whose spatial components are Maxwell's stress tensor $T_{ij} = (1/4\pi)(E_i E_j + B_i B_j) - \delta_{ij}\omega$; the mixed space and time components are $T_{i4} = -ic(g^e)_j$, $T_{4j} = -i\mathscr{S}_j/c$; the time component is $T_{44} = \omega$. The tensor $T_{\mu v}$'s properties of symmetry and tracelessness were Lorentz-invariant, and Planck's theorem of the inertia of energy was considered to be a consequence of the former property. Since the quantities f_μ and $T_{\mu v}$ had the proper properties under Lorentz transformation, then Eq. (12.82) expressed laws of nature that were valid in every inertial reference system. Consequently if a physical system was in equilibrium in one inertial reference system, it was in equilibrium in every inertial system – that is, equilibrium could not be transformed away.

At the end of Planck's lecture Minkowski, who two days earlier had delivered his lecture "Space and Time," made a comment whose content emphasized both that he grasped the mathematical meaning of Planck's proposal and that he interpreted Einstein's 1905 paper on electrodynamics within the context of an electromagnetic world-picture: "In my view the law of momentum is obtained from the energy law; namely, in Lorentz's theory of space and time the energy law depends on the reference system. We write the energy law for every possible reference system so that we have many equations, and in those are contained the law of momentum." To which Planck replied with his characteristic emphasis on empirical data: "True. But I consider the independence from the reference system not as an established physical result, but rather as a hypothesis which I nevertheless consider as promising, but by no means proven. It remains to examine if these relations truly exist in nature. We can only find that out by experiment and hopefully is the time not distant when we will know."

12.5.8. Some Essential Results of Max von Laue: In Particular on Elementary-Particle Theory in 1911

Von Laue [(1911a, e), (1912b)] demonstrated explicitly the advantages and deep meaning of Minkowski's stress-tensor formalism, particularly for developing a relativistic theory of deformable bodies, and for elucidating the state of elementary-particle theory.

Referring to Minkowski's $T_{\mu v}$ as the "world–tensor (*Welttensor*)," von Laue developed the relation between its elements in two inertial reference systems

$$T_{\mu v} = \sum_{\alpha,\beta=1}^{4} a_{\mu\alpha} a_{v\beta} T'_{\alpha\beta} \qquad (12.83)$$

where the quantities $a_{\mu v}$ are the elements of the Lorentz transformation matrix

Eq. (7.38) as (taking the primed system to be the body's rest system):[26]

$$g_x = (v/c^2)\gamma^2 (T'_{44} - T'_{11}), \tag{12.84}$$

$$g_y = - (v/c^2)\gamma T'_{12}, \tag{12.85}$$

$$g_z = - (v/c^2)\gamma T'_{13}, \tag{12.86}$$

$$T_{22} = T'_{22}, \tag{12.87}$$

$$T_{33} = T'_{33}, \tag{12.88}$$

$$T_{12} = \gamma T'_{12}, \tag{12.89}$$

$$T_{13} = \gamma T'_{13}, \tag{12.90}$$

$$T_{11} = \gamma^2 [(v^2/c^2)T'_{44} - T'_{11}], \tag{12.91}$$

$$T_{44} = \gamma^2 [T'_{44} - (v^2/c^2)T'_{11}]. \tag{12.92}$$

In electromagnetic theory, the stresses T_{ij} were the Maxwell stresses; however, as von Laue emphasized, the beauty of Minkowski's tensor formalism was that the world–tensor could include the properties of a physical system under any sort of forces — elastic, chemical or mechanical. One of von Laue's major contributions to relativity theory was his proposal that the state of a physical system could be described completely through the equation

$$f_\mu = \sum_v \frac{\partial}{\partial x_v} \left[\sum T_{\mu v} \right] \tag{12.93}$$

in which $\sum T_{\mu v}$ contained the world–tensors for mechanical, chemical, thermal, elastic, etc., processes. Like Planck, von Laue used a heuristic interpretation of the principle of relativity to extend a formalism far beyond its originally intended boundaries — in this case, Minkowski's world–tensor for electromagnetic processes.

The cases of T_{ij} of interest to us here are also those on which von Laue focused — that is, where the pressures are uniform so that $T_{ii} = T'_{ii} = T_0$ and $T_{ij} = T'_{ij} = 0$ ($i \neq j$). Hence, Eqs. (12.84) and (12.92) can be replaced by

$$g_x = (v/c^2)\gamma^2(\omega' - T_0), \tag{12.94}$$

$$\omega = \gamma^2(\omega' - (v^2/c^2)T_0) \tag{12.95}$$

where ω (ω') is an energy density in K (k). Von Laue's integration over the system's finite volume V in K yielded

$$G_x = \frac{v}{c^2} \frac{(E' + p_0 V')}{\sqrt{1 - v^2/c^2}}, \tag{12.96}$$

$$E = \frac{E' + p_0 V'(v^2/c^2)}{\sqrt{1 - v^2/c^2}}, \tag{12.97}$$

where we have used $\gamma \, dV = dV'$, and set $T_0 = -p_0$, since p_0 was the stress leading to work of compression that increased the system's energy by the amount $p_0 \, dV$; the Eqs. (12.96) and (12.97) had been deduced by Planck (1907a) and Einstein (1907e) in less general ways: not yet in possession of Minkowski's world-tensor formalism, Planck (1907a) had spent many pages deducing these two equations for energy and momentum by direct Lorentz-transformation of the terms in the principle of least action; Einstein (1907e) had analyzed the case of a deformable envelope surrounding electrical charges.

Von Laue (1911a, e) went on to point out that a body could be described as moving like a "point mass" if its internal state, characterized by the quantities (E', p_0, V') remained unchanged. In this case, from Eq. (12.96) the transverse mass is

$$m_T = \frac{G_x}{v} = \left[\frac{E' + p_0 V'}{c^2} \right] \bigg/ \sqrt{1 - v^2/c^2} \qquad (12.98)$$

and the longitudinal mass is

$$m_L = \frac{\partial G_x}{\partial v} = \left(\frac{E' + p_0 V'}{c^2} \right) \bigg/ (1 - v^2/c^2)^{3/2}. \qquad (12.99)$$

Since the quantity $(E' + p_0 V')/c^2$ was the body's rest mass, then von Laue found that he could explain Planck's seemingly unusual result in Eq. (12.71) as "finding its clarification in the inertia of energy": according to Eq. (12.96) a body in inertial motion and undergoing changes in its internal coordinates that lead to a measurable force in K can be represented by

$$F_x = \frac{dG_x}{dt} = \frac{dG_x}{dt'} \frac{dt'}{dt} \qquad (12.100)$$

or

$$F_x = \frac{v}{c^2} \left[\frac{dE'}{dt'} + \frac{dp_0}{dt'} V' + p_0 \frac{dV'}{dt'} \right]. \qquad (12.101)$$

According to the conservation of energy

$$\frac{dE'}{dt'} = T' \frac{dS}{dt'} - p_0 \frac{dV'}{dt'}, \qquad (12.102)$$

then substituting Eq. (12.102) into (12.101) yields

$$F_x = \frac{v}{c^2} \left[V' \frac{dp_0}{dt'} + T' \frac{dS}{dt'} \right] \qquad (12.103)$$

which is Planck's Eq. (12.71) for the case in which no external forces act in k, and yet a force was measured in K that served to maintain the inertial motion of k. The force arose from a change of state of the body in k.

This was one among several key differences between the relativity and classical mechanics which von Laue believed were rooted in the relativistic notion of the particle. He was struck particularly by the transition from the mechanics of deformable bodies to mass points because Newtonian mechanics made the opposite transition. Von Laue sought to prove that this could not be the case in relativity theory, and he showed that investigations into the relativistic dynamics of continuous media provided a response to Ehrenfest's 1907 query to Einstein of whether the relativistic mechanics could determine the parallel translation of a deformable electron. First von Laue proved that the "mass point" was inconsistent with the relativity theory. For an adiabatic-isobaric system — i.e., a system satisfying also the quasi-stationary motion, see Planck's Eq. (12.71) — the statement of energy conservation is

$$\boldsymbol{v} \cdot \frac{d\boldsymbol{G}}{dt} = \frac{dE}{dt} + p_0 \frac{dV}{dt}. \tag{12.104}$$

Since according to the relativity theory a moving particle's volume measured in K is a function of its relative velocity, then energy conservation cannot be satisfied without the term $p_0 \, dV/dt$. Therefore, a particle in relativity theory cannot be compressed into a point particle, although it may be possible to describe its trajectory as generated by its center of mass. In fact, the term $p_0 \, dV/dt$ exactly cancels the term $m_T/4$ in Eq. (1.195), and so a charged particle's kinematical properties could be deduced without the quasi-stationary approximation of Lorentzian electrodynamics. In addition, the aura of arbitrariness was removed from the Poincaré stress.

Then von Laue (1911a) pursued analysis of the electron's electromagnetic properties according to the theory of relativity by calculating the world–tensor for the electron's self-electromagnetic fields. The vanishing of the sum of the world–tensor's diagonal elements yielded the result

$$T'_{11} = -\tfrac{1}{3} T'_{44}. \tag{12.105}$$

Owing to the isotropy of the self-stress in the electron's rest system, the off-diagonal elements of the Maxwell stress tensor vanish and Eqs. (12.84) and (12.92) become

$$g^e_x = \frac{4}{3} \frac{v}{c^2} \gamma^2 \, w^{e'}, \tag{12.106}$$

$$w^e = \gamma^2 \, w^{e'} \left(1 + \frac{1}{3} \frac{v^2}{c^2}\right). \tag{12.107}$$

Integrating Eqs. (12.106) and (12.107) over the electron's volume in K yields

$$G^e_x = \frac{4}{3} \frac{v}{c^2} \gamma \int w^{e'} \, dV', \tag{12.108}$$

$$E_T^e = \gamma \left[1 + \frac{1}{3}\frac{v^2}{c^2} \right] \int w^{e\prime}\, dV', \tag{12.109}$$

where, as in Eq. (1.196), E_T^e is the electron's total energy arising from its self-electromagnetic fields. Since $\int w^{e\prime}\, dV' = (1/8\pi) \int E'^2\, dV' = m_0^e c^2$ then

$$G_x^e = \tfrac{4}{3} m_0^e v / \sqrt{1 - v^2/c^2}, \tag{12.110}$$

$$E_T^e = m_0^e c^2 \left(1 + \frac{1}{3}\frac{v^2}{c^2} \right) \Big/ \sqrt{1 - v^2/c^2}. \tag{12.111}$$

Eq. (12.110) displayed the collinearity of the spherical electron's velocity and momentum, but Eqs. (12.110) and (12.111), as von Laue recognized, could not be associated with a moving particle. Poincaré's method for replacing the term $\gamma m_0^e (v^2/3)$ with $m_0^e c^2 \gamma/3$ was to introduce a counterterm whose energy had the proper Lorentz transformation properties for ensuring that the longitudinal mass derived from the electron's energy or Lagrangian was the same as that derived from its momentum. Eq. (12.111) is equivalent to Eq. (1.196). *Mathematically* the reason for the incorrectness of Eq. (12.111) was that, according to Minkowski's space-time formalism, where quantities transform under the Lorentz group, the Poynting vector was not the proper sort of quantity to the associated with a particle possessing mass; that is, in four-dimensional space-time, the Poynting vector was not a vector but an element of the world–tensor. *Physically* Eq. (12.111) could not be associated with an electron because attempting to deduce the dynamics of a deformable electron from its self-fields was like trying to study the static equilibrium of a fluid without assessing the forces exerted by the walls of its container. Von Laue took the electron to be in equilibrium in every inertial reference system under the influence of both mechanical and electromagnetic forces; that is, he ensured the vanishing of f_μ by adding onto the electromagnetic world–tensor $T_{\mu\nu}^e$ a mechanical world–tensor $T_{\mu\nu}^m$, thereby closing the system. According to this procedure, von Laue wrote

$$\sum_{\nu=1}^{4} \frac{\partial}{\partial x_\nu} (T_{\mu\nu}^e + T_{\mu\nu}^m) = 0. \tag{12.112}$$

From the transformation properties of $T_{\mu\nu}^e + T_{\mu\nu}^m$ the electron's total energy and total x-component of momentum were

$$E_T = E_T^e + E^m, \tag{12.113}$$

$$G_T = G_x^e + G_x^m, \tag{12.114}$$

where G_x^m, G_x^e and E^m, E_T^e transform like the quantities in Eqs. (12.110) and (12.111), respectively, e.g., as usual,

$$E^m = \gamma(E^{m\prime} + (v^2/c^2) p_0^m V'), \tag{12.115}$$

$$G_x^m = \gamma\,(v/c^2)(E^{m\prime} + p_0^m V'). \tag{12.116}$$

Since the mechanical stress acted inward in order to prevent the deformable electron from exploding in its rest system, i.e., (see Eq. (1.208) and Chapter 1, footnote 67)

$$p_0^m V' = T^{e'} V' \tag{12.117}$$

then p_0^m was the Poincaré stress p_0^c.[27] Consequently, E_T and G_T from Eqs. (12.113) and (12.114) become, respectively,

$$E_T = \gamma(E^{e'} + E^{m'}), \tag{12.118}$$

$$G_T = (v/c^2)\gamma\,(E^{e'} + E^{m'}) \tag{12.119}$$

and although none of the quantities E^m, E_T^e, G_x^m, G_x^e, taken separately, possessed the proper Lorentz transformation properties to be associated with a particle, E_T and G_T describe a particle with rest mass M_0, where

$$M_0 = (E^{e'} + E^{m'})/c^2. \tag{12.120}$$

According to the electromagnetic world-picture, continued von Laue,

$$G_x^m = 0 \tag{12.121}$$

and therefore [see Eq. (1.209)]

$$p_0^m V' = - E^{m'} = - \tfrac{1}{3}m_0^e c^2; \tag{12.122}$$

the equations for E_T and G_T become

$$E_T = \frac{4}{3}\frac{m_0^e c^2}{\sqrt{1 - v^2/c^2}}, \tag{12.123}$$

$$G_T = \frac{4}{3}\frac{m_0^e v}{\sqrt{1 - v^2/c^2}}. \tag{12.124}$$

These results could be obtained in the electromagnetic world-picture only by using the quasi-stationary approximation. Indeed, substituting Eq. (12.122) into Eq. (12.115) yields

$$E^m = (m_0^e c^2/3)\sqrt{1 - v^2/c^2} \tag{12.125}$$

which is the amount of nonelectrical energy that Abraham claimed was necessary to prevent the moving deformable electron from exploding (see Eq. (1.210)). Von Laue demonstrated that in relativity theory the energy in Eq. (12.115) appeared naturally through demanding axiomatically that the electron be a closed system. The additional elastic stress served to maintain the electron's equilibrium in its rest system, as well as to cause the electron to undergo a Lorentz contraction relative to K. *Consequently it was within the space-time formalism of Minkowski that the Poincaré stress could be understood.*

(Assuming that Einstein specialized Eqs. (12.55) and (12.57) to Lorentz's electron, we may now note that Einstein went astray in (1907d) by having taken account only of the term $\gamma\beta^2 p_0^m V'$ in Eq. (12.115), which is the energy $-\beta^2\gamma m_0^e c^2/3$. Inclusion of the two terms in Eq. (12.115) with E_T^e yielded the correct $\frac{4}{3}$-factor.)

Von Laue next invoked the sharpest possible statement within the relativity theory for the electron's stability — the electron was a "perfectly static system." Consequently the electron's energy and momentum obtained from the world-tensor formalism *ab initio* contained contributions from every force necessary for its equilibrium. From Eqs. (12.84)–(12.92) this condition meant that the volume integral over every stress vanished; Gustav Mie (1913) referred to this condition as "von Laue's theorem."

The resulting equations for the electron's energy and momentum were identical to those obtained from assuming quasi-stationary motion, and also to those pertaining to a point particle:[28]

$$G_x = (v/c^2) E_T' / \sqrt{1 - v^2/c^2},\qquad(12.126)$$

$$E_T = E_T' / \sqrt{1 - v^2/c^2},\qquad(12.127)$$

where E_T' is the particle's rest-mass energy, and Eq. (12.127) is equivalent to Eq. (12.34). Thus, as concerns the extended electron, von Laue was led to conclude that "we can never draw from considerations based upon quasi-stationary acceleration any consequences concerning the shape, [the] distribution of charge, or the electromagnetic portion of its total momentum."[29] Von Laue directed this statement also at Eqs. (12.118) and (12.119) because they could be satisfied by many different models of the electron. For all intensive purposes this result sounded the death knell for the electromagnetic world-picture as it was conceived of by Abraham, Lorentz and Poincaré;[30] it also provided further support for von Laue's contention that, according to the relativity theory, the mechanics of particles could be deduced from the mechanics of continuous media, but the converse was not true.

But a complete reply to Ehrenfest's question required that von Laue make still another major contribution to the relativity theory, in addition to the axiomatic assertion that physical systems be closed: namely, his explication of the properties of the flow of momentum density.[31] This contribution was based on a prerelativistic result of Abraham [see Eq. (1.126)] which found its most general expression in the *kinematics* of special relativity: a body's momentum need not be collinear with its velocity, and Planck's "law of inertia" generalized this result to the flow of both momentum and energy densities; from Eqs. (12.62)–(12.64) also, it followed that to move a body in the x-direction required forces in y- and z-directions. These results of Abraham and Planck found their expression in von Laue's Eqs. (12.84)–(12.86). But, continued von Laue, the noncollinearity between momentum and velocity was actually not so peculiar, because, for example, a rotating axle transmitted energy down a drive shaft that was normal to the axle's angular velocity.[32]

Armed with these results von Laue could write that Einstein had been correct in 1907 when he asserted that one could not begin replying to Ehrenfest's query until a framework had been formulated for introducing forces to maintain electrodynamical equilibrium. According to von Laue, Ehrenfest's query required no reply at all, because the extended particle undergoing quasi-stationary motion could be treated as a point particle. But in reply to Ehrenfest, the relativistic dynamics of elastic bodies stated axiomatically that stresses existed to maintain a body of any shape in uniform translation. Furthermore, although certain calculational conveniences were introduced in the assumption that the electron had a spherical shape in its rest system as both Lorentz and Born had assumed, e.g., the collinearity of momentum and velocity, this assumption was unnecessary because a flow of momentum density normal to the particle's direction of motion served to prevent it from exploding; if these momentum currents could one day be calculated from a constructive theory of the electron, they would have to transform like the elements of a second-rank tensor. All these consequences stemmed from von Laue's axiomatic claim that the relativity theory had to be in agreement with the electron's stability. This was consistent with Einstein's interpretation of the relativity theory as a theory of principle.[33]

FOOTNOTES

1. His (1907e) proof went as follows. Consider that k is the instantaneous rest system for an electrically charged body of charge density ρ', as measured in k. Then in k the body's total charge e' is

$$e' = \int \rho' \, d\xi \, d\eta \, d\zeta. \tag{A}$$

In order to determine the total charge in K "at a particular time t," Einstein transformed the charge's volume element $d\xi \, d\eta \, d\zeta$ to K,

$$e' = \int \rho' J \, dx \, dy \, dz, \tag{B}$$

where J is the Jacobian for transformation from k to K — that is, $J = \gamma$. Consequently, Eq. (A) becomes

$$e' = \int \rho' \gamma \, dx \, dy \, dz. \tag{C}$$

Consequently the condition for charge invariance, i.e.,

$$e' = \int \rho \, dx \, dy \, dz = e, \tag{D}$$

is that the charge densities in k and K be related through the equation

$$\rho = \rho' \gamma, \tag{E}$$

which Einstein had deduced independently by demanding that the Maxwell-Lorentz equations be Lorentz covariant [see Eq. (12.5) for $u_x = v$]. For the general case where k is not the charged body's instantaneous rest system, the Jacobian of transformation from k to K can be deduced from the Lorentz invariance of volume elements in four-dimensional space as follows. Since

$$dx\,dy\,dz\,dt = d\xi\,d\eta\,d\zeta\,d\tau \qquad\qquad (F)$$

and we can substitute from Eq. (6.52) for $d\tau$ the expression

$$d\tau = \gamma(1 - u_x v/c^2)\,dt \qquad\qquad (G)$$

to obtain

$$dx\,dy\,dz\,dt = \gamma(1 - u_x v/c^2)\,d\xi\,d\eta\,d\zeta\,dt$$

or

$$dx\,dy\,dz\,dt = \gamma(1 - u_x v/c^2)J\,dx\,dy\,dz\,dt$$

therefore

$$J = [\gamma(1 - u_x v/c^2)]^{-1}$$

which becomes $J = \gamma$ for $u_x = v$. Consequently for the general case

$$\int \rho'\,d\xi\,d\eta\,d\zeta = \int \rho'\,[\gamma(1 - u_x v/c^2)]^{-1}\,dx\,dy\,dz$$

or

$$\rho'\,[\gamma(1 - u_x v/c^2)]^{-1} = \rho$$

which is Eq. (12.5).

2. The lack of historic credibility of Whittaker's (1973, vol. 2) well-known Chapter 2, "The Relativity Theory of Poincaré and Lorentz," has been demonstrated effectively by Holton (1973a) — see also Miller (1973) for comments on Whittaker (1973) and Guillaume (1924).

3. For simplicity Einstein assumed that when $x = y = z = t = 0$, then $\tau = \xi = \eta = \zeta = 0$, so that the transformation equations relating k and K were Eqs. (6.52)–(6.55). Einstein transformed the right-hand side of Eqs. (12.12)–(12.14) to K by application of Eqs. (9.24)–(9.29), and then he used relativistic transformations for relating the electron's acceleration between K and its instantaneous rest system k; that is,

$$a_\xi = \gamma^3 a_x, \qquad\qquad (A)$$

$$a_\eta = \gamma^2 a_y, \qquad\qquad (B)$$

$$a_\zeta = \gamma^2 a_z, \qquad\qquad (C)$$

and with these results Einstein could write the electron's equations of motion relative to K. From incorrect formulae for velocities Lorentz (1904c) managed to arrive at Eqs. (A)–(C) by restricting himself to the electron's instantaneous rest system. Poincaré (1906) deduced Eqs. (A)–(C) from their general expression in Σ'. In Einstein's view when k is not the instantaneous rest system then Eqs. (A)–(C) are

$$a_\xi = \frac{a_x}{\gamma^3(1 - u_x v/c^2)^3}, \tag{D}$$

$$a_\eta = \frac{a_y}{\gamma^2(1 - u_x v/c^2)^2} + \frac{u_y a_x v/c^2}{\gamma^2(1 - u_x v/c^2)^3}, \tag{E}$$

$$a_\zeta = \frac{a_z}{\gamma^2(1 - u_x v/c^2)^2} + \frac{u_z a_x v/c^2}{\gamma^2(1 - u_x v/c^2)^3}, \tag{F}$$

and Eqs. (D)–(F) are mathematically identical to those used by Poincaré (1906).

4. Lewis and Tolman (1909) considered such results as the variation of mass with velocity as transcending electromagnetic theory. Tolman (1911b) derived it from a *Gedanken* experiment based upon conservation of momentum and the principle of relativity. Observers in K and k throw identical balls at one another, and the balls' trajectories are normal to the relative velocity between k and K. Each observer analyzes the change in velocity of his ball owing to its elastic collision with the other one. Tolman (1912a) extended this demonstration to inelastic collisions, and emphasized the inappropriateness of the terms transverse and longitudinal mass; rather, the quantity $m_0\gamma$ "is best suited for THE mass of a moving body" (emphasis in original).

5. In fact, since the first pages of the 1906 version bear the inscription, "*Stampato il 14 dicembre 1905*," then Poincaré must have read Kaufmann's (1905) and then added this final paragraph just before his (1906) went to press.

6. In the paper that followed Bucherer's, Classen (1908) reported on an improved version of Simon's experiment for ε/μ_0^e—for example, Classen placed the anode and cathode within the uniform magnetic field of a solenoid, thereby preventing the cathode rays from encountering the fringing field of the solenoid as was the case in Simon's experiment. Classen obtained $\varepsilon/\mu_0^e = 1.7728 \times 10^7$, in good agreement with the values of Bestelmeyer and Bucherer. In fact, Classen's measured charge-to-mass ratio is only 1% higher than the currently accepted value of 1.7589×10^7.

7. As Poincaré wrote (1912a), further weight was added to Bucherer's data because they disconfirmed Bucherer's theory of the electron. This is a curious statement because Bucherer had discarded his own theory of the electron about two years previously, and in 1908 compared his data only with the theories of Abraham and Lorentz. Poincaré never elevated the principle of relativity to a convention because of disagreement with the measured value of the result of his Lorentz-covariant gravitational theory for the advance of Mercury's perihelion [see Poincaré (1912a)].

8. By 1908 many physicists considered the free positively-charged hydrogen ion to be the positive electron [see Lorentz (1909)]. This free specie of the electron eliminated a *unitarity* that was disturbing to most physicists who preferred a dualistic theory of electricity that contained free positive and negative charges [see Kaufmann (1901a)].

9. With encouragement from Planck, E. Hupka (1910) empirically investigated Einstein's second proposal of 1905, and found agreement with the "Lorentz-Einstein theory" rather than Abraham's theory, which predicted a potential P of

$$P \sim \frac{1}{v/c}\left[\ln\left(\frac{1 + v/c}{1 - v/c}\right) - 2\right].$$

This experiment was subjected to a detailed criticism by Heil (1910) who claimed that the 10% accuracy of Hupka's potential difference rendered the data inconclusive concerning the "Lorentz-Einstein relativity theory". As of mid-1910 Laub (1910) could only concur with Heil.

10. Kaufmann's contributions were "*Die Kathodenstrahlen*" and "*Die Röntgenstrahlen*." The topic of β-rays was reserved for the papers on radioactivity. As far as I know after 1908 Kaufmann ceased to publish on β-ray deflection experiments. In 1908 he moved from Bonn to Königsberg where he became Professor of experimental physics. At Königsberg his research interests were centered on the electromagnetic properties of bulk matter and the design of apparatus for electrical measurements. Kaufmann retired from Königsberg in 1935, and remained a guest professor at Freiberg until his death on 1 January 1947.

11. Ives' (1952) analysis of Einstein's (1905e) derivation of the mass-energy equivalence revealed a logical inconsistency – Einstein had assumed the result to be proven at the outset. For discussion of Ives' paper see Jammer (1961).

12. From the "Maxwell-Lorentz equations" Einstein deduced that

$$\int x\rho[\boldsymbol{v} \cdot \boldsymbol{E}]\, dV + \frac{d}{dt}\left\{\frac{1}{8\pi}\int x(E^2 + B^2)\, dV\right\} = \frac{c}{8\pi}\int (\boldsymbol{E} \times \boldsymbol{B})_x\, dV, \qquad (A)$$

where x was the center-of-mass coordinate of the ensemble of particles, each with a charge density ρ, and the center of mass moved with the velocity v; E and B were the external fields acting on each particle owing to the other particles; the integrals extended over all space. Since $\rho[\boldsymbol{v} \cdot \boldsymbol{E}] = \boldsymbol{f} \cdot \boldsymbol{v}$ (where f was the Lorentz force density), then the first term in Eq. (A) was the rate at which the electromagnetic field supplied energy density to each particle; Einstein replaced the first term in Eq. (A) with

$$\int x[\boldsymbol{f} \cdot \boldsymbol{v}]\, dV = \sum x_v\, dE_v/dt = c^2 \sum x_v\, dm_v/dt, \qquad (B)$$

where m_v is the mass of one of the particles. Next Einstein associated a mass density ρ_e of the electromagnetic field with the field's energy density $(E^2 + B^2)/8\pi$, and he replaced the second term with $d\{c^2 \int x\rho_e\, dV\}/dt$. With these substitutions Einstein's Eq. (A) became

$$\sum x_v \frac{dm_v}{dt} + \frac{d}{dt}\int x\rho_e\, dV = \frac{1}{4\pi c}\int (\boldsymbol{E} \times \boldsymbol{B})_x\, dV. \qquad (C)$$

Then, like Poincaré, Einstein demonstrated that

$$\frac{d}{dt}\left[\frac{1}{4\pi c}\int (\boldsymbol{E} \times \boldsymbol{B})_x\, dV\right] = -\int f\, dV = -\sum m_v \frac{d^2 x_v}{dt^2}; \qquad (D)$$

since the integrals extend over all space, the contributions from the Maxwell stress tensor vanish. Suppose, Einstein continued, that in Eq. (D) mass and energy are taken as independent; then the x-component of the resultant force is

$$\frac{d}{dt}\left[\sum m_v \frac{dx_v}{dt}\right] = \sum m_v \frac{d^2 x_v}{dt^2}, \qquad (E)$$

and Einstein obtained Poincaré's 1900 result

$$\frac{1}{4\pi c}\int (E \times B)_x + \sum m_v \frac{dx_v}{dt} = \text{constant.} \tag{F}$$

Einstein's combining Eqs. (C) and (F) yielded

$$\sum x_v \frac{dm_v}{dt} + \sum m_v \frac{dx_v}{dt} + \frac{d}{dt}\int x \frac{\rho_e}{4\pi} dV = \text{constant.} \tag{G}$$

But, he continued, reintroducing the dependence of mass upon energy vitiated extraction of the expected result from Eq. (G) — that is, conservation of motion of the x-coordinate; for in this case Eq. (E) should have been written for each mass m_v as

$$\frac{d}{dt}\left(m_v \frac{dx_v}{dt} \right) - m_v \frac{d^2 x_v}{dt^2} = \frac{dm_v}{dt}\frac{dx_v}{dt} \tag{H}$$

where

$$\frac{dm_v}{dt}\frac{dx_v}{dt} = \frac{1}{c^2}\int \rho \frac{dx_v}{dt}(v \cdot E)\,dV, \tag{I}$$

and so the additional term was of second order in the ratio of velocities to c^2.

13. In the late nineteenth and early twentieth centuries various experimenters believed they had discovered discrepancies in the law of conservation of mass in chemical reactions, i.e., Lavoisier's law. The discrepancies turned out to have resulted from experimental errors [see Jammer (1961)].

14. But contrary to Ives (1952) and Whittaker (1973, vol. 2), Poincaré did not discuss the proportionality of the mass of the "*fluid fictif*" to the fluid's energy density divided by c^2. In fact, in the course of rejecting his early effort to save the principle of action and reaction in Lorentz's theory, Poincaré (1908a) wrote that light "energy has no mass."

15. But the cancellation required by Hasenöhrl could never be exact: he needed the term $(1 + \frac{2}{3}\beta^2)^{-3/4}$ to be unity. For $\beta \ll 1$, this quantity becomes $1 - \frac{1}{2}\beta^2$, and the term $-\frac{1}{2}\beta^2$ vanished upon assuming that the radiation cavity underwent a Lorentz contraction.

16. The controversy ended in a stalemate because by suitably redefining his kinematical and thermodynamic variables along the lines suggested by Planck's (1907a), Hasenöhrl's and von Mosengeil's results became *equivalent mathematically*. Hasenöhrl's reputation as a brilliant physicist (he was at Solvay in 1911), coupled with the fact that he was killed in World War I, made him ideal material for propaganda against Einstein. Thus, Lenard (1930) wrote that Hasenöhrl was the discoverer of the mass-energy equivalence. For further discussion see Born (1969) and Frank (1953).

Mostly owing to Germany's defeat in World War I, Lenard's antisemitism and extreme nationalism came into the open. He was an early supporter of Hitler, subsequently becoming Hitler's authority in science. Many of Lenard's former students joined him in reorganizing science under National Socialism, among them was August Becker whose (1905) research was mentioned earlier. Stark too had early espoused the cause of National Socialism but a courageous stand by Max von Laue, in particular, forced Stark into retirement prior to World War II. During the first decade of the 20th century Lenard and Stark had been among the leaders in German science: Lenard was the Nobel laureate in 1905 (see Chapter 1, footnote 38); Stark was awarded the Nobel Prize in 1919 for having discovered the Doppler effect in fast moving hydrogen ions (canal rays) and the splitting of spectral lines in an electric field (Stark effect). Their correspondence reveals that both men had been early supporters of Einstein. For example, we recall the Einstein-Stark correspondence of 1907. Stark considered the transverse Doppler effect of the radiation from canal rays to have been proof of the relativity theory. For biographical information on Stark see Hermann (1975), and see Beyerchen (1977) for a discussion of the state of German science under National Socialism.

17. Planck's antipositivistic view and its appeal to Einstein in the years after 1909 is discussed in Holton (1973d).

18. Pursuing his program of deducing certain results from the more general view of mechanics, Tolman (1913) deduced Planck's Eqs. (12.62)–(12.64) from defining force as the rate of change of a mechanical momentum mv where $m = m_0$. In (1911a) he had succeeded for the case in which k was the moving body's instantaneous rest system. Tolman's investigations on the nature of force in relativistic mechanics were highly regarded. For example, he was the first to calculate the force on a moving test charge due to a moving source charge [Tolman (1913)]. In (1911a) he had performed the calculation by direct substitution of the electromagnetic fields arising from a moving charge, and restricted the test charge's motion to be along the y-axis of K. However, the credit is Poincaré's (1906) for having been the first to deduce from the Lorentz transformation the field of a uniformly moving charge.

19. The pressure contributes to the principle of least action a term $(p' \, dV') \, dt'$. Since $dV' \, dt'$ is a Lorentz-invariant quantity, then so is p'.

20. For cavity radiation $p' = \frac{1}{3}aT'^4$ (where a is a constant) and $E' = aV'T'^4$ (according to the Stefan-Boltzmann law), where primed quantities refer to k. Thus, in K the momentum of the cavity radiation is

$$G = \frac{\gamma v}{c^2}\left(aV'T'^4 + \frac{1}{3}aT'^4V'\right) \qquad (A)$$

(since $G'_{y'} = G'_{z'} = 0$ because k is the cavity's rest system). Then G becomes

$$G = \frac{\gamma v}{c^2}\frac{4}{3}aV'T'^4$$

$$= \gamma\frac{v}{c^2}\frac{4}{3}a\gamma V\gamma^4 T^4 \quad \text{(since } V' = \gamma V \text{ and } T' = \gamma T\text{)}$$

$$= \gamma^6\frac{v}{c^2}\frac{4}{3}aVT^4$$

$$= \frac{4}{3}\frac{aVT^4}{c^2}\frac{1}{(1 - v^2/c^2)^3}v. \qquad (B)$$

This led Planck to write that whereas G/v is the transverse mass of the cavity radiation, there are two longitudinal masses: one for constant entropy and volume, and another for constant entropy and pressure, i.e., $(\partial G/\partial v)_{S,V}$ or $(\partial G/\partial v)_{S,P}$ respectively. The result (B) had been obtained by Hasenöhrl from the Lorentz theory of radiation by the laborious process of considering Abraham's (1904b) methods for the relative and absolute rays.

21. This statement confused Tolman (1911b) who interpreted it as Einstein's having removed all physical content from Planck's Eqs. (12.62)–(12.64) unless the forces were electromagnetic.

22. In fact, Einstein went on to reverse Planck's procedure by demonstrating that an action principle followed from the relativistic expressions for force, pressure, momentum, volume and energy.

23. The other two examples were: (1) A moving system consisting of radiation within a closed massless inflexible envelope, on which no external forces acted. From Eqs. (12.69) and (12.72) the system's energy and momentum were $E = \gamma E_0$ and $G = vE/c^2$, where E_0 was the system's energy relative to comoving axes. (2) The envelope in case (1) was flexible and so the radiation pressure had to be balanced by an exterior pressure of amount $p_0 = E_0/3c^2$. From Eqs. (12.69) and (12.72) Einstein obtained

$$E = \gamma E_0 \left(1 + \frac{1}{3} \left(\frac{v}{c} \right)^2 \right)$$

$$G = \frac{4}{3} \frac{v}{c^2} E,$$

and he made no comment on the equivalence of these results to the case of a deformable electron without the Poincaré stress.

24. Einstein (1907e) used a *Gedanken* experiment to demonstrate the means for extending the principle of relativity to reference systems in relative acceleration: a reference system Σ_1 moves in the direction of its x-axis with a constant acceleration a relative to a reference system Σ_2 that remains at rest. In Σ_2 there is a constant gravitational field in which all bodies fall with an acceleration $-a$. Assuming the equivalence between inertial and gravitational masses, Newton's laws predict the same motions in Σ_1 and Σ_2; hence, Einstein concluded – "we have no cause from our experience to distinguish between Σ_1 and Σ_2; thus in what follows we shall accept the full physical equivalence of the gravitational field and the corresponding acceleration of the coordinate system." Thus, the widened principle of relativity asserts that no experiment can distinguish between the effects of a uniform acceleration and a constant gravitational field. Consequently, acceleration lost its absoluteness and with the gravitational field became a relative quantity. Einstein next used the special relativity theory as a guide in order to seek the new definition of simultaneity in a uniformly accelerating reference system Σ. He obtained an equation for time that incorporated the physical equivalence of a uniform acceleration and a constant gravitational field [Einstein (1912b) referred to this equivalence as the "principle of equivalence"]: time was a function of the gravitational potential and thus varied with position. This result led Einstein to predict an increase in the wavelength of light emitted from atoms on the sun compared with those on the earth, owing to the sun's larger gravitational field.

Although, as Einstein emphasized (1907e), the principle of equivalence was known to hold for mechanical processes, the generalized principle of relativity embraced all physical processes and so also should the principle of equivalence. In order to seek the consequences of these assertions, Einstein next exploited the new principle of relativity's "heuristic value." Working to order $a\xi/c^2$, where ξ is an element of distance in Σ, Einstein applied his new time coordinate to transform the Maxwell-Lorentz equations from a Galilean accelerated reference system into Σ (precisely analogous, incidentally, to Lorentz's methods of transformation during 1892–1904). Demanding that the Maxwell-Lorentz equations be covariant in Σ meant that Σ's acceleration should be replaced by a constant gravitational field with potential $\Phi = a\xi$, thereby rendering Σ into a rest system. Two results followed: (1) The velocity of light c was no longer constant; rather, c depended on the potential Φ. (2) As he had done earlier in (1907e) for special relativity, Einstein next deduced from the covariant Maxwell-Lorentz equations in Σ with sources, the equation for conservation of energy. In this result Einstein identified an energy of position E in the gravitational field; according to his earlier results E/c^2 is a mass – the gravitational mass. But from the "physical equivalence" of a uniform acceleration and a homogeneous gravitational field, E/c^2 must also be the inertial mass – this was a "very remarkable result," wrote Einstein. He could go no further in 1907 than this circular argument, which served to reveal no internal inconsistencies in relating a constant acceleration with a constant gravitational field. Furthermore, Einstein's 1907 results stressed that enlarging the principle of relativity could cause him to stray far from the comfortable concepts of the special theory, i.e., the measuring rods and clocks, for the second axiom had to be dropped. Just how wide was the gap between the special theory and a generalized theory of relativity became clear by 1912.

As Einstein (1919) recalled [quoted in Holton (1973c)], the *Gedanken* experiment in his (1907e) was a version of the one that led to his discovery of "physical equivalence."

> When, in the year 1907, I was working on a summary essay concerning the special theory of relativity for the *Jahrbuch für Radioaktivität und Elektronik*, I had to try to modify Newton's theory of gravitation in such a way that it would fit into the theory [of relativity]. Attempts in this direction showed the possibility of carrying out this enterprise, but they did not satisfy me because they had to be supported by hypotheses without physical basis. At that point, there came to me the happiest thought of my life, in the following form:
>
> Just as is the case with the electric field produced by electromagnetic induction, the gravitational field has similarly only a relative existence. *For if one considers an observer in free fall, e.g. from the roof of a house, there exists for him during his fall no gravitational field – at least in his immediate vicinity.* (Italics in original.)

As Einstein (1911a) wrote, "Galileo's principle ... is one of the most universal which the observation of nature has yielded; but in spite of that the law has not found any place in the foundations of our edifice of the physical universe." Einstein's treatment of noninertial reference systems is another example of the power of his visual thinking, coupled with a strong sense of aesthetics. Although Poincaré (1902) had also noted the tacit assumption of the equality of inertial and gravitational masses, he saw no connection between acceleration and gravitation.

25. An important early indirect test of the mass-energy equivalence was its use by Sommerfeld [(1915), (1916)] for calculating the fine-structure in the hydrogen atom spectrum. In 1932–1934 the mass-energy equivalence was confirmed in the nuclear physics experiments of Cockcroft-Walton, Bainbridge, Blackett-Occhialini, and Klemperes. In a note added to the 1958 English translation of his *Theory of Relativity*, published originally in German in 1921, Pauli wrote:

> Today the equivalence of mass and energy (inertia of energy) postu-
> lated by Einstein is one of the most certain foundations of nuclear physics.

For further historically-oriented discussions of the mass-energy equivalence see Jammer (1961) and Siegel (1978).

26. Minkowski (1908a) had developed only the transformation properties of a space-time vector of the first kind.

27. However, von Laue (1911a) continued, this result was not unique because the pressure p_0^m was not determined by surface forces only; yet this "has no influence upon the result."

28. The power of relativity theory to discuss a moving particle as if it were a point had also impressed Wien in 1909, although Wien did not develop this property as did von Laue. This property of relativity theory, however, in conjunction with Bucherer's (1908a) data, led Wien to add a lengthy footnote approving of relativity theory to the (1909) text of the lecture he had given to the 1906 *Naturforscherversammlung*.

29. Therefore, statements in certain recent texts to the effect that the $\frac{4}{3}$-factor mars the Lorentz covariance of the Lorentz-Einstein theory are incorrect [e.g., Röhrlich (1965), and Panofsky and Phillips (1961)]. My (1973) discusses the post-1911 part of this story, here I shall review the status of the $\frac{4}{3}$-factor during the period covered in this book.

In the three models of the electron the $\beta = 0$ limit of m_L and m_T is

$$m_L = m_T = \mu_0^e = \frac{2}{3}(e^2/R) = \frac{4}{3}m_0^e$$

where m_0^e is the electron's mass due only to its self-coulomb field, i.e., $m_0^e = E_0^e/c^2$. Consequently, the energy of a resting electron is greater than its electrostatic mass. Neither Abraham, nor Bucherer nor Lorentz questioned this point. Whereas the $\frac{4}{3}$-factor appeared naturally in the electron theories of Abraham and Bucherer, Poincaré had to insert an internal adhesive stress into the Lorentz theory that served both to bind the Lorentz electron and to ensure the Lorentz theory's Lorentz covariance. In short, the Poincaré stress insures that the $\frac{4}{3}$-factor appears in the momentum and energy of Lorentz's electron. For example, from either Eqs. (1.196) or (12.111), when $\beta = 0$ then the Lorentz electron's total energy is E_0^e instead of $\frac{4}{3}E_0^e$, which is the case when the Poincaré stress is included [see Eqs. (1.211) or (12.123)]. Consequently, the $\frac{4}{3}$-factor's origin resides in the adhesive force postulated for the purpose of preventing the electron from exploding in its rest system. Abraham *ab initio* postulated that his electron was rigid. Bucherer and Langevin chose a constant-volume electron so that its total energy was accounted for only by Maxwell-Lorentz electrodynamics, and Poincaré invented an interior adhesive stress which imposed on the Maxwell-Lorentz electrodynamics the additional restriction of being Lorentz covariant. We recall (Section 12.5.3) Einstein's (1907d) demonstration for an internal adhesive stress based on the relativity of simultaneity; he went on to prove that this stress guaranteed a description of a moving charged body that was both Lorentz covariant and independent of the particle's orientation in its rest system.

30. However, this was not the case. I shall survey some post-1912 views on an electromagnetic foundation for physics, and take the opportunity to fill in, with broad strokes, Abraham's activities during 1908–1922.

In *The Theory of Electrons* (1909) Lorentz proved that although the Poincaré stress explained the electron's stability for changes of volume, additional stresses were necessary to maintain stability for changes of shape. In a later essay Lorentz (1917) suggested introducing these additional stresses if necessary. He continued in the essay of 1917:

> We shall always understand by "mass," the "mass of Minkowski," which is constant for each body, or particle, and independent of the choice of the system of coordinates. It is considered to be a measure of quantity of matter [and Lorentz wrote the electron's mass as $\mu_0^e = 2e^2/(3Rc^2)$].

By 1921 problems with the quantum theory of radiation led Lorentz at the 1921 Solvay Conference to discuss the question, "Can we maintain Maxwell's equations?" Furthermore, Lorentz continued, according to the theory of relativity every system's mass varies in the same manner, and consequently it was unnecessary to assume that the electron's mass was entirely electromagnetic. This was, of course, von Laue's theorem.

Until his death in 1912 Poincaré continued to pursue an electromagnetic world-picture. In the face of the difficulty pointed out by Lorentz in 1909, in 1912 Poincaré wrote, "is Lorentz's electron stable? That is a very difficult question and we make no response here" (1912a). In a lecture presented 11 April 1912 before the French Society of Physics, Poincaré said:

> It is to the self-induction of the convection currents produced by the movements of these electrons [in Rutherford's atom] that the atom which is made up of them owes its apparent inertia which we call its mass.

In 1914 Rutherford, himself, wrote that "the exceedingly small dimensions found for the atomic nucleus add weight to the suggestion that the hydrogen nucleus is the *positive electron*, and that its mass is entirely electromagnetic in origin" (italics in original).

After the appearance of Bucherer's data, Abraham focused on further explicating Minkowski's theory of the electrodynamics of moving bulk matter to which he made several lasting contributions, e.g., the construction of a symmetric energy-momentum tensor (1909a). He believed that from Minkowski's theory could emerge some sort of ether-based "electromagnetic mechanics" (1909b).

In the Preface to the 1914 edition of *Theorie der Elektrizität*, vol. 2, Abraham registered his opinion that present data had not yet decided the issue between the various electron theories. His extensive listing of experiments did not include Bucherer's 1908 experiment.

Abraham was intrigued by Einstein's (1911a) calculation of the bending of light near the sun, which Einstein had accomplished by assuming that the velocity of light is a function of the gravitational potential. After all, Abraham had always been critical of Einstein's second axiom of the special theory. Abraham (1912a, b) included the gravitational potential into the framework of Minkowski's theory, and was able to reproduce Einstein's (1911a) result for the bending of light. This line of research, in the footsteps of Poincaré, toward an "electromagnetic theory of gravitation" [Abraham (1914a)] brought Abraham into a direct controversy with Einstein, which is outside the scope of this book. Its suffices to discuss Abraham's 1914 paper that summarized his opposition to the special theory of relativity and the still evolving generalized theory of relativity. The Lorentz and Einstein theories, Abraham wrote, were "essentially identical because Einstein's rods demonstrate Lorentz's contraction and Einstein's clocks indicate Lorentz's local time." Then as Kaufmann had asserted to Planck in the discussion session to Planck's (1906b), Abraham emphasized the special theory's restriction to inertial reference systems while the "field theory provides a logical *Weltbild* that embraces a wide circle of facts . . . it includes accelerated and curvilinear motion." Abraham's preference for theories that embraced the widest class of phenomena was not unlike Einstein's criterion in his (1907e) for preferring the relativity theory over the seemingly empirically vindicated theories of Abraham and Bucherer. But it is the following spirited tirade that is indicative of how hard physical and philosophical predilections die, particularly Abraham's predilection for Mach's philosophy of science, which had also attracted Einstein as a young man [Abraham (1914a)]:

> The theory of relativity to which Minkowski later gave an appropriate mathematical form attracted the attention of many people to the new mechanics. The overthrow of the fundamental conceptions of kinematics and dynamics surprised those who had not followed the historical evolution of those problems which we have just sketched. The apparent generality of the solution of the problem of space and time met the desire of that period for unifying and synthesizing science. That is why the theory of relativity excited the young devoted to the study of mathematical physics; under the influence of that theory they filled the halls and corridors of the universities. On the other hand, the physicists of the former generation, whose philosophy was formed under the influence of Mach and Kirchhoff, remained for the most part skeptical of the audacious innovators who allowed themselves to rely upon a small number of experiments, still debated by specialists, to overthrow the fundamental tests of every physical measurement.

After thoroughly lambasting Einstein's current work on the general theory, Abraham was willing to grant special relativity theory a place in history because of its "criticism of the concepts of space and time."

In a short sequel paper (1914b), Abraham took Einstein to task for having not yet replied. Reverting to the acid tone that had caused him to have become a persona non grata in academic circles in Germany, Abraham wrote that he did not particularly care whether the majority of theorists agreed with the relativistic notions of space and time; however, he believed that if a vote were taken special relativity would not win a majority. In any case, Abraham concluded, "Einstein himself would have to vote against it because in the 1913 theory of relativity he dropped the postulate of the constancy of the velocity of light...".

Consequently, unlike Lorentz in 1906, we cannot even loosely pidgeonhole Abraham into that class of scientists referred to by certain modern philosophers of science as naive falsificationists. Even at the 1906 discussion that followed Planck's critique of Kaufmann's data [see Planck (1906b)], Abraham had placed greater emphasis on what he deemed to have been the fundamental problems with Lorentz's theory of the electron; namely that the deformable electron's stability was ensured by adhesive forces which did not square with Abraham's view of the electromagnetic world-picture. Then, in his paper of 1914, Abraham stressed his preference for theories with the widest possible explanatory power. In summary, despite the empirical data that most physicists took as having conclusively disconfirmed his theory, Abraham refused to permit these empirical data to decide the issue between the competing electron theories. It turned out that Abraham was correct, but by 1938 the problem of the correct form of m_T was moot.

In their obituary notice for Abraham, Born and von Laue recalled their last conversation with him at a meeting in 1920. Abraham had no objections against the "logical inclusiveness" of Einstein's generalized theory of relativity, and he "acknowledged" and admired its accomplishments. On the other hand, they reported, Abraham hoped for its empirical disproof. Born and von Laue aptly wrote that Abraham will be remembered in the "history of physics as the completor of classical electrodynamics." See Miller (1973) for further discussion of post-1912 attempts at an electromagnetic world-picture based in part on the special theory of relativity.

31. Von Laue (1949) is a well-known popular exposition of these results.

32. Although von Laue did not pursue this analysis further, he had also succeeded in explaining Einstein's (1907d) "remarkable result": The impulses applied simultaneously to the rod in its rest system alter the rod's internal state thereby increasing its energy relative to K. Consequently Einstein's Eq. (12.52) can be obtained from Eq. (12.112).

Thus, Einstein's (1907d) "unknown *Qualität*" from his "remarkable result" is a current of energy resulting from a change in the rod's internal state.

33. So also, continued von Laue (1911a), was the negative result of Trouton and Noble accounted for, because the system of electromagnetic field and supported capacitor plates was a perfectly static system; that is, the system was in equilibrium under the action of the electromagnetic field and the elastic forces stressing the capacitor in its rest system. In K the elastic stress produced a momentum flow that gave rise to a torque cancelling the one from the moving capacitor's electromagnetic field. Von Laue (1912b) contains the details missing from his (1911a). Within a different framework, Lorentz (1904c) had also provided a full *explanation* of Trouton and Noble's result that was based on elastic forces transforming like electromagnetic forces.

The notion of energy and momentum flow also enabled von Laue to resolve the following problem: the point of intersection of a right-angled lever is fixed to k's origin, and the rod's arms are along the x'- and y'-axes; forces of equal magnitude act on the ends of the arms and normal to the x'- and y'-axes (the direction of the force normal to the $x'(y')$-axis is in the direction of the positive $y'(x')$-axis). The lever is in equilibrium in k, but it is easy to show that an observer in K calculates that a torque acts on the lever. The problem was originally stated by Lewis and Tolman (1909) who erroneously found no torque relative to K; Sommerfeld brought their paper to von Laue's attention [von Laue (1911c)]. Applying the special relativity theory as a theory of principle, von Laue demanded that the lever be in equilibrium relative to K as well; thus he was led to ensure the system's closure with the flow of a momentum density in the proper direction to cancel any torques arising from the relativistic behavior of the rod's dimensions and of the forces acting on it [von Laue (1911a, c, d)]. See also Epstein (1911), and for further discussion Pauli (1958). Misunderstandings of the content of Einstein's special theory of relativity as a theory of principle, has led to a so-called "right-angled lever paradox" which, from time to time, surfaces in the modern literature.

EPILOGUE

EINSTEIN: THE YEARS 1905–1915

> In a memoir published four years ago [1907e] I tried to answer the question whether the propagation of light is influenced by gravitation. I return to this theme, because my previous presentation of the subject does not satisfy me . . . .
>
> A. Einstein (1911a)

During 1905–1911 Einstein's principal preoccupation was with the structure of radiation. By 1909 he had convinced himself that a granular structure for light had to emerge from a field-theoretical description of electromagnetism [see McCormmach (1970) and Einstein (1972)]. Toward this goal he attempted to modify the Maxwell-Lorentz equations into covariant nonlinear equations whose solutions contain light quanta and stable electrons as knots, i.e., singularities, in the electromagnetic field.

Although this program was unsuccessful, Einstein (1909b) maintained the necessity of altering electrodynamics to accommodate the results of cavity radiation: Einstein's predilection for a field-theoretical route for resolving the "disturbing duality" in the Maxwell-Lorentz theory ran counter to the goal of an electromagnetic world-picture, whose hallmark was the coexistence of particle and field.

Then there was the still unresolved problem of generalizing the special relativity theory, on which Einstein did not publish between 1908 and 1911, although he was working on it (as he said in unpublished essays). By 1912 Einstein (1934) had concluded that the equivalence of physical laws in accelerated reference systems required a special mathematics with which he was unfamiliar. The principle of equivalence led to such severe distortions of measuring rods that the notion of distance in noninertial systems could not easily be defined, and the dependence of time on the gravitational potential seemed to preclude a well defined notion of simultaneity. On taking up a post at the ETH in 1912, Einstein learned from his old friend Grossman, also on the faculty, just the mathematics he needed.

The description of distances in curved spaces of any number of dimensions had been initiated in 1827 by Gauss, and then developed further by Ricci, Riemann and Levi-Civita. The curvature of Gaussian coordinate systems is determined by quantities that are indigenous to these systems. Since gravity acts on all bodies in the same way, Einstein asserted that the quantities characterizing the four-dimensional curved space described gravity, and he referred to these ten quantities as the metric tensor. The enormous distortions of measuring rods in noninertial reference systems led Einstein to take a step away from a Machian view that was more drastic than those in the 1905 relativity paper – he (1913) shifted physical reality away from the space-time intervals to the metric tensor; in the case of no gravitational field, the curved space-time becomes the flat space-time of special relativity. Since matter is the source of gravitational fields, the physical characteristics of space are conditioned by matter.

Thus, Einstein went beyond the contemporaneous philosophical view, e.g. Poincaré's, that there was no relation between matter and space. The sheer intellectual virtuosity of this step, as well as Einstein's assertion that every sort of acceleration should be included in the principle of equivalence, concerned many physicists. For example, Abraham (1914a) wrote: "Even if such a theory were admissable from the mathematical point of view, it would not be true from the view of physics."

Looking back on electrodynamics in 1905, we can conjecture that its ambiguities would have been removed eventually without Einstein. The requisite mathematical formalism already existed. Refinements of already performed ether-drift experiments, with additional null experiments, would have given physicists such as Langevin, Lorentz and Poincaré the license to assert that the Maxwell-Lorentz equations had exactly the same form on the moving earth as they had relative to the ether (perhaps further fine tuning of the Maxwell-Lorentz theory would also have been required). The result of these endeavors could only have been a consistent theory of electromagnetism, and not a special theory of relativity. In his obituary to Langevin, Einstein wrote (1947):

> It appears to me as a foregone conclusion that he would have developed
> the Special Relativity Theory, had that not been done elsewhere.

What greater honor could have been bestowed on Langevin's memory? Yet Langevin's 1911 exposition of relativity theory reveals that Langevin's view of relativity physics differed fundamentally from Einstein's. Langevin developed Einstein's special relativity theory with an ether whose purpose was to have been the active agent that caused physical effects such as time dilation. Yet without Einstein, Langevin may have advanced to a new view of electrodynamics that did not contain an ether and was therefore capable of further generalization. However, general relativity could not have evolved without Einstein because its formulation required the folding together of physics and geometry. This was a double-edged masterstroke, for Einstein broke sharply with both the physics and philosophy of his day.

For Einstein the year 1915 was another Annus Mirabilis because he completed the general relativity theory.[1] Yet the disturbing duality between particle and field remained, and so it was natural for him to proceed to an attempt at a unified theory of the gravitational and electromagnetic fields from which matter could be deduced. However, the unified-field theory, and thus a fully relativistic *Weltbild*, eluded Einstein. Could it have been that for the final step he lacked further insights such as the *Gedanken* experiments of 1895 and 1907? That part of modern intellectual history is, however, beyond the scope we set for ourselves. My goal here has been to analyze key contributions from early years of Einstein's scientific career, so that we may better appreciate the emergence of a style of thinking that, ever since, has characterized 20th-century physics.

FOOTNOTE

1. Einsteins's (1917b) extended the general relativity theory to considerations of cosmology. Guided by current data Einstein demanded that the gravitational-field equations describe a static, closed and homogeneous universe; this required his adding a term containing the "cosmological constant." He was uncomfortable with the cosmological constant because it was "gravely detrimental to the formal beauty of the theory" (1919b). In order to fix this constant Einstein boldly shifted from considerations on the nature of the cosmos to the universe in the small, i.e., to elementary particles. In (1919b) he replaced the Poincaré stress with a gravitational stress whose magnitude could be determined from calculating the cosmological constant in a region of space that contained only the gravitational and electromagnetic fields [see also Einstein (1921a)]. Thus, the cosmological constant became merely a constant of integration and not a "universal constant" indicative of any "fundamental law." Einstein found that three fourths of the charged particle's energy could be ascribed to its self-electromagnetic field, and one fourth to the binding gravitational field. However, his theory of the elementary particle did not posses unique solutions. Incidentally, by 1929 Edwin Hubble's observations of extra-galactic nebulae revealed that the universe was expanding; consequently, Einstein could omit the cosmological constant, thereby restoring the general relativity theory's "formal beauty".

APPENDIX

This Appendix contains my translation of Einstein's 1905 special relativity paper as published in the *Annalen der Physik*.* It should be noted that it differs from previous (and in places quite unacceptable) English language translations. In order to indicate places in the original *Annalen* particle that contain misprints and a point of possible confusion for the modern reader, I have added footnotes in square brackets that contain my initials (A.I.M.). Einstein's own four footnotes in the original paper are given here without square brackets. I have flagged typographical errors in the original *Annalen* version which were carried over into the Teubner edition and, with additional ones from that edition, went into the Dover reprint volume *The Principle of Relativity*. I have also included in square brackets Arnold Sommerfeld's footnotes which appeared in the Dover reprint volume and in its original German version – I have added the initials A.S. to these footnotes. I have changed only Einstein's notation for the velocity of light in vacuum V to the symbol c, which was already becoming standard in 1905.

The history of the Teubner reprint volume is of some interest to the theme of this book. In the "Foreword" to the 1913 Teubner edition, the mathematician Otto Blumenthal wrote that he meant the book – for which he acted as editor – to be a sequel to the highly successful 1909 Teubner reprint of Minkowski's lecture "Space and Time." Blumenthal attributed to Arnold Sommerfeld the suggestion that the "fundamental original works on the principle of relativity should be collected."

Blumenthal and Sommerfeld decided on the following format for the 1913 edition: (1) An excerpt from Lorentz's (1895) book that discussed the Michelson-Morley experiment, without mentioning that it was a miniscule portion of a highly successful work (see Section 1.6); (2) Lorentz's (1904c) paper on his theory of the electron, without alerting the reader that the last portion had been omitted; (3) Einstein's special relativity paper; (4) Einstein's (1905e) derivation of the mass-energy equivalence; (5) Minkowski's lecture "Space and Time," with useful notes by Sommerfeld; (6) two selections from Lorentz's Wolfskehl lectures.

Of this cut-and-paste collection Blumenthal grandly wrote: "Therefore this small volume contains a collection of documents on the history of the principle of relativity." It has taken a great deal of work by modern historians of science to straighten out Blumenthal's "history," that made it appear that the

---

* Passages from Einstein's paper that are cited in the text are keyed to the line numbers in the margin – see the Author's Notes to the Reader.

Michelson-Morley experiment was a prerequisite for Einstein's publication on the special relativity theory.

ON THE ELECTRODYNAMICS OF MOVING BODIES

By A. Einstein

That Maxwell's electrodynamics — the way in which it is usually understood — when applied to moving bodies, leads to asymmetries which do not appear to be inherent in the phenomena is well known. Consider, for example, the reciprocal electrodynamic interaction of a magnet and a conductor. The observable phenomenon here depends only on the relative motion of the conductor and the magnet, whereas the customary conception draws a sharp distinction between the two cases in which either the one or the other of these bodies is in motion. For if the magnet is in motion and the conductor at rest, there arises in the neighborhood of the magnet an electric field with a certain definite energy, producing a current at the places where parts of the conductor are situated. But if the magnet is at rest and the conductor in motion, no electric field arises in the neighborhood of the magnet. In the conductor, however, we find an electromotive force, to which in itself there is no corresponding energy, but which gives rise — assuming equality of relative motion in the two cases discussed — to electric currents of the same path and intensity as those produced by the electric forces in the former case.

Examples of this sort, together with the unsuccessful attempts to discover any motion of the earth relatively to the "light medium," lead to the conjecture that to the concept of absolute rest there correspond no properties of the phenomena, neither in mechanics, nor in electrodynamics, but rather that as has already been shown to quantities of the first order, for every reference system in which the laws of mechanics are valid*, the laws of electrodynamics and optics are also valid.

We will raise this conjecture (whose intent will from now on be referred to as the "Principle of Relativity") to a postulate, and moreover introduce another postulate, which is only apparently irreconcilable with the former: light is always propagated in empty space with a definite velocity c which is independent of the state of motion of the emitting body. These two postulates suffice in order to obtain a simple and consistent theory of the electrodynamics of moving bodies taking as a basis Maxwell's theory for bodies at rest. The introduction of a "luminiferous ether" will prove to be superfluous because the view here to be developed will introduce neither an "absolutely resting space" provided with special properties, nor associate a velocity-vector with a point of empty space in which electromagnetic processes occur.

The theory to be developed is based — like all electrodynamics — on the kinematics of the rigid body, since the assertions of any such theory concern

---

[* The preceeding memoir by Lorentz was not at this time known to the author. (A.S.)]

the relationships between rigid bodies (coordinate systems), clocks, and electromagnetic processes. Insufficient consideration of this circumstance is the root of the difficulties with which the electrodynamics of moving bodies presently has to contend.

I. KINEMATICAL PART

§1. Definition of Simultaneity

Let us consider a coordinate system in which the equations of Newtonian mechanics hold.* For precision of demonstration and to distinguish this coordinate system verbally from others which will be introduced later, we call it the "resting system."

If a material point is at rest relatively to this coordinate system, its position can be defined relative to it by rigid measuring rods employing the methods of Euclidean geometry, and can be expressed in Cartesian coordinates.

If we wish to describe the *motion* of a material point, we give the values of its coordinates as functions of the time. Now we must bear carefully in mind that a mathematical description of this kind has no physical meaning unless we are quite clear as to what we will understand by "time". We have to take into account that all our judgments in which time plays a role are always judgments of *simultaneous events*. If, for instance, I say, "That train arrives here at 7 o'clock," I mean something like this: "The pointing of the small hand of my watch to 7 and the arrival of the train are simultaneous events."**

It might appear possible that all the difficulties concerning the definition of "time" can be overcome by substituting "the position of the small hand of my watch" for "time." In fact such a definition is satisfactory when we are concerned with defining a time exclusively for the place where the watch is located; but it is no longer satisfactory when we have to connect in time series of events occurring at different places, or – what comes to the same thing – to evaluate the times of events occurring at places remote from the watch. We could in principle content ourselves to time events by using an observer located at the origin of the coordinate system, and equipped with a clock, who coordinates the arrival of the light signal originating from the event to be timed and traveling to his position through empty space, to be timed with the hands of his clock. Yet as we know from experience, this coordination has the disadvantage that it is not independent of the standpoint of the observer with the clock. We arrive at a much more practical arrangement by means of the following considerations.

[* i.e. to the first approximation. (A.S.)]

** We shall not here discuss the inexactitude which lurks in the concept of simultaneity of two events at (approximately) the same place, which must be removed through introducing an abstract concept.

If at the point A of space there is a clock, an observer at A can time the events in the immediate vicinity of A by coordinating the positions of the hands which are simultaneous with these events. If there is at the space point B another clock — and we wish to add, "a clock being of exactly the same characteristics as the one at A" — then it is possible for an observer at B to time the events in the immediate neighborhood of B. But, without further definitions it is not possible to compare, in respect with time, an event at A with an event at B. Thus far we have defined only an "A time" and a "B time", but no common "time" for A and B. The latter time can now be defined by requiring that by definition the "time" necessary for light to travel from A to B be identical to the "time" necessary to travel from B to A. Let a ray of light start at the "A time" t_A from A toward B, let it at the "B time" t_B be reflected at B in the direction of A, and arrive again at A at the "A time" t'_A. The two clocks run in synchronization by definition if

$$t_B - t_A = t'_A - t_B. \tag{§1.1}$$

We assume this definition of synchronization to be free of any possible contradictions, applicable to arbitrarily many points, and that the following relations are universally valid: —

1. If the clock at B synchronizes with the clock at A, the clock at A synchronizes with the clock at B.

2. If the clock at A synchronizes with the clock at B and also with the clock at C, the clocks at B and C also synchronize with each other.

Thus with the help of certain (imaginary) physical experiments we have defined what is to be understood by synchronous stationary clocks located at different places, and have clearly obtained a definition of "simultaneous," or "synchronous," and of "time." The "time" of an event is the reading simultaneous with the event of a clock at rest and located at the position of the event, this clock being synchronous, and indeed synchronous for all time determinations, with a specified clock at rest.

In addition, in agreement with experience we further require that the quantity

$$\frac{2AB}{t'_A - t_A} = c, \tag{§1.2}$$

be a universal constant (the velocity of light in empty space).

It is essential to have time defined by means of clocks at rest in a resting system, and the time now defined being appropriate to the resting system we call "the time of the resting system."

§2. On the Relativity of Lengths and Times

The following considerations are based on the principle of relativity and on the principle of the constancy of the velocity of light. We define these two principles thus —

1. The laws by which the states of physical systems undergo changes are
independent of whether these changes of state are referred to one or the other
of two coordinate systems moving relatively to each other in uniform
translational motion.

2. Any ray of light moves in the "resting" coordinate system with the definite
velocity c, which is independent of whether the ray was emitted by a resting or
by a moving body. Consequently,

$$\text{velocity} = \frac{\text{light path}}{\text{time interval}}$$

where time interval is to be understood in the sense of the definition in §1.

Consider a rigid rod at rest whose length is l when measured by a measuring-
rod which is also at rest. We now imagine the axis of the rod lying along the x-
axis of the resting coordinate system, and that a uniform motion of parallel
translation with velocity v along the x-axis in the direction of increasing x is
then imparted to the rod. We now inquire as to the length of the *moving* rod,
which we imagine to be determined by means of the following two
operations: —

(a) The observer moves together with the given measuring-rod and the rod to
be measured, and measures the length of the rod directly by superposing the
measuring-rod, in just the same way as if the rod to be measured, observer and
measuring rod were at rest.

(b) By means of clocks at rest set up in the resting system and synchronized in
accordance with §1, the observer ascertains at what points of the resting system
the two ends of the rod to be measured are located at a definite time t. The
distance between these two points, measured by the measuring-rod already
employed, which in this case is at rest, is also a length which may be designated
"the length of the rod."

According to the principle of relativity the length to be discovered by the
operation (a) — we will call it "the length of the rod in the moving
system" — must be equal to the length l of the rod at rest.

The length to be discovered by the operation (b) we will call "the length of
the (moving) rod in the resting system," which we shall determine on the basis
of our two principles. We shall find that it differs from l.

Current kinematics assumes tacitly that the lengths determined by these two
operations are precisely equal, or in other words, that a moving rigid body at
the instant of time t may in geometrical respects be perfectly represented by *the
same* body *at rest* in a definite position.

We imagine further that at the two ends A and B of the rod, clocks are placed
which synchronize with the clocks of the resting system — that is, that their
indications correspond at any instant to the "time of the resting system" at the
places where they happen to be. Consequently these clocks are "synchronous
in the resting system."

We imagine further that with each clock there is a moving observer, and that these observers apply to both clocks the criterion established in §1 for the synchronization of two clocks. Let a ray of light depart from A at the time* t_A, let it be reflected at B at the time t_B, and reach A again at the time t'_A. Taking into consideration the principle of the constancy of the velocity of light we find that

$$t_B - t_A = \frac{r_{AB}}{c - v} \quad \text{and} \quad t'_A - t_B = \frac{r_{AB}}{c + v}$$

where r_{AB} denotes the length of the moving rod — measured in the resting system. Observers moving with the moving rod would thus find that the two clocks were not synchronous, while observers in the resting system would declare the clocks to be synchronous.

Thus we see that we can attribute no *absolute* meaning to the concept of simultaneity, but that two events which, examined from a coordinate system, are simultaneous, can no longer be interpreted as simultaneous events when examined from a system which is in motion relatively to that system.

§3. Theory of the Transformation of Coordinates and Times from a Resting System to another System in Uniform Motion of Translation Relatively to the Former

Let us in "resting" space take two systems of coordinates, i.e., two systems, each of three rigid material lines, perpendicular to one another, and originating from a point. Let the X-axes of the two systems coincide, and their Y- and Z-axes, respectively, be parallel. Let each system be provided with a rigid measuring-rod and a number of clocks, and let the two measuring-rods, and likewise all the clocks of the two systems, be in all respects identical.

Now to the origin of one of the two systems (k) let a (constant) velocity v be imparted in the direction of the increasing x of the other resting system (K), and let this velocity be communicated to the coordinate axes, the relevant measuring-rod, and the clocks. To any time t of the resting system K there then will correspond a definite position of the axes of the moving system, and from reasons of symmetry we are entitled to assume that the motion of k may be such that the axes of the moving system are at the time t (this "t" always denotes a time of the resting system) parallel to the axes of the resting system.

We now imagine space to be measured from the resting system K by means of the measuring-rod at rest, and also from the moving system k by means of the measuring-rod moving with it; thus we determine the coordinates x, y, z, and ξ, η, ζ, respectively. Further, let the time t of the resting system be determined for all points thereof at which there are clocks by means of light

---

* "Time" here denotes "time of the resting system" and also "position of hands of the moving clock located at the place under discussion."

signals in the manner indicated in §1. Similarly let the time τ of the moving system be determined for all points of the moving system at which there are clocks at rest relatively to that system by applying the method, given in §1, of exchanging light signals between the points at which the latter clocks are located.

To any system of values x, y, z, t, which completely defines the place and time of an event in the resting system, there belongs a system of values ξ, η, ζ, τ, determining that event relatively to the system k, and our task is now to find the system of equations connecting these quantities.

In the first place it is clear that the equations must be *linear* on account of the properties of homogeneity which we attribute to space and time.

If we set $x' = x - vt$, it is clear that a point at rest in the system k belongs to a system of values x', y, z, independent of time. We first define τ as a function of x', y, z, and t. To do this we have to express in equations that τ is nothing else than the collection of the data of clocks at rest in system k, which have been synchronized according to the rule given in §1.

From the origin of system k let a ray be emitted at the time τ_0 along the X-axis to x', and at the time τ_1 be reflected back to the coordinate origin, arriving there at the time τ_2; thus we must have $\frac{1}{2}(\tau_0 + \tau_2) = \tau_1$, or, by inserting the arguments of the function τ and applying the principle of the constancy of the velocity of light in the resting system : —

$$\frac{1}{2}\left[\tau(0, 0, 0, t) + \tau\left(0, 0, 0, t + \frac{x'}{c - v} + \frac{x'}{c + v}\right)\right] = \tau\left(x', 0, 0, t + \frac{x'}{c - v}\right).$$

$$[\text{§3.1}]$$

Hence, if x' be chosen infinitesimally small,

$$\frac{1}{2}\left(\frac{1}{c - v} + \frac{1}{c + v}\right)\frac{\partial \tau}{\partial t} = \frac{\partial \tau}{\partial x'} + \frac{1}{c - v}\frac{\partial \tau}{\partial t}, \qquad [\text{§3.2}]$$

or

$$\frac{\partial \tau}{\partial x'} + \frac{v}{c^2 - v^2}\frac{\partial \tau}{\partial t} = 0. \qquad [\text{§3.3}]$$

It is to be noted that instead of the origin of the coordinates we might have chosen any other point for the point of origin of the ray, and the equation just obtained is therefore valid for all values of x', y, z.

A similar analysis — applied to the H- and Z-axes* — taking into account that when considered from the resting system light propagates along these axes with the velocity $\sqrt{(c^2 - v^2)}$, yields the results

_____

[* Einstein's letters H and Z are the Greek upper case symbols for eta and zeta, and refer to axes in the system k. For consistency in notation, in line 39 Einstein should have employed Ξ instead of X. Since in line 90 below appears Ξ, then his first use of X was an oversight. (A.I.M.)]

$$\frac{\partial \tau}{\partial y} = 0, \qquad \frac{\partial \tau}{\partial z} = 0.$$

Since τ is a *linear* function, it follows from these equations that

$$\tau = a\left(t - \frac{v}{c^2 - v^2}x'\right) \qquad\qquad [\S 3.4]$$

where a is a function $\phi(v)$ at present unknown, and where for brevity it is assumed that at the origin of k, $\tau = 0$, when $t = 0$.

With the help of this result we easily determine the quantities ξ, η, ζ by expressing in equations that light (as required by the principle of the constancy of the velocity of light, in combination with the principle of relativity) is also propagated with velocity c when measured in the moving system. For a ray of light emitted at the time $\tau = 0$ in the direction of the increasing ξ

$$\xi = c\tau \quad \text{or} \quad \xi = ac\left(t - \frac{v}{c^2 - v^2}x'\right). \qquad\qquad [\S 3.5]$$

But the ray moves relatively to the initial point of k, when measured in the resting system, with the velocity $c - v$, so that

$$\frac{x'}{c - v} = t. \qquad\qquad [\S 3.6]$$

If we insert this value of t in the equation for ξ, we obtain

$$\xi = a\frac{c^2}{c^2 - v^2}x'. \qquad\qquad [\S 3.7]$$

In an analogous manner we find, by considering rays moving along the two other axes, that

$$\eta = c\tau = ac\left(t - \frac{v}{c^2 - v^2}x'\right) \qquad\qquad [\S 3.8]$$

when

$$\frac{y}{\sqrt{(c^2 - v^2)}} = t, \qquad x' = 0. \qquad\qquad [\S 3.9]$$

Thus

$$\eta = a\frac{c}{\sqrt{(c^2 - v^2)}}y \quad \text{and} \quad \zeta = a\frac{c}{\sqrt{(c^2 - v^2)}}z. \qquad\qquad [\S 3.10]$$

Substituting for x' its value, we obtain

$$\tau = \phi(v)\beta(t - vx/c^2), \qquad\qquad [\S 3.11]$$

$$\xi = \phi(v)\beta(x - vt), \qquad\qquad\qquad \text{[§3.12]}$$

$$\eta = \phi(v)y, \qquad\qquad\qquad\qquad \text{[§3.13]}$$

$$\zeta = \phi(v)z, \qquad\qquad\qquad\qquad \text{[§3.14]}$$

where

$$\beta = \frac{1}{\sqrt{(1 - v^2/c^2)}}, \qquad\qquad\qquad \text{[§3.15]}$$

and ϕ is an as yet unknown function of v. If no assumption whatever be made as to the initial position of the moving system and as to the zero point of τ, an additive constant is to be placed on the right side of each of these equations.

We now have to prove that any ray of light, measured in the moving system, is propagated with the velocity c, if, as we have assumed, this is the case in the resting system; for we have not as yet furnished the proof that the principle of the constancy of the velocity of light is compatible with the principle of relativity.

At the time $t = \tau = 0$, when the coordinate origins of the two systems coincide, let a spherical wave be emitted from a source at the origin of both systems, and be propagated with the velocity c in system K. If (x, y, z) is a point just reached by this wave, then

$$x^2 + y^2 + z^2 = c^2 t^2. \qquad\qquad\qquad \text{[§3.16]}$$

Transforming this equation with the aid of our equations of transformation we obtain after a simple calculation

$$\xi^2 + \eta^2 + \zeta^2 = c^2 \tau^2. \qquad\qquad\qquad \text{[§3.17]}$$

The wave under consideration is therefore no less a spherical wave with velocity of propagation c when considered in the moving system. Consequently our two fundamental principles are compatible with each other.*

In the equations of transformation which have been developed there enters an unknown function ϕ of v, which we will now determine.

For this purpose we introduce a third system of coordinates K′, which relatively to the system k is in a state of parallel translational motion parallel to the Ξ-axis such that its coordinate origin moves with velocity $-v$ on the Ξ-axis relative to k. At the time $t = 0$ let all three origins coincide, and when $t = x = y = z = 0$ let the time t' of the system K′ be zero. We call the coordinates, measured in the system K′, x', y', z', and by a twofold application of our transformation equations, we obtain

---

[* The equations of the Lorentz transformation may be more simply deduced directly from the condition that in virtue of those equations the relation $x^2 + y^2 + z^2 = c^2 t^2$ shall have as its consequence the second relation $\xi^2 + \eta^2 + \zeta^2 = c^2 \tau^2$. (A.S.)]

$$t' = \phi(-v)\beta(-v)(\tau + v\xi/c^2) = \phi(v)\phi(-v)t, \qquad [\S 3.18]$$

$$x' = \phi(-v)\beta(-v)(\xi + v\tau) \quad = \phi(v)\phi(-v)x, \qquad [\S 3.19]$$

$$y' = \phi(-v)\eta \qquad\qquad\quad = \phi(v)\phi(-v)y, \qquad [\S 3.20]$$

$$z' = \phi(-v)\zeta \qquad\qquad\quad = \phi(v)\phi(-v)z. \qquad [\S 3.21]$$

Since the relations between x', y', z' and x, y, z do not contain the time t, the systems K and K' are at rest with respect to one another, and it is clear that the transformation from K to K' must be the identity transformation. Thus

$$\phi(v)\phi(-v) = 1. \qquad [\S 3.22]$$

We now inquire into the meaning of $\phi(v)$. We fix our attention on that part of the H-axis of system k which lies between $\xi = 0, \eta = 0, \zeta = 0$ and $\xi = 0, \eta = l, \zeta = 0$. This part of the H-axis is a rod moving perpendicularly to its axis with velocity v relatively to system K, whose ends have in K the coordinates

$$x_1 = vt, \qquad y_1 = \frac{l}{\phi(v)}, \qquad z_1 = 0 \qquad [\S 3.23]$$

and

$$x_2 = vt, \qquad y_2 = 0, \qquad z_2 = 0. \qquad [\S 3.24]$$

The length of the rod measured in K is therefore $l/\phi(v)$; and this gives us the meaning of the function $\phi(v)$. From reasons of symmetry it is now evident that the length of a given rod moving perpendicularly to its axis, measured in the resting system, can depend on only the velocity and not on the direction and the sense of the motion. Thus, the length of the moving rod measured in the resting system does not change when v and $-v$ are interchanged. Consequently $l/\phi(v) = l/\phi(-v)$, or

$$\phi(v) = \phi(-v). \qquad [\S 3.25]$$

It follows from this relation and the previous one that $\phi(v) = 1$, so that the transformation equations which have been found become

$$\tau = \beta(t - vx/c^2), \qquad [\S 3.26]$$

$$\xi = \beta(x - vt), \qquad [\S 3.27]$$

$$\eta = y, \qquad [\S 3.28]$$

$$\zeta = z, \qquad [\S 3.29]$$

where

$$\beta = 1/\sqrt{(1 - v^2/c^2)}. \qquad [\S 3.30]$$

§4. *Physical Meaning of the Equations Obtained Concerning Moving Rigid Bodies and Moving Clocks*

We consider a rigid sphere* of radius R, at rest relatively to the moving system k, and whose center is at the coordinate origin of k. The equation of the surface of this sphere moving relatively to the system K with velocity v is

$$\xi^2 + \eta^2 + \zeta^2 = R^2. \qquad [\S4.1]$$

The equation of this surface is expressed in x, y, z at the time $t = 0$ as

$$\frac{x^2}{(\sqrt{(1 - v^2/c^2)})^2} + y^2 + z^2 = R^2. \qquad [\S4.2]$$

A rigid body which, measured in a state of rest, has the form of a sphere, therefore has in a state of motion – viewed from the resting system – the form of an ellipsoid of revolution with the axes

$$R\sqrt{(1 - v^2/c^2)},\ R,\ R. \qquad [\S4.3]$$

Thus, whereas the Y- and Z-dimensions of the sphere (and therefore of every rigid body of arbitrary form) do not appear modified by the motion, the X dimension appears shortened in the ratio $1 : \sqrt{(1 - v^2/c^2)}$, i.e., the greater the value of v, the greater the shortening. For $v = c$ all moving objects – viewed from the "resting" system – shrivel up into plane figures. For velocities greater than that of light our deliberations become meaningless; we shall, however, find in what follows that the velocity of light in our theory plays the role, physically, of an infinitely great velocity.

It is clear that the same results hold good of bodies at rest in the "resting" system, viewed from a system in uniform motion.

Further, we imagine one of the clocks which are qualified to mark the time t when at rest relatively to the resting system, and the time τ when at rest relatively to the moving system, to be located at the origin of the coordinates of k, and so adjusted that it marks the time τ. What is the rate, of this clock, when viewed from the resting system?

Between the quantities, x, t and τ, which refer to the position of the clock, the equations clearly hold – and

$$\tau = \frac{1}{\sqrt{(1 - v^2/c^2)}}(t - vx/c^2)$$

and

$$x = vt.$$

Therefore,

* That is, a body possessing spherical form when examined at rest.

$$\tau = t\sqrt{(1 - v^2/c^2)} = t - (1 - \sqrt{(1 - v^2/c^2)})t$$

whence it follows that the time marked by the clock (viewed in the resting system) is slow by $1 - \sqrt{(1 - v^2/c^2)}$ seconds per second, or – neglecting magnitudes of fourth and higher order – by $\frac{1}{2}v^2/c^2$ seconds.

From this there ensues the following peculiar consequence. If at the points A and B of K there are stationary clocks which, viewed in the resting system, are synchronous; and if the clock at A is moved with the velocity v along the line AB to B, then on its arrival at B the two clocks no longer synchronize, but the clock moved from A to B lags behind the other which has remained at B by $\frac{1}{2}tv^2/c^2$ seconds (up to magnitudes of fourth and higher order), t being the time required to move the clock from A to B.

It is at once apparent that this result still holds good if the clock moves from A to B in any polygonal line, and also when the points A and B coincide.

If we assume that the result proved for a polygonal line is also valid for a continuously curved line, we obtain the theorem: If one of two synchronous clocks at A is moved in a closed curve with constant velocity until it returns to A, the journey lasting t seconds, then the clock that moved runs $\frac{1}{2}tv^2/c^2$ seconds slower than the one that remained at rest. Thus we conclude that a balance-clock* at the equator must go more slowly, by a very small amount, than a precisely similar clock situated at one of the poles under otherwise identical conditions.

§5. The Theorem of Addition of Velocities

In the system k moving along the X-axis of the system K with velocity v, let a point move in accordance with the equations

$$\xi = w_\xi\tau, \qquad \eta = w_\eta\tau, \qquad \zeta = 0,$$

where w_ξ and w_η denote constants.

Required: the motion of the point relatively to the system K. If with the help of the transformation equations developed in §3 we introduce the quantities x, y, z, t into the quations of motion of the point, we obtain

$$x = \frac{w_\xi + v}{1 + vw_\xi/c^2}\,t,$$

$$y = \frac{\sqrt{(1 - v^2/c^2)}}{1 + vw_\xi/c^2}\,w_\eta t,$$

$$z = 0.$$

Thus the law of the parallelogram of velocities is valid according to our theory

---

[* Not a pendulum-clock, which is physically a system to which the earth belongs. This case had to be excluded. (A.S.)]

only to a first approximation. We set*

$$V^2 = \left(\frac{dx}{dt}\right)^2 + \left(\frac{dy}{dt}\right)^2,$$

$$w^2 = w_\xi^2 + w_\eta^2,$$

$$\alpha = \tan^{-1} w_\eta/w_\xi,$$

α is then to be looked upon as the angle between the velocities v and w. After a
simple calculation we obtain

$$V = \frac{\sqrt{[(v^2 + w^2 + 2vw\cos\alpha) - (vw\sin\alpha)^2/c^2]}}{1 + vw\cos\alpha/c^2}$$

It is worthy of remark that v and w enter into the expression for the resultant
velocity in a symmetrical manner. If w also has the direction of the (Ξ-axis) we
get

$$V = \frac{v + w}{1 + vw/c^2}.$$

It follows from this equation that from a composition of two velocities which
are less than c, there always results a velocity less than c. For if we set $v = c - \kappa$,
$w = c - \lambda$, κ and λ being positive and less than c, then

$$V = c\frac{2c - \kappa - \lambda}{2c - \kappa - \lambda + \kappa\lambda/c} < c.$$

It follows, further, that the velocity of light c cannot be altered by
composition with a velocity less than that of light. For this case we obtain

$$V = \frac{c + w}{1 + w/c} = c.$$

We might also have obtained the formula for V, for the case when v and w have
the same direction, by compounding two transformations in accordance with
§3. If in addition to the systems K and k figuring in §3 we introduce still another
system of coordinates k′ moving parallel to k, its initial point moving on the Ξ-
axis with the velocity w, we obtain equations between the quantities x, y, z, t
and the corresponding quantities of k′, which differ from the equations found
in §3 only in that the place of "v" is taken by the quantity

$$\frac{v + w}{1 + vw/c^2};$$

[* Here the *Annalen* version contains the following typographical error:

$$\alpha = \tan^{-1} w_y/w_x$$

(A.I.M.).]

from which we see that such parallel transformations – necessarily – form a group.

We now have deduced the essential theorems of the kinematics corresponding to our two principles, and we proceed to exhibit their application to electrodynamics.

II. ELECTRODYNAMICAL PART

§6. *Transformation of the Maxwell-Hertz Equations for Empty Space. On the Nature of the Electromotive Forces Occurring in a Magnetic Field During Motion*

Let the Maxwell-Hertz equations for empty space be valid for the resting system K, so that we have

$$\frac{1}{c}\frac{\partial X}{\partial t} = \frac{\partial N}{\partial y} - \frac{\partial M}{\partial z}, \qquad \frac{1}{c}\frac{\partial L}{\partial t} = \frac{\partial Y}{\partial z} - \frac{\partial Z}{\partial y},$$

$$\frac{1}{c}\frac{\partial Y}{\partial z} = \frac{\partial L}{\partial z} - \frac{\partial N}{\partial x}, \qquad \frac{1}{c}\frac{\partial M}{\partial t} = \frac{\partial Z}{\partial x} - \frac{\partial X}{\partial z},$$

$$\frac{1}{c}\frac{\partial Z}{\partial t} = \frac{\partial M}{\partial x} - \frac{\partial L}{\partial y}, \qquad \frac{1}{c}\frac{\partial N}{\partial t} = \frac{\partial X}{\partial y} - \frac{\partial Y}{\partial x}, \qquad [\S6.1]$$

where (X, Y, Z) denotes the electric-force vector and (L, M, N) the magnetic-force vector.

If we apply to these equations the transformation developed in §3, by referring the electromagnetic processes to the coordinate system there introduced, moving with the velocity v, we obtain the equations

$$\frac{1}{c}\frac{\partial X}{\partial \tau} = \frac{\partial}{\partial \eta}\left\{\beta\left(N - \frac{v}{c}Y\right)\right\} - \frac{\partial}{\partial \zeta}\left\{\beta\left(M + \frac{v}{c}Z\right)\right\}, \quad [\S6.2]$$

$$\frac{1}{c}\frac{\partial}{\partial \tau}\left\{\beta\left(Y - \frac{v}{c}N\right)\right\} = \frac{\partial L}{\partial \zeta} \qquad\qquad - \frac{\partial}{\partial \xi}\left\{\beta\left(N - \frac{v}{c}Y\right)\right\}. \quad [\S6.3]$$

$$\frac{1}{c}\frac{\partial}{\partial \tau}\left\{\beta\left(Z + \frac{v}{c}M\right)\right\} = \frac{\partial}{\partial \xi}\left\{\beta\left(M + \frac{v}{c}Z\right)\right\} - \frac{\partial L}{\partial \eta}, \qquad [\S6.4]$$

$$\frac{1}{c}\frac{\partial L}{\partial \tau} = \frac{\partial}{\partial \zeta}\left\{\beta\left(Y - \frac{v}{c}N\right)\right\} - \frac{\partial}{\partial \eta}\left\{\beta\left(Z + \frac{v}{c}M\right)\right\}, \quad [\S6.5]$$

$$\frac{1}{c}\frac{\partial}{\partial \tau}\left\{\beta\left(M + \frac{v}{c}Z\right)\right\} = \frac{\partial}{\partial \xi}\left\{\beta\left(Z + \frac{v}{c}M\right)\right\} - \frac{\partial X}{\partial \zeta}, \qquad [\S6.6]$$

$$\frac{1}{c}\frac{\partial}{\partial \tau}\left\{\beta\left(N - \frac{v}{c}Y\right)\right\} = \frac{\partial X}{\partial \eta} \qquad\qquad - \frac{\partial}{\partial \xi}\left\{\beta\left(Y - \frac{v}{c}N\right)\right\}, \quad [\S6.7]$$

where

$$\beta = 1/\sqrt{(1 - v^2/c^2)}.$$

Now the principle of relativity requires that if the Maxwell-Hertz equations for empty space hold in system K, they also hold in system k. In other words, the vectors of the electric and the magnetic force (X', Y', Z') and (L', M', N') of the moving system k, which are defined by their ponderomotive effects on electric or magnetic substances, respectively, satisfy the following equations: —

$$\frac{1}{c}\frac{\partial X'}{\partial \tau} = \frac{\partial N'}{\partial \eta} - \frac{\partial M'}{\partial \zeta}, \qquad \frac{1}{c}\frac{\partial L'}{\partial \tau} = \frac{\partial Y'}{\partial \zeta} - \frac{\partial Z'}{\partial \eta},$$

$$\frac{1}{c}\frac{\partial Y'}{\partial \tau} = \frac{\partial L'}{\partial \zeta} - \frac{\partial N'}{\partial \xi}, \qquad \frac{1}{c}\frac{\partial M'}{\partial \tau} = \frac{\partial Z'}{\partial \xi} - \frac{\partial X'}{\partial \zeta},$$

$$\frac{1}{c}\frac{\partial Z'}{\partial \tau} = \frac{\partial M'}{\partial \xi} - \frac{\partial L'}{\partial \eta}, \qquad \frac{1}{c}\frac{\partial N'}{\partial \tau} = \frac{\partial X'}{\partial \eta} - \frac{\partial Y'}{\partial \xi}.$$

Evidently the two systems of equations found for system k must express exactly the same thing, since both systems of equations are equivalent to the Maxwell-Hertz equations for system K. Since, further, the equations of the two systems agree, with the exception of the symbols for the vectors, it follows that the functions occurring in the systems of equations at corresponding places must agree, with the exception of a factor $\psi(v)$, which is common for all functions of the one system of equations, and is independent of ξ, η, ζ and τ but depends upon v. Thus we have the relations

$$X' = \psi(v)X, \qquad\qquad L' = \psi(v)L,$$

$$Y' = \psi(v)\beta\left(Y - \frac{v}{c}N\right), \quad M' = \psi(v)\beta\left(M + \frac{v}{c}Z\right),$$

$$Z' = \psi(v)\beta\left(Z + \frac{v}{c}M\right), \quad N' = \psi(v)\beta\left(N - \frac{v}{c}Y\right).$$

If we now form the reciprocal of this system of equations, firstly by solving the equations just obtained, and secondly by applying the equations to the inverse transformation (from k to K), which is characterized by the velocity $-v$, it follows, when we consider that the two systems of equations thus obtained must be identical, that $\psi(v)\psi(-v) = 1$. Further, from reasons of symmetry* $\psi(v) = \psi(-v)$, and therefore

$$\psi(v) = 1,$$

---

* If, for example, $X = Y = Z = L = M = 0$, and $N \neq 0$, then from reasons of symmetry it is clear that when v changes sign without changing its numerical value, Y' must also change sign without changing its numerical value.

and our equations assume the form

$$X' = X, \qquad\qquad L' = L,$$

$$Y' = \beta\left(Y - \frac{v}{c}N\right), \quad M' = \beta\left(M + \frac{v}{c}Z\right),$$

$$Z' = \beta\left(Z + \frac{v}{c}M\right), \quad N' = \beta\left(N - \frac{v}{c}Y\right).$$

As to the interpretation of these equations we make the following remarks: Let a point charge of electricity have the magnitude "one" when measured in the resting system K, i.e., let it when at rest in the resting system exert a force of one dyne upon an equal quantity of electricity at a distance of one cm. By the principle of relativity this electric charge is also of the magnitude "one" when measured in the moving system. If this quantity of electricity is at rest relatively to the resting system, then by definition the vector (X, Y, Z) is equal to the force acting upon it. If the quantity of electricity is at rest relatively to the moving system (at least at the relevant instant), then the force acting upon it, measured in the moving system, is equal to the vector (X', Y', Z'). Consequently the first three equations above allow themselves to be expressed in words in the following two ways: —

1. If a unit electric point charge is moving in an electromagnetic field, there acts upon it, in addition to the electric force, an "electromotive force" which, if we neglect the terms multiplied by the second and higher powers of v/c, is equal to the vector-product of the velocity of the charge and the magnetic force, divided by the velocity of light. (Old manner of expression.)

2. If a unit electric point charge is moving in an electromagnetic field, the force acting upon it is equal to the electric force which is present at the position of this unit charge, and which we determine by transformation of the field to a coordinate system at rest relatively to the electrical unit charge. (New manner of expression.)

The analogy holds with "magnetomotive forces." We see that in the developed theory the electromotive force plays only the part of an auxiliary concept, which owes its introduction to the circumstance that electric and magnetic forces do not exist independently of the state of motion of the coordinate system.

Furthermore it is clear that the asymmetry mentioned in the introduction as arising when we consider the currents produced by the relative motion of a magnet and a conductor, now disappears. Likewise, questions as to the "seat" of electrodynamic electromotive forces (unipolar machines) become meaningless.

§7. Theory of Doppler's Principle and of Aberration

In the system K, very far from the coordinate origin, let there be a source of electrodynamic waves, which in a part of space containing the coordinate

origin may be represented to a sufficient degree of approximation by the
equations

$$X = X_0 \sin \Phi, \qquad L = L_0 \sin \Phi,$$
$$Y = Y_0 \sin \Phi, \qquad M = M_0 \sin \Phi,$$
$$Z = Z_0 \sin \Phi, \qquad N = N_0 \sin \Phi,$$

where

$$\Phi = \omega \left\{ t - \frac{1}{c}(lx + my + nz) \right\}.$$

Here (X_0, Y_0, Z_0) and (L_0, M_0, N_0) are the vectors defining the amplitude of
the wave-train, and l, m, n the direction-cosines of the wave-normals. We seek
the characteristics of these waves, when they are examined by an observer at
rest in the moving system k.

Applying the transformation equations found in §6 for electric and magnetic
forces, and those found in §3 for the coordinates and the time, we obtain
directly

$$X' = X_0 \sin \Phi', \qquad\qquad L' = L_0 \sin \Phi',$$
$$Y' = \beta(Y_0 - vN_0/c) \sin \Phi', \qquad M' = \beta(M_0 + vZ_0/c) \sin \Phi',$$
$$Z' = \beta(Z_0 + vM_0/c) \sin \Phi', \qquad N' = \beta(N_0 - vY_0/c) \sin \Phi',$$

$$\Phi' = \omega' \left\{ \tau - \frac{1}{c}(l'\xi + m'\eta + n'\zeta) \right\}$$

where

$$\omega' = \omega\beta(1 - lv/c),$$

$$l' = \frac{l - v/c}{1 - lv/c},$$

$$m' = \frac{m}{\beta(1 - lv/c)},$$

$$n' = \frac{n}{\beta(1 - lv/c)}.$$

From the equation for ω' it follows that if an observer is moving with
velocity v relatively to an infinitely distant source of light of frequency ν, in such
a way that the connecting line "light source-observer" makes the angle ϕ with
the velocity of the observer referred to a coordinate system which is at rest
relatively to the source of light, the frequency ν' of the light perceived by the
observer is given by the equation

$$v' = v \frac{1 - \cos\phi \cdot v/c}{\sqrt{(1 - v^2/c^2)}}.$$

This is Doppler's principle for arbitrary velocities. When $\phi = 0$ the equation assumes the simple form

$$v' = v \sqrt{\frac{1 - v/c}{1 + v/c}}.$$

We see that, in contrast with the customary view, when $v = -c$, $v' = \infty$.*

If we call the angle between the wave-normal (direction of the ray) in the moving system and the connecting line "light source-observer" ϕ', the equation for l' becomes

$$\cos\phi' = \frac{\cos\phi - v/c}{1 - \cos\phi \cdot v/c}.$$

This equation expresses the law of aberration in its most general form. If $\phi = \frac{1}{2}\pi$, the equation becomes simply

$$\cos\phi' = -v/c.$$

We still have to find the amplitude of the waves, as it appears in the moving system. If we call the amplitude of the electric or magnetic force A or A' respectively, accordingly as it is measured in the resting system or in the moving system, we obtain

$$A'^2 = A^2 \frac{(1 - \cos\phi \cdot v/c)^2}{1 - v^2/c^2},$$

for $\phi = 0$ this equation becomes

$$A'^2 = A^2 \frac{1 - v/c}{1 + v/c}.$$

From the equations developed here, it follows that to an observer approaching a source of light with the velocity c, this source of light must appear of infinite intensity.

[* Here the *Annalen* version has the following typographical error: "for $v = -\infty$, $v = \infty$." In the desk copy of Einstein's own set of reprints, which was presented to Professor Gerald Holton at the completion of his work on a *catalogue raisonné* of the Einstein papers in August 1964, there are some additions and corrections in Einstein's own hand. On this page the corrections consist of the deletion of the $-\infty$ (in $v = -\infty$) and its replacement by V which is the symbol that Einstein used for the velocity of light in vacuum c; and the replacement in line 17 of the phrase "connecting line 'light source-observer'" by the phrase "direction of motion." (A.I.M.)]

§8. *Transformation of the Energy of Light Rays. Theory of the Pressure of*   1
Radiation Exerted on Perfect Reflectors

Since $A^2/8\pi$ equals the energy of light per unit volume, according to the
principle of relativity, we have to regard $A'^2/8\pi$ as the energy of light per unit
volume in the moving system. Thus A'^2/A^2 would be the ratio of the "measured   5
in motion" to the "measured at rest" energy of a given light complex, if the
volume of a light complex were the same, whether measured in K or in k. This
is, however, not the case. If l, m, n are the direction-cosines of the wave-normals
of the light in the resting system, no energy passes through the surface elements
of a spherical surface moving with the velocity of light: —   10

$$(x - lct)^2 + (y - mct)^2 + (z - nct)^2 = R^2.$$

We may say, therefore, that this surface permanently encloses the same light
complex. We inquire into the amount of energy enclosed by this surface, from
the viewpoint of the system k — that is, the energy of the light complex relatively
to the system k.

The spherical surface — viewed in the moving system — is an ellipsoidal   15
surface, whose equation at the time $\tau = 0$, is

$$(\beta\xi - l\beta\xi v/c)^2 + (\eta - m\beta\xi v/c)^2 + (\zeta - n\beta\xi v/c)^2 = R^2. \qquad [\S8.1]$$

If S is the volume of the sphere, and S' the volume of the ellipsoid, then by a
simple calculation

$$\frac{S'}{S} = \frac{\sqrt{1 - v^2/c^2}}{1 - \cos\phi \cdot v/c}. \qquad [\S8.2]$$

Thus, if we call the light energy enclosed by this surface E when it is measured in
the resting system, and E' when measured in the moving system, we obtain   20

$$\frac{E'}{E} = \frac{A'^2 S'}{A^2 S} = \frac{1 - \cos\phi \cdot v/c}{\sqrt{(1 - v^2/c^2)}}, \qquad [\S8.3]$$

and, when $\phi = 0$, this formula simplifies into

$$\frac{E'}{E} = \sqrt{\frac{1 - v/c}{1 + v/c}}. \qquad [\S8.4]$$

It is remarkable that the energy and the frequency of a light complex vary
with the observer's state of motion in accordance with the same law.

Now let the coordinate plane $\xi = 0$ be a perfectly reflecting surface, at which
the plane waves considered in the previous section are reflected. We seek the   25
pressure of light exerted on the reflecting surface, and the direction, frequency,
and intensity of the light after reflection.

Let the incident light be defined by the quantities A, $\cos\phi$, v (referred to
system K). Viewed from k the corresponding quantities are

$$A' = A \frac{1 - \cos \phi \cdot v/c}{\sqrt{(1 - v^2/c^2)}}, \qquad [\S 8.5]$$

$$\cos \phi' = \frac{\cos \phi - v/c}{1 - \cos \phi \cdot v/c}, \qquad [\S 8.6]$$

$$v' = v \frac{1 - \cos \phi \cdot v/c}{\sqrt{(1 - v^2/c^2)}}. \qquad [\S 8.7]$$

Referring the process to system k, we obtain for the reflected light

$$A'' = A', \qquad [\S 8.8]$$

$$\cos \phi'' = -\cos \phi', \qquad [\S 8.9]$$

$$v'' = v'. \qquad [\S 8.10]$$

Finally, by transforming back to the resting system K, we obtain for the reflected light*

$$A''' = A'' \frac{1 + \cos \phi'' \cdot v/c}{\sqrt{(1 - v^2/c^2)}} = A \frac{1 - 2\cos \phi \cdot v/c + v^2/c^2}{1 - v^2/c^2}, \qquad [\S 8.11]$$

$$\cos \phi''' = \frac{\cos \phi'' + v/c}{1 + \cos \phi'' \cdot v/c} = -\frac{(1 + v^2/c^2)\cos \phi - 2v/c}{1 - 2\cos \phi \cdot v/c + v^2/c^2}, \qquad [\S 8.12]$$

$$v''' = v'' \frac{1 + \cos \phi'' \cdot v/c}{\sqrt{(1 - v^2/c^2)}} = v \frac{1 - 2\cos \phi \cdot v/c + v^2/c^2}{1 - v^2/c^2}. \qquad [\S 8.13]$$

The energy (measured in the resting system) incident per unit time upon the unit of surface of the mirror is evidently $A^2(c\cos\phi - v)/8\pi$. The energy leaving per unit time the unit of surface of the mirror is $A'''^2(-c\cos\phi''' + v)/8\pi$. According to the principle of energy, the difference of these two expressions is the work done per unit time by the pressure of light. If we express this work as the product Pv, where P is the pressure of light, we obtain

$$P = 2 \cdot \frac{A^2}{8\pi} \frac{(\cos \phi - v/c)^2}{1 - v^2/c^2}. \qquad [\S 8.14]$$

In agreement with experiment and with other theories, to a first approximation we obtain

$$P = 2 \cdot \frac{A^2}{8\pi} \cos^2 \phi. \qquad [\S 8.15]$$

---

* In the *Annalen* version the Eq. [§8.13] is incorrect because the part containing the frequency and angle in K has a denominator of $(1 - v/c)^2$.

All problems in the optics of moving bodies can be solved by the method here employed. The essential point is that the electric and magnetic force of the light, which is influenced by a moving body, be transformed to a coordinate system at rest relative to the body. By this means all problems in the optics of moving bodies are reduced to a series of problems in the optics of bodies at rest.

§9. Transformation of the Maxwell-Hertz Equations when Convection-Currents are Taken into Account

We start from the equations

$$\frac{1}{c}\left\{\frac{\partial X}{\partial t}+u_x\rho\right\}=\frac{\partial N}{\partial y}-\frac{\partial M}{\partial z},\qquad \frac{1}{c}\frac{\partial L}{\partial t}=\frac{\partial Y}{\partial z}-\frac{\partial Z}{\partial y},$$

$$\frac{1}{c}\left\{\frac{\partial Y}{\partial t}+u_y\rho\right\}=\frac{\partial L}{\partial z}-\frac{\partial N}{\partial x},\qquad \frac{1}{c}\frac{\partial M}{\partial t}=\frac{\partial Z}{\partial x}-\frac{\partial X}{\partial z},$$

$$\frac{1}{c}\left\{\frac{\partial Z}{\partial t}+u_z\rho\right\}=\frac{\partial M}{\partial x}-\frac{\partial L}{\partial y},\qquad \frac{1}{c}\frac{\partial N}{\partial t}=\frac{\partial X}{\partial y}-\frac{\partial Y}{\partial x},$$

where

$$\rho=\frac{\partial X}{\partial x}+\frac{\partial Y}{\partial y}+\frac{\partial Z}{\partial z}$$

denotes 4π times the charge density, and (u_x, u_y, u_z) the velocity-vector of the charge. If we imagine the electrical substances to be coupled in an unchanging manner to small rigid bodies (ions, electrons), these equations are the electromagnetic basis of Lorentz's electrodynamics and optics of moving bodies.

Let these equations be valid in the system K, and using the transformation equations given in §§3 and 6, transform them to the system k. We then obtain the equations

$$\frac{1}{c}\left\{\frac{\partial X'}{\partial \tau}+u_\xi\rho'\right\}=\frac{\partial N'}{\partial \eta}-\frac{\partial M'}{\partial \zeta},\qquad \frac{1}{c}\frac{\partial L'}{\partial \tau}=\frac{\partial Y'}{\partial \zeta}-\frac{\partial Z'}{\partial \eta},$$

$$\frac{1}{c}\left\{\frac{\partial Y'}{\partial \tau}+u_\eta\rho'\right\}=\frac{\partial L'}{\partial \zeta}-\frac{\partial N'}{\partial \xi},\qquad \frac{1}{c}\frac{\partial M'}{\partial \tau}=\frac{\partial Z'}{\partial \xi}-\frac{\partial X'}{\partial \zeta},$$

$$\frac{1}{c}\left\{\frac{\partial Z'}{\partial \tau}+u_\zeta\rho'\right\}=\frac{\partial M'}{\partial \xi}-\frac{\partial L'}{\partial \eta},\qquad \frac{1}{c}\frac{\partial N'}{\partial \tau}=\frac{\partial X'}{\partial \eta}-\frac{\partial Y'}{\partial \xi},$$

where

$$u_\xi=\frac{u_x-v}{1-u_xv/c^2},$$

$$u_\eta = \frac{u_y}{\beta(1 - u_x v/c^2)},$$

$$u_\zeta = \frac{u_z}{\beta(1 - u_x v/c^2)},$$

and

$$\rho' = \frac{\partial X'}{\partial \xi} + \frac{\partial Y'}{\partial \eta} + \frac{\partial Z'}{\partial \zeta}$$

$$= \beta(1 - u_x v/c^2)\rho.$$

Since — as follows from the addition theorem of velocities (§5) — the vector (u_ξ, u_η, u_ζ) is nothing else than the velocity of the electric substances, measured in the system k. Consequently we have proved that, on the basis of our kinematical principles, the electrodynamic foundation of Lorentz's theory of the electrodynamics of moving bodies is in agreement with the principle of relativity.

In passing, it may be remarked that the following important theorem may easily be deduced from the developed equations: If an electrically charged body is in motion anywhere in space without altering its charge when regarded from a system of coordinates moving with the body, its charge also remains constant — when viewed from the "resting" system K.

§10. Dynamics of the (Slowly Accelerated) Electron

Let there be in motion in an electromagnetic field a point particle possessing an electric charge ε (in the sequel this electrically charged particle is called an "electron"). We assume for its law of motion the following: If at a certain instant of time the electron is at rest, then in the next instant of time the electron's motion is described by the equations

$$\mu \frac{d^2 x}{dt^2} = \varepsilon X,$$

$$\mu \frac{d^2 y}{dt^2} = \varepsilon Y,$$

$$\mu \frac{d^2 z}{dt^2} = \varepsilon Z,$$

where x, y, z denote the electron's coordinates, and μ the mass of the electron, as long as its motion is slow.

Now, secondly, let the velocity of the electron at a given instant of time be v. We seek the electron's law of motion in the immediately following instant of time.

Without loss of generality, we may and will assume that the electron, at the moment when we focus our attention on it, is at the coordinate origin, and moves with the velocity v along the X-axis of the system K. It is then clear that at the designated instant ($t = 0$) the electron is at rest relatively to a coordinate system k which is in motion with a velocity v parallel to the X-axis.

From the above assumption, in combination with the principle of relativity, it is clear that in the immediately following time (for small values of t) the electron, considered from the system k, moves in accordance with the equations

$$\mu \frac{d^2\xi}{d\tau^2} = \varepsilon X',$$

$$\mu \frac{d^2\eta}{d\tau^2} = \varepsilon Y',$$

$$\mu \frac{d^2\zeta}{d\tau^2} = \varepsilon Z',$$

in which the symbols ξ, η, ζ, τ, X', Y', Z' refer to the system k. If, further, we decide that when $t = x = y = z = 0$ then $\tau = \xi = \eta = \zeta = 0$, the transformation equations of §§3 and 6 hold, so that we have

$$\xi = \beta(x - vt), \quad \eta = y, \quad \zeta = z, \quad \tau = \beta(t - vx/c^2),$$

$$X' = X, \quad Y' = \beta(Y - vN/c), \quad Z' = \beta(Z + vM/c).$$

With the help of these equations we transform the above equations of motion from system k to system K, and obtain

$$\left.\begin{array}{l} \dfrac{d^2x}{dt^2} = \dfrac{\varepsilon}{\mu\beta^3} X, \\[2mm] \dfrac{d^2y}{dt^2} = \dfrac{\varepsilon}{\mu\beta}\left(Y - \dfrac{v}{c}N\right), \\[2mm] \dfrac{d^2z}{dt^2} = \dfrac{\varepsilon}{\mu\beta}\left(Z + \dfrac{v}{c}M\right) \end{array}\right\} \quad \cdots (A)$$

Taking the customary point of view we now inquire as to the "longitudinal" and the "transverse" mass of the moving electron. We write the equations (A) in the form

$$\mu\beta^3 \frac{d^2x}{dt^2} = \varepsilon X = \varepsilon X',$$

$$\mu\beta^2 \frac{d^2y}{dt^2} = \varepsilon\beta\left(Y - \frac{v}{c}N\right) = \varepsilon Y',$$

$$\mu\beta^2 \frac{d^2z}{dt^2} = \varepsilon\beta\left(Z + \frac{v}{c}M\right) = \varepsilon Z',$$

and remark firstly that $\varepsilon X'$, $\varepsilon Y'$, $\varepsilon Z'$ are the components of the ponderomotive force acting upon the electron, as viewed in a system moving at this moment with the same velocity as the electron. (This force might be measured, for example, by a spring balance at rest in the last-mentioned system.) Now if we call this force simply "the force acting upon the electron,"* and maintain the equation — mass $\times$ acceleration $=$ force — and if we also decide that the accelerations are to be measured in the resting system K, we obtain from the above equations

$$\text{Longitudinal mass} = \frac{\mu}{(\sqrt{1 - v^2/c^2})^3},$$

$$\text{Transverse mass} = \frac{\mu}{1 - v^2/c^2}.$$

Naturally, with a different definition of force and acceleration we would obtain other values for the masses. This shows us that in comparing different theories of the motion of the electron we must proceed very cautiously.

We remark that these results as to the mass are also valid for ponderable material points, because a ponderable material point can be made into an electron (in our sense of the word) by the addition of an *arbitrarily small* electric charge.

Next we determine the electron's kinetic energy. If an electron that was initially at rest at the coordinate origin of the system K moves along the X-axis under the influence of an electrostatic force X, it is clear that the energy obtained from the electrostatic field is $\int \varepsilon X \, dx$. As the electron is to be slowly accelerated, and consequently may not emit any energy in the form of radiation, the energy obtained from the electrostatic field must be equal to the electron's energy of motion W. Bearing in mind that during the whole process of motion which we are considering, the first of the equations (A) applies, we obtain

$$W = \int \varepsilon X \, dx = \mu \int_0^v \beta^3 v \, dv$$

$$= \mu c^2 \left\{ \frac{1}{\sqrt{1 - v^2/c^2}} - 1 \right\}.$$

[* The definition of force here given is not advantageous, as was first shown by M. Planck. It is more to the point to define force in such a way that the laws of momentum and energy assume the simplest form. (A.S.)]

Thus, when $v = c$, W becomes infinite. Velocities greater than that of light have — as in our previous results — no possibility of existence.

This expression for the kinetic energy must also, by virtue of the argument stated above, apply to ponderable masses as well.

We will now enumerate the properties of the electron's motion which result from the system of equations (A), and are accessible to experiment.

1. From the second equation of the system (A) it follows that when $Y = Nv/c$, an electric force Y and a magnetic force N have an equally strong deflective action on an electron moving with the velocity v. Thus we see that it is possible by our theory to determine the velocity of the electron from the ratio of the magnetic deflection A_m to the electric deflection A_e, for any velocity, by applying the law

$$\frac{A_m}{A_e} = \frac{v}{c}.$$

This relationship may be tested experimentally, since the velocity of the electron can be directly measured, e.g. by means of rapidly oscillating electric and magnetic fields.

2. From the derivation for the electron's kinetic energy, it follows that between the potential difference traversed and the acquired velocity v of the electron there must be the relationship

$$P = \int X\,dx = \frac{\mu}{\varepsilon}c^2\left\{\frac{1}{\sqrt{1 - v^2/c^2}} - 1\right\}.$$

3. We calculate the radius of curvature R of the electron's path when a magnetic force N is present (the only deflective force) acting perpendicularly to the electron's velocity. From the second of the equations (A) we obtain

$$-\frac{d^2y}{dt^2} = \frac{v^2}{R} = \frac{\varepsilon}{\mu}\frac{v}{c}N\sqrt{1 - \frac{v^2}{c^2}}$$

or

$$R = \frac{\mu c^2}{\varepsilon} \cdot \frac{v/c}{\sqrt{(1 - v^2/c^2)}} \cdot \frac{1}{N}.$$

These three relationships are a complete expression for the laws according to which, by the theory here advanced, the electron must move.

In conclusion I wish to say that in working at the problem here dealt with I have had the loyal assistance of my friend and colleague M. Besso, and that I am indebted to him for several valuable suggestions.

BIBLIOGRAPHY

See the *Author's Notes to the Reader* for the code used in this listing. For brevity, in such journals as *Ann. Phys.* and *Philos. Mag.* I have omitted the series number, and in the case of *Ann. Phys.* the editor's name as well.

Secondary Sources – cf. pp. 434 – 440.

Primary Sources

Abraham, Max (1875–1922)

1902a   Dynamik des Electrons, *Göttinger Nachr.*, 20–41 (1902).

1902b   Prinzipien der Dynamik des Elektrons, *Phys. Z.*, *4*, 57–63 (1902).

1903   Prinzipien der Dynamik des Elektrons, *Ann. Phys.*, *10*, 105–179 (1903).

1904a   Die Grundhypothesen der Elektronentheorie, *Phys. Z.*, *5*, 576–579 (1904).

1904b   Zur Theorie der Strahlung und des Strahlungsdruckes, *Ann. Phys.*, *14*, 236–287 (1904).

1904c   *Theorie der Elektrizität: Einführung in die Maxwellsche Theorie der Elektrizität* (Leipzig: Teubner, 1904; 2nd ed., 1907). Revision of Föppl (1894).

1905   *Theorie der Elektrizität: Elektromagnetische Theorie der Strahlung* (Leipzig: Teubner, 1905; 2nd ed., 1908; 3rd ed., 1914); 1904c and this book are a two-volume set.

1909a   Zur Elektrodynamik bewegter Körper, *Rend. Circ. Mat. Palermo*, *28*, 1–28 (1909).

1909b   Zur elektromagnetischen Mechanik, *Phys. Z.*, *10*, 737–741 (1909).

1910   Die Bewegungsgleichungen eines Massenteilchens in der Relativtheorie, *Phys. Z.*, *11*, 527–530 (1910).

1912a   Zur Theorie der Gravitation, *Phys. Z.*, *13*, 1–14 (1912).

1912b   Das Elementargesetz der Gravitation, *Phys. Z.*, *13*, 4–5 (1912).

1914a   Die neue Mechanik, *Scientia*, *15*, 10–29 (1914).

1914b   Sur le probleme de la relativité, *Scientia*, *16*, 101–103 (1914).

Airy, George Biddell (1801–1892)

1871   On a supposed alteration in the amount of Astronomical Aberration of Light, produced by the passage of the Light through a considerable thickness of Refracting Medium, *Proc. R. Soc. London*, *20*, 35–39 (1871).

Arago, François (1786–1853)

1810   Mémoire sur la vitesse de la lumière, *C. R. Acad. Sci.*, *36*, 38–49 (1853). (Arago delayed publishing his results of 1810.)

Arnold, F.

1895   Über die unipolare Induktion an Wechselstrommaschinen mit ruhenden Wickelungen, *Elektrot. Z.*, *16*, 136–140 (1895).

Barnett, Samuel Johnson (1873–1956)

1912   On Electromagnetic Induction and Relative Motion, *Phys. Rev.*, *35*, 323–336 (1912).

1918   On Electromagnetic Induction and Relative Motion. II., *Phys. Rev.*, *12*, 95–114 (1918).

Becker, August

1905   Messungen an Kathodenstrahlen, *Ann. Phys.*, *17*, 381–470 (1905).

Becker, Richard (1887–1955)

1932   Unipolar-Induktion als Folge des relativistischen Zeitbegriffs, *Naturwissenschaften*, *51*, 917–919 (1932).

Bergson, Henri (1859–1941)

1922   *Durée et Simultanéité: A Propos de théorie d'Einstein* (1st ed., Paris: Librarie Felix Alcan, 1922; 2nd ed., 1923; 3rd ed., 1925; 4th ed., 1929), translation of 4th ed. by Leon Jacobson with an introduction by H. Dingle (New York: Bobbs-Merrill, 1965).

Bestelmeyer, Adolf Christoph Wilhelm (1875–1954?)
  1907   Spezifische Ladung und Geschwindigkeit der durch Röntgenstrahlen erzeugten
          Kathodenstrahlen, *Ann. Phys.*, *22*, 429–447 (1907).
  1909   Bemerkung zu der Abhandlung Herrn A. H. Bucherer's 'Die experimentelle
          Bestätigung des Relativitätsprinzips', *Ann. Phys.*, *30*, 166–174 (1909).
Boltzmann, Ludwig (1844–1906)
  1891   *Vorlesungen über Maxwell's Theorie der Elektricität und des Lichtes* (2 vols.; Leipzig:
          Barth, vol. I, 1891; vol. II, 1893).
  1897   *Vorlesungen über die Prinzipe der Mechanik* (Leipzig: Barth, 1897); see *Ludwig
          Boltzmann: Theoretical Physics and Philosophical Problems* (B. McGuiness, ed.)
          (Boston: Reidel, 1974), translated by R. Foulkes, where pp. 223–254 is a translation of
          the Preface of Boltzmann's *Vorlesungen*.
Born, Max (1882–1970)
  1909a   Die träge Masse und das Relativitätsprinzip, *Ann. Phys.*, *28*, 571–584 (1909).
  1909b   Die Theorie des starren Elektrons in der Kinematik des Relativitätsprinzips, *Ann. Phys.*,
          30, 1–56 (1909).
  1910a   Eine Ableitung der Grundgleichungen für die elektromagnetischen Vorgänge in
          bewegten Körpern vom Standpunkte der Elektronentheorie, *Math. Ann.*, *68*, 526–551
          (1909–1910).
  1910b   Über die Definition des starren Körpers in der Kinematik des Relativitätsprinzips, *Phys.
          Z.*, *11*, 233–234 (1910).
  1910c   Zur Elektrodynamik bewegter Körper, *Verh. D. Phys. Ges.*, *8*, 457–467 (1910).
  1910d   Zur Kinematik des starren Körpers im System des Relativitätsprinzips, *Göttinger
          Nachr.*, 161–179 (1910).
  1911   Elastizitätstheorie und Relativitätsprinzip, *Phys. Z.*, *12*, 569–575 (1911).
  1920   *Einstein's Theory of Relativity* (New York: Dutton, n.d.), translated from the third
          German edition of 1922 by H. L. Brose.
  1923   Max Abraham, *Phys. Z.*, *24*, 49–53 (1923). With M. von Laue.
  1969   *Physics in my Generation* (New York: Springer-Verlag, 1969).
Brace, Dewitt Bristol (1859–1905)
  1904   On Double Refraction in Matter moving through the Aether, *Philos. Mag.*, *7*, 317–329
          (1904).
  1905a   The Negative Results of Second and Third Order Tests of the 'Aether Drift,' and
          Possible First Order Methods, *Philos. Mag.*, *10*, 71–80 (1905),
  1905b   The Aether 'Drift' and Rotary Polarization, *Philos. Mag.*, *10*, 383–396 (1905).
  1905c   A Repetition of Fizeau's Experiment on the Change Produced by the Earth's Motion on
          the Rotation of the Refracted Ray, *Philos. Mag.*, *10*, 591–599 (1905).
Bradley, James (1692–1762)
  1728   A new Apparent Motion discovered in the Fixed Stars; its Cause assigned; the Velocity
          and Equable Motion of Light deduced, *Proc. Roy. Soc. London*, *35*, 308–321 (1728).
Bucherer, Alfred Heinrich (1863–1927)
  1904   *Mathematische Einführung in die Elektronentheorie* (Leipzig: Teubner, 1904).
  1905   Das deformierte Elektron und die Theorie des Elektromagnetismus, *Phys. Z.*, *6*,
          833–834 (1905).
  1907a   On a New Principle of Relativity in Electromagnetism, *Philos. Mag.*, *13*, 413–421 (1907).
  1907b   The Action of Uniform Electric and Magnetic Fields on Moving Electrons, *Philos.
          Mag.*, *13*, 721 (1907).
  1908a   Messungen an Becquerelstrahlen. Die experimentelle Bestätigung der Lorentz-
          Einsteinschen Theorie, *Phys. Z.*, *9*, 755–762 (1908).
  1908b   On the Principle of Relativity and on the Electromagnetic Mass of the Electron. A Reply
          to Mr. Cunningham, *Philos. Mag.*, *15*, 316–318 (1908).
  1909   Die experimentelle Bestätigung des Relativitätsprinzips, *Ann. Phys.*, *28*, 513–536 (1909).
Budde, Emil Arnold (1842–1921)
  1880   Das Clausius'sche Gesetz und die Bewegung der Erde im Raume, *Ann. Phys.*, *10*,

553–560 (1880); II, *Ibid.*, *12*, 644–647 (1881).

Classen, Johannes Wilhelm (1864–1928)
1908   Eine Neubestimmung von ε/μ für Kathodenstrahlen, *Phys. Z.*, *9*, 762–765 (1908).

Cohn, Emil (1854–1944)
1900   Über die Gleichungen der Electrodynamik für bewegte Körper, in *Recueil de travaux offerts par les auteurs à H. A. Lorentz* (The Hague: Nijhoff, 1900), pp. 516–523.
1902   Über die Gleichungen des elektromagnetischen Feldes für bewegte Körper, *Ann. Phys.*, *7*, 29–56 (1902).
1904a   Zur Elektrodynamik bewegter Systeme, *Berl. Ber.*, *40*, 1294–1303 (1904).
1904b   Zur Elektrodynamik bewegter Systeme. II, *Berl. Ber.*, *40*, 1404–1416 (1904).
1913   *Physikalisches über Raum und Zeit* (2 vols.; Leipzig: Teubner, 1913).

Comstock, Daniel Frost (1885–?)
1910   A Neglected Type of Relativity, *Phys. Rev.*, *30*, 267 (1910).

Des Coudres, Theodor (1862–1926)
1889   Über das Verhalten des Lichtäthers bei den Bewegungen der Erde, *Ann. Phys.*, *38*, 71–79 (1889).
1900   Zur Theorie des Kraftfeldes electrischer Ladungen, die sich mit Überlichtgeschwindigkeit bewegen, in *Recueil de travaux offerts par les auteurs à H. A. Lorentz* (The Hague: Nijhoff, 1900), pp. 652–664.

Cramp, W. and E. H. Norgrove
1936   Some Investigations on the Axial Spin of a Magnet and on the Laws of Electromagnetic Induction, *J. Inst. Elec. Eng.*, *78*, 481–491 (1936); the correspondence in *Ibid.*, *79*, 344–348 (1936).

Crocker, F. B. and C. H. Parmly
1894   Unipolar Dynamos for Electric Light and Power, *Amer. Inst. Elec. Eng.*, *11*, 406–429 (1894).

Drude, Paul (1863–1906)
1900   *The Theory of Optics* (New York: Dover, 1919), translated from the first German edition of 1900 by C. R. Mann and R. A. Millikan.

Ehrenfest, Paul (1880–1933)
        [Ehrenfest's papers are reprinted in *P. Ehrenfest Collected Scientific Papers* (M. J. Klein, ed.) (Amsterdam: North-Holland, 1959).]
1906   Zur Stabilitätsfrage bei den Bucherer-Langevin-Elektronen, *Phys. Z.*, *7*, 302–303 (1906). Reprinted in *Papers*, pp. 117–118.
1907   Die Translation deformierbarer Elektronen und der Flächensatz, *Ann. Phys.*, *23*, 204–205 (1907). Reprinted in *Papers*, pp. 144–145.
1909   Gleichförmige Rotation starrer Körper und Relativitätstheorie, *Phys. Z.*, *10*, 918 (1909). Reprinted in *Papers*, p. 154.
1910   Zu Herrn v. Ignatowskys Behandlung der Bornschen Starrheitsdefinition, *Phys. Z.*, *11*, 1127–1129 (1910). Reprinted in *Papers*, pp. 156–158.
1911   Zu Herrn v. Ignatowskys Behandlung der Bornschen Starrheitsdefinition, II, *Phys. Z.*, *12*, 412–413 (1911). Reprinted in *Papers*, pp. 159–160.
1912   Zur Frage nach der Entbehrlichkeit des Lichtäthers, *Phys. Z.*, *13*, 317–319 (1912). Reprinted in *Papers*, pp. 303–305.

Eichenwald, Aleksandr Aleksandrovich (1864–1944)
1903   Über die magnetischen Felde, *Ann. Phys.*, *11*, 1–30 and 421–441 (1903).

Einstein, Albert (1879–1955)
1901   Folgerungen aus den Kapillaritätserscheinungen, *Ann. Phys.*, *4*, 513–523 (1901).
1902a   Thermodynamische Theorie der Potentialdifferenz zwischen Metallen und vollständig dissoziierten Lösungen ihrer Salze, und eine elektrische Methode zur Erforschung der Molekularkräfte, *Ann. Phys.*, *8*, 798–814 (1902).
1902b   Kinetische Theorie des Wärmegleichgewichtes und des zweiten Hauptsatzes der Thermodynamik, *Ann. Phys.*, *9*, 417–433 (1902).
1903   Theorie der Grundlagen der Thermodynamik, *Ann. Phys.*, *11*, 170–187 (1903).

1904   Allgemeine molekulare Theorie der Wärme, *Ann. Phys.*, *14*, 354–362 (1904).

1905a   Eine neue Bestimmung der Moleküldimensionen, Inaugural-Dissertation, Zürich Universität.

1905b   Über einen die Erzeugung und Verwandlung des Lichtes betreffenden heuristischen Gesichtspunkt, *Ann. Phys.*, *17*, 132–148 (1905), translated by A. B. Arons and M. B. Peppard, *Am. J. Phys.*, *33*, 367–374 (1965).

1905c   Die von der molekularkinetischen Theorie der Wärme geforderte Bewegung von in ruhenden Flüssigkeiten suspendierten Teilchen, *Ann. Phys.*, *17*, 549–560 (1905). Reprinted in *A. Einstein, Investigations on the Theory of the Brownian Movement* (New York: Dover, 1956), translated by A. D. Cowper, with notes by R. Furth.

1905d   Zur Elektrodynamik bewegter Körper, *Ann. Phys.*, *17*, 891–921 (1905). Reprinted in *H. A. Lorentz, A. Einstein, H. Minkowski, Das Relativitätsprinzip, eine Sammlung von Abhandlungen* (Leipzig: Teubner, 1st ed., 1913; 2nd and 3rd enlarged eds., 1919, 1923), translated from the edition of 1923 by W. Perrett and G. B. Jeffery as *The Principle of Relativity: A Collection of Original Memoirs on the Special and General Theories of Relativity by H. A. Lorentz, A. Einstein, H. Minkowski and H. Weyl* (London: Metheuen, 1923); the Metheuen version was reprinted (New York: Dover, n.d.). Hereafter the Dover reprint volume is designated as *PRC*. Einstein's (1905d) is on pp. 37–65 of *PRC*. A facsimile of Einstein (1905d) is in *Selected Papers on Relativity Theory During 1905–1911: Einstein, Ehrenfest, Planck, von Laue, Langevin* (A. I. Miller, ed.) (New York: Arno Press, 1981). See the Appendix on pp. 391 − 415 for the version referred to in this book.

1905e   Ist die Trägheit eines Körpers von seinem Energieinhalt abhängig?, *Ann. Phys.*, *18*, 639–641 (1905). Reprinted in *PRC*, pp. 69–71, where the volume number is stated incorrectly and the title misspelled.

1906a   Eine neue Bestimmung der Moleküldimensionen, *Ann. Phys.*, *19*, 289–306 (1906). Reprinted in the reprint volume in 1905c, pp. 19–35.

1906b   Theorie der Lichterzeugung und Lichtabsorption, *Ann. Phys.*, *20*, 199–206 (1906).

1906c   Prinzip von der Erhaltung der Schwerpunktsbewegung und die Trägheit der Energie, *Ann. Phys.*, *20*, 627–633 (1906).

1906d   Eine Methode zur Bestimmung des Verhältnisses der transversalen und longitudinalen Masse des Elektrons, *Ann. Phys.*, *21*, 583–586 (1906).

1907a   Plancksche Theorie der Strahlung und die Theorie der spezifischen Wärme, *Ann. Phys.*, *22*, 180–190 (1907).

1907b   Möglichkeit einer neuen Prüfung des Relativitätsprinzips, *Ann. Phys.*, *23*, 197–198 (1907).

1907c   Bemerkung zur Notiz des Herrn P. Ehrenfest: Translation deformierbarer Elektronen und der Flächensatz, *Ann. Phys.*, *23*, 206–208 (1907).

1907d   Die vom Relativitätsprinzip geforderte Trägheit der Energie, *Ann. Phys.*, *23*, 371–384 (1907).

1907e   Relativitätsprinzip und die aus demselben gezogenen Folgerungen, *Jahrb. Radioakt.*, *4*, 411–462 (1907); *5*, 98–99 (Berichtigungen).

1908a   Elektromagnetische Grundgleichungen für bewegte Körper, *Ann. Phys.*, *26*, 532–540 (1908). With J. Laub.

1908b   Die im elektromagnetischen Felde auf ruhende Körper ausgeübten ponderomotorischen Kräfte, *Ann. Phys.*, *26*, 541–550 (1908). With J. Laub.

1909a   Zum gegenwärtigen Stande des Strahlungsproblems, *Phys. Z.*, *10*, 185–193 (1909).

1909b   Entwicklung unserer Anschauungen über das Wesen und die Konstitution der Strahlung, *Phys. Z.*, *10*, 817–825 (1909).

1910   Principe de relativité et ses conséquences dans la physique moderne, *Arch. Sci. Phys. et Nat.*, *29*, 5–28 and 125–244 (1910).

1911a   Einfluß der Schwerkraft auf die Ausbreitung des Lichtes, *Ann. Phys.*, *35*, 898–908 (1911). Reprinted in *PRC*, pp. 99–108.

1911b   Zum Ehrenfestschen Paradoxon, *Phys. Z.*, *12*, 509–510 (1911).

1911c   Relativitätstheorie, *Nat. Ges. Zürich, Viers.*, *56*, 1–14 (1911).

1912a   Lichtgeschwindigkeit und Statik des Gravitationsfeldes, *Ann. Phys.*, *38*, 355–369 (1912).

1912b   Theorie des statischen Gravitationsfeldes, *Ann. Phys.*, *38*, 443–458 (1912).

1913   Entwurf einer verallgemeinerten Relativitätstheorie und eine Theorie der Gravitation, I. Physikalischer Teil von A. Einstein. II. Mathematischer Teil von M. Grossmann (Leipzig: Teubner, 1913) and *Z. Math. und Phys.*, *62*, 225–261 (1913).

1915a   Zur allgemeinen Relativitätstheorie, *Verh. D. Phys. Ges.*, 778–786 and 799–801 (1915).

1915b   Das Relativitätsprinzip, in *Kultur der Gegenwart: Physik* (E. Warburg, ed.) (Leipzig: Teubner, 1915), pp. 703–713.

1916   Grundlage der allgemeinen Relativitätstheorie, *Ann. Phys.*, *49*, 769–822 (1916). Reprinted in part in *PRC*, pp. 111–173.

1917a   *Relativity, The Special and the General Theory* (Braunschweig: Vieweg, 1917; New York: Holt, 1920), translated from the fifth German Edition by R. W. Lawson.

1917b   Kosmologische Betrachtungen zur allgemeinen Relativitätstheorie, *Berl. Ber.*, 142–152 (1917). Reprinted in *PRC*, pp. 177–188.

1918a   Dialog über Einwände gegen die Relativitätstheorie, *Naturwissenschaften*, 697–702 (1918).

1918b   Motiv des Forschens. Lecture delivered in honour of Max Planck's sixtieth birthday in 1918, and reprinted with the incorrect title "Principles of Research" in A. Einstein, *Essays in Science* (New York: Philosophical Library, 1934), translated by A. Harris, pp. 1–5. Hereafter *Essays in Science* is designated as *ES*.

1918c   Prinzipielles zur allgemeinen Relativitätstheorie, *Ann. Phys.*, *55*, 241–244 (1918).

1919a   What is the Theory of Relativity, written for the London Times, 28 November 1919. Versions appear in A. Einstein, *Ideas and Opinions* (New York: Bonanza Books, n.d.), pp. 227–232; A. Einstein, *Out of My Later Years* (Totowa, New Jersey: Littlefield Adams and Co., 1967), pp. 54–57. Hereafter *Ideas and Opinions* is designated as *IO*.

1919b   Spielen Gravitationsfelder im Aufbau der materiellen Elementarteilchen eine wesentliche Rolle?, *Berl. Ber.*, 349–356 (1919). Reprinted in *PRC*, pp. 191–198.

1920   Relativity and the Ether. Lecture presented on 27 October 1920 at Leiden University, and reprinted in *ES*, pp. 98–111.

1921a   *The Meaning of Relativity: Four Lectures Delivered at Princeton University*, May 1921 (5th ed.; Princeton: Princeton University Press, 1970), translated by E. P. Adams.

1921b   Geometry and Experience. Lecture presented on 27 January 1921 to the Prussian Academy of Sciences, and reprinted in *IO*, pp. 232–246.

1922   Einstein and the Philosophies of Kant and Mach, *Nature*, 112–253 (1923).

1923   Fundamental Ideas and Problems of the Theory of Relativity. Lecture delivered on 11 July 1923 to the Nordic Assembly of Naturalists at Gothenburg, in acknowledgement of the Nobel Prize. Reprinted in *Nobel Lectures, Physics: 1901–1921* (New York: Elsevier, 1967), pp. 479–490.

1934   Notes on the Origin of the General Theory of Relativity. Reprinted in *ES*, pp. 78–84.

1936   Physics and Reality, *Journal of the Franklin Institute*, *221*, 313–347 (1936). Reprinted in *IO*, pp. 290–323.

1944   Remarks on Bertrand Russell's Theory of Knowledge, in *The Philosophy of Bertrand Russell* (P. A. Schilpp, ed.) (Evanston: The Library of Living Philosophers, 1944), pp. 277–291. Reprinted in *IO*, pp. 18–24.

1946   Autobiographical Notes, in *Albert Einstein: Philosopher-Scientist* (P. A. Schilpp, ed.) (Evanston: The Library of Living Philosophers, 1949), pp. 2–94.

1949   Paul Langevin, in Memoriam. Reprinted in *IO*, pp. 210–211.

1949   Reply to Criticisms, in *Albert Einstein: Philosopher-Scientist* (P. A. Schilpp, ed.) (Evanston: The Library of Living Philosophers, 1949), pp. 665–688.

1956   *Albert Einstein: Lettres à Maurice Solovine* (Paris: Gauthier-Villars, 1956).

1957   H. A. Lorentz, His Creative Genius and His Personality, in *H. A. Lorentz, Impressions of His Life and Work* (G. L. de Haas-Lorentz, ed.) (Amsterdam: North-Holland, 1957). Reprinted in *IO*, pp. 73–76.

1972   *Albert Einstein – Michele Besso: Correspondance 1903–1955* (Paris: Hermann, 1972),
        translated into French by P. Speziali who also supplied notes and an introduction.
Epstein, Paul Sophus (1883–1966)
    1911   Über relativische Statik, *Ann. Phys.*, *36*, 779–798 (1911).
Faraday, Michael (1791–1867)
    1932   *Faraday's Diary* (6 vols.; London: G. Bell, 1932).
    1965   *Experimental Researches in Electricity* (3 vols.; New York: Dover, 1965).
FitzGerald, George Francis (1851–1901)
    1882   On Electromagnetic Effects due to the Motion of the Earth, *Trans. R. Dublin Soc.*, *1*,
            319–324 (1882).
Fizeau, Hippolyte (1819–1896)
    1851   Sur les hypothèses relatives à l'éther lumineux, et sur une expérience qui paraît
            démontrer que le mouvement des corps change la vitesse avec laquelle la lumière se
            propage dans leur intérieur, *C. R. Acad. Sci.*, *33*, 349–355 (1851).
    1860   Sur un méthode propre à rechercher si l'azimut de polarisation du rayon réfracte est
            influence par le mouvement du corps réfringent, *C. R. Acad. Sci.*, *49*, 717–723 (1859);
            also in *Ann. Chim. Phys.*, *58*, 129–163 (1860).
Föppl, August (1854–1924)
    1891   Über magnetische Ströme, *Elektrot. Z.*, *12*, 203–205 (1891).
    1894   *Einführung in die Maxwell'sche Theorie der Elektricität* (Leipzig: Teubner, 1894).
Frank, Philipp (1884–1966) and H. Rothe
    1911   Über die Transformation der Raum-Zeitkoordinaten von ruhenden auf bewegte
            Systeme, *Ann. Phys.*, *34*, 825–855 (1911).
    1912   Zur Herleitung der Lorentztransformation, *Phys. Z.*, *13*, 750–753 (1912).
Fresnel, Augustin (1788–1827)
    1818   Lettre d'Augustin Fresnel à François Arago, sur l'influence du mouvement terrestre
            dans quelques phénomènes d'optique, *Ann. Chim. Phys.* IX, 57 (1818); translated in
            Schaffner (1972), pp. 125–135. Reprinted in A. Fresnel, *Oeuvres complètes* (3 vols.;
            Paris: Imprimerie Impériale; vol. I, 1866; vol. II, 1868; vol., III, 1870), II, pp. 627–636.
Gans, Richard (1880–1954)
    1905   Zur Elektrodynamik in bewegten Körpern, *Ann. Phys.*, *16*, 516–534 (1905).
Goldstein, Eugen (1850–1930)
    1876   Vorläufige Mittheilungen über elektrische Entladungen in verdünnten Gasen, *Berl.
            Monatsber.*, 279–245 (1876)
Grotrian, Otto (1847–1921)
    1901   Elektrometrische Untersuchungen über unipolare Induction, *Ann. Phys.*, *6*, 794–817
            (1901).
Guye, Charles-Eugène (1866–1942) and Charles Lavanchy
    1916   Vérification expérimentale de la fomule de Lorentz-Einstein par les rayons cathodiques
            de grande vitesse, *Arch. Sci. Phys. Mat.*, *41*, 286, 353 and 441 (1916).
Hasenöhrl, Friedrich (1874–1915)
    1904   Zur Theorie der Strahlung in bewegten Körpern, *Ann. Phys.*, *15*, 344–370 (1904).
    1905   Zur Theorie der Strahlung in bewegten Körpern. Berichtigung, *Ann. Phys.*, *16*, 589–592
            (1905).
Heaviside, Oliver (1850–1925)
    1889   On the electromagnetic effects due to the motion of electrifications through a dielectric,
            Philos. Mag., *27*, 324–339 (1889).
    1894   *Electrical Papers* (2 vols.; London: MacMillan, 1894).
    1925   *Electromagnetic Theory* (3 vols.; London: MacMillan, vol. I: 1893, 1922, 1925; vol. II:
            1899, 1922, 1925; vol. III: 1912, 1922, 1925).
Heil, W.
    1910   Diskussion der Versuche über die träge Masse bewegter Elektronen, *Ann. Phys.*, *31*,
            519–546 (1910); Zur 'Diskussion der Versuche über die träge Masse bewegter
            Elektronen', *Ibid.*, *33*, 403–413 (1910).

von Helmholtz, Hermann (1821–1894)
  1874   Über die Theorie der Elektrodynamik, *Borchardt's J. Math.*, *78*, 273–324 (1874).
  1875   Versuche über die im ungeschlossenen Kreise durch Bewegung inducirten elektromotorischen Kräfte, *Monatsberichte Akademie der Wissenschaften zu Berlin*, 400–418 (1875).
  1894a   The Origin and Correct Interpretation of our Sense Impressions, *Z. Psychologie und Physiologie der Sinnesorgane*, VII, 81–96 (1894), translated in *Selected Writings of Hermann von Helmholtz* (R. Kahl, ed.) (Middletown, CT: Wesleyan University Press, 1971), pp. 501–572.
  1894b   Folgerungen aus Maxwell's Theorie über die Bewegungen des reinen Aethers, *Ann. Phys.*, *53*, 135–143 (1894).
  1897   *Vorlesungen über die Elektromagnetische Theorie des Lichts* (Leipzig: Leopold Voss, 1897).
Herglotz, Gustav (1881–1953)
  1903   Zur Elektronentheorie, *Göttinger Nachr.*, 357–382 (1903).
  1910   Über den vom Standpunkt des Relativitätsprinzips aus als 'starr' zu bezeichnenden Körper, *Ann. Phys.*, *31*, 393–415 (1910).
  1911   Über die Mechanik des deformierbaren Körpers vom Standpunkte der Relativitätstheorie, *Ann. Phys.*, *36*, 493–533 (1911).
Hertz, Heinrich Rudolf (1857–1894)
  1880   On Induction in Rotating Spheres, Inaugural-Dissertation in Berlin, 15 March 1880. Reprinted in Hertz (1895), pp. 35–126.
  1884   On the Relations between Maxwell's Fundamental Electromagnetic Equations and the Fundamental Equations of the Opposing Electromagnetics, *Ann. Phys.*, *23*, 84–103 (1884). Reprinted in Hertz (1895), pp. 273–290.
  1890a   On the Fundamental Equations of Electromagnetics for Bodies at Rest, *Ann. Phys.*, *40*, 577 (1890). Reprinted in Hertz (1892), pp. 195–240.
  1890b   On the Fundamental Equations of Electromagnetics for Bodies in Motion, *Ann. Phys.*, *41*, 369 (1890). Reprinted in Hertz (1892), pp. 241–268.

          Hertz's scientific and philosophic writings are in a three volume collection that was edited by P. Lenard. For a biographical sketch of Hertz see McCorrmach (1972). The three volumes are:

  1892   *Electric Waves* (Leipzig: Teubner, 1892; London: MacMillan, 1893; New York: Dover, 1962), translated by D. E. Jones.
  1894   *The Principles of Mechanics* (Leipzig: Teubner, 1894; London: MacMillan, 1893; New York: Dover, 1956), translated by D. E. Jones and J. T. Walley, with a Preface by Hermann von Helmholtz. The Dover edition contains an Introduction by R. S. Cohen.
  1895   *Miscellaneous Papers* (Leipzig: Teubner, 1895; London: MacMillan, 1896), translated by D. E. Jones and G. A. Schott.
Hicks, William M. (1850–1934)
  1902   On the Michelson-Morley Experiment relating to the Drift of the Ether, *Philos. Mag.*, *3*, 9–42 (1902).
Hilbert, David (1862–1943)
  1909–1910
          Hermann Minkowski, *Math. Ann.*, *68*, 445–471 (1909–1910).
Hittorf, Johann Wilhelm (1824–1914)
  1869   Über die Elektricitätsleitung der Gase, *Ann. Phys.*, *136*, 1–31 (1869).
Hoek, Martinus (1834–1873)
  1868   Détermination de la vitesse avec laquelle est entraînée une onde traversant un milieu en mouvement, *Arch. Néerl.*, *3*, 180–185 (1868).
Hupka, Erich (1884–1919)
  1910   Beitrag zur Kenntnis der trägen Masse bewegter Elektronen, *Ann. Phys.*, *31*, 169–204 (1910).

von Ignatowski, Waldemar (1875–?)
    1910a Der starre Körper und das Relativitätsprinzip, *Ann. Phys.*, *33*, 607–630 (1910).
    1910b Einige allgemeine Bemerkungen zum Relativitätsprinzip, *Phys. Z.*, *11*, 972–976 (1910).
    1911a Zur Elastizitätstheorie vom Standpunkte des Relativitätsprinzips, *Phys. Z.*, *12*, 164–169 (1911).
    1911b Über Überlichtgeschwindigkeiten in der Relativitätstheorie, *Phys. Z.*, *12*, 776–778 (1911).

Ives, Herbert Eugene (1882–1953)
    1937 Light Signals on Moving Bodies as Measured by Transported Rods and Clocks, *J. Opt. Soc. Am.*, *27*, 177–180 (1937). With G. R. Stilwell.
    1938 An Experimental Study of the Rate of a Moving Atomic Clock, *J. Opt. Soc. Am.*, *28*, 215–226 (1938).
    1952 Derivation of the Mass-Energy Relation, *J. Opt. Soc. Am.*, *42*, 520–543 (1952).

Jochmann, Emil Carl Gustav Georg (1833–1871)
    1864 Über die durch einen Magnet in einem rotirenden Stromleiter inducirten elektrischen Ströme, *J. R. Angew. Math.*, *63*, 158–178 (1864), translated in *Philos. Mag.*, *27*, 506–528 (1864).

Jones, R. Clark
    1939 On the Relativistic Doppler Effect, *Phys. Rev.*, *29*, 337–339 (1939).

Kaufmann, Walter (1871–1947)
    1897a Die magnetische Ablenkbarkeit der Kathodenstrahlen und ihre Abhängigkeit vom Entladungspotential, *Ann. Phys.*, *61*, 544–552 (1897).
    1897b Über die Deflexion der Kathodenstrahlen, *Ann. Phys.*, *62*, 588–598 (1897).
    1897c Nachtrag zu der Abhandlung 'Die magnetische Ablenkbarkeit der Kathodenstrahlen etc.,' *Ann. Phys.*, *62*, 596–598 (1897).
    1898 Die magnetische Ablenkbarkeit electrostatisch beeinflußter Kathodenstrahlen, *Ann. Phys.*, *65*, 431–439 (1898).
    1901a Die Entwicklung des Elektronenbegriffs, *Phys. Z.*, *3*, 9–15 (1901), translated in *Electrician*, *48*, 94–97 (1901).
    1901b Methode zur exakten Bestimmung von Ladung und Geschwindigkeit der Becquerelstrahlen, *Phys. Z.*, *2*, 602–603 (1901).
    1901c Die magnetische und electrische Ablenkbarkeit der Becquerelstrahlen und die scheinbare Masse der Elektronen, *Göttinger Nachr.*, 143–155 (1901).
    1902a Über die elektromagnetische Masse des Elektrons, *Göttinger Nachr.*, 291–303 (1902).
    1902b Die elektromagnetische Masse des Elektrons, *Phys. Z.*, *4*, 54–57 (1902).
    1903 Über die 'Elektromagnetische Masse' der Elektronen, *Göttinger Nachr.*, 90–103, 148 – Berichtigung (1903).
    1905 Über die Konstitution des Elektrons, *Berl. Ber.*, *45*, 949–956 (1905).
    1906 Über die Konstitution des Elektrons, *Ann. Phys.*, *19*, 487–553 (1906); Nachtrag zu der Abhandlung: 'Über die Konstitution des Elektrons', *Ann. Phys.*, *20*, 639–640 (1906).
    1907 Bemerkungen zu Herrn Plancks: 'Nachtrag zu der Besprechung der Kaufmannschen Ablenkungsmessungen', *Verh. D. Phys. Ges.*, *9*, 667–673 (1907).
    1908 Erwiderung an Herrn Stark, *Verh. D. Phys. Ges.*, *10*, 91–95 (1908).

Kennard, Earle Hesse (1885–1968)
    1912 Unipolar Induction, *Philos. Mag.*, *23*, 937–941 (1912).
    1913 The Effect of Dielectrics on Unipolar Induction, *Phys. Rev.*, *1*, 355–359 (1913).
    1917 On Unipolar Induction: Another Experiment and its Significance as Evidence for the Existence of the Aether, *Philos. Mag. 33*, 179–190 (1917).

Kennedy, Roy J. and Edward M. Thorndike
    1932 Experimental Establishment of the Relativity of Time, *Phys. Rev.*, *42*, 400–418 (1932).

Kirchhoff, Gustav (1824–1887)
    1876 *Vorlesungen über mathematische Physik: Mechanik* (Leipzig: Teubner, 1876).

Langevin, Paul (1872–1946)
    1905 La physique des électrons. Lecture delivered on 22 September 1904 at the International

Congress of Arts and Science at St. Louis, Missouri, and published in *Rev. Générale Scis. Pures Appl.*, *16*, 257–276 (1905).

1911   L'évolution de l'espace et du temps. Lecture delivered on 10 April 1911 at the Philosophy Congress at Bologna, and published in *Scientia*, *10*, 31–54 (1911).

1912   *La Théorie du rayonnement et les quanta: Rapports et discussions de la réunion tenue à Bruxelles, du 30 octobre au 3 novembre 1911* (Paris: Gauthier-Villars, 1912) (P. Langevin and M. de Broglie, eds.).

Larmor, Joseph (1857–1942)

1884   Electromagnetic Induction in Conducting Sheets and Solid Bodies, *Philos. Mag.*, *17*, 1–23 (1884). Reprinted in J. Larmor, *Mathematical and Physical Papers* (2 vols.; Cambridge: Cambridge University Press, 1929), vol. I, pp. 8–28.

1895   A Dynamical Theory of the Electric and Luminiferous Medium – Part II, *Philos. Trans. R. Soc. London*, *186*, 695–742 (1895). Reprinted in *Mathematical and Physical Papers*, vol. I, pp. 543–598.

1897   A Dynamical theory of the Electric and Luminiferous Medium – Part III: Relations with Material Media, *Philos. Trans. R. Soc. London*, *190*, 205–300 (1897). Reprinted in *Mathematical and Physical Papers*, vol. II, pp. 11–132.

1900   *Aether and Matter* (Cambridge: Cambridge University Press, 1900).

1904   On the Ascertained Absence of Effects of Motion through the Aether, in Relation to the Constitution of Matter, and the Fitzgerald-Lorentz Hypothesis, *Philos. Mag.*, *7*, 621–625. Reprinted in *Mathematical and Physical Papers*, vol. II, pp. 274–280.

Laub, Jakob (1879–1962)

1907   Zur Optik der bewegten Körper, *Ann. Phys.*, *23*, 738–744 (1907).

1908   Zur Optik der bewegten Körper. II, *Ann. Phys.*, *25*, 175–184 (1908).

1910   Über die experimentellen Grundlagen des Relativitätsprinzips, *Jahrb. Radioakt.*, *7*, 405–463 (1910).

von Laue, Max (1879–1960)

1907   Die Mitführung des Lichtes durch bewegte Körper nach dem Relativitätsprinzip, *Ann. Phys.*, *23*, 989–990 (1907).

1909   Die Wellenstrahlung einer bewegten Punktladung nach dem Relativitätsprinzip, *Ann. Phys.*, *28*, 436–442 (1909).

1911a   *Das Relativitätsprinzip* (Braunschweig: Vieweg, 1911).

1911b   Zur Diskussion über den starren Körper in der Relativitätstheorie, *Phys. Z.*, *12*, 85–87 (1911).

1911c   Ein Beispiel zur Dynamik der Relativitätstheorie, *Verh. D. Phys. Ges.*, *9*, 513–518 (1911).

1911d   Bemerkungen zum Hebelgesetz in der Relativitätstheorie, *Phys. Z.*, *12*, 1008–1010 (1911).

1911e   Zur Dynamik der Relativitätstheorie, *Ann. Phys.*, *35*, 524–542 (1911).

1912a   Zwei Einwände gegen die Relativitätstheorie und ihre Widerlegung, *Phys. Z.*, *13*, 118–120 (1911).

1912b   Zur Theorie des Versuches von Trouton und Noble, *Ann. Phys.*, *38*, 370–384 (1912).

1912c   On the conception of the current of energy, *Proc. R. Acad. Amsterdam*, *14*, 825–831 (1912).

1913   Das Relativitätsprinzip, in *Jahrbücher der Philosophie* (Berlin: Mittler & Sohn, 1913), pp. 99–128.

1949   Inertia and Energy, in *Albert Einstein: Philosopher-Scientist* [see Einstein (1946)], pp. 501–533.

Lebedew, Petr N. (1866–1912)

1901   Untersuchung über die Druckkräfte des Lichtes, *Ann. Phys.*, *6*, 433–458 (1901).

Lecher, Ernst (1856–1926)

1895   Eine Studie über unipolare Induction. *Ann. Phys.*, *54*, 276–304 (1895).

Lenard, Philipp (1862–1947)

1898   Über die electrostatischen Eigenschaften der Kathodenstrahlen, *Ann. Phys.*, *64*, 279–289 (1898).

1900 Erzeugung von Kathodenstrahlen durch ultraviolettes Licht, *Ann. Phys.*, *2*, 359–379 (1900).

1905 On Cathode Rays. Lecture delivered on 28 May 1906 in acknowledgement of the Nobel Prize. Reprinted in *Nobel Lectures, Physics: 1901–1921* (New York: Elsevier, 1967), pp. 101–138.

1930 *Große Naturforscher: Eine Geschichte der Naturforschung in Lebensbeschreibungen* (Munich: Lehmans, 1930).

Lewis, Gilbert N. (1875–1946)

1908 A Revision of the Fundamental Laws of Matter and Energy, *Philos. Mag.*, *16*, 705–717 (1908).

1909 The Principle of Relativity, and Non-Newtonian Mechanics, *Philos. Mag.*, *18*, 510–523 (1909). With R. C. Tolman.

Lorentz, Hendrik Antoon (1853–1928)

 [Most of Lorentz's published papers are in H. A. Lorentz, *Collected Papers* (9 vols.; The Hague: Nijhoff, 1935–1939).]

1886 De l'influence du mouvement de la terre sur les phénomènes lumineux, *Versl. Kon. Akad. Wetensch. Amsterdam*, *2*, 297 (1886). Reprinted in *Collected Papers*, *4*, pp. 153–214.

1892a La théorie électromagnétique de Maxwell et son application aux corps mouvants, *Arch. Néerl.*, *25*, 363 (1892). Reprinted in *Collected Papers*, *2*, pp. 164–343.

1892b The relative Motion of the Earth and the Ether, *Versl. Kon. Akad. Wetensch. Amsterdam*, *1*, 74 (1892). Reprinted in *Collected Papers*, *4*, pp. 219–223.

1892c On the Reflection of light by Moving bodies, *Versl. Kon. Akad. Wetensch. Amsterdam*, *1*, 28 (1892). Reprinted in *Collected Papers*, *4*, pp. 215–218.

1895 Versuch einer Theorie der elektrischen und optischen Erscheinungen in bewegten Körpern (Leiden: Brill, 1895). Reprinted in *Collected Papers*, *5*, pp. 1–137.

1897 Concerning the problem of the dragging along of the ether by the earth, *Versl. Kon. Akad. Wetensch. Amsterdam*, *6*, 266 (1897). Reprinted in *Collected Papers*, *4*, pp. 237–244.

1898 Optical Phenomena Connected with the Charge and Mass of the Ions. I., *Versl. Kon. Akad. Wetensch. Amsterdam*, *6*, 506 (1898); II., *Ibid.*, *6*, 555 (1898). Reprinted in *Collected Papers*, *3*, pp. 17–39.

1899a Théorie simplifiée des phénomènes électriques et optiques dans des corps en mouvement, *Versl. Kon. Akad. Wetensch. Amsterdam*, *7*, 507 (1899). Reprinted in *Collected Papers*, *5*, pp. 139–155.

1899b La théorie de l'aberration de Stokes dans l'hypothèse d'un éther n'ayant pas partout la même densité, *Versl. Kon. Akad. Wetensch. Amsterdam*, *7*, 523 (1899). Reprinted in *Collected Papers*, *4*, pp. 245–251.

1900 Considérations sur la pesanteur, *Versl. Kon. Akad. Wetensch. Amsterdam*, *8*, 603 (1900). Reprinted in *Collected Papers*, *5*, pp. 198–215.

1901a De electronentheorie. Lecture delivered to the Nederl. Natuur-en Geneeskundig Congres on 2 April 1901. Reprinted in *Collected Papers*, *9*, pp. 102–111.

1901b Über die scheinbare Masse der Ionen, *Phys. Z.*, *2*, 78 (1901). Reprinted in *Collected Papers*, *3*, pp. 113–116.

1901c Boltzmann's and Wien's laws of Radiation, *Versl. Kon. Akad. Wetensch. Amsterdam*, *9*, 572 (1901). Reprinted in *Collected Papers*, *6*, pp. 280–292.

1902 The Rotation of the Plane of Polarization in Moving Media, *Versl. Kon. Akad. Wetensch. Amsterdam*, *10*, 793 (1902). Reprinted in *Collected Papers*, *5*, pp. 56–166.

1904a Maxwell's elektromagnetische Theorie, in *Encykl. Math. Wiss.*, *13*, 63–144 (1904).

1904b Weiterbildung der Maxwellschen Theorie. Elektronentheorie, in *Encykl. Math. Wiss.*, *14*, 145–288 (1904).

1904c Electromagnetic Phenomena in a System Moving with any Velocity Less than that of Light, *Proc. R. Acad. Amsterdam*, *6*, 809 (1904). Reprinted in *Collected Papers*, *5*, pp. 172–197, and in part in *PRC*, pp. 11–34.

1904d Remarque au sujet d'induction unipolaire, *Arch. Néerl.*, *9*, 380 (1904). Reprinted in

Collected Papers, 3, pp. 177–179.

1904e   Ergebnisse und Probleme der Elektronentheorie, *Elektrotechn. Verein zu Berlin, 1904* (Berlin: Springer-Verlag, 1905). Reprinted in *Collected Papers*, 8, pp. 79–124.

1909   *The Theory of Electrons* (Leiden: Brill, 1909; rev. ed., 1916; New York: Dover, 1952).

1910   Alte und neue Fragen der Physik *Phys. Z.*, *11*, 1234–1257 (1910). Reprinted in *Collected Papers*, 7, pp. 205–257.

1910–1912

    Lectures on Theoretical Physics, vol. III (London: MacMillan, 1931), translated by L. Silberstein ar.d A. P. H. Trivelli. (Lorentz's lectures at Leiden during 1910–1912 with additional material by A. D. Fokker.)

1917   The Connection Between Momentum and the Flow of Energy, Remarks Concerning the Structure of Electrons and Atoms, *Versl. Kon. Akad. Wetensch. Amsterdam*, *26*, 981 (1917). Reprinted in *Collected Papers*, 5, pp. 314–329.

1922   *Problems of Theoretical Physics* (New York: Dover, 1967). (A course of lectures delivered by Lorentz in the beginning of 1922 at the California Institute of Technology.)

1923   Notes sur la Théorie des Electrons, in *Institut International de Physique Solvay – Atomes et Electrons: Rapports et Discussions du Conseil de Physique tenu à Bruxelles du 1ᵉʳ au 6 avril 1921* (Paris: Gauthier-Villars, 1923), pp. 1–35.

Mach, Ernst (1838–1916)

(1889, 1897, 1960)

    Die Mechanik in ihrer Entwicklung historisch-kritisch dargestellt (Leipzig: F. A. Brockhaus, 1883, 1888, 1897), translated from the sixth German edition of 1908 with revisions through the ninth German edition by T. J. McCormmack as *The Science of Mechanics: A Critical and Historical Account of its Development* (LaSalle, IL: Open Court, 1960).

1910   Die Leitgedanken meiner naturwissenschaftlichen Erkenntnislehre und ihre Aufnahme durch die Zeitgenossen, *Phys. Z.*, *11*, 599–606 (1910). Reprinted in Toulmin (1970), pp. 30–43.

1921   *The Principles of Physical Optics* (Leipzig: Barth, 1921; London: Metheuen, 1926), translated by J. S. Anderson and A. F. A. Young.

1943   *Popular Scientific Lectures* (LaSalle, IL: Open Court, 1943), translated by T. J. McCormmack.

1959   *The Analysis of Sensations* (New York: Dover, 1959), translated by C. M. Williams.

1976   *Knowledge and Error* (Boston: Reidel, 1976), translated by T. J. McCormmack and P. Foulkes with an Introduction by E. Hiebert.

Mascart, Eleuthère E. N. (1837–1908)

1872   Sur les modifications qu'éprouve la lumière par suite du mouvement de la source lumineuse et du mouvement de l'observateur, *Ann. Ecole Norm.*, *1*, 157–214 (1872); *Ibid.*, *3*, 157–214 (1874).

Maxwell, James Clerk (1831–1879)

1873   *Treatise on Electricity and Magnetism* (2 vols.; Oxford: Oxford University Press, 1873; reprinted from the abridged 3rd ed.; New York: Dover, 1954).

1878   Ether, in Maxwell (1952), pp. 763–775.

1952   *The Scientific Papers of James Clerk Maxwell* (2 vols.; Cambridge: Cambridge University Press, 1890; New York: Dover, 1952) (W. D. Niven, ed.)

McCrea, W. H.

1957   Reply to Professor Dingle, *Nature*, *177*, 784–785 (1957).

Michelson, Albert Abraham (1852–1931)

1881   The Relative Motion of the Earth and the Luminiferous Ether, *Amer. J. Sci.*, *22*, 120–129 (1881). Reprinted in Swenson (1972), pp. 249–258.

1886   Influence of Motion of the Medium on the Velocity of Light, *Amer. J. Sci.*, *31*, 377–386 (1886). With Edward W. Morley. Reprinted in Swenson (1972), pp. 261–270.

1887   On the Relative Motion of the Earth and the Luminiferous Ether, *Amer. J. Sci.*, *34*, 333–345 (1887). With Edward W. Morley. Reprinted in Swenson (1972), pp. 273–285.

Mie, Gustave (1868–1957)
    1913   Grundlagen einer Theorie der Materie, *Ann. Phys.*, *40*, 1–66 (1913).

Minkowski, Hermann (1864–1909)
    1907   Das Relativitätsprinzip. Lecture delivered to the Math. Ges. Göttingen on 5 November 1907, and published in *Ann. Phys.*, *47*, 927–938 (1915).
    1908a Die Grundgleichungen für die elektromagnetischen Vorgänge in bewegten Körpern, *Göttinger Nachr.*, 53–111 (1908).
    1908b Raum und Zeit. Lecture delivered to the 80th *Naturforscherversammlung* at Cologne on 21 September 1908 in *Phys. Z.*, *20*, 104–111 (1909). Reprinted in *PRC*, pp. 75–91.

Morton, William Blair (1868–1949)
    1896   Electro-Magnetic Theory of Moving Charges, *Philos. Mag.*, *41*, 488–494 (1896).

von Mosengeil, Kurd
    1907   Theorie der stationären Strahlung in einem gleichförmig bewegten Hohlraum, *Ann. Phys.*, *22*, 867–904 (1907).

Neumann, G.
    1914   Die träge Masse schnell bewegter Elektronen, *Ann. Phys.*, *45*, 529–579 (1914).

Newton, Isaac (1642–1727)
    1687   *Philosophiae naturalis principia mathematica* (Berkeley: University of California Press, 1934), translated by A. Motte and revised by F. Cajori.

Nichols, Ernest Fox (1869–1924) and Gordon Ferry Hull
    1903   The Pressure of Radiation, *Astron. J.*, *17*, 315–351 (1903), translated as Über Strahlungsdruck, *Ann. Phys.*, *12*, 225–263 (1903).

Noether, Fritz (1884–?)
    1910   Zur Kinematik des starren Körpers in der Relativitätstheorie, *Ann. Phys.*, *31*, 919–944 (1910).

Otting, G.
    1939   Der quadratische Dopplereffekt, *Phys. Z.*, *40*, 681–687 (1939).

Pauli, Wolfgang (1900–1958)
    1958   *Theory of Relativity*, translated by G. Field (New York: Pergamon, 1958). This is a translation of Pauli's article Relativitätstheorie, in *Encykl. Math. Wiss.*, *19* (Leipzig: Teubner, 1921).

Pegram, George B. (1876–1958)
    1917   Unipolar Induction and Electron Theory, *Phys. Rev.*, *10*, 591–600 (1917).

Perrin, Jean (1870–1942)
    1895   Nouvelles propriétés des rayons cathodiques, *C. R. Acad. Sci.*, *121*, 1130–1134 (1895).

Petzoldt, Josef (1862–1929)
    1914   Die Relativitätstheorie der Physik, *Z. Positivische Philos.*, *1*, 1–56 (1914).

Planck, Max Karl Ludwig (1858–1947)
    1896   Gegen die neuere Energetik, *Ann. Phys.*, *57*, 72–78 (1896).
    1900a Über eine Verbesserung der Wienschen Spektralgleichung, *Verh. D. Phys. Ges.*, *2*, 202–204 (1900), translated in ter Haar (1967), pp. 79–81.
    1900b Zur Theorie des Gesetzes der Energieverteilung im Normalspektrum, *Verh. D. Phys. Ges.*, *2*, 237–245 (1900), translated in ter Haar (1967), pp. 82–90.
    1901   Über das Gesetz der Energieverteilung im Normalspektrum, *Ann. Phys.*, *4*, 553–563 (1901).
    1906a Das Prinzip der Relativität und die Grundgleichungen der Mechanik, *Verh. D. Phys. Ges.*, *4*, 136–141 (1906).
    1906b Die Kaufmannschen Messungen der Ablenkbarkeit der β-Strahlen in ihrer Bedeutung für die Dynamik der Elektronen, *Phys. Z.*, *7*, 753–761 (1906); published without the discussion session in *Verh. D. Phys. Ges.*, *8*, 418–432 (1906).
    1906c *Vorlesungen über die Theorie der Wärmestrahlung* (Barth: Leipzig, 1906); second revised edition of 1913 translated by M. Masius as *Theory of Radiation* (New York: Dover 1959).

1907a Zur Dynamik bewegter Systeme, *Berl. Ber.*, *13*, 542–570 (1907); also in *Ann. Phys.*, *26*, 1–34 (1908).

1907b Nachtrag zu der Besprechung der Kaufmannschen Ablenkungsmessungen, *Verh. D. Phys. Ges.*, *9*, 301–305 (1907).

1908 Bemerkungen zum Prinzip der Aktion und Reaktion in der allgemeinen Dynamik, *Phys. Z.*, *9*, 828–830 (1908).

1909 Die Einheit des physikalischen Weltbildes, *Phys. Z.*, *10*, 62–75 (1909). Reprinted in Toulmin (1970), pp. 3–27.

1910a Gleichförmige Rotation und Lorentz-Kontraktion, *Phys. Z.*, *11*, 294 (1910).

1910b Zur Machschen Theorie der physikalischen Erkenntnis, *Phys. Z.*, *11*, 1186–1190 (1910). Reprinted in Toulmin (1970), pp. 45–52.

1915 Das Prinzip der kleinsten Wirkung, in *Kultur der Gegenwart*: *Physik* (E. Warburg, ed.) (Leipzig: Teubner, 1915), pp. 692–702. Reprinted as The Principle of Least Action, in M. Planck, *A Survey of Physical Theory* (New York: Dover, 1960), translated by R. Jones and D. H. Williams, pp. 69–81.

1931 Maxwell's Influence in Germany, in *James Clerk Maxwell*: *A Commemoration Volume, 1831–1931* (Cambridge: Cambridge University Press, 1931).

Plücker, Julius (1801–1868)

1858 Über die Einwirkung des Magneten auf die elektrischen Entladungen in verdünnten Gasen, *Ann. Phys.*, *103*, 88–106, 151–157; *104*, 113–128, 622–630; *105*, 67–84.

Poincaré, Henri (1854–1912)

[Most of Poincaré's published papers are in *Oeuvres de Henri Poincaré* (11 vols.; Paris: Gauthier-Villars, 1934–1953).]

1887 Sur les hypothèses fondamentales de la Géométrie, *Bull. Soc. Math. France*, *15*, 203–216 (1887).

1891 Les Géométries non-euclidiennes, *Rev. Générale Sci. Pures Appl.* *2*, 769–774 (1891).

1895 A propos de la théorie de M. Larmor, *L'Eclairage électrique*, *3*, 5–13 and 289–295 (1895); Ibid., *5*, 5–14 and 385–392 (1895). Reprinted in *Oeuvres de Henri Poincaré*, *9*, pp. 396–426. Hereafter vol. 9 is referred to as *Oeuvres*.

1898 La mesure du temps, *Rev. Mét. Mor.*, *6*, 371–384 (1898); translated in V.S., pp. 26–36.

1899 Des fondements de la Géométrie: A propos d'un Livre de M. Russell, *Rev. Mét. Mor.*, *7*, 251–279 (1899). This is a book review of Bertrand Russell's *An Essay on the Foundation of Geometry* (Cambridge: Cambridge University Press, 1897; New York: Dover, 1952). Pages 265–269 of Poincaré's review constitute pages 75–79 of Chapter V, "Experiment and Geometry", of Poincaré's (1902).

1900a Sur les rapports de la Physique expérimentale et de la Physique mathématique, *Rapports presentés au Congrès international de Physique réuni à Paris en 1900* (4 vols.; Paris: Gauthier-Villars, 1900), vol. 1, pp. 1–29. Reprinted in German as Über die Beziehungen zwischen der experimentellen und der mathematischen Physik, *Phys. Z.*, *2*, 166–171, 182–186 and 196–201 (1900–1901). This lecture forms the substance of Chaps. IX and X of Poinaré's (1902), pp. 140–182.

1900b La théorie de Lorentz et le principe de réaction, in *Recueil de travaux offerts par les auteurs à H. A. Lorentz* (The Hague: Nijhoff, 1900), pp. 252–278. Reprinted in *Oeuvres*, pp. 464–488.

1900c Sur les principes de la Mécanique, *Bibliothèque du Congres International de Philosophie tenu à Paris du 1ᵉʳ au 5 août 1900* (Paris: Colin, 1901), pp. 457–494; an expanded version of this paper is presented on pp. 123–139 of Poincaré's (1902).

1900d Sur l'induction unipolaire, *L'Eclairage électrique*, *23*, 41–53 (1900).

1901 *Electricité et Optique* (Paris: Gauthier-Villars, 1901). (Poincaré's lectures at the Sorbonne from 1888, 1890 and 1899.)

1902 *Science and Hypothesis* (New York: Dover, 1952), translator unknown. This is a translation of Poincaré's, *La Science et l'Hypothèse* (Paris: Ernest Flammarion, 1902).

1904 L'état actuel et l'avenir de la Physique mathématique. Lecture delivered on 24 September 1904 to the International Congress of Arts and Science, Saint Louis,

Missouri, and published in *Bull. Sci. Mat.*, *28*, 302–324. Reprinted on pp. 91–111 of Poincaré's (1905b).

1905a   Sur la dynamique de l'électron, *C. R. Acad. Sci.*, *140*, 1504–1508 (1905). Reprinted in *Oeuvres*, pp. 489–493.

1905b   *The Value of Science* (New York: Dover, 1958), translated by George Bruce Halsted. This is a translation of Poincaré's *La Valeur de la Science* (Paris: Ernest Flammarion, 1905).

1906   Sur la dynamique de l'électron, *Rend. Circ. Mat. Palermo*, *21*, 129–175 (1906). Reprinted in *Oeuvres*, pp. 494–550.

1908a   *Science and Method* (New York: Dover, n.d.), translated by Francis Maitland. This is a translation of Poincaré's *Science et Méthode* (Paris: Ernest Flammarion, 1908).

1908b   La dynamique de l'électron, *Rev. Générale Sci. Pures Appl.*, *19*, 386–402 (1908). Reprinted in *Oeuvres*, pp. 551–586. Excerpts from this paper have been translated in Book III, pp. 199–250 of Poincaré's (1908a).

1909   La mécanique nouvelle, the last of Poincaré's six Wolfskehl lectures given at Göttingen in 1909 and published in *Sechs Vorträge über ausgewählte Gegenstände aus der reinen Mathematik und mathematischen Physik* (Leipzig: Teubner, 1910).

1912a   La dynamique de l'électron (Paris: Dumas, 1913).

1912b   Les rapports de la matière et de l'éther, *J. Phys. Théor. Appl.*, *2*, 347 (1912). Reprinted in Chapter VII, pp. 89–101 of Poincaré (1913).

1912c   L'éspace et le temps. Lecture delivered on 4 May, 1912 at the University of London. Reprinted in Chapter II, pp. 15–24 of Poincaré's (1913).

1913   *Mathematics and Science: Last Essays*, (New York: Dover, 1963), translated by J. W. Bolduc. This is a translation of Poincaré's, *Dernières Pensées* (Paris, Ernest Flammarion, 1913).

1924   *Le mécanique nouvelle* (Paris: Gauthier-Villars, 1924).

Precht, Julius (1871 – 1942)

1906   Strahlungsenergie von Radium, *Ann. Phys.*, *21*, 595–601 (1906).

Preston, S. Tolver (1844 – ?)

1885a   On some Electromagnetic Experiments of Faraday and Plücker, *Philos. Mag.*, *19*, 131–140 (1885).

1885b   On some Electromagnetic Experiments, continued, – No. II, Diverse views on Faraday, Ampère, and Weber, *Philos. Mag.*, *19*, 215–218 (1885).

1891   The Problem of the Behavior of the Magnetic Field about a Revolving Magnet, *Philos. Mag.*, *31*, 100–102 (1891).

Lord Rayleigh [J. W. Strutt] (1842–1919)

1902   Does Motion through the Aether cause Double Refraction?, *Philos. Mag.*, *4*, 678–683 (1902).

Richardson, Owen W. (1879–1959)

1916   *The Electron Theory of Matter* (Cambridge: Cambridge University Press, 1916).

Ritz, Walther (1878–1909)

1908a   Recherches critiques sur l'Electrodynamique générale, *Ann. Chim. Phys.*, *13*, 145 (1908). Reprinted in W. Ritz, *Oeuvres* (Paris: Gauthier-Villars, 1911), pp. 317–426.

1908b   Recherches critiques sur les théories électrodynamiques de Cl. Maxwell et de H. A. Lorentz, *Arch. Sci. Phys. et Nat.*, *26*, 209 (1908). Reprinted in *Oeuvres*, pp. 422–446.

1908c   Rôle de l'éther en physique, *Scientia*, III, 260 – 274 (1908). Reprinted in *Oeuvres*, pp. 447–461.

1908d   Über ein neues Gesetz der Serienspektren, *Phys. Z.*, *9*, 521 (1908). Reprinted in *Oeuvres*, pp. 141–162.

1909   Zum gegenwärtigen Stande des Strahlungsproblems, *Phys. Z.*, *10*, 323–324 (1909). Reprinted in *Oeuvres*, pp. 507–508. With A. Einstein.

Röntgen, Wilhelm Konrad (1845–1923)

1888   Über die durch Bewegung eines im homogenen elektrischen Felde befindlichen Dielektrums hervorgerufene elektrodynamische Kraft, *Ann. Phys.*, *35*, 246–283 (1888).

Runge, Carl (1856–1927)
  1903   Über die elektromagnetische Masse der Elektronen, *Göttinger Nachr.*, 326–330 (1903).
Russell, Bertrand (1872–1970)
  1897   *An Essay on the Foundations of Geometry* (Cambridge: Cambridge University Press,
          1897; New York: Dover, 1952).
Rutherford, Ernest (1871–1937)
  1899   Uranium Radiation and the Electrical Conduction produced by it, *Philos. Mag.*, 47,
          109–163 (1899).
  1914   The structure of the Atom, *Philos. Mag.*, 27, 488 (1914).
Schwarzschild, Karl (1873–1916)
  1903   Zur Elektrodynamik. III. Über die Bewegung des Elektrons, *Göttinger Nachr.*, 245–278
          (1903).
Searle, George Frederick Charles (1864– ?)
  1897   On the motion of an electrified ellipsoid, *Philos. Mag.*, 44, 329–341 (1897).
von Siemens, Werner (1816–1892)
  1881   Die dynamoelektrische Maschine, *Elektrot. Z.*, 2, 89–95 (1881).
  1893   *Personal Recollections* (New York: Appleton, 1893), translated by W. C. Coupland.
Silberstein, Ludwik (1872–1948)
  1914   *The Theory of Relativity* (London: MacMillan, 1914).
Simon, S.
  1899   Über das Verhältnis der elektrischen Ladung zur Masse der Kathodenstrahlen, *Ann.
          Phys.*, 69, 589–611 (1899).
Sommerfeld, Arnold (1868–1951)
          [Most of Sommerfeld's scientific papers have been reprinted in *Gesammelte Schriften*
          (4 vols.; Braunschweig: Vieweg, 1968).]
  1904a   Allgemeine Untersuchung des Feldes eines beliebig bewegten Elektrons, *Göttinger
          Nachr.*, 99–130 (1904). Reprinted in *Schriften*, II, pp. 39–70.
  1904b   Grundlagen für eine allgemeine Dynamik des Elektrons, *Göttinger Nachr.*, 363–439
          (1904). Reprinted in *Schriften*, II, pp. 71–147.
  1904c   Simplified Deduction of the Field and the Forces of an Electron, moving in any given
          way, *Proc. R. Acad. Amsterdam*, 7, 346–367 (1904).
  1905   Über Lichtgeschwindigkeits- und Überlichtgeschwindigkeits-Elektronen, *Göttinger
          Nachr.*, 201–235 (1905). Reprinted in *Schriften*, II, pp. 148–182.
  1907   Ein Einwand gegen die Relativtheorie der Elektrodynamik und seine Beseitigung, *Phys.
          Z.*, 8, 841–842 (1907). Reprinted in *Schriften*, II, pp. 183–184.
  1909   Über die Zusammensetzung der Geschwindigkeiten in der Relativtheorie, *Phys. Z.*, 10,
          826–829 (1909). Reprinted in *Schriften*, II, pp. 185–188.
  1910a   Zur Relativitätstheorie I: Vierdimensionale Vektoralgebra, *Ann. Phys.*, 32, 749–776
          (1910).
  1910b   Zur Relativitätstheorie II: Vierdimensional Vektoralgebra, *Ann. Phys.*, 33, 649–689.
          Both papers of 1910 are reprinted in *Schriften*, II, pp. 189–257.
  1911   Das Plancksche Wirkungsquantum and seine allgemeine Bedeutung für die Molekular-
          physik, *Phys. Z.*, 12, 1057–1069 (1911).
  1915   Die Feinstruktur der wasserstoff- und wasserstoffähnlichen Linien, *Münchener
          Berichte*, 459–500 (1915).
  1916   Zur Quantentheorie der Spektrallinien, *Ann. Phys.*, 51, 1–94, 125–167 (1916).
  1948   Philosophie und Physik seit 1900, *Naturwissenschaftliche Rundschau*, 1, 97–100 (1948).
          Reprinted in *Schriften*, IV, pp. 640–643.
c. 1950   Autobiographische Skizze, in *Schriften*, IV, pp. 673–682 (F. Bopp, ed.).
  1964   *Electrodynamics* (New York: Dover, 1964), translated by E. G. Ramberg.
Stark, Johannes (1874–1957)
  1906   Über die Lichtemission der Kanalstrahlen in Wasserstoff, *Ann. Phys.*, 401–456 (1906).
  1908   Bemerkung zu Herrn Kaufmanns Einwand 'Antwort auf einen Einwand von Herrn
          Planck', *Verh. D. Phys. Ges.*, 10, 14–16 (1908).

Starke, Hermann (1874–1960)
   1903   Über die elektrische und magnetische Ablenkung schneller Kathodenstrahlen, *Verh. D. Phys. Ges.*, *1*, 241–250 (1903).
   1934   Über die Bestimmung der Massenveränderlichkeit des Elektrons an schnellen Kathodenstrahlen, *Ann. Phys.*, *21*, 67–88 (1934). With M. Nacken.
Stead, G. and H. Donaldson
   1910   The Problem of Uniform Rotation Treated on the Principle of Relativity, *Philos. Mag.*, *115*, 92–95 (1910).
Stokes, George Gabriel (1819–1903)
          [Many of Stokes' published papers are in G. G. Stokes, *Mathematical and Physical Papers* (5 vols.; Cambridge: Cambridge University Press, 1880–1905).]
   1845   On the Aberration of Light, *Philos. Mag.*, *27*, 9 (1845). Reprinted in *Mathematical and Physical Papers*, vol. I., pp. 134–140.
   1846   On Fresnel's Theory of the Aberration of Light, *Philos. Mag.*, *28*, 76 (1846). Reprinted in *Mathematical and Physical Papers*, vol. I, pp. 141–147.
Stoney, George Johnstone (1826–1911)
   1894   Of the 'Electron' or Atom of Electricity, *Philos. Mag.*, *38*, 418–420 (1894).
Sutherland, William (1859–1911)
   1899   Cathode, Lenard and Röntgen Rays, *Philos. Mag.*, *47*, 269–284 (1899).
Swann, William Francis Gray (1884–1962)
   1920   Unipolar Induction, *Phys. Rev.*, *15*, 365–398 (1920).
Szarvassi, Arthur
   1909   Die Theorie der elektromagnetischen Erscheinungen in bewegten Körpern und das Energieprinzip, *Phys. Z.*, *10*, 811–813 (1909).
Tate, J. T.
   1922   Unipolar Induction, *Bull. Nat. Res. Council*, *4*, 75–95 (1922).
Thomson, Joseph John (1856–1940)
   1880   On Maxwell's Theory of Light, *Philos. Mag.*, *9*, 284–291 (1880).
   1881   On the electric and magnetic effects produced by motion of electrified bodies, *Philos. Mag.*, *11*, 229–249 (1881).
   1889   On the Magnetic Effects produced by Motion in the Electric Field, *Philos. Mag.*, *28*, 1–14 (1889).
   1893   *Recent Researches in Electricity and Magnetism: Intended as a Sequel to Professor Clerk Maxwell's Treatise on Electricity and Magnetism* (Oxford: Clarendon Press, 1893).
   1897   Cathode Rays, *Philos. Mag.*, *44*, 293–316 (1897).
   1898   On the Charge of Electricity produced by Röntgen rays, *Philos. Mag.*, *46*, 528–545 (1898).
   1899   On the Masses of the Ions in Gases at Low Pressures, *Philos. Mag.*, *48*, 547–567 (1899).
   1906   Carriers of negative electricity. Lecture delivered on 11 December 1906 in acknowledgement of the Nobel Prize. Reprinted in *Nobel Lectures, Physics: 1901–1921* (New York: Elsevier, 1967), pp. 141–155.
Tolman, Richard Chase (1881–1948)
   1910   The Second Postulate of Relativity, *Phys. Rev.*, 291 (1910); *Ibid.*, *31*, 26–40 (1910).
   1911a Note on the Derivation from the Principle of Relativity of the Fifth Fundamental Equation of the Maxwell-Lorentz Theory, *Philos. Mag.*, *21*, 296–301 (1911).
   1911b Non-Newtonian Mechanics: The Direction of Force and Acceleration, *Philos. Mag.*, *22*, 458–463 (1911).
   1912a Non-Newtonian Mechanics, The Mass of a Moving Body, *Philos. Mag.*, *23*, 375–380 (1912).
   1912b Some Emission theories of Light, *Phys. Rev.*, *35*, 136–143 (1912).
   1913   Non-Newtonian Mechanics: Some Transformation Equations, *Philos. Mag.*, *25*, 150–157 (1913).
   1917   *The Theory of the Relativity of Motion* (Berkeley: University of California Press, 1917).

Trocheris, M. G.
  1949   Electrodynamics in a Rotating Frame of Reference, *Philos. Mag.*, *40*, 1143–1154 (1949).
Trouton, Frederick Thomas (1863–1922) and H. R. Noble
  1903   The Mechanical Forces Acting on a Charged Electric Condenser moving through Space, *Philos. Trans. R. Soc. London*, *202*, 165–181 (1903).
Uppenborn, F.
  1885   Über Unipolarmaschinen, *Centralblatt für Elektrotechnik*, *7*, 324–329 (1885).
Varičak, V.
  1911   Zum Ehrenfestschen Paradoxon, *Phys. Z.*, *12*, 169 (1911).
Veltmann, Wilhelm (1832–1902)
  1870a   Fresnel's Hypothese zur Erklärung der Aberrationserscheinungen, *Astron. Nachr.*, *75*, 145–150 (1870).
  1870b   Über die Fortpflanzung des Lichtes in bewegten Medien, *Astron. Nachr.*, *76*, 129–144 (1870).
  1873   Über die Fortpflanzung des Lichtes in bewegten Medien, *Ann. Phys.*, *150*, 497–535 (1873).
Voigt, Woldemar (1850–1919)
  1887   Über das Doppler'sche Prinzip, *Göttinger Nachr.*, 14 (1887). Reprinted in *Phys. Z.*, *16*, 381–386 (1915).
Van der Waals, J. D., Jr.
  1912   Energy and Mass. II, *Proc. R. Acad. Amsterdam*, *14*, 821–831 (1912).
Weber, C. L.
  1895   Über unipolare Induktion, *Elektrot. Z.*, *16*, 513–514 (1895).
Weber, Wilhelm (1804–1891)
  1841   Unipolare Induktion, *Ann. Phys.*, *52*, 353–386 (1841).
Wiechert, Emil (1861–1928)
  1898   Hypothesen für eine Theorie der elektrischen und magnetischen Erscheinungen, *Göttinger Nachr.*, 87–106 (1898).
  1900   Elektrodynamische Elementargesetze, in *Recueil de travaux offerts par les auteurs à H. A. Lorentz* (The Hague: Nijhoff, 1900), pp. 549–573. Reprinted in *Ann. Phys.*, *4*, 667–669 (1901).
  1911   Relativitätsprinzip und Aether. I, *Phys. Z.*, *12*, 689–707 (1911); II, *Ibid.*, *12*, 737–758 (1911).
  1915   Die Mechanik im Rahmen der allgemeinen Physik, in *Die Kultur der Gegenwart: Physik* (E. Warburg, ed.) (Leipzig: Teubner, 1915), pp. 1–78.
Wiedemann, Gustav Heinrich (1826–1899)
  1885   *Die Lehre von der Elektricität* (4 vols.; Braunschweig: F. Vieweg und Sohn, 1885).
Wien, Wilhelm (1864–1928)
  1898   Untersuchungen über die electrische Entladung in verdünnten Gasen, *Ann. Phys.*, *65*, 440–452 (1898).
  1900   Über die Möglichkeit einer elektromagnetischen Begründung der Mechanik, in *Recueil de travaux offerts par les auteurs à H. A. Lorentz* (The Hague: Nijhoff, 1900), pp. 96–107. Reprinted in *Ann. Phys.*, *5*, 501–513 (1901).
  1904a   Über die Differentialgleichungen der Elektrodynamik für bewegte Körper. I. *Ann. Phys.*, *13*, 641–662 (1904); II., *Ibid.*, 663–668 (1904).
  1904b   Erwiderung auf die Kritik des Hrn. Abraham, *Ann. Phys.*, *14*, 635–637 (1904).
  1909   *Über Elektronen* (Leipzig: Teubner, 1909).
  1911   On the laws of Thermal Radiation. Lecture delivered on 11 December 1911 in acknowledgement of the Nobel Prize. Reprinted in *Nobel Lectures, Physics: 1901–1921* (New York: Elsevier, 1967), pp. 271–289.
Young, Thomas (1773–1829)
  1804   Experiments and Calculations relative to physical optics, *Proc. Roy. Soc. London*, *94*, 1–16 (1804).

Zahn, C. T. and A. A. Spees
    1938a An Improved Method for the Determination of the Specific Charge of Beta-Particles,
           Phys. Rev., *53*, 357–364 (1938).
    1938b The Specific Charge of Disintegration Electrons From Radium E, *Phys. Rev.*, *53*,
           365–373 (1938).
    1938c A critical analysis of the classical experiments on the variation of electron mass, *Phys.
           Rev.*, *53*, 511–521 (1938).
Zeeman, Pieter (1865–1943)
    1897   On the Influence of Magnetism on the Nature of the Light emitted by Substance, *Philos.
           Mag.*, *43*, 226–239 (1897).
    1914–1915
           Fresnel's coefficient for light of different colours, *Proc. R. Acad. Amsterdam*, *17*,
           445–451 (1914); 2nd part in *Ibid.*, *18*, 398–408 (1915).

Secondary Sources

Anderson, David L.
    1966   Resource Letter ECAN-1 on the Electronic Charge and Avogadro's Number, *Am. J.
           Phys.*, *34*, 2–8 (1966).
Arnheim, Rudolf
    1971   *Visual Thinking* (Berkeley: University of California Press, 1971).
Becker, Richard and Fritz Sauter
    1964   *Electromagnetic Fields and Interactions* (2 vols.; New York: Blaisdell, 1964).
Bell, Eric Temple
    1945   *The Development of Mathematics* (New York: McGraw-Hill, 1945).
Bernstein, Jeremy
    1973   *Einstein* (New York: Viking Press, 1973).
Beyerchen, Alan D.
    1977   *Scientists Under Hitler: Politics and the Physics Community in the Third Reich* (New
           Haven: Yale University Press, 1977)
Biquard, Pierre
    1969   *Paul Langevin: scientifique, éducateur, citoyen* (Paris: Seghers, 1969)
Blackmore, John
    1972   *Ernst Mach: His Life, Work, and Influence* (Berkeley: University of California Press,
           1972).
Bork, A. M.
    1966a Physics Just Before Einstein, *Science*, *152*, 597–603 (1966).
    1966b The 'Fitzgerald' Contraction Hypothesis, *ISIS*, *57*, 199–207 (1966).
Bowman, Peter
    1978   Is Signal Synchrony Independent of Transport Synchrony?, *Philos. Sci.*, *45*, 309–311
           (1978).
Brecher, Kenneth
    1977   Is the Velocity of Light Independent of its Source, *Phys. Rev. Letters*, *39*, 1051–1054
           (1977).
Bridgman, Percy W.
    1949   Einstein's Theories and the Operational Point of View, in *Albert Einstein: Philosopher-
           Scientist* (P. A. Schilpp, ed.) (La Salle, IL: Open Court, 1949), pp. 333–354.
Brush, Stephen G.
    1967   Note on the History of the Fitzgerald-Lorentz Contraction, *ISIS*, *54*, 230–232 (1967).
    1976   *The Kind of Motion We Call Heat* (2 vols.; Amsterdam: North-Holland, 1976).
Builder, Geoffrey
    1957   The Resolution of the Clock Paradox. *Austr. J. Phys.*, *10*, 246–262 (1957).
    1958   Ether and Relativity. *Austr. J. Phys.*, *11*, 279–297 (1958).

Campbell, John T.
1973   Kaufmann, Walter, in *Dictionary of Scientific Biography* (C. C. Gillispie, ed.), VII (New York, Scribner, 1973), pp. 263–265.

Cullwick, E. G.
1959   *Electromagnetism and Relativity: With Particular Reference to Moving Media and Electromagnetic Induction* (1st ed., New York: Wiley, 1957; 2nd ed., 1959).

Darboux, Gaston
1913   Elogie de Henri Poincaré, in *Oeuvres de Henri Poincaré* (11 vols.; Paris: Gauthier-Villars, 1934–1954), vol. 2, pp. vii–lxxii.

Darwin, C. G.
1957   The Clock Paradox in Relativity, *Nature*, 976–977 (1957).

Ellis, B. and P. Bowman
1967   Conventionality in Distant Simultaneity. *Philos. Sci.*, *34*, 116–136 (1967).

Everitt, C. W. F.
1974   Maxwell, James Clerk, in *Dictionary of Scientific Biography*, (C. C. Gillispie, ed.), IX (New York: Scribner, 1974), pp. 198–230; also published as *James Clerk Maxwell* (New York: Scribner, 1975).

Flückiger, Max
1974   *Albert Einstein in Bern* (Bern: Haupt, 1974).

Forman, Paul
1975   Sommerfeld, Arnold, in *Dictionary of Scientific Biography* (C. C. Gillispie, ed.), XII (New York: Scribner, 1975), pp. 525–532. With A. Hermann.

Fox, J. G.
1965   Evidence Against Emission Theories, *Am. J. Phys.*, *33*, 1–17 (1965).

Frank, Philipp
1947   *Einstein: Sein Leben und seine Zeit* (New York: Knopf, 1947), translated by G. Rosen, and edited and revised by S. Kusaka (New York: Knopf, 1953).

French, Anthony P.
1966   *Special Relativity* (New York: Norton and Company, 1966).

Galison, Peter
1979   Minkowski's Space-Time: From Visual Thought to the Absolute World, *Hist. Studies Phys. Scis.*, *10*, 85–121 (1979).

Giannoni, Carlo
1978   Relativistic Mechanics and Electrodynamics with One-way Velocity Assumptions, *Philos. Sci.*, *45*, 17–46 (1978).

Gillispie, Charles Coulston
1960   *The Edge of Objectivity: An Essay in the History of Scientific Ideas* (Princeton: Princeton University Press, 1960).

Goldberg, Stanley
1967   Henri Poincaré and Einstein's Theory of Relativity, *Am. J. Phys.*, *35*, 934–944 (1967).
1969   The Lorentz Theory of Electrons and Einstein's Theory of Relativity, *Am. J. Phys.*, *37*, 498–513 (1969).
1970a   Poincaré's Silence and Einstein's Relativity: The Role of Theory and Experiment in Poincaré's Physics, *Brit. J. Hist. Sci.*, *5*, 73–84 (1970).
1970b   The Abraham Theory of the Electron: The Symbiosis of Experiment and Theory, *Arch. Hist. Exact Scis.*, *7*, 7–25 (1970).
1976   Max Planck's Philosophy of Nature and His Elaboration of the Special Theory of Relativity, *Hist. Studies Phys. Scis.*, *7*, 125–160 (1976).

Grünbaum, Adolf
1973   *Philosophical Problems of Space and Time* (Dordrecht: Reidel, 1973).

Guillaume, Charles-Edouard
1924   Introduction to Poincaré's (1924).

Gunter, P. A. Y.
1969   *Bergson and the Evolution of Physics* (P. A. Y. Gunter, ed.) (Knoxville: University of

Tennessee Press, 1969).

ter Haar, D.

1967   *The Old Quantum Theory* (D. ter Haar, ed.) (New York: Pergamon Press, 1967).

De Haas-Lorentz, G. L.

1957   Reminiscences, in *H. A. Lorentz, Impressions of His Life and Work*, (G. L. de Haas-Lorentz, ed.) (Amsterdam: North-Holland, 1957), pp. 15–47, 82–120.

Hahn, Roger

1970   Arago, Dominique François Jean, in *Dictionary of Scientific Biography* (C. C. Gillispie, ed.), I (New York: Scribner, 1970), pp. 200–203.

Heilbron, John

1976   Thomson, Joseph John, in *Dictionary of Scientific Biography* (C. C. Gillispie, ed.) XII (New York: Scribner, 1976), pp. 362–372.

Heimann, Peter M.

1970   Maxwell and Modes of Consistent Representation, *Arch. Hist. Exact Scis.*, 6, 171–213 (1970).

1971   Maxwell, Hertz and the Nature of Electricity, *ISIS*, 62, 149–157 (1971).

Hermann, Armin

1966   Albert Einstein und Johannes Stark: Briefwechsel und Verhältnis der beiden Nobelpreisträger, *Sudhoffs Archiv für Geschichte der Medizin und der Naturwissenschaften der Pharmazie und der Mathematik*, 50, 267–285 (1966).

1971   *The Genesis of Quantum Theory (1899–1913)* (Cambridge: MIT Press, 1971), translated by C. W. Nash.

1973   Lenard, Philipp, in *Dictionary of Scientific Biography* (C. C. Gillispie, ed.) VIII (New York: Scriber, 1973), pp. 180–183.

1975   Stark, Johannes, in *Dictionary of Scientific Biography* (C. C. Gillispie, ed.) XII (New York: Scribner, 1975), pp. 613–616.

Hiebert, Erwin N.

1973   Mach, Ernst, in *Dictionary of Scientific Biography* (C. C. Gillispie, ed.) VII (New York: Scribner, 1973), pp. 595–607.

Hirosige, Tetu

1966   Electrodynamics before the Theory of Relativity, *Jpn. Studies Hist. Sci.*, 5, 1–49 (1966).

1969   Origins of Lorentz' Theory of Electrons and the Concept of the Electromagnetic Field, *Hist. Studies Phys. Scis.*, 1, 151–209 (1969).

1976   The Ether Problem, the Mechanistic Worldview, and the Origins of the Theory of Relativity, *Hist. Studies Phys. Scis.*, 7, 3–82 (1976).

Hoffmann, Banesh

1972   *Albert Einstein, Creator and Rebel* (New York: Viking, 1972). With the collaboration of Helen Dukas.

1979   *Albert Einstein, The Human Side* (Princeton: Princeton University Press, 1979). With Helen Dukas.

Holton, Gerald

1962   Resource Letter SRT-1 on Special Relativity Theory, *Am. J. Phys.*, 30, 462–469 (1962).

1973   *Thematic Origins of Scientific Thought: Kepler to Einstein* (Cambridge, MA: Harvard University Press, 1973) – the essay's original dates of publication are given in parenthesis;
   (a) On the Origins of the Special Theory of Relativity, pp. 165–183 (1962).
   (b) Poincaré and Relativity, pp. 185–195 (1964).
   (c) Influences on Einstein's Early Work, pp. 197–217 (1967).
   (d) Mach, Einstein, and the Search for Reality, pp. 219–259 (1968).
   (e) Einstein, Michelson, and the 'Crucial' Experiment, pp. 261–352 (1969).
   (f) On Trying the Understand Scientific Genius, pp. 353–380 (1971).

1978   Subelectrons, presuppositions and the Millikan-Ehrenhaft dispute, in G. Holton, *The Scientific Imagination: Case Studies* (Cambridge, U.K.: Cambridge University Press, 1978), pp. 25–83.

1979   Einstein's Scientific Program: The Formative Years. Lecture presented on 5 March
        1979 at the Einstein Centennial Celebration, the Institute for Advanced Study, 4–9
        March 1979. Scheduled to appear in the *Proceedings*.

Infeld, Leopold

1950   *Albert Einstein: His Work and Influence on Our World* (New York: Scribner, 1950).

Jammer, Max

1957   *Concepts of Force* (Cambridge, MA: Harvard University Press, 1957).

1961   *Concepts of Mass* (Cambridge, MA: Harvard University Press, 1961).

1966   *The Conceptual Development of Quantum Mechanics* (New York: McGraw-Hill, 1966).

Kahan, Théo

1959   Sur les Origines de la théorie de la relativité restreinte, *Rev. Hist. Scis.*, *13*, 159–165
        (1959).

Kangro, Hans

1976   Wien, Wilhelm Carl Werner Otto Franz, in *Dictionary of Scientific Biography* (C. C.
        Gillispie, ed.) XIV (New York: Scribner, 1976), pp. 337–342.

Kittel, Charles

1974   Larmor and the Prehistory of the Lorentz Transformations, *Am. J. Phys.*, *42*, 726–729
        (1974).

Klein, Martin J.

1962   Max Planck and the Beginnings of the Quantum Theory, *Arch. Hist. Exact Scis.*, *1*,
        459–479 (1962).

1963   Planck, Entropy, and Quanta, 1901–1906, *The Natural Philosopher* (New York:
        Blaisdell), *1*, 83–108 (1963).

1964   Einstein and the Wave-Particle Duality, *The Natural Philosopher* (New York: Blaisdell)
        3, 3–49 (1964).

1965   Einstein, Specific Heats and the Early Quantum Theory, *Science*, *148*, 173–180 (1965).

1967   Thermodynamics in Einstein's Thought, *Science*, *157*, 509–516 (1967).

1970   *Paul Ehrenfest: vol. 1, The Making of a Theoretical Physicist* (Amsterdam: North-
        Holland, 1970).

1972   Mechanical Explanation at the End of the Nineteenth Century, *Centaurus*, *17*, 58–82
        (1972).

1975   Einstein on Scientific Revolutions, *Vistas in Astron.*, *17*, 114–120 (1975).

1977   Some Unnoticed Publications by Einstein, *ISIS*, *68*, 601–604 (1977). With A. Needell.

Kollros, Louis

1956   Albert Einstein en Suisse Souvenirs, in *Jubilee of Relativity Theory* (Basel: Birkhäuser-
        Verlag, 1956), pp. 271–281.

von Kossel, W.

1947   Walter Kaufmann, *Naturwissenschaften*, *34*, 33–34 (1947).

Kuhn, Thomas S.

1978   *Black-Body Theory and the Quantum Discontinuity, 1894–1912* (Oxford: Oxford
        University Press, 1978).

Lebon, Ernest

1912   *Savants du Jour: Henri Poincaré* (2nd ed.; Paris: Gauthier-Villars, 1912).

Marder, L.

1971   *Time and the Space Traveller* (Philadelphia: University of Pennsylvania Press, 1971).

McCormmach, Russell

1967   Hertz, Heinrich Rudolf, in *Dictionary of Scientific Biography* (C. C. Gillispie, ed.) (New
        York: Scribner, 1972), IV, pp. 340–350.

1970a   H. A. Lorentz and the Electromagnetic View of Nature, *ISIS*, *58*, 37–55 (1970).

1970b   Einstein, Lorentz and the Electromagnetic View of Nature, *Hist. Studies Phys. Scis.*, *2*,
        41–87 (1970).

1973   Lorentz, H. A., in *Dictionary of Scientific Biography* (C. C. Gillispie, ed.) (New York:
        Scribner, 1973), VIII, pp. 487–500.

1976   Editor's Foreword, in *Hist. Studies Phys. Scis.*, *7*, xi-xxxv (1976).

Merz, John Theodore
   1965   *A History of European Scientific Thought in the Nineteenth Century* (4 vols., 1904–1912;
          New York: Dover, 1965) and vols. 3 and 4 of this set are entitled, *A History of European
          Thought in the Nineteenth Century.*

Miller, Arthur I.
   1973   A Study of Henri Poincaré's "Sur la Dynamique de l'Electron", *Arch. Hist. Exact Scis.,*
          10, 207–328 (1973).
   1974   On Lorentz's Methodology, *Brit. J. Philos. Sci., 25*, 29–45 (1974).
   1975a   Book review of Adolf Grünbaum's, *The Philosophical Problems of Space and Time,*
          ISIS, 66, 590–594 (1975).
   1975b   Albert Einstein and Max Wertheimer: A Gestalt Psychologist's View of the Genesis of
          Special Relativity Theory, *History of Science, 13*, 75–103 (1975).
   1976   On Einstein, Light Quanta, Radiation and Relativity in 1905, *Am. J. Phys., 44*, 912–923
          (1976).
   1977a   Reply to Adolf Grünbaum's 'Remarks on Arthur I. Miller's Review of *Philosophical
          Problems of Space and Time*', *ISIS, 68*, 449–450 (1977).
   1977b   On Unipolar Dynamos, The Electrical Industry and Relativity Theory. Lecture
          presented on 27 April 1977 at the *Technology Studies Seminar Series*, Massachusetts
          Institute of Technology. Scheduled to appear in *Ann. Sci.*
   1977c   The Physics of Einstein's Relativity Paper of 1905 and the Electromagnetic World-
          Picture of 1905, *Am. J. Phys., 45*, 1040–1048 (1977).
   1978a   Visualization Lost and Regained: The Genesis of the Quantum Theory in the Period
          1913–1927, in *On Aesthetics in Science* (J. Wechsler, ed.) (Cambridge, MA.: MIT Press,
          1978), pp. 72–102.
   1978b   Reply to 'Some New Aspects of Relativity: Comments on Zahar's Paper', *Brit. J. Philos.
          Sci., 29*, 252–256 (1978).
   1979a   On Some Other Approaches to Electrodynamics in 1905. Lecture presented on 5 March
          1979 at the Einstein Centennial Celebration, the Institute for Advanced Study, 4–9
          March 1979. Scheduled to appear in the *Proceedings.*
   1979b   Poincaré and Einstein: A Comparative Study. Scheduled to appear in vol. XXXI of the
          Boston Studies Philos. Sci.
   1979c   The Special Relativity Theory: Einstein's Response to the Physics of 1905. Lecture
          presented on 14 March 1979 at the Einstein Centennial Symposium in Jerusalem, Israel,
          14–23 March 1979. Scheduled to appear in the *Proceedings.*

Møller, C.
   1972   *The Theory of Relativity* (Oxford: Oxford University Press, 1952; 2nd enlarged ed.,
          1972).

Nye, Mary Jo
   1972   *Molecular Reality: A Perspective on the Scientific Work of Jean Perrin* (New York:
          Elsevier, 1972).

Panofsky, Wolfgang and Melba Phillips
   1961   *Classical Electricity and Magnetism* (Reading: Addison-Wesley, 1961).

Prokhovnik, S. J.
   1967   *The Logic of Special Relativity* (Cambridge, U.K.: Cambridge University Press, 1967).

Purcell, Edward M.
   1975   *Electricity and Magnetism: Berkeley Physics Course* (New York: McGraw-Hill, 1975).

Pyenson, Lewis
   1977   Hermann Minkowski and Einstein's Special Theory of Relativity, *Arch. Hist. Exact
          Scis., 17*, 71–95 (1977).

Lord Rayleigh [J. W. Strutt]
   1942   *Sir J. J. Thomson* (Cambridge, U.K.: Cambridge University Press, 1942).

Reichenbach, Hans
   1958   *The Philosophy of Space and Time* (New York: Dover, 1958), translated by M.
          Reichenbach and J. Freund.

1969   *Axiomatization of the Theory of Relativity* (Berkeley: University of California Press, 1969), translated and edited by M. Reichenbach.

Reid, Constance
1970   *Hilbert* (New York: Springer, 1970).

Reiser, Anton
1930   *Albert Einstein, A Biographical Portrait* (New York: A & C Boni, 1930).

Resnick, Robert
1968   *Introduction to Special Relativity* (New York: Wiley, 1968).
1978   *Physics* (2 vols.; New York: Wiley, 1978). With D. Halliday.

Robertson, Howard P.
1949   Postulate versus Observation in the Special Theory of Relativity, *Rev. Mod. Phys.*, *21*, 378–382 (1949).

Röhrlich, Fritz
1965   *Classical Charged Particles* (Reading, MA.: Addison-Wesley, 1965).

Rosenfeld, Léon
1952   The Velocity of Light and the Evolution of Electrodynamics, *Nuovo Cimento*, *4*, 1630–1669 (1957).

Rosser, W. G. V.
1964   *An Introduction to the Theory of Relativity* (London: Butterworths, 1964).

Rossi, Bruno and D. B. Hall
1941   Rate of Decay of Mesotrons with Momentum, *Phys. Rev.*, *59*, 223–228 (1941).

Salmon, W. C.
1975   *Space, Time and Motion: A Philosophical Introduction* (Encino, CA: Dickenson, 1975)

Sauter, Joseph
1965   *Erinnerungen an Albert Einstein*. This phamphlet (unpaginated) was published in 1965 by the Patent Office in Bern, and contains documents pertaining to Einstein's years at that office as well as a note by Sauter.

Schaffner, Kenneth
1969   The Lorentz Theory of Relativity, *Am. J. Phys.*, *37*, 498–513 (1969).
1972   *Nineteenth-Century Aether Theories* (New York: Pergamon, 1972).

Schiff, Leonard I.
1939   A Question in General Relativity, *Proc. Nat. Acad. Sci. USA*, *25*, 391–395 (1939).

Schlomka, T. and G. Schenkel
1949   Relativitätstheorie und Unipolarinduktion, *Ann. Phys.*, *5*, 57–62 (1949).

Seelig, Carl
1952   *Albert Einstein und die Schweiz* (Zürich: Europa-Verlag, 1952).
1954   *Albert Einstein: Eine dokumentarische Biographie* (Zürich: Europa-Verlag, 1954).

Shankland, Robert S.
1963   Conversations with Albert Einstein, *Am. J. Phys.*, *31*, 47–57 (1963).
1964   The Michelson-Morley Experiment, *Sci. Am.*, November (1964), pp. 107–114.

Siegel Daniel
1978   Classical-Electromagnetic and Relativistic Approaches to the Problem of Nonintegral Atomic Masses, *Hist. Studies Phys. Scis.*, *9*, 323–360 (1978).

Siemens, Georg
1957   *History of the House of Siemens*, (2 vols.; Munich: Alber, 1957), translated by A. F. Rodger.

Süsskind, Charles
1972   Heaviside, Oliver, in *Dictionary of Scientific Biography* (C. C. Gillispie, ed.) VI (New York: Scribner, 1972), pp. 211–212.

Swenson, Loyd, Jr.
1972   *The Ethereal Aether* (Austin: University of Texas Press, 1972).

Terrell, James
1959   Invisibility of the Lorentz Contraction. *Phys. Rev.*, *116*, 1041–1045 (1959).

Teske, A. A.
  1975   Smoluchowski, Marian, in *Dictionary of Scientific Biography* (C. C. Gillispie, ed.) (New York: Scribner, 1975), XII, pp. 496–498.
Tonnelat, Marie-Antoinette
  1971   *Histoire du principe de relativité* (Paris: Flammarion, 1971).
Toulmin, Stephen
  1970   *Physical Reality* (S. Toulmin, ed.). (New York: Harper Torchbooks, 1970).
Weidener, R. T. and R. L. Sells
  1975   *Elementary Physics, Classical and Modern* (Boston: Allyn and Bacon, 1975).
Weill-Brunschvicg, Adrienne
  1973   Langevin, Paul, in *Dictionary of Scientific Biography* (C. C. Gillispie, ed.) VIII (New York: Scribner, 1973), pp. 9–14.
Weinberg, Steven
  1972   *Gravitation and Cosmology* (New York: Wiley, 1972).
Weinstein, Roy
  1960   Observation of Length by a Single Observer, *Am. J. Phys.*, *28*, 607–610 (1960).
Weiss, Pierre
  1912   Prof. Dr. Heinrich Friedr. Weber, 1843–1912, *Schweizerische Naturf. Ges. Verh.*, *95*, 44–53 (1912).
Weisskopf, Victor F.
  1960   The Visual Appearance of Rapidly Moving Objects, *Phys. Today*, *13*, 24–27 (1960).
Wertheimer, Max
  1959   *Productive Thinking* (1st ed., New York: Harper, 1945; enlarged ed., 1959).
Whitrow, Gerald J.
  1961   *The Natural Philosophy of Time* (London: Nelson, 1961).
Whittaker, Sir Edmund
  1973   *A History of the Theories of Aether and Electricity* (vol. 1, London: Nelson, 1910, revised and enlarged edition, 1953; vol. 2, London: Nelson, 1953, 2 vols.; reprinted by New York: Humanities Press, 1973).
Williams, L. Pearce
  1971   *Michael Faraday: A Biography* (New York: Basic Books, 1971).
Winnie, J. A.
  1970   Special Relativity without one-way velocity assumptions, *Philos. Sci.*, *37*, 81–99; II, 223–238 (1970).
  1972   The Twin-Rod Thought Experiment, *Am. J. Phys.*, *40*, 1091–1094 (1972).
Woodruff, Arthur E.
  1968   The Contributions of Hermann von Helmholtz to Electrodynamics, *ISIS*, *59*, 300–311 (1968).
Yildiz, Asim
  1966   Electromagnetic Cavity Resonances in Accelerated Systems, *Phys. Rev.*, *146*, 947–954 (1966). With C. H. Tang. II, *Nuovo Cimento*, *61*, 1–11 (1969).

INDEX

Numbers set in *italics* designate those page numbers on which the complete literature citations are given.